Observing Wildlife in Tropical Forests

Volume 1
A Geosemeiotic Approach

Nils Lindahl Elliot

Delome Publications
Bristol, England

Revised edition published 2019
Delome Publications

Except for the quotation of short passages for the purpose of criticism and review, no part of the following work may be reprinted or reproduced or utilised in any form or by any electronic, mechanical, or other means, now known or hereafter invented, including photocopying and recording, or in any information storage or retrieval system, without express permission in writing from the author.

ISBN 978-1-9161626-1-7 (Pbk)
ISBN 978-1-9161626-2-4 (Ebk)

The cover and any images in this book are included in the above copyright unless otherwise specified. Every effort has been made to trace and recognise the rights of other copyright holders. If any copyright holders have been inadvertently overlooked, the author/publisher would be pleased to include any necessary credits in any subsequent edition of this work.

For Laura and Isabel

Contents

Preface

The photograph on the cover of this book shows visitors next to the 'Big Tree', a gargantuan kapok that stood on Barro Colorado Island, Panama, until 2013 (the picture itself was taken in 2011). Scientists believe that by the time that it fell, this *Ceiba pentandra* was hundreds of years old, and may well have been the single-stemmed tree with the largest crown in the world. Alas, none of the people standing in front of the kapok – or indeed any of the rest of the tourists whom I joined during ethnographic research on Barro Colorado – travelled to Barro Colorado expressly to see this or any other of the island's floristic wonders. While many visitors were interested in seeing Barro Colorado's rain forest, most wanted to see the island's fauna – in particular, what this volume will describe as its *charismatic* fauna.

The eagle-eyed reader will spot some dark patches directly above the visitor with the white hat, in the sunlit area on the right of the tree. The patches are actually bats (probably *Saccopteryx leptura*, also known as the lesser white-lined bat). Bat scientists and more generally the fans of chiropterans might well believe that winged mammals *do* have a certain 'charisma', but for a majority of the visitors who participated in my research on Barro Colorado, it was only fauna such as monkeys, poisonous snakes or sloths that really mattered.

At first, my main interest in researching this kind of observational preference was a comparatively simple one: to investigate if it was linked to the representations of the nature media, the latter being my shorthand for a field devoted to the popular representation of wildlife via genres such as wildlife documentaries on TV. However, by the end of my fieldwork on Barro Colorado, I became aware that I had, in an academic sense, failed to see the wood for the trees. The focus on the mediazation of nature led me to overlook the significance of wildlife observation in its own right. It also led me to take for granted Barro Colorado as a remarkable site in which to study wildlife observation amongst tourists.

So about a year after I first conducted fieldwork on Barro Colorado, I decided to follow up the initial research with a much broader project. The new project had two aims: the first, which is developed in this volume, was to produce a theoretical approach capable of explaining key aspects of wildlife observation amongst tourists visiting tropical forests, as well as other kinds of ecosystems. The second aim, which is the subject of a second volume (in preparation), was to engage in a genealogical inquiry, which is to say a 'history of the present' of wildlife observation amongst tourists on Barro Colorado Island. Both aspects of the investigation involve what I describe as a *geosemeiotics* of wildlife observation – a transdisciplinary inquiry that articulates geographic, semiotic, ecological and socio-anthropological perspectives on wildlife observation.

As the list above begins to suggest, the following pages cover a lot of ground. This being the case, one of the biggest challenges in writing this book was to find a *modus presentandi* that might make the work more inclusive for all readers, but especially for the visitors, park staff, and scientists with whom I had discussions during the ethnographic research, and who more often than not expected my research to adhere to positivist principles. Given these expectations, it became clear that the investigation should be presented by way of a hybrid genre: one that might combine the conventions associated with a research monograph, with those of a *propaedeutic* work, i.e. a text that goes to some length to introduce, contextualise and justify theoretical choices.

There is clearly a tension between meeting the requirements of each of the mentioned genres. From the point of view of a research monograph, a certain speed is required to cover new ground, and readers familiar with the background theory may become impatient, or feel patronised if the writing goes too slowly; to be sure, several of the scholars that this volume will cite would probably dismiss out of hand anything like a propaedeutic style of writing. Certainly one risk is that, in so far as the writing seeks to make complex theories more accessible, it might inadvertently invite readers to assume that those theories are no more than a matter of common sense. Conversely, from a propaedeutic perspective, if a perceived 'need for speed' or 'due precision' sacrifices adequate contextualisation, then those new to the theories might well feel that the work excludes them.

There is probably no happy medium and certainly no one size fits all. This being the case, I've chosen to err on both ends of the scale, if that is possible. In some places, readers will be asked to quite rapidly appraise the ideas proposed by some of the most abstruse thinkers in the critical social sciences and humanities. That said, the writing across the work as a whole will proceed relatively slowly in so far as it will take the time to introduce in considerable detail numerous concepts, theories, and their paradigms (not least those that this study will oppose).

I've spoken of speeds, but it is also pertinent to say something about extension. In an increasingly Twitterised world, few if any readers are likely to have the time to read a study as long as the present one (to be sure, does anyone read academic *books* nowadays?). With this constraint in mind, the present and the following volume are divided into relatively independent parts (one might almost say 'books within books'). So long as the reader starts with the introduction of each volume, then the parts that follow may be read as stand-alone sections, or as parts of the sequences suggested by each volume's table of contents. This kind of flexibility undoubtedly makes for some annoying repetition (including the bibliographic references, which start anew with each part). However, hopefully it will make it easier to 'dip in' to those aspects of the work that seem most relevant to each reader. The obvious caveat is that reading one or two parts will not necessarily give the reader a sense of the whole.

A word, finally, about the mode of publication. I've decided to self-publish this book via Amazon KDP. The name I've chosen for the imprint involves a Peircian wink: in Peirce, a delome is a type of sign that represents its object in its character as sign. Put more simply, it is another term for argument.

Acknowledgements

The research for this volume began in 2007, in relatively conventional academic circumstances. However, those circumstances changed quite drastically soon after I started the ethnographic fieldwork on Barro Colorado. The reader will find a detailed account of the changes elsewhere;[1] here it suffices to say that, having just begun the investigation in Panama, it became clear that a new rector (vice-chancellor) at my university was determined to impose a crudely neoliberal intellectual property (IP) policy on the institution, and this had direct implications for my research. The university proposed to give itself the right to sell a database that I'd generated via research funded by the UK's Economic and Social Research Council (ESRC). The research in Panama might be similarly affected unless I resigned more or less immediately.

Of course, this kind of IP policy has long applied to researchers doing work for big pharma, IT, artificial intelligence, and many other technical-vocational fields. However, at the time it was unheard of for a university in the UK to try to commercialise publicly funded research in the critical social sciences and humanities – let alone research about science and environmental communication of the kind I had engaged in. In this context, it became apparent that any academic unwilling to have their teaching or research sold for purely commercial purposes had no business working at the institution. So I refused to sign on the dotted line of the new policy, and left.

A few months after I resigned, the vice-chancellor who imposed this transformation became embroiled in a financial scandal that reached the UK's national media. This was followed by his sudden resignation, and a speedy departure for another post, where further scandals would emerge. Two months after these developments took place, the global financial crisis of 2007-2009 erupted, and in the years that followed, what many of us had interpreted as a purely self-serving experiment by a runaway privateer turned out to be a not-so-dry run for what became known as the marketisation of higher education across much of the UK. This was a shift that was a long time in the making, but it gained a certain urgency when an unholy alliance of greedy university managers, conservative politicians and the heads of global educational and private equity corporations

---

[1] See Nils Lindahl Elliot, 'New Labour's Skills Policy at the Intersection of Business and Politics', *Policy Futures in Education*, 7 (2009), 297-312. The following newspaper articles, published between 2007 and 2016, offer additional insights on the events in question: George Monbiot, 'These Men Would've Stopped Darwin', *Guardian* online, 11 May 2009, <http://www.theguardian.com/commentisfree/2009/may/11/science-research-business> [Accessed 30 December 2016]; and Robin McKie, 'Scientists Attack Their "Muzzling" by Government', in *The Observer* online, 20 February 2016 <https://www.theguardian.com/science/2016/feb/20/scientists-attack-muzzling-government-state-funded-cabinet-office>[Accessed 21 February 2016].

decided to use the financial crisis to try to cash in on higher education in England, Wales and Northern Ireland.

The silver lining in this cloud of academic neoliberalisation was that leaving formal academia opened up two marvellous opportunities. The first was that I could become a researcher completely free of the notorious red tape introduced to British universities by successive Thatcherite ministers (it really ought to be called 'blue tape'). The second was that in 2011, after I presented the most practical findings of the initial ethnographic research in Panama, I was able to get a new full-time position, this time as the stay-at-home parent for my baby daughter. From that point onwards, the wealth of research time was more than replaced with the richness of a new life, sleeping eyes and wrinkled hands, and then a growing state of amazement at the world the personification of what this volume will describe as *firstness*.

If I was able to complete the research in these circumstances, it was thanks to the support of Laura Walder. In recognition of this, but also to express my gratitude for all the love, generosity, and the discoveries that began a day late in the spring of 2011, I dedicate the pages that follow to her and to Isabel. I would also like to thank Dawn and David, Gunnar and Ann, Hans and Rosa for all the ways in which they too, supported my decision to leave formal academia.

The fieldwork on Barro Colorado was made possible by the Smithsonian Tropical Research Institute (STRI). I am very grateful for the interest shown in my first research proposals by Ira Rubinoff, then the director and now emeritus director of STRI. Dr Rubinoff recognised the significance for biological reserves of the mediazation of wildlife observation, and agreed to provide workspace, room and board on Barro Colorado, as well as access to STRI's Visitors Programme during the ethnographic fieldwork.

STRI's Visitors Programme was led by Oris Acevedo, STRI's Science Coordinator on BCI, with the assistance of her deputy, Belkys Jiménez, and I am very grateful for the support that both provided via that programme. I would also like to thank Beth King, STRI's Science Interpretation Officer, for her support; I am particularly grateful for her liaison role with STRI when the project was getting under way.

Once the ethnographic observations began, STRI's Visitors Programme's excellent guides not only agreed to allow me to accompany their groups, but provided a useful foundation for my later investigations into Barro Colorado's ecology. Despite the inevitably awkward, to not say intrusive nature of participant observation, the members of the programme were always supportive, and for this I am particularly grateful.

Thanks are also due for the visitors who agreed to participate in the research during their tours. They answered detailed questionnaires at the end of long, and hot walks through Barro Colorado's tropical forest, and shared many insightful comments about their experiences on the trail. Beyond expressing my gratitude for their support, I hope that the following chapters will go some way in answering their questions about the nature of the research.

Throughout my stay on Barro Colorado, STRI's residential support staff constituted the *sine qua non* of everyday research on the island. I hope that Volume 2 of this study will go some ways towards acknowledging this work, which has always played a key, if often unrecognised role in the investigations conducted by all researchers on Barro Colorado.

On the Panamanian mainland, I am also very grateful for the assistance provided by Gregorio Marín, whose friendship and knowledge of Panama made more meaningful the many voyages along the beautiful Galliard highway that links Panama City with the Smithsonian's pier at Gamboa.

Back in Bristol, I would like to thank my former head of school, Jane Arthurs, for facilitating a research sabbatical during which I could travel to Barro Colorado. I would also like to thank the then-new dean of my faculty, Paul Gough, for trying to find a way around the university's IP policies.

My research benefited from the expertise of several scientists whose research I would also like to acknowledge. On Barro Colorado, I am very grateful for the knowledge offered by Egbert Leigh, W. Douglas Robinson, Jorge Ventocilla and Jackie Willis. In Bristol, I would like to thank Hannah ter Hofstede and Holger Goerlitz.

In a fit of first-time parent – or perhaps it is former lecturer – optimism, I tried to begin to publish this research on the centenary of the first official crossing of the Panama Canal. At the time, my daughter was about to become eligible for state-supported early learning, and I imagined that a couple of hours on weekday mornings would be a kind of research Shangri-La – one that would enable the publication, even if only by instalments, of what I hoped might become an edited collection about wildlife observation in tropical forests. Those plans came to naught, but I'd like to thank Michelle Henning and Carmen Alfonso for sharing their critical skills in a manner that greatly improved the introductory remarks for that attempt. Their suggestions live on in this iteration of the research, if only because both helped me to see the foibles of my first introduction for the work. Thanks are also due to Laura Walder, who proofread parts of the later drafts of the book. It goes almost without saying that any remaining mistakes in the work are very much my own.

The design of the book's cover is the work of Clare Devonport at Morph Brands. The cover has somehow made the volume really *feel* like a book. Clare also suggested the burnt orange for the font in my previous book's cover, and I am very grateful for her contribution to both books.

I would also like to acknowledge an academic form of support which occurred as I was completing this study, and for which I am especially thankful. Timo Maran, the head of the Department of Semiotics at the University of Tartu, Estonia, invited me to present my research as part of the Jakob von Uexküll Lectures in November 2018. Between 1892 and 1902, Uexküll, a biologist, was a regular visitor of the *Stazione Zoologica* in Naples, and this, coupled to his role as a key precursor of what is today known as biosemiotics, meant that this study could be presented in a particularly appropriate academic context.

I would like to acknowledge, finally, the contribution to my research of Gunther Kress, who passed away just as I was publishing an ebook version of this volume. Gunther was one of the founders of a properly *social* semiotics, and before that of a critical form of linguistic discourse analysis. During my doctoral research, Gunther's example of an explicitly interdisciplinary integration of semiotic, sociological, and anthropological forms of analysis offered a welcome alternative to media studies and cultural studies. I'd like to think that the spirit of that kind of approach, and of Gunther's remarkably concrete pedagogies of theory, live on in the present study.

N.L.E., Bristol, September 2019

A friend of mine, in consequence of a fever, totally lost his sense of hearing. He had been very fond of music before his calamity; and, strange to say, even afterwards would love to stand by the piano when a good performer played. So then, I said to him, after all you can hear a little. Absolutely not at all, he replied; but I can feel the music all over my body. Why, I exclaimed, how is it possible for a new sense to be developed in a few months! It is not a new sense, he answered. Now that my hearing is gone I can recognize that I always possessed this mode of consciousness, which I formerly, with other people, mistook for hearing.

Charles Sanders Peirce,
Immortality in the Light of Synechism

TV showing BBC *Planet Earth II* Series.
Series is © Copyright BBC 2016

Prologue

The bokeh sink in slow motion, polygons of rain that crash into a glistening ensemble of plants and animals in the understory of a neotropical forest. Amid the exuberant greenery, lantern flies lurk and katydids launch as a strawberry poison dart frog vies with a bromeliad for the title of Most Crimson Object. A red-eyed tree frog, shown at first in a cryptic flatness, is pictured jumping up, also in slow motion, revealing unexpected hues on its underside.

Even as this scene unfolds, a soundtrack starts with the musical equivalent of a rainstorm about to break. The soundtrack then dramatises the fall of each drop with a percussive lashing, its melodic motion mirroring the pitter-patter of rain in cascading quavers: G#, G, F, C. The cascade recedes to give way to an ascending chord progression of minims played initially by a piano, and then sung by a sublimely synthesised choir: F, G, G#.

This scene is not, so far as I know, the product of a drug-fuelled hallucination, though at least some of the bokeh may well have been added by digital means. It is instead just one of many similarly extraordinary montages included in the BBC's *Planet Earth II* series, which premiered at around the time I was finishing an early draft of this volume. Like the original *Planet Earth* (which was broadcast shortly before I *began* the research), this was a very successful TV series, at least if one goes by the ratings: apparently each of its programmes attracted well over 11 million viewers per episode in the UK alone. The first episode made headlines in news media across the world when one of its scenes 'went viral'; the scene showed how Galápagos racer snakes attacked hatchling marine iguanas, and some observers went so far as to describe the scene as the best wildlife TV sequence ever produced.

What lies behind the success of blockbuster wildlife TV series of this kind? There can be little doubt that at least part of the answer may be found in that high modern predilection for the solace of an Edenic Nature – a nature that ostensibly continues to exist as it did before the fall of industrialisation, or the rise of the consumer societies that followed. In the case of *Planet Earth II*, it seems that the light, or perhaps one should say the chiaroscuro of that myth provided at least some TV audiences with relief from the political developments that began to unfold in the US in November 2016.

The importance of such a hermeneutic notwithstanding, it would be a mistake to overlook the programme-makers' ability to represent wildlife in the most viscerally compelling of ways. Series producers have long suggested that the secret of this interpellation lies in the skills of the cinematographers. But of course, other production roles are also important. What would the programmes be without, say, the magic of montage, the sublime (and sublimating) soundscapes, or the managerial but still creative skills of the executive producers? To be sure, the producers would not be able to hire the best craftspeople, or indeed crisscross the world and use equipment on the sharpest edge of technological innovation were it not for the vast budgets that they command, especially when making the so-called 'Blue-Chip' TV series – those that, like *Planet Earth*, have the highest production values.

Here again, *Planet Earth II* offers a useful example. The series will have cost many millions to produce,[1] and was shot using Ultra High Definition (4K) and Hybrid Log-Gamma (HLG) technologies, the latter being what by the middle of the 2010s was a new standard that allowed for a greater contrast range, and the representation of colours hitherto not technically reproducible by BBC producers. Perhaps to whet audiences' appetite for these new technologies, in the UK the BBC used its iPlayer platform to make available, in full 4K/HLG glory, two scenes from the full *Planet Earth II* series: the one described above (albeit with a different soundtrack), and another one featuring a jaguar stalking capybaras. Not coincidentally, both of the scenes selected to wow audiences were shot in neotropical forests, and both were described by the science and technology website Gizmodo as being 'jaw-dropping'.[2] The BBC itself suggested that they showed shades of green and a red 'never before seen on a TV'.[3]

Why dwell on such scenes, let alone technicalities which will themselves no doubt soon be superseded? This study starts from a contradiction that is somehow made more tangible by the idea of 4K bokeh falling wetly in a Hybrid Log-Gamma'd forest. Leading producers of wildlife TV have long claimed, not without some validity, that their marvellous representations portray 'nature itself'. As one pioneering wildlife TV producer once put it (perhaps inadvertently candidly), 'Nature as it really exists is our line of business.'[4] Yet to boast that certain colours have never before been seen on TV is to admit, in so many words, that previous programmes did not quite portray Nature, or at any rate what passes for Nature, 'as it really exists' – at least not as it might really exist for a human observer standing and looking in a tropical forest, *in situ*. Until now, the Nature portrayed was, shall we say, N – CC(g,r...), i.e. Nature minus Certain Colours such as the greens and reds made visible by HLG technologies. My use of a quasi-mathemati-

---

[1] At the time I was writing this prologue, the precise figure had not been advertised, but in 2006 the first Planet Earth cost a reported £8 million to produce. Daniele Alcinii, 'BBC's return to Earth with "Planet Earth II"', *Realscreen*, 25 October 2016 <http://realscreen.com/2016/10/25/bbcs-return-to-earth-with-planet-earth-ii/>[Accessed 24 August 2017].

[2] James O Malley, 'The BBC is Test Streaming 4K HDR Planet Earth Footage from Today', *Gizmodo*, 8 December 2017, <http://www.gizmodo.co.uk/2016/12/the-bbc-is-test-streaming-4k-hdr-planet-earth-footage-from-today/>[Accessed 24 August 2017].

[3] BBC Research and Development, '4K UHD Trial of Planet Earth II in HDR', 8 December 2016, *BBC online* <http:// www.bbc.co.uk/2016-12-hdr-4k-uhd-iplayer-trial-planet-earth>[Accessed 24 August 2017].

[4] Colin Willock, *The World of Survival* (London: André Deutch, 1978), p. 41.

cal form of description is meant to acknowledge, with a wink, that some readers might find this objection to be little more than a purely technical limitation: wildlife TV really does equal the capital N of Nature, give or take a few shades of green (or red). And yet, when one starts to think more carefully about all that is left out of series such as *Planet Earth II*, it becomes apparent that rather more than a technical limitation is at stake.

One can begin by pointing to the long list of ecological phenomena that might have, indeed perhaps *ought* to have been included, but have been systematically left out of this, and most of the other BBC wildlife series. What is at issue is not so much the myriad species that have never been represented, but for example the remarkable bias against the narrative foregrounding of the flora, or the discussion of truly synecological aspects of the relatively small number of ecosystems that the producers have long favoured (itself a form of bias). And that is to say nothing about the genre's fundamental omission: a critical consideration of the myriad ways in which already by 1979 – the year in which the first BBC wildlife 'blockbuster', *Life on Earth*, was broadcast – modern cultural activities had long since begun to transform most if not all biomes across the planet.

The not unreasonable rejoinder might well be that a TV programme is not to be confused with an ecology textbook, let alone a pamphlet of the kind once produced by environmentalist NGOs. But while that argument is certainly a valid one in itself – clearly, there *are* differences in genres – it fails to explain why wildlife TV producers – and in fairness, the producers and editors of many other genres also devoted to 'Nature' – have devoted so little effort to developing more compelling accounts of, say, those keystone species that are almost never represented. Why not more programmes, for example, about the humble agouti? After all, many producers have argued that their magnificent images play a major role in modern conservationist discourse by 'showing what there is to save'. If so, why not spend more time – and money – showing more of those organisms that really must be saved if certain ecosystems are to survive? And that is to say absolutely nothing about the aspect not just of tropical forests, but of the entire planet which nature series such as *Planet Earth* have either entirely ignored, or tiptoed around: the rising catastrophe that is anthropogenic climate change.

Anyone unpersuaded by this kind of critique might wish to consider the long list of things that *are* included in the TV series, but which no observer, human or otherwise, would be likely to see with their own eyes. As a first example, consider the wondrous phenomenon of bokeh-rain. Alternatively, imagine what it would be like to see, 'with your own eyes', an amphibian launching in slow motion. I for one have never seen or heard bokeh-rain pelting surfaces, floral or amphibian, to the tune of cascading quavers.

Readers may infer from these remarks that I am either a hyper-realist, or a closet constructionist. Actually, and despite the risk of stating the obvious, I have no doubt that *Planet Earth II does* portray real frogs, even as I welcome the many aesthetic twists and turns in this and all other wildlife TV series. That said, perhaps few readers would disagree with the proposition that we humans do not look at the fauna in rain forests (or more generally at objects in any other context), via two-dimensional, 16:9 screens with expertly intensified scenes of the kind made available by TV shows.

Or do 'we'?

Introduction to Volume 1

Something falls heavily to the ground to one side of the trail. At first we think it's a branch, but a writhing motion in the ankle-deep leaf litter suggests otherwise: the fallen branch is actually a slender but bright green snake – a parrot snake (most likely *Leptophis ahaetulla occidentalis*). The snake has managed to catch a turnip-tail gecko (*Thecadactylus rapicauda*), but both have fallen out of a tree in the process. As we watch transfixed, the snake starts to devour the unlucky lizard alive. Soon the gecko's head is in the snake's mouth, but in an almost caricatural attempt to prevent the unpreventable, the gecko manages to use its hind feet to get a hold on the ophidian's head. As our cameras click and whir, the lizard disappears into what looks like an impossibly narrow gullet. In the end, the gecko returns to the tree from whence it has fallen – but this time as an outline of its prior existence in the distended body of the snake.

This event occurred whilst a group of tourists was visiting Barro Colorado Island, a 15.6 km² (approx. 6 sq. mile) biological reserve surrounded by the Gatún Lake, a vast reservoir in the northern half of the Panama Canal (see Figure 1, next page). First established as a natural park in 1923, when the Panama Canal was part of a US colonial enclave known as the Canal Zone, the island is today the best preserved part of the 54 km² (21 sq. mile) Barro Colorado Nature Monument, a Panamanian national park and site of special scientific interest which comprises Barro Colorado itself, several islets, and the five mainland peninsulas that surround the island.

Starting in early 2007, I engaged in participant observations with visitors as they toured Barro Colorado. The visitors who partook in the research were on day trips offered by the Visitors Programme of the Smithsonian Tropical Research Institute (STRI), an overseas bureau of the Smithsonian Institution in Washington DC which manages the Barro Colorado biological reserve. The Smithsonian first began to administer the biological reserve and its field station in 1946, when Barro Colorado was still a part of the Canal Zone. In 1966, scientists working at the field station persuaded the Smithsonian to allow them to establish a semi-independent bureau in order to secure funding for their activities on Barro Colorado, and this led to the establishment of STRI. When in 1977 presidents Omar Torrijos and Jimmy Carter signed the treaties that began to devolve the Canal Zone to the Republic of Panama, STRI negotiated a parallel agreement with the Panamanian authorities that allowed it to remain as the official custodian of Barro Colorado, and of the then newly created Barro Colorado Nature Monument (cf. Vol. 2).

Snake-lizard predations must have occurred countless times on Barro Colorado. It is, however, unlikely that many people have witnessed them in the manner that the group of visitors mentioned above did. Indeed, the visitors were probably doubly lucky: lucky to encounter the strikingly beautiful specimen of *Leptophis ahaetulla*, and even luckier to witness a part of the predation. So it was somewhat surprising to discover that at least one family came away from the tour with a sense of disappointment. According to the answers its members provided in a survey administered at the end of the tour, the family was displeased at the overall dearth of sightings of what may be described as the *charismatic* fauna on the island. As far as the family members were concerned, seeing a skinny snake eating a small lizard was no substitute for a fuller repertoire of encounters involving animals such as monkeys, the larger poisonous snakes, sloths, etc.

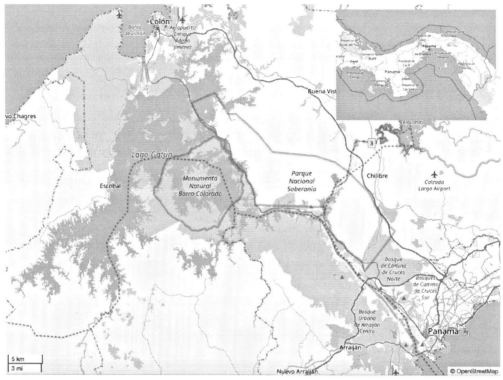

Figure 1. Barro Colorado Island in the Panama Canal. The line surrounding the island shows the boundaries of the Barro Colorado Nature Monument. © OpenStreetMap.org contributors.[1]

By the time that I completed the ethnographic fieldwork on Barro Colorado, it had become clear that this kind of disappointment was by no means a matter of an isolated response or two. On the contrary, a majority of the tourists – and especially those who were first-time visitors to a tropical forest – came to Barro Colorado with high expectations regarding the number and type of wildlife encounters they would be able to experience. When such expectations failed to come to fruition, many of the visitors left the forest with a sense of disillusion – and this despite the best efforts of the STRI guides that led each tour.

Why the disappointment? One obvious answer may be found in the forest itself. Unlike, say, a savannah, where at least the larger animals might be spotted

---

[1] Map used in accordance with the provisions of the Open Data License. See <http://www.openstreetmap.org/copyright>.

from afar, the floristic structure of a tropical forest often creates an environment in which it is *bound* to be difficult to view creatures – even those that are relatively large, and only a few meters away. Then again, many species in tropical forests have evolved in such a way as to make themselves even more difficult to detect; if crypsis does not make the creatures nearly invisible, then sheer scarcity may ensure that the chances of an encounter are slim. Unfortunately for tourists, even when animals *are* spotted, many try to put as much distance as possible between themselves and would-be predators, reducing encounters to little more than a glimpse.

Important as these proximate causes of visibility, or rather, *in*-visibility may be, they do not explain why it is that many of the visitors arrived on Barro Colorado expecting to experience something quite different. Nor do they explain why many visitors had a wish list of creatures they hoped to observe. To account for these aspects, a different set of issues would have to be considered and researched, and that is precisely what my ethnographic research on Barro Colorado set out to do.

The fieldwork for that research began from a relatively simple abduction (or what is more commonly described as a *hypothesis*[2]), and with it, a distal cause of disappointment: that especially first-time visitors to a tropical forest might well expect to experience the *in situ* equivalent of the representations of 'jungle' and wildlife generated by certain media of mass communication.

I will return to this abduction in a moment. First it should be explained that it was grounded in research conducted earlier with visitors in zoos and wildlife parks.[3] That research showed that the media of mass communication played an important role in shaping many of the visitors' interpretations of displays with wild, or quasi-wild, animals. Simplifying greatly, the zoo research, which also included an ethnographic dimension, revealed that younger children often interpreted the displays with reference to story books and fictional films. By contrast, adults more often tended to interpret the displays with reference to wildlife television. For example, before Pixar/Disney's *Finding Nemo* premiered in 2003, almost none of the visitors who partook in my zoo research paid much attention to

---

[2] Abduction is the term used by Charles Sanders Peirce to refer to 'some belief, active or passive, formulated or unformulated, that has just been broken up… The mind seeks to bring the facts, as modified by the new discovery, into order; that is, to form a general conception embracing them… This synthesis suggesting a new conception or hypothesis, is the Abduction.' Charles Sanders Peirce, 'Sundry Logical Conceptions', in Nathan Houser (General Ed.), *The Essential Peirce, Volume 2 (1893-1913)* (Bloomington: Indiana University Press, 1998), 267-288, p. 287.

[3] See Nils Lindahl Elliot, 'The New Zoos: Science, Media and Culture' at <http://nilslindahl.net/projects/new-zoos/>[Accessed 30 December 2016]. See also the Economic and Social Research Council (ESRC) press release at Eureka Alert, 'Is that lizard a basilisk? How Harry Potter sparks interest in the reptile house', 15 August 2005, < http://www.eurekalert.org/pub_releases/2005-08/esr-itl081505.php>[Accessed 30 December 2016].

the common clownfish (*Amphiprion ocellaris*) displayed at one zoo; after the film appeared, families almost literally queued up to 'find Nemo'. Or to offer an example that centres on adults, many visitors assumed that naturalistic, and immersive zoo enclosures were automatically superior from the point of view of the welfare of the animals on display. It was no coincidence that the design of these displays echoed the techniques of wildlife observation employed by the *nature media* – my shorthand for a field of mass communication devoted to the production and dissemination of factual (which is to really to say documentary) representations of wildlife and their habitats. These and numerous other examples suggested that the visitors had internalised at least some aspects of the predominant 'ways of seeing' employed by nature media representations. When the visitors came face to face with real animals in zoos, many judged the quality of the displays on the basis of the extent to which the enclosures mimicked at least aspects of the media viewing experience. The media had, however inadvertently, provided visitors with certain criteria of/for evaluating the naturalism of the displays.

Elsewhere I have described this kind of transposition or translation as a media-based *dynamic of transmediation*: simplifying somewhat, someone interprets something in one context, but using interpretive keys transferred, mostly if not entirely implicitly, from another context.[4] We can say provisionally that no interpretive process can do without this kind of dynamic: at least if one goes by conventional semiotic wisdom, to know what one thing is and is not, one must know another. That said, *media*-based dynamics of transmediation entail a more specific, and in some sense unnecessary frame: they start from the characteristic representations associated with one or another genre of mass communication. As such, media-based dynamics of transmediation can be said to be a result of relatively recent social and cultural developments – in particular, what the sociologist John B. Thompson has described as the *mediazation* of modern cultures. According to Thompson, mediazation is 'the general process by which the transmission of symbolic forms becomes increasingly mediated by the technical and institutional apparatuses of the media industries'.[5] Symbolic forms is Thompson's concept for actions, utterances, images and texts produced by subjects and recognised by them and others as meaningful constructs.[6] While the expression 'transmission of symbolic forms' suggests a reference to the mechanical aspects of mass communication, Thompson is in fact referring to any process which entails the social production, circulation, or reception of said forms.

---

[4] Nils Lindahl Elliot, *Mediating Nature* (London: Routledge, 2006), p. 63.

[5] John B. Thompson, *The Media and Modernity: A Social Theory of the Media* (Cambridge: Polity Press, 1995), p. 46. See also John B. Thompson, *Ideology and Modern Culture* (Cambridge: Polity, 1990), p. 3-4.

[6] John B. Thompson, *Ideology and Modern Culture*, p. 59.

Returning to the research in Barro Colorado Island, the findings of the zoo research, as coupled to my wider investigations into modern environmentalism and the media[7] gave me reason to believe that if, as expected, adult tourists on Barro Colorado *did* engage in media-based dynamics of transmediation, they might well draw on wildlife documentaries and other genres of the nature media. In effect, I proposed that the cultural authority of the nature media was such that the field's representations might well work to shape tourists' dynamics of transmediation, and with them, their anticipated ways of seeing wildlife on Barro Colorado Island. More specifically, I proposed that nature media representations might well have unexpected pedagogic consequences in so far as they might inadvertently teach, and at least some audiences might themselves unconsciously learn, particular *techniques of wildlife observation* – what I will define provisionally as structured, and structuring ways of perceiving, conceiving and interpreting wild animals – or at any rate, animals represented as being wild. On Barro Colorado, it was these techniques, or aspects of such techniques, which I suspected audiences might employ to interpret and evaluate actual encounters with wildlife during tours of the island.

During the months that I spent on Barro Colorado, participant observations enabled me to accompany the visitors as they walked on Barro Colorado's trails, and to investigate their *in situ* responses to the island, its forest and its wildlife. Surveys administered at the end of each visit provided a means of obtaining detailed background information about each visitor, including markers of economic, social and cultural capital, past and present tourist itineraries, as well as a mapping of media habits. The surveys also provided an opportunity to ask questions about visitor expectations, and the extent to which the tour of Barro Colorado met those expectations.

I will say a lot more about the methods and the results of this research in Volume 2. Here it suffices to explain that the research uncovered two kinds of evidence that suggested that many of the tourists *did* interpret Barro Colorado's forest and its wildlife with reference to the nature media field. On the one hand, the investigation documented numerous instances in which visitors made *in situ* references to nature media genres. For example, one visitor interpreted an encounter on Barro Colorado with an American crocodile (*Crocodylus acutus*) by referring to TV images of the Nile crocodile (*Crocodylus niloticus*) in eastern Africa. Even as the visitor gazed upon the crocodile in Barro Colorado, he described how Nile crocodiles ambushed Common Wildebeest (*Connochaetes taurinus*) – and this despite the fact that he'd never been to Africa.

On the other hand, a less direct but arguably more strategic kind of evidence involved a pattern of correspondence between the nature media field's characteristic ways of representing wildlife in tropical forests, and self-reported visitor ex-

---

[7] See Nils Lindahl Elliot, *Mediating Nature*.

pectations as to what wildlife they hoped to observe, and *how* they hoped to observe it on Barro Colorado. The pattern in question involved several aspects.

First, many of the visitors expected to find a forest that was teeming not just with wildlife, but with certain kinds of creatures – what I described earlier as the 'charismatic' fauna, a category which included the larger mammals (e.g. island's four diurnal species of monkey, or either of its two sloth species) and reptiles (e.g. the crocodiles found in the region, and the island's snakes – especially the pit vipers, *Crotalinae/Viperidae*). As the following chapters will make clear, the concept of a *charismatic* fauna is not meant to suggest that there is an innate attraction between humans and some species of animals – for example, the kind of relation hypothesised by some evolutionary psychologists. Instead, the attraction is best understood as a matter of a certain 'following' – one which involves an affinity to the forms of the creatures themselves, but where the affinity is at least partly constructed via relations of authority instituted, amongst other sources, by nature media presenters, and by certain genres' characteristic forms of interpellation (discursive, visual, audio-visual, etc.).

Second, most visitors also expected not only to *see* the charismatic species, but to have *close* encounters with them. Wildlife ought to be experienced via the *in situ* equivalent of close-ups, and ideally, via close-ups involving the kind of visual drama so often privileged by the nature media. It might seem that the family group's response to the encounter with the parrot snake and the gecko contradicts this point; in fact, what this family and many other visitors seemed to desire were tours *punctuated* by this kind of encounter, and ideally with larger and better known species. This expectation may be interpreted as a desire for the tour to be structured in much the same way as wildlife TV: as a succession of encounters involving what wildlife TV producers once described as 'Hey Mays' – moments of audio-visual drama and delight that, in the gendering discourse of the producers, might lead the male partners in households to shout 'Hey May come and see this…'.[8]

Third, many of the visitors further reported that they expected to experience such scenes in a lush *jungle*. This being the case, many tourists expressed surprise that Barro Colorado's forest was so dry, and that the ground was often covered by what seemed like an autumnal leaffall (my initial research took place during Barro Colorado's dry season, which is when most international tourists visit the island). Few realised that the category of jungle is often used in an imprecise manner – or indeed, that the also popular category of *rain* forest is itself a very general one that may lead one to overlook important variations in tropical forests.

And yet, even as many of the tourists expected to find a jungle, almost none of those who partook in the research had much interest in what many ecologists might regard as Barro Colorado's true wonder: its flora, and especially its re-

---

[8] See Nils Lindahl Elliot, 'Signs of Anthropomorphism: the Case of Natural History Television Documentaries', in *Social Semiotics*, 11(2001), 289-305, p. 302.

markable trees. As I began to explain in the Preface, the cover of this volume shows a massive kapok (*Ceiba pentandra*) which at the time was possibly the single-stemmed tree with the largest crown in the world, with a spread of over 60 meters (more than 200 feet).[9] While this and other giants pointed out by STRI's guides certainly generated some interest amongst many of the tourists, by and wide the flora was regarded as little more than a kind of *backdrop* for what ought to be the main act – numerous and close-up encounters with monkeys, sloths, poisonous snakes, etc.

As Volume 2 will explain, this pattern mirrors quite precisely the representational tendencies found especially in wildlife documentaries on television, and many an article about wildlife in the *National Geographic*. In these and in other genres, the nature media field has historically reduced tropical forests to jungles even as its producers have treated the flora as little more than a visual setting for the dramatic appearance of a changing 'menu' of charismatic fauna.

When considered in relation to visitors' direct references to the nature media, and to statistical analyses of visitors' media use, the correspondence between representational strategies and visitor responses suggested that many of the tourists were indeed engaging in nature media-based dynamics of transmediation: they were attempting, however inadvertently, to project onto Barro Colorado's forest at least some aspects of the nature media field's characteristic ways of observing wildlife.

Now the findings of the initial fieldwork on Barro Colorado might well have brought the research to its planned conclusion, as per the original research proposal. Instead, the fieldwork on Barro Colorado served as a prelude to a more ambitious investigation. The more I engaged in participant observation with visitors on Barro Colorado, the more it became apparent that the encounters with wildlife were themselves at least as deserving of research as were any 'media effects'. Then again, the more surveys I read, the more it became apparent that visitor responses to Barro Colorado were part of a much broader matrix of tourist experiences, however much the tourists' observations did indeed reflect the influence of their own media use.

A similar change of perspective occurred with regard to the geography of Barro Colorado. The design of the initial ethnographic research reduced Barro Colorado to something like a convenient location for an investigation that might well have taken place in virtually any other biological reserve, or indeed in any other kind of ecosystem frequented by tourists, and represented by the nature media. However, the more I learned about Barro Colorado's environmental history, the more it became apparent that the island and its environs invited critical attention in their own right.

---

[9] STRI, 'Farewell Big Tree', in *STRI News* 21 June 2013, <https://www.stri.si.edu/sites/strinews/PDFs/STRINews_Jun21_2013.pdf>[Accessed 1 August 2016].

So it is that, after completing the analysis of the initial fieldwork, I decided to follow up the initial investigation with a new project – one that sought to develop a theoretical approach, and with it a form of analysis capable of explaining the interrelation between the ecological, geographic, social and cultural aspects of wildlife observation amongst tourists visiting tropical forests. This approach should then constitute a part of the basis for a genealogical inquiry into the emergence of tourism on Barro Colorado Island, and with it particular techniques of observation developed beyond the island.

The present volume is about the first part of this project, and so the rest of this introduction will provide an overview of the critical approach to wildlife observation amongst tourists visiting tropical forests. Volume 2 will then provide a separate introduction to the genealogical inquiry.

Now anyone unfamiliar with the literature might be forgiven for imagining that a wealth of ethnographic research has been published on tourist experiences of tropical forests; after all, tropical forests have long been the object of scientific, as well as popular cultural fascinations. More importantly, over the last few decades, tropical forests have become popular destinations for millions of tourists the world over.

In fact, while many excellent studies have been produced about ecotourism, and efforts have been made to develop a more specialised body of knowledge that focusses on wildlife tourism, by 2008 (when the research for the new project started) there was a dearth of ethnographic research about tourism in tropical forests. Perhaps more importantly, by the end of the 2010s, no one appeared to have bothered to develop, and to publish a detailed theoretical account of *wildlife observation* amongst tourists.

This meant that I had to look elsewhere for theoretical inspiration. An obvious place to start were the many positivist treatises that have been produced about *perception* – especially *visual* perception. And indeed, in the first part of this volume I will consider three theories which are representative of three fields that might contribute to research about the perception of wild animals in tropical forests. The theories in question are those of David Marr, a neuroscientist who develops what remains one of the more influential versions of a computational/neuroscientific approach to visual perception;[10] James J. Gibson, a psychologist who develops a very different, but also influential theory known as the ecological approach to visual perception;[11] and Edward O. Wilson, the famous myrmecologist, sociobiologist and conservationist who proposes what he describes as the

---

[10] David Marr, *Vision: A Computational Investigation into the Human Representation and Processing of Visual Information* (San Francisco: W. H. Freeman, 1982).

[11] James J. Gibson, *The Perception of the Visual World* (Boston: Houghton Mifflin, 1950); James J. Gibson, *The Senses Considered as Perceptual Systems* (Boston: Houghton Mifflin, 1966), and James J. Gibson, *The Ecological Approach to Visual Perception* (London: Psychology Press, 2015 [1979]).

'biophilia hypothesis'[12] – strictly speaking not a theory of *perception*, or indeed by Wilson's own account, a *theory* of anything – but an approach which nonetheless makes important claims about how people perceive nonhuman animals.

While some aspects of each of these approaches can make a contribution to the study, I will argue that their authors' thought is vitiated by a reliance on positivist philosophy, and unhelpfully reductive forms of conceptualisation: Marr focusses entirely on the central nervous system and the organs of visual perception; Wilson argues that any regularities in perception, conception or interpretation must ultimately be explained from an adaptationist perspective; and Gibson believes that even if perception cannot be reduced to a matter of organismic functions or adaptations, it is necessary to adopt a radically empiricist approach that excludes representation, memory and institutional forces.

To be clear, it of course *must* be true that our human capacity to observe the world is at once enabled and constrained by certain physiological faculties, aspects of which are bound to have been affected by evolutionary processes. To refer to just two rather obvious examples, in general we humans cannot hear sounds much above or below a range of 20 to 20,000 Hertz (and indeed those of us who are older will probably be unable to hear sounds much above 15K Hertz, if that). And when it comes to using our visual capacities, the unaided eyes and mind will only be able to see objects that reflect light within the so-called visible spectrum, i.e. about 390 to 770 nanometers. Characteristics of this kind must be an outcome of ancient evolutionary dynamics, and certainly make a difference when it comes to trying to observe, say, an ocelot (*Leopardus pardalis*) in a forest such as Barro Colorado's. The ocelot, whose ethology has itself undergone an evolutionary process, is not only likely to have a much better sense of smell and hearing than we humans do (and perhaps better night vision as well), but is bound to be much faster and quieter. Especially if an inexperienced tourist and a forest-savvy ocelot (as opposed to, say, a kitten) were to be left to their own devices in a patch of neotropical forest, there can be little doubt as to which would be most likely to detect the other first, or which would more easily evade detection. In such a context it would be absurd to deny, or minimise the importance of the different species' ethologies, conceived at least partly along traditional adaptationist lines.

The last point notwithstanding, the expression 'left to their own devices' is a misleading one in so far as it suggests that each species has no more than its own purely natural 'devices', e.g. its own evolved ethological mechanisms which automatically determine, *a priori*, how its members will respond to a certain set of events. At least in the case of humans, that perspective is manifestly mistaken if only because humans can turn to additional devices such binoculars, camera traps, radio collars etc. in order to locate just about any creature found on the sur-

---

[12] See Edward O. Wilson, *Biophilia: The Human Bond with Other Species* (Cambridge, MA: Harvard University Press, 1984). See also Stephen R. Kellert and Edward O. Wilson (Eds.), *The Biophilia Hypothesis* (Washington D.C.: Island Press, 1993).

face of the planet. Are we really expected to believe that such devices are themselves no more than an evolutionary adaptation, whose fundamental nature was determined in the Pleistocene?

Then again, an adaptationist perspective is also likely to overlook the fact that at least some aspects of our sensory dispositions are learned. As the ecological psychologist Eleanor Gibson has pointed out,[13] what we perceive, and how we perceive it is a matter of actively, if unselfconsciously *acquired* dispositions – learned ways of looking at, touching, smelling or otherwise sensing our surroundings. Such dispositions must be acquired by each generation, and indeed by each *individual* from the earliest infancy – and this always in intimate relation to a particular set of environmental possibilities and constraints.

A perspective such as Gibson's is a very welcome antidote to the biological reductionism associated with adaptationist perspectives. But ecological psychologists tend to overlook the relation between perceptual learning, and broader socio-cultural mediations: as scholars in the critical social sciences and humanities have long argued, the uses of any evolved physiological capacities are at least partly determined by the workings of institutions, discourses, genres, and cultural formations more generally. A good example of this verity are the findings of my own initial ethnographic research on Barro Colorado, which as I noted above, revealed the extent to which many tourists had, in effect, interiorised media-based techniques of observation which they then sought to transmediate to the actual forests.

In this volume, I will thus suggest that any researcher interested in studying wildlife observation needs to explain rather more than the overtly ecological, physiological or indeed psychological determinants of observation. The problem is not to exclude the mentioned aspects, but rather to articulate them as part of an approach that acknowledges the importance of social and cultural determinations – and this without treating *any* of the aspects as the 'ultimate cause' of what tourists (or any other group) perceive, conceive and interpret when they encounter wildlife in tropical forests.

The issues just raised begin to explain why the present investigation employs the concept of *observation*, and not that of perception. While the two are often used interchangeably, perception tends to be associated with positivist conceptions, or at least with approaches that focus on the relation between (human) bodies and whatever is perceived. By contrast, the concept of observation facilitates the linking of the *here-and-now* of more or less direct perception in a certain context with the *there-and-then* of social and cultural determinations. According to the dictionary definition of the term, observation is the process or action of attending to

---

[13] See Eleanor J. Gibson, *Principles of Perceptual Learning and Development* (New York: Century Psychology Series, 1969). See also Eleanor J. Gibson and Anne D. Pick, *An Ecological Approach to Perceptual Learning and Development* (Oxford: Oxford University Press, 2000).

something or someone carefully, and this typically to obtain information. The connotations of the underlying verb (*to observe*) go beyond 'direct perception' to suggest a process whereby something is registered as being *significant*, and as being worthy of being watched carefully and attentively.[14] This implies in turn a particular *motivation* and with it a purposefulness – and thereby a process that is not only structured or directed by the most immediate or ostensive elements of an encounter with wildlife (or any other kind of entity).

In the aforementioned definition, the expression 'watched' suggests a purely visual process, but another reason for choosing observation over other alternatives such as the 'gaze', 'ways of looking', or indeed 'visual perception' is that it can also be used to refer to beyond-visual ways of attending to something. As an example, one might consider the classification of the sciences proposed by the philosopher Charles Sanders Peirce, whose semiotic theory I will introduce in Part 2. According to Peirce, the classification of each kind of scientific inquiry can be based on each discipline's characteristic form of observation. Peirce notes that even mathematics '... is observational, in so far as it makes constructions in the imagination according to abstract precepts, and then observes these imaginary objects, finding in them relations of parts not specified in the precept of construction. This is truly observation, yet certainly in a very peculiar sense; and no other kind of observation would at all answer the purpose of mathematics'.[15]

Thus far I've focussed on the meanings of observation that have to do with attention. But another advantage of observation is that it has additional meanings which allow one to shift the register, and in so doing to acknowledge other closely related dimensions or aspects of observation/perception. For instance, an observation is also a remark, statement, or comment that is based on something that one has seen, heard or otherwise noticed,[16] and I will make the case in this volume that wildlife observation more often than not goes with such observations. To this meaning we may add another noted by the cultural theorist and historian of observation Jonathan Crary: the Latin root *observare* also refers to conforming one's action, or complying – as in observing rules, codes, regulations and practices.[17] For example, in the case of scientific practice, this and the connotations that I mentioned earlier come together when the researcher 'makes observations', in the sense that the individual not only *takes* note, but *makes* notes that translate

---

[14] Oxford English Dictionary, 90th edition, <https://en.oxforddictionaries.com> [Last accessed 1 August 2017].

[15] Charles Sanders Peirce, 'Detailed Classification of the Sciences', in *Collected Papers of Charles Sanders Peirce*, 8 Vols., Charles Hartshorne, Paul Weiss, and Arthur Burks (eds.) (Cambridge: Harvard University Press, 1931-1958), Volume 1, p. 110.

[16] Oxford English Dictionary, 90th edition, <https://en.oxforddictionaries.com> [Last accessed 1 August 2017].

[17] Jonathan Crary, *Techniques of the Observer* (Cambridge, MA: MIT Press, 1990), pp. 5-6.

what has been attended to into a set of themselves encoded and encoding statements. The observational process is, to an extent, itself rule-bound.

The possibility of shifting the register of observation in these ways is useful for at least three reasons. The first is that I will argue that wildlife observation amongst tourists is a complex practice that entails multiple aspects and dimensions, including ones of the kinds made explicit by the different meanings of 'observation'. As part of this complexity, a second reason is that observational practices are always *multimodal* – i.e. they involve more than one channel of impression or expression. The last point is particularly pertinent in the case of wildlife observation in tropical forests, which typically entails beyond-visual channels, and so also beyond-visual perceptual systems. Finally, at least what I will describe as *mediate* wildlife observation is likely to be disciplined in a variety of ways, and by different regimes (cultural/human but also beyond-human).

With these kinds of considerations in mind, Parts 2 to 4 of this volume will develop what I will describe as a *geosemeiotic* approach to wildlife observation. The etymology of *geo-* goes back to the Ancient Greek *gē*, which means earth. For its part, semeiotic refers to the theory of signs (as I will explain at a later stage, the extra 'e' in semeiotic is not a typo; it is meant to signal a particular approach to semiotic theory). So geosemeiotic can be taken to mean an approach concerned with 'signs of the earth' – though as I will explain in the chapters that follow, perhaps a better account would be signs of (and by) the *world*. In the remainder of this introduction, I would like to offer an overview of what this approach entails as applied to wildlife observation amongst tourists visiting tropical forests.

It is pertinent to begin by explaining why I speak of a theoretical *approach*, as opposed to a theory, full stop. This choice is meant to signal that my aim is not to produce a 'general theory' of wildlife observation that explains what causes people to observe wildlife, or that explains which variables are most important when it comes to predicting how someone is likely to engage in wildlife observation. One of the reasons for this is that I do not believe that wildlife observation should be explained in mechanistic terms. On the contrary, I will argue that, even if linear causality is involved in certain aspects of the process, other forms of causality, and indeed other forms of *relation* are also involved which cannot be reduced to causation.

In keeping with this perspective, I will not be setting out to find out why people engage in wildlife observation. Posed in this manner, this question suggests that there is just one reason, one motivation, and my research suggests that in fact, different groups of people approach wildlife observation in at least partly different ways. While it is of course possible to argue that there must be a shared valorisation of *wild* life, and of the activity of observing wild animals in their own habitats, the more interesting questions have to do with what might be described as the *intríngulis* of wildlife observation amongst tourists – *intríngulis* being the Spanish language word for a non-obvious or hidden complexity.

So when I speak of a 'theoretical approach', I refer mainly to explaining what kinds of elements are involved in wildlife observation, and what happens to those elements when tourists engage in wildlife observation whilst visiting tropical forests. I will, in this sense, aim to problematise, and to consider some alternatives to the largely tacit *ontologies* that inform everyday ways of conceiving wildlife observation, and observation more generally. As the philosopher Manuel DeLanda puts it, 'A philosopher's ontology is the set of entities he or she assumes to exist in reality, the types of entities he or she is committed to assert actually exist'.[18]

As the following chapters will make clear, this kind of approach is not 'subjective' or 'introspective', as per some positivist scholars' dismissal of any research not based on 'hard science'. Nor does it simply reject anything like causality, or indeed empirical criteria regarding research about wildlife observation. Those who eschew causation *tout court*, or who reject empirical research or indeed scientific realism without further ado, run the risk of relativism, idealism, or what is equally problematic, a rampant *culturalism*. As the cultural critic Terry Eagleton puts it, culturalism is the doctrine that everything in human affairs is a matter of culture.[19] Culturalism is the obverse of *biologism*, the stance adopted by those who attempt to reduce everything in human affairs to a transcendental biology. Culturalism is no better than biologism; on the contrary, when culturalism takes over, then it becomes rather difficult to explain why academics' bodies (as distinct from some of their/our theories) don't simply float away.

I've now said something about what the research is *not* setting out to do. How, then, will it approach wildlife observation? As I noted above, the research will propose a geosemeiotic approach that is itself *transdisciplinary*. Until the last quarter of the twentieth century, research in both the natural and the social sciences was more often than not strongly grounded in particular disciplines. While there were many exceptions, the rule was that researchers produced research under the aegis of theoretical and methodological regimes that were conceived as being more or less monadic. This being the case, the researcher was *either* a biologist, or an economist, or a sociologist, etc.

Actually, a case can be made that no discipline is ever truly monadic, and that on the contrary, what *becomes* a discipline is in fact a certain *set* of approaches that are recontextualised as part of one ostensibly homogeneous field of inquiry. From this perspective, disciplines involve not just a certain approach, but the *disciplining* of conceptions deemed to be, or *not* to be, compatible with those generated in the own field.

Especially during the last quarter of the twentieth century (in some cases much earlier), some of the more adventurous researchers began to see the benefits of

---

[18] Manuel DeLanda, *Intensive Science and Virtual Philosophy* (London: Continuum, 2002), p. 2.

[19] Terry Eagleton, 'Culture and Nature', in *The Idea of Culture* (Oxford: Blackwell 2000), 87-111, p. 91.

loosening disciplinary boundaries, and of combining approaches based in wholly or partly different disciplines; this to explain mutually recognised phenomena, even if the precise matrixes of concepts and methodologies of explanation differed across the combined disciplines. Interdisciplinary approaches resulted from these conjunctions, and in many contexts eventually began to displace more narrowly disciplinary orientations (though again, there would of course be significant exceptions, and even today, some degree of disciplinarity often still plays an important role in teaching and learning, the organisation of university departments, and academic careers).

Important and useful as interdisciplinarity may be, *trans*disciplinary research entails a qualitatively different form of investigation. A first way of conceiving transdisciplinarity involves much the kinds of principles put forward by Basarab Nicolescu, and with him Lima de Freitas and Edgar Morin as part of what has become known as the *Charter of Transdisciplinarity*.[20] To paraphrase some of those principles, a transdisciplinary approach rejects reductionist perspectives on human practice; recognises the existence of different levels of reality regulated by different kinds of logic or interrelation; presupposes a dialogue between the natural sciences, the humanities and the social sciences; and most importantly, suggests that '[t]he keystone of [t]ransdisciplinarity is the semantic and practical unification of the meanings that traverse and lie *beyond* different disciplines'.[21] Put more simply, transdisciplinary research goes in search of problems, answers or solutions which completely evade the radar of disciplinary or even interdisciplinary orientations.

Of course, this account of transdisciplinarity entails a certain ethic, and as part of that, a *disciplining* approach in its own right. While this kind of acknowledgement does not deny the validity of Nicolescu *et al.'s* proposals, it may be suggested that what starts out as a revolutionary 'transdisciplinary' approach may in time come to be, or seem, like a not-so-new disciplinarity in its own right. And indeed, academic research has come a long way since the *Charter* was proposed, and so perhaps some modifications are now required with respect to the stated values.

A first point is that, if it is important to avoid reductionist perspectives vis-a-vis *humans*, the same is true for non-human, or as I will prefer to describe them, *beyond-human* creatures and entities more generally. A second suggestion is that, more than unification, *tout court*, what is at stake is perhaps the *recognition* of meanings which cannot, and indeed should not be simply unified even by a per-

---

[20] See Basarab Nicolescu, Limas de Freitas, and Edgar Morin, 'The Charter of Transdisciplinarity', *Inters.org: Encyclopedia of Religion & Science*, 1994 <http://www.inters.org/Freitas-Morin-Nicolescu-Transdisciplinarity>[Last accessed 25 August 2017].

[21] Basarab Nicolescu, Limas de Freitas, and Edgar Morin, 'The Charter of Transdisciplinarity', italics added to the original.

spective that cuts across multiple disciplines. What is required instead is an articulation that is capable of recognising continuities, even as it makes place(s) for what Part 3 will characterise as a philosophy of difference, and with it, a logic of assemblage. A third suggestion is that, given the extraordinary inroads made by interdisciplinary research, today transdisciplinarity probably also involves something like 'trans-interdisciplinarity', and that will certainly be true for this study.

So why a transdisciplinary account of wildlife observation amongst tourists in tropical forests? First, and as I began to suggest earlier, wildlife observation involves a multiplicity of aspects, some of which may be adequately explained via more traditional disciplinary forms of investigation, but others which can only be really understood via multiple disciplinary/interdisciplinary perspectives. So it is, for example, that the explanation of observational practice *qua* practice requires semiotic, psychological, sociological and anthropological theories.

Then again, and second, investigating wildlife observation requires a mode of inquiry that can explain what happens not just to the humans, but what happens when human and beyond-human aspects not only 'come together', but are intertwined in ways that defy the conventional classification of 'natural' or 'cultural', and so the time-honoured distribution of academic labour across the physical and social sciences.

This point requires elucidation. On the one hand, and perhaps most obviously, if at least *in situ* wildlife observation involves *encounters* with actual fauna, then it involves both human and beyond-human elements – including human and beyond-human *bodies*. If the researcher approaches such encounters from *either* a social science *or* a physical science perspective, then one or another kind of entity, one or another kind of body is likely to be excluded from the analysis. To be sure, a similar point can be made even with respect to the role of the human body itself; if, as has historically been the case, the researcher has to choose between the mind or the body, then key aspects of observational practice amongst humans are also likely to be excluded.

On the other hand, even a comprehensively interdisciplinary approach may fail to recognise those elements that are not best characterised as being *either* 'human' *or* 'non-human'. If this is true for the encounters as encounters, it is now also true, to an extent, for the forests themselves. Today all tropical forests betray at least a degree of anthropogenic change, and this is likely to generate what can be described as a nature-culture *hybridity* that resists any reduction along purely natural or cultural lines. Barro Colorado is a particularly obvious case in point: while the island is widely and accurately regarded as having one of the Republic of Panama's best conserved tropical forest fragments, the island has an environmental history that forces a reconsideration of any naively naturalistic account of its nature. Between 1910 and 1913, US engineers dammed the Chagres River near its mouth on the Caribbean in order to create the Gatún Lake, a part of the Panama Canal waterway. The new reservoir transformed Barro Colorado and some other hilltops into islands, and in so doing drastically changed the topography of the former Chagres River Valley. A decade or so after the flooding was complete, scientists set up a research station on the island, and began conducting research that entailed varying degrees of intervention vis-a-vis the island's wildlife, and its ecology more generally. Over the following century, this research station grew in size and international reputation, to the point that by 2007, when I began my research, its scientists could boast that Barro Colorado's forest was the most intensively researched tropical forest in the world. The infrastructure required to support that research, which included almost 40 kilometres (approx. 25 miles) of trails cut through the forest, would have provoked numerous, if more often than not subtle changes to the island's ecology.

While Barro Colorado's environmental history is undoubtedly unique in some respects, a case can be made that dynamics of hybridisation have occurred in tropical forests right across the planet. On the one hand, most forests are undergoing what are often accelerating dynamics of fragmentation.[22] On the other hand, the onset of an increasingly catastrophic climate change is causing transformations in tropical forests.[23] More generally, it seems very clear now that human activity has modified biomes in subtle and not-so-subtle ways right across the world, making it an illusion to speak of Edenic 'last wildernesses'. Indeed, by the beginning of the twenty-first century, there was probably no part at least of the *surface* of the planet that did not involve a degree of culture-nature hybridity, and this is one reason why so many scholars now speak of an 'Anthropocene' age.

In this volume I will thereby argue that the researcher must go beyond the time-honoured boundary between the social and the natural worlds, between the physical and the social sciences, but this in a way that does not simply reinstate the nature-culture dichotomy by other means. To return to the *Charter of Transdisciplinarity*, the problem is recognise and explain meanings (and to be sure, not just *meanings*) that traverse or lie *beyond* the great nature-culture divide.

With these considerations in mind, I will approach wildlife observation by way of three sets of interdisciplinary perspectives: semeiotic-phenomenological, sociological-philosophical, and ecological-geographic. The volume will begin by presenting each of these perspectives separately, but will then bring them together in a concluding chapter. The following is a brief account of each of the perspectives.

---

[22] See for example, William F. Laurance, 'Beyond Island Biogeography Theory: Understanding Habitat Fragmentation in the Real World', in Jonathan B. Losos & Robert E. Ricklefts (Eds.) *The Theory of Island Biogeography Revisited* (Princeton: Princeton University Press, 2010), 214-236.

[23] See for example, William F. Laurance, 'The World's Tropical Forests Are Already Feeling the Heat', Yale Environment 360, 2 May 2011, <https://e360.yale.edu/features/the_-worlds_tropical_forests_are_already_feeling_the_heat>[Accessed 5 July 2017]. See also William F. Laurance, J.L.C. Camargo, Philip Fearnside, T. Lovejoy, G. Williamson, R.C.G. Mesquita, C.F.J. Meyer, Paulo Bobrowiec, and Susan Laurance, 'An Amazonian Rainforest and its Fragments as a Laboratory for Global Change', *Biological Reviews*, 2017, 000-000.

Semeiotic-Phenomenological. Whatever else they entail, encounters with wild animals involve a process whereby someone detects and interprets *signs* produced by the animals and the forest (strictly speaking, the signs are 'co-produced', but I will have the opportunity to consider these and other complexities at a later stage). I will thereby devote Part 2 to developing a semiotics of wildlife observation. Simplifying somewhat, semiotics may be defined provisionally as the theory, or set of theories with which to explain the structuring of signs. In the critical social sciences and in Continental philosophy, semiotics has long been associated with the work of Ferdinand de Saussure (1857-1913), the founder of modern linguistics and the scholar behind the well-known characterisation of signs as conjunctions of signifieds and signifiers.[24] Saussure's semiotic theory has been used to explain how all manner of objects and events are socially and culturally constructed. After acknowledging the importance of this approach, Part 2 will explain why the research will *not* adopt a Saussurean perspective, and why it will turn instead to the work of Charles S. Peirce (1839-1914).[25] Peirce (pronounced as 'purse') was the founder of a very different approach to semiotics (or sem*e*iotics, as he spelled it, and as I myself will spell it when employing a Peircian perspective). Peirce's semeiotic rejects what I will characterise as the nominalism that haunts structuralist and poststructuralist semiotics. He adopts instead a realist perspective on the relations between signs and the cosmos. Accordingly, he makes the case that there may be a direct link – what he described as an indexical link – between objects and the forms by which those objects either express themselves, or are represented via additional objects, additional signs. Peirce's realism no doubt reflects the fact that, even as he was and remains one of North America's leading philosophers, by the early twentieth century he was also a world-class mathematician, astrophysicist, and geodesist. As it happens, Charles Peirce's father, Benjamin Peirce, the founding director of Harvard's mathematics department, was part of a group of notables that helped to set up the Smithsonian Institution in Washington D.C. – the parent institution of Barro Colorado's 'custodian', the Smithsonian Tropical Research Institute (STRI).

This kind of approach is more appropriate in the context of a study of wildlife observation for at least two reasons. First, it allows for signs and semeiotic processes (what semioticians describe as 'semiosis') that are not 'human-made'. The wildlife also produce signs, and this entails an agency which is denied, however implicitly, when the focus is entirely on social or culturally constructed meanings. Second, that selfsame agency may mean that beyond-human objects may have a certain capacity to determine how human groups interpret wild animals and their

---

[24] Ferdinand de Saussure, *Course in General Linguistics*, Charles Bally and Albert Sechehaye (eds.), translated by R. Harris (London: Duckworth 1983 [1916]).

[25] See for example, Charles Sanders Peirce, *Collected Papers of Charles Sanders Peirce*, 8 Vols., Charles Hartshorne, Paul Weiss, and Arthur Burks (eds.) (Cambridge: Harvard University Press, 1931-1958).

surroundings. I say a 'certain capacity' because of course, if this point is taken too far, it denies the importance of social and cultural processes in the interpretation of wildlife and their habitats. But here again, the Peircian perspective offers a good alternative to Saussurean or post-Saussurean approaches in so far as it allows for semeiotic regimes that combine indexical and symbolic, 'direct' and associative forms of meaning-making.

Thus far I have highlighted the *semeiotic* contribution that Peirce can make, but in Part 2 I will suggest that Peirce can make two further contributions to a study of wildlife observation. The first involves his phenomenology, or what can be described, provisionally, as a kind of philosophy of the fundamental aspects of consciousness (at a later stage it will be explained that Peirce's phenomenology in some respects goes far beyond *consciousness*). The second, related contribution involves Peirce's pragmaticist philosophy – a philosophy that, simplifying again, investigates not just the 'representational' role of signs, but their, or their actors' real capacity to make, express, and in the process *transform* the world. Taken together, these aspects of Peirce's thought, which are intimately related to his semeiotic theory, allow the analyst to develop a remarkably rich and subtle characterisation of the ways in which forms determine how both humans and beyond-humans perceive and interpret each other.

Sociological-Philosophical. Wildlife observation, amongst tourists or any other kind of group, typically involves practices which are mediated by institutions, fields, codes, discourses, genres, techniques and technologies, and more generally the ensembles of such elements which constitute cultural formations. In Part 3, I will thereby begin by employing what can be broadly characterised as a cultural sociological approach in order to consider the nature of such mediations. In doing so, I will draw on the work of scholars such as Pierre Bourdieu,[26] Michel Foucault,[27] and Raymond Williams.[28] Key to the analysis will be an account of how cultural formations may structure particular techniques of observation, and

---

[26] See for example Pierre Bourdieu and Loïc J. D. Wacquant, *An Invitation to Reflexive Sociology* (Cambridge: Polity Press, 1992); Pierre Bourdieu, *The Logic of Practice,* translated by Richard Nice (Cambridge: Polity Press, 1990[1980]); and Pierre Bourdieu, *Language and Symbolic Power,* edited by John B. Thompson and translated by Gino Raymond and Matthew Adamson (Cambridge: Polity Press, 1991).

[27] See for example Michel Foucault, *The Order of Things: An Archeology of the Human Sciences* (London: Routledge, 1986 [1966]); and Michel Foucault, *The Archeology of Knowledge*, translated by A. M. Sheridan Smith (New York: Barnes & Noble, 1993 [1969]). See also the later account of discourse in 'The Order of Discourse', included as an appendix in Michel Foucault, *The Archeology of Knowledge*, pp. 215-238.

[28] See for example, Raymond Williams, *Problems in Materialism and Culture* (London: Verso, 1980); Raymond Williams, *Culture* (London: Fontana Press, 1981); and Raymond Williams, *Keywords: A Vocabulary of Culture and Society,* revised edition (London: Fontana Press, 1983).

how individuals and groups may interiorise those techniques and employ them to interpret tropical forests and their wild denizens.

If, however, I say that I will *begin* by adopting a cultural sociological perspective, it is because Part 3 will also consider perspectives which take critical distance from the kind of conceptions proposed by the earlier cultural sociology. In particular, I will consider two interrelated sets of issues which have been raised by philosophers over the last two or so decades: those to do with the role of the body (or 'embodiment'), and those to do with a philosophy of difference.

Let us begin with the question of the body. It might be assumed that in the case of a human observer as in the case of any beyond-human animal, the own body is entirely a matter of a physical organism with certain parts and functions. For the cultural sociologists as for the philosophers and interdisciplinary forms of research that they draw on, this is an unacceptably reductive conception in so far as it overlooks the many ways in which socio-cultural processes determine how bodies are not only interpreted, but also lived and even shaped. According to poststructuralist philosophers such as Michel Foucault[29] and Judith Butler,[30] sociobiological and neuroscientific approaches to the body overlook the importance of sociality and culture. Especially in the writing of post-Foucaultian authors such as Butler, it is argued that a biological notion of embodiment is a figment of a discursive imagination.

There is much to be said for this kind of approach in so far as it alerts us to the problems with either naively organismic, or crudely biologist conceptions of the body. However, the risk is that it goes so far that it implicitly or explicitly denies embodiment as a more-than-discursive, indeed more-than-cultural reality with materialities which are not themselves purely or even mainly social or cultural (we return to the critique of culturalism). For this reason, in Part 3 I will problematise approaches such as Foucault's and Butler's via the writings of critical realist scholars such as Toril Moi,[31] and before her existentialist and phenomeno-

---

[29] See for example, Michel Foucault, *History of Sexuality, Volume I: An Introduction*, translated by Robert Hurley (New York: Pantheon Books, 1978[1976]). See also, Michel Foucault, 'Nietzsche, Genealogy, History', translated by Donald F. Brouchard and Sherry Simon, in James D. Faubion (Ed.), *Aesthetics, Method, and Epistemology: Essential Works of Foucault 1954-1984, Volume 2.* (Harmondsworth: Penguin, 1998 [1971]), 369-392.

[30] See for example, Judith Butler, *Gender Trouble: Feminism and the Subversion of Identity* (London: Routledge, 1990).

[31] See for example, Toril Moi, 'What is a Woman? Sex, Gender and the Body in Feminist Theory', in *What is a Woman and Other Essays* (Oxford: Oxford University Press, 2000), 3-120.

logical philosophers such as Simone de Beauvoir[32] and Maurice Merleau-Ponty.[33] In their different ways, the mentioned scholars acknowledge embodiment in ways that avoid the idealism that creeps into many poststructuralist accounts. As Toril Moi puts it, 'The widespread tendency to criticize anyone who thinks that biological facts exist for their "essentialism" or "biologism" is best understood as a *recoil* from the thought that biological facts can ground social values. Instead of denying that biological facts ground any such thing, ... poststructuralists prefer to deny that there *are* biological facts independent of our social and political norms'.[34] Moi argues that, rather than erase biological facts, it is necessary to acknowledge their materiality. This can be done without returning to biologism. One of way of doing so is via the thought of existentialist and phenomenological philosophers such as Simone de Beauvoir and Maurice Merleau-Ponty, who consider the body not just in situation, but *as* situation – a proposal that I consider in some detail in Part 3.

Let us now turn to the issues raised by a philosophy of difference. After recognising the contribution of the phenomenological philosophers, I turn to the philosopher Gilles Deleuze,[35] and later to Deleuze writing with the philosopher and activist Félix Guattari.[36] Deleuze and Guattari take issue not just with the culturalism of social theory, but with its tendency to take for granted an *identitary* ontology of all manner of bodies. Simplifying somewhat, such an ontology hinges on the notion that a body is something other than itself: x is y. This stance is arguably the predominant one in modern cultures, and can be illustrated in a number of ways. I noted above that Butler argues that bodies are not so much what they are in themselves, as the outcomes of discursive regimes. In the context of so-called identity politics, groups base their politics on an identification with one or another cause, itself grounded in a single, or relatively narrowly defined issue. An identitary perspective also prevails in scientific research apparently a world a way

---

[32] See Simone de Beauvoir, *The Second Sex,* translated by Constance Borde and Sheila Malovany-Chevallier (New York: Vintage Books, 2009 [1949]).

[33] See for example Maurice Merleau-Ponty, *Phenomenology of Perception,* translated by Colin Smith (London: Routledge, 2002 [1945]).

[34] Toril Moi, 'What is a Woman?', p. 41.

[35] See for example Gilles Deleuze, *Difference and Repetition,* translated by Paul Patton (New York: Columbia University Press, 1994 [1968]); Gilles Deleuze, *Expressionism in Philosophy: Spinoza,* translated by Martin Joughin (New York: Zone Books, 1990 [1968]); and Gilles Deleuze, *Spinoza: Practical Philosophy,* translated by Robert Hurley (San Francisco: City Light Books, 1988 [1970]).

[36] See for example Gilles Deleuze and Félix Guattari, *Anti-Oedipus: Capitalism and Schizophrenia,* translated by Robert Hurley, Mark Seem, and Helen R. Lane (Minneapolis, MN: University of Minnesota Press, 1983[1972]); and Gilles Deleuze and Félix Guattari, *A Thousand Plateaus: Capitalism and Schizophrenia,* translated by Brian Massumi (London: Continuum, 1988[1980]).

from such politics; this in so far as, for example, organisms are automatically approached as being versions of *x*, *y*, or *z species*, themselves regarded as being essential *to* the individual organism.

By contrast, and simplifying again, Deleuze argues that a body is itself before it is anything else; certainly it is not a kind of derivative of an essentialised *type*. Whereas it would appear that the identitary approach is all about recognising difference, from a Deleuzian perspective it may be argued that difference is actually denied in quite a fundamental way in so far as entities are defined by likening them to something else. If we are truly to recognise the uniqueness, and with it the positive difference of a body, it is necessary to adopt what is in some respects a radically materialist, and empiricist approach that starts from the *haecceity* i.e. the 'thisness' or 'itselfness' of actual bodies. This kind of ontology – a differential ontology – is what is required for a true 'philosophy of difference'.

Of course, no body is ever alone, and one first way in which Deleuze mediates his radically monadic account is by way of the concept of *affect*: what a body, what an *individual* can do always *also* depends at least in part on another body's, another individual's capacities; together, they work, however independent or unselfconsciously, to increase or decrease each other's affections. Then again, individuals don't simply work in twos, or in dyadic formations. What a body can do typically depends on numerous other bodies working as some kind of 'system', or better yet, some kind of *ensemble*. This is what Deleuze, and Deleuze and Guattari describe as an *assemblage*. But here the key insight is that assemblages are not wholes made up of wholly *dependent* parts; the parts continue to be relatively independent.

This may all seem very abstract until one considers the following: any encounter with wildlife involves a complex interrelation between the haecceity of the individual bodies (it is *this* person or *that* parrot snake, and it is *Barro Colorado's* forest and not tropical forests in general); the affects between any two bodies (e.g. the capacity of a tourist and an observed animal to detect each other's presence); and the *assemblage* of those bodies as part of a larger group of bodies whose logic of interrelation is not itself organismic. So for example, whether a tourist and an ocelot detect each other will depend on each individual, but also on the floristic structure of the forest, or the prevailing wind direction, neither of which is dependent on either the visitor or the ocelot. Or to offer a different example, a person who travels to a tropical forest for leisure purposes is not an organic part of that forest; to be sure, the same point can be made about a person flying in a jet on their way to the tourist destination (there is a crowd of people, most of whom are unrelated to each other). Then again, both the forest and the jetliner clearly affect what the individual can and cannot do, at least when they are in one or the other place. So again, the problem is to explain at once the haecceity of the bodies, and affects between bodies, and the logic(s) of assemblage.

As I will explain in Part 3, as part of such an explanation, there can be no simple distinction between what counts as a cultural and a biological body, a cultural and a biological assemblage. On the contrary, especially but not only in the case of wildlife observation, there is clearly a 'mixture' that denies any effort to categorise the different actors, the different parts, as being one or the other.

Ecological-Geographic. Of course, *in situ* wildlife observation always entails a certain ecology, and a certain geography. And indeed, wildlife observation as a tourist practice is all about going to see (or otherwise sense) wild animals in their own environments. This is a point that seems at once so obvious, and so fundamental as to render secondary the rest of the aspects or dimensions involved in wildlife observation – so much so that a case might be made that the study really ought to *begin* with the ecological and geographic aspects of wildlife observation, especially in so far as it is concerned with tourist practices in a particular kind of biome (tropical forests).

There are, however, at least two reasons for delaying the consideration of these aspects until the last full part of Volume 1. The first is that tourists' own habitual modes and techniques of observation typically develop before they reach the tropical forests. From a tourist perspective, the own embodied biography and the acquired techniques of observation come *before* the physical geography and ecology of the tropical forests; indeed, in so far as these techniques betray the influence of the media and other socio-cultural forces, then it is perhaps hardly surprising that, as suggested earlier, many tourists appear to come away with a deep sense of disappointment over what they actually *can* experience *in situ*.

A second, related reason for delaying the geographical analysis is that what one can or cannot observe is not *only* determined by the immediate, physical environment that one finds oneself in, nor by the morphology or ethology of the wild animals. Put more technically, there are good reasons to avoid both an ecological and a geographic determinism when it comes to wildlife observation.

The last points notwithstanding, it would be absurd to overlook the real significance of what might be termed the physics (in the sense of the physical properties) of tropical forests and their flora. When tourists visit a tropical forest, what fauna and flora they encounter *will* of course depend at least in part on the ecology and physical geography of the location (we return by this route to affects and to assemblage). If a similar principle applies to any biome, any ecosystem, it appears to be a particularly important one in the case of tropical forests, where a combination of the floristic structure and the ethology of the fauna almost guarantees that certain creatures will be difficult to find and/or observe.

With these two sets of considerations in mind, I will begin Part 4 with an account of the physical geography, and the ecology of tropical forests that is based on research conducted in the physical sciences. As part of this account, I will refer, amongst others, to the research of the ecologists Egbert G. Leigh,[37] Leslie Holdridge[38] and John A. Endler[39] to describe aspects such as the floristic structure of tropical forests, and some of the zoological aspects that most clearly work to determine what tourists can and cannot observe whilst engaging in *in situ* wildlife observation in those forests. An account of these kinds of determinations will also serve to specify the reference to tropical forests, a category that may seem self-explanatory but actually is as broad as it is potentially vague. As Part 4 will explain, the kind of forest that this volume will refer to may be more precisely characterised as the *moist* forests found in the lowland areas (i.e. < 1000 metres above sea level) of the Neotropics.

The importance of this 'physics' notwithstanding, and in keeping with the issues I raised above, in Part 4 I will suggest that an eco-geographic approach must include additional geographic perspectives that both problematise, and go beyond the physics of tropical moist forests. The second perspective that I will consider is the research conducted by human geographers, and more specifically the kind of analysis afforded by a subfield of human geography once known as the new cultural geography. This perspective builds on many of the same assumptions, scholars, and paradigms associated with the earlier cultural sociology. Its fundamental concepts were developed in the 1980s by scholars such as Denis Cosgrove and Stephen Daniels,[40] and Doreen Massey.[41] For the mentioned scholars, the experience of places and spaces must be understood from the perspective of cultural signifying systems, and the social orders which such systems at once produce and reproduce. This being the case, one cannot treat landscapes or wildernesses (or indeed other analogous categories such as 'the tropics') as if they were entirely objective entities that exist as they appear to exist, regardless of their

---

[37] See for example, Egbert Giles Leigh, Jr, *Tropical Forest Ecology* (Oxford: Oxford University Press, 1999) and Egbert G. Leigh Jr, A. Stanley Rand, & Donald M. Windsor (Eds.) *The Ecology of a Tropical Forest: Seasonal Rhythms and Long Term Changes* 2nd edition (Washington D.C.: Smithsonian, 1996).

[38] See for example Leslie Holdridge, 'Determination of World Plant Formations From Simple Climatic Data', *Science*, 105:2727 (1947), 367-368. See also Leslie Holdridge, *Life Zone Ecology*, Revised Edition (San José, Costa Rica: Tropical Science Centre, 1967).

[39] See for example John A. Endler, 'Interactions Between Predators and Prey', in John R. Krebs and Nicholas B. Davies (Eds.) *Behavioural Ecology: An Evolutionary Approach*, 3rd edition, (Oxford: Blackwell Scientific, 1991), 169-196.

[40] See for example, Denis Cosgrove and Stephen Daniels, 'Introduction: Iconography and Landscape', in Denis Cosgrove and Stephen Daniels (Eds.) *The Iconography of Landscape: Essays on the Symbolic Representation, Design, and Use of Past Landscapes* (Cambridge: Cambridge University Press, 1988), 1-10. See also Denis Cosgrove, *Geography and Vision: Seeing, Imagining, and Representing the World* (London: I.B. Tauris, 2008).

[41] See for example, Doreen Massey, *Space, Place and Gender* (Cambridge: Polity Press, 1994), and Doreen Massey, *For Space* (London: Sage, 2005).

representation. On the contrary, from the neo-Kantian perspective of the new cultural geography, a landscape is an entirely human and humanising phenomenon, one that may refer to a beyond-human set of bodies, but which is itself first and foremost a symbolic form.

Useful as the new cultural geography is in providing a kind of antidote to reductive approaches to ecology and geography, it too, may be critiqued in so far as its scholars also err on the side of culturalism, and with it the kind of idealism that may result from an over-emphasis on human consciousness. In a shift that suggests a number of parallels to the one I described in Part 3, Part 4 will thereby consider more recent forms of scholarship which entail a variation on the theme of what may be described as the new materialism, and the neo-empiricism introduced in Part 3. I will begin with the proposals of Science and Technology Studies (STS), and Actor-Network-Theory (ANT) – especially the research of both fields' leading exponent, Bruno Latour.[42] Latour's work, like that of Deleuze and Guattari, has been hugely influential in the critical social sciences and humanities. In particular, Latour's proposals regarding the hybridity of networks involving humans and beyond-human objects offers a number of important insights for the present study. Amongst other proposals, Latour vindicates the importance of including all sorts of 'things' as part of the analysis, including techniques and technologies of the kind employed by scientists. According to Latour, all such techniques and technologies, all such *things* work in networks to construct particular senses of reality. A parallel can be established, in this sense, between the Deleuzian/Guattarian concept of assemblage, and Latour's concept of network. One important difference is that Latour takes constructivism to a radically nominalist extreme, and this will be one of several aspects of his research that I will critique.

Above I noted that Latour's research has been highly influential across the critical social sciences, and indeed in the early 2000s a new form of human geography emerged that came to be known as 'hybrid geographies', which had Latourian ANT as one of its main inspirations. The founder, and leading exponent of the new geography was the human geographer Sarah Whatmore.[43] Simplifying somewhat, Whatmore's hybrid geography researches what might be described as the 'real' interface between the referents of the categories of culture and nature – an interface for which the nature-culture dualism either doesn't apply, or makes it difficult to even begin to understand dynamics which escape the conventional dichotomy.

---

[42] See for example, Bruno Latour, *We Have Never Been Modern*, translated by Catherine Porter (Cambridge, MA: Harvard University Press, 1993[1991]); Bruno Latour, *Politics of Nature: How to Bring the Sciences Into Democracy*, translated by Catherine Porter (Cambridge, MA: Harvard University Press, 2004); and Bruno Latour, *Reassembling the Social: An Introduction to Actor-Network-Theory* (Oxford: Oxford University Press, 2005).

[43] See for example Sarah Whatmore, *Hybrid Geographies: Natures Cultures Places* (London: Sage, 2002).

I will be particularly interested in considering the implications for the present study of what Whatmore describes as topologies of wildlife. Such topologies explain what happens when wild animals come to be incorporated in the kinds of networks generated by zoos, conservationist institutions, and wild animal farms. Whatmore employs ANT, but also posthumanist feminist theory, biophilosophy, and animal geographies to show how the animals in effect become something other than the *wild* animals they are conceived as when they enter the networks instituted by the mentioned genres, institutions, and fields. I will suggest that this kind of approach provides important insights for tourism – especially in the case of the more overtly intrusive forms of tourism, as well as conservationist research of the kind that is often engaged by scientists in biological reserves such as Barro Colorado's.

Significant as ANT and hybrid geographies may be to a study of wildlife observation, in the last chapter in Part 4 I will argue that neither of these approaches engages in sufficient detail with what some philosophers have described as the 'question of the animal'. If one doesn't accept the now centuries-old opposition and hierarchy established between humans and beyond-human animals, then how should animals be conceived? To answer this question, I will draw on three theoretical approaches. To begin with, I will refer to the work of the biologist, and proto-biosemiotician Jakob von Uexküll,[44] whose influential concept of *umwelt* posits a kind of surrounding world or 'bubble' for different animals, and which constitutes the basis for their engagement with the world. Uexküll's approach allows us to move beyond the anthropocentrism associated with the earlier forms of cultural sociology, for which human consciousness and human practices more generally are all that really matters.

Thereafter, I will return to Deleuzian and Guattarian writing in order to consider how their philosophy allows us to conceive animals. Key to this will be the notion that animals, like bodies more generally, are always 'becoming'; they are not fixed entities, or indeed essences defined by type (e.g. species), but entities with trajectories that are constantly re- and de-territorialising. What is true for individual animals and indeed for the populations that constitute species is also true for their habitats. In her later work, the geographer Doreen Massey, whom as I noted is associated with the earlier cultural geography, illustrates this point beautifully when she notes that Skiddaw, a 'not pretty' but impressive block of a mountain in the English Lake District – a seemingly 'unmovable' and 'timeless' mountain – was actually once south of the equator. As Massey puts it, '[t]he bit that we know today as the slates of Skiddaw crossed the equator about 300 mil-

---

[44] See for example Jakob von Uexküll, *Theoretical Biology*, translated by D. L. MacKinnon (New York: Harcourt Brace, 1926). See also Jakob von Uexküll, *A Foray Into the Worlds of Animals and Humans* with *A Theory of Meaning*, translated by Josep D. O'Neil (Minneapolis: University of Minnesota Press, 2010 [1934 & 1940, respectively]).

lion years ago'.[45] A similar point might be made about what we now know as the Panamanian isthmus, which once was a sea, but which some three million years ago began to change into the land bridge that enabled a certain biogeographic circulation between the northern and southern landmasses known as North and South America.

The third approach that I will consider is that of the biologist and feminist philosopher Donna Haraway,[46] who in her work on 'companion species'[47] takes issue with Deleuze and Guattari's perspective on wild animals, and introduces additional layers and elements of analysis by considering all those aspects of animals which tend to be either overlooked, or dismissed by both scientific, and traditional philosophical analysis: 'Can animals play? Or work? And even, can I learn to play with *this* cat? Can I, the philosopher, respond to an invitation or recognize one when it is offered? ... And what if the question of how animals engage *one another's* gaze *responsively* takes center stage for people?'[48]

These approaches to animality (and to ecology and geography more generally) take the analysis back to the kinds of issues raised in Part 3 with respect to the haecceity of bodies and encounters with wildlife. But they also reveal further problems with the kind of approaches that I consider in Part 1. A positivist, and sociobiological approach to wildlife encounters of the kind advocated by Edward O. Wilson (and considered in Part 1) emphasises the transcendental role of ancient adaptations; for Wilson, as for other sociobiologists, what I described earlier with Eleanor Gibson as perceptual learning, or indeed what might be described with a sociologist like Pierre Bourdieu as a *habitus* – a socially acquired disposition – are of secondary importance when compared to a template ostensibly established for humans in the Pleistocene. So, for example, Wilson argues that we humans have an ancient fear and fascination of snakes, which can only really be confirmed by cultural experience. By contrast, a stance such as Deleuze and Guattari's insists on the importance for humans (as for all other creatures) of an ongoing, and always transformative relationality – but one which starts from real individuals, and their own situations. From this perspective, how we humans react to animals depends not so much on adaptations, but on each individual's at once biological, sociological, and geographic trajectory, or what I will also describe as a dynamic situation which may shaped by aspects such as media use, prior experiences with certain animals, and of course social relations more generally, but which can never be reduced to any such 'factors'. The same principle applies to

---

45 Doreen Massey, *For Space*, p. 133-134.

46 See for example Donna Haraway, *Primate Visions: Gender, Race, and Nature in the World of Modern Science* (New York: Routledge, 1989), and *Simians, Cyborgs, and Women: The Reinvention of Nature* (London: Free Association Books, 1991).

47 Donna J. Haraway, *When Species Meet* (Minneapolis: University of Minnesota Press, 2008).

48 Donna J. Haraway, *When Species Meet*, p. 22.

the beyond-human animals themselves; the members of a species that may have, as 'members' of that species, a shared ethology. But each individual's response is never the outcome of a kind of transcendental species logic; as I will show, beyond a contextual logic, there are always also those aspects which have to do with what I will describe as a principle of immanence: simplifying greatly, each animal is itself before it is something else.

* * *

Having provided an overview of the different approaches, I would now like to say something about the logic that informs my recontextualisation of the different theories. As I have already explained, the book is divided into four parts, each of which explores a different dimension, or a set of dimensions of wildlife observation of the kind engaged by tourists in tropical forests. I have structured the different parts in such a way that each starts from a concrete example of encounters with wildlife, and then uses that example to consider various theoretical issues. In general, I start from a kind of theoretical common sense, and then problematise the initial approach via more and more complex perspectives.

In so doing, I bring together the work of scholars whose proposals may entail significant theoretical discontinuities. For example, while many seem to believe that Deleuzian philosophy goes hand in hand with Latour's Actor-Network-Theory, the former perspective is broadly realist, while the latter is radically nominalist. Conversely, in some cases research that I will present in different parts of the volume may entail important continuities. This is particularly evident in the case of the chapters that focus on the older forms of cultural sociology (Part 3) and the once new cultural geography (Part 4); while it is possible to treat each of these quite separately, in practice scholars in the latter field initially drew on the research from the former field (or at any rate, from cultural studies). Then again, the work of some of the authors that will be presented in this volume refuses any kind of classification even along interdisciplinary lines – the most obvious case in point, the research of Deleuze, and of Deleuze and Guattari, which goes far beyond what I will nevertheless describe as sociological-philosophical perspectives.

As this description begins to make clear, the reader will not find in this work the kind of study that appears to stick seamlessly to one theory. On the contrary, for the reasons that I explained in my account of transdisciplinarity, in this study I opt for an approach that is deliberately polyvalent. Deleuze famously suggested that it is possible to treat the work of other scholars in much the way that one does a toolbox,[49] picking and using the conceptual tools that are most appropriate to the own research, and changing them to suit the own project. In this study I will be doing just that with aspects of Deleuze and Guattari's philosophy (and the theories of a good many other scholars as well). In so doing I will strive to point out the most obvious discontinuities between theories, as well as the more important changes introduced vis-a-vis the borrowed concepts.

For these and other reasons that will become clear in due course, the reader will notice that as the chapters within each section progress, the analyses will end up in places that are, conceptually and methodologically speaking, quite different from the ones from which each part begins. The same point can be made about the book as a whole. In so far as Parts 2 to 4 develop particular perspectives with respect to wildlife observation amongst tourists, it will be necessary to draw together the different parts of the analysis to offer a fully transdisciplinary perspective on wildlife observation in tropical forests. That will be the task of the last full chapter of this volume, in which I will suggest how a geosemeiotic approach might be developed via a series of homologies between the key concepts developed in each of the different sections of the work.

This volume ends with a postscript that acknowledges the elephant in the transdisciplinary room: catastrophic climate change. The research for this book began at a time when it was still possible to pretend that climate change was a problem for 'future generations'. In 2017 and 2018, when I was completing this volume, it became abundantly clear that climate change is now an emergency that not only cannot be denied (unless one lies to oneself or to others), but which demands immediate action from each and all of us. In the postscript I offer at once a *mea culpa,* and some suggestions for ways of engaging with the catastrophe that are consistent, I hope, with the approach that I develop across the following chapters.

---

[49] Gilles Deleuze, 'Intellectuals and Power', in *Desert Island and Other Texts (1953-1974),* (Cambridge: MA: MIT Press, 2004), 206-213, p. 208. In this text, which records an interview with Michel Foucault, Deleuze says: 'Yes, that's what a theory is, exactly like a tool box. It has nothing to do with the signifier ... A theory has to be used, it has to work. And not just for itself.'

PART 1

Perceptions of Wildlife
Positivist Perspectives

Our guide stops and points at the base of a tree: crawling along the trunk are several enormous ants. The guide explains that these are bullet ants (*Paraponera clavata*), also known in the Spanish language as the *Hormiga Veinticuatro* – the 24 (Hour) Ant. The vernacular names of the ants refer to the excruciatingly painful, and long-lasting effects of the species' sting. The guide further explains that bullet ants tend to forage high up in trees, so their stings are unlikely to be a problem for anyone walking along tropical forest trails. To be sure, even if bullet ants *were* to be encountered at the base of a tree (which is where the species often builds its nests), the species' large size – typically about 30 mm, or more than one inch in length – should make the ants relatively easy to spot. After taking the requisite pictures, our group moves on.

I later learn that one scientist, Justin O. Schmidt, has described the bullet ants' sting as the most painful of any produced by the hymenopteran insects.[1] Schmidt should know: his 'Schmidt Sting Pain Index' attempts to develop a scale of the relative pain inflicted by the venom of the different members of the order, which includes ants, wasps and bees. To develop the scale, Schmidt has endured the stings of many such insects; so he must be (or at least must *have been*) an expert in pain. From an orthodox scientific perspective, the idea of generating such an index based on one individual's experiences is problematic, to say the least.[2] However, Schmidt has also partaken in the more conventionally scientific analysis of the chemistry of *P. clavata's* venom. The venom contains three fractions that block synaptic transmission in other insects; unfortunately for humans, two of the fractions may also act as agonists on vertebrates, which are presumably more of a threat to the ants. One of the fractions contains kinin, a polypeptide that produces vasodilation and the contraction of smooth muscle. The other fraction, which the

---

[1] Justin O. Schmidt, 'Hymenoptera Venoms: Striving Toward the Ultimate Defense Against Vertebrates' in D. L. Evans and J. O. Schmidt (Eds.), *Insect Defenses: Adaptive Mechanisms and Strategies of Prey and Predators* (Albany: State University of New York Press, 1990), 387–419.

[2] I use the term 'orthodox' advisedly. About a year after I completed the first draft of this chapter, the *Guardian* published an article by Justin O. Schmidt in which he explained that 'working as a chemical ecologist, I was still fascinated by the defence mechanisms of wasps, bees and ants. As part of my job studying interactions between insects and their environment, I did fieldwork that involved collecting different species, providing endless opportunities to get stung. It occurred to me that while the venom could be analysed, it was harder to make comparisons between the type of pain each insect caused, which seemed crucial in understanding how and why different types of sting had evolved'. See Justin O. Schmidt, 'Experience: I Have Been Stung by 150 Species of Insect', in *Guardian* online, 11 May 2018 <https://www.theguardian.com/environment/2018/may/11/experience-i-have-been-stung-by-150-species-of-insect>[Accessed 12 May 2018].

researchers suggest is the most active from a neurotoxic perspective, is what they call a *poneratoxin*, a peptide that causes paralysis and pain.[3]

Research of the kind engaged by Schmidt and his colleagues tells us about bullet ants and the effects of the species' venom. But what happens when a person *observes* a bullet ant, or indeed any other creature, in a tropical forest? To my knowledge, no researcher has taken the time to explain this kind of activity as it occurs amongst tourists visiting a tropical forest (indeed much the same point might be made about other biomes). This raises the question: how best to theorise such a practice?

In keeping with the propaedeutic aims described in the Preface, in this first part of the study I will offer an account of several paths *not* taken in this study: what I will characterise as three *positivist* perspectives on perception. From a positivist perspective, even if there are vast differences between ants and human beings, there is, in principle, no reason to approach the study of either species (or indeed of any other kind of organism) according to fundamentally different forms of scientific practice. At least some of the research must of course be species-specific, but generally speaking, anyone interested in studying the ways in which ants or humans relate to their environment should employ much the same epistemological criteria. Indeed, however much *observing* a bullet ant, and being *stung* by one entail rather different kinds of events, both sets of events involve, broadly speaking, much the same physical forces and dynamics. As we have seen, bullet ant venom acts by way of peptides and polypeptides that react with the neurophysiological function of vertebrates; by a similar token, if we are able to *see* bullet ants, it is because light enters our pupils and provokes a different set of neurological reactions, *but neurological reactions nevertheless.*

In the following four chapters I will explain what's wrong with this kind of approach. To this end, Chapter 1 will provide an introduction to positivist philosophy. Chapter 2 will consider Edward O. Wilson's biophilia hypothesis and sociobiological theory of gene-culture coevolution. Chapter 3 will then turn to the neuropsychological theory of perception proposed by David Marr. Finally, Chapter 4 will consider the proposals of the ecological psychologist James Gibson.

---

[3] See Tom Piek, Alain Duval, Bernard Hue, Henk Karst, Bruno Lapied, Piet Mantel, Terumi Nakahima, Marcel Pelhate, & Justin O. Schmidt, 'Poneratoxin, a Novel Peptide Neurotoxin From the Venom of the Ant *Paraponera clavata*', *Comparative Biochemistry and Physiology Part C: Comparative Pharmacology*, 99(1991), 487-495. For an updated account, see the more recent Stephen R.Johnson, Hillary G.Rikli, Justin O.Schmidt and M. Steven Evans, 'A Reexamination of Poneratoxin from the Venom of the Bullet Ant *Paraponera clavata*', *Peptides*, 98 (December 2017), 51-62.

1

On the Rules of Positivism

In the critical social sciences and humanities, to say that someone, or someone's research is 'positivist' is more often than not a way of suggesting that the investigator adheres to overly restrictive parameters, that they have a tendency to simplify complex dynamics on the basis of a reductionist approach, and that they engage in politically conservative forms of research. It goes almost without saying that such accusations are not always fair, and indeed they may say more about the accuser's ideology than about the quality of the criticised research.

To be sure, there was once a time when positivism was equated with a progressive disposition. Indeed, across much of the nineteenth century and the early twentieth century, precisely the values that would later be criticised for their conservatism were regarded as a sign of a liberal, or in some cases even a socialist politics. It says something about our times that, at a historical point when the forces of fascism are once again on the rise, those who would normally critique positivist forms of research may find themselves in the ironic position of defending positive science against no science at all.

The development of an explicitly positivist philosophy is often attributed to Auguste Comte (1798-1857), an influential nineteenth century philosopher and the founder of sociology as a scientific discipline.[1] Comte developed his philosophy at a time when France was experiencing the aftermath of the French Revolution and the Napoleonic Wars. This was a time of epochal transformations generated by the shattering of the autocratic orders of old, and the rise of cultural formations strongly influenced by industrialisation and an incipient modern science. In this context, Comte proposed a new way of studying the social world, but also what he unabashedly described as a new world *religion* – one that he believed was more fit for the times, and which he thought should be based on a social evolutionist account of intellectual development, or what Comte more simply described as 'mind'.

According to this account, mind progresses across three stages of development: the *theological*, dominated by the presupposition that all phenomena are produced by the immediate action of supernatural beings; the *metaphysical*, a transitional phase that is only a modification of the theological in so far as it substitutes a variety of abstract forces for the supernatural beings (so, for example, 'Nature' is substituted for 'God'); and finally, the scientific or *positive*, when the mind is finally able to devote itself to finding the laws that govern phenomena, viz.,

---

[1] See Auguste Comte, *The Positive Philosophy of August Comte, Vols I & II*, translated and abridged by Harriet Martineau (London: Kegan Paul, Trench, Trübner & Co, 1893[1830-1842]).

what Comte describes as 'their invariable relations of succession and resemblance'.[2]

As Comte puts it in *A General View of Positivism*, in the society made possible by positive mind, it is necessary to 'generalize our scientific conceptions' and to 'systematize the art of social life'.[3] The discipline that ought to lead such practices is what Comte characterises as a new *physique sociale*, or what he also describes as *sociologie*: a science that should build on the methods of the physical sciences, whose development Comte analyses as part of an itself influential survey of the state of positive mind. In Comte's account, sociology will become the 'queen of the sciences'; and indeed, in keeping with his evolutionist inclinations, Comte sees a progression within science itself: if the positive, or scientific mind began with mathematics, astronomy and physics, and then progressed to chemistry and to biology, henceforth sociology will constitute the culmination of positive mind in so far as it would be devoted to explaining that most difficult of subjects, the physics of human social practices.

Comte was brought up as a Roman Catholic, but in keeping with the liberal currents of his time, he is critical of the Church and is an advocate of a completely secular state. He nevertheless proposes that, in the society made possible by positive mind, a 'spiritual power' is necessary to 'direct the future of society by means of education'; as a 'supplementary part' of that education, this spiritual power should 'pronounce judgement upon the past' and 'the relations between different classes'.[4] Indeed, the new religion should be led by 'intellectual and moral directors' whom Comte conceives as priest-like scientists. The new scientific clerics should employ their superior knowledge and the methods of positive science in order to correct any inclinations amongst the population to regress to the theological and metaphysical errors of old.

As this account begins to make clear, Comte's social physics was a forerunner of *policy science*:[5] if the task of sociology is to discover the immutable laws of social physics, the scientific intelligentsia should employ the resulting knowledge not only to predict, but where necessary also to *correct* prevailing practices.

In our own time, one with a generalised scepticism regarding any non-technical role for scientists in society, a utopian society led by priest-like scientists may seem bizarre, to say the least. However, Comte's writing proved to be extraordinarily influential. Amongst other things, throughout the second half of the nineteenth century and beyond, Comtian positivism provided an ironically spiritual guide for the establishment of modern states across much of the world. To men-

---

[2] Auguste Comte, *The Positive Philosophy of August Comte, Vol. I*, p. 2.

[3] Auguste Comte, *A General View of Positivism*, translated by J. H. Bridges (Cambridge: Cambridge University Press, 2009 [1844]), p. 3.

[4] Auguste Comte, *The Positive Philosophy of August Comte, Vol. I*, pp. 114-115.

[5] See Brian Fay, *Social Theory and Political Practice* (London: Allen and Unwin, 1975).

tion just one famous example, Brazilian disciples of Comte founded the Republic of Brazil in 1891 and declared the national motto to be 'Order and Progress' (*Ordem e Progresso*) in an explicit reference to Comte's own 'Order and Progress, and Above All Else Love'. As Comte explains it, 'The Positivist regards artificial Order in Social phenomena, as in all others, as resting necessarily upon the Order of nature, in other words, upon the whole series of natural laws.[…] But Order has to be reconciled with Progress…'; '[i]n Sociology the correlation assumes this form: order is the condition of all Progress; Progress is always the object of Order. Or, to penetrate the question still more deeply, Progress may be regarded simply as the development of Order; for the order of nature necessarily contains within itself the germ of all possible progress'.[6] Especially in the later work of Comte, love comes into the picture as the basis for the new positivist religion, whose adherents should express a love of *humanity*.

I will return to questions of order and progress in Chapter 2. For now, it should be noted that, for all of Comte's influence, positivism as we know it today probably owes at least as much to the rise of logical positivism in the Europe of the 1920s and 30s, and more specifically in Vienna. I would now like to turn to this kind of positivism, which entails some continuities but also important discontinuities with respect to Comtian thought.

At the time of the First Austrian Republic (1919-1934), Austrian politics were dominated by the conflict generated by the Allied Power's imposition of the Treaty of Saint Germain, which effectively transformed Austria into a rump state. In this context, Vienna, formerly an imperial capital, underwent a remarkable transformation: it became something akin to a proto-welfare state-within-a-state. While the Austrian state was dominated by parties to the right and far-right – with the active support of the Roman Catholic Church – in Vienna itself a democratically elected, and socialist government promoted a social welfarism that was still decades away in the rest of western Europe. For the Church and for fascist politicians alike, the 'Red Vienna' was thereby a social experiment that must be stopped at all costs, including violent attacks by right-wing paramilitaries. The attacks were resisted by left-wing groups in what became an increasingly murderous political context across the 1920s and 1930s.

Even as Vienna became a battleground for the far-right and left, a group of scientists-cum-philosophers began to meet regularly in the University of Vienna, and to discuss ways of using science to oppose the re-emergence of what they too, described as theological and metaphysical thought. Throughout the 1920s, the group referred to itself as the Ernst Mach Association, after the famous physicist; however, in 1929 its members renamed themselves as the 'Vienna Circle' (*Wiener Kreis*), and today the members of the group are more often referred to by that name, or as the logical positivists.

---

[6] Auguste Comte, *The Positive Philosophy of August Comte, Vol. I*, pp. 115-116.

Part 1 Positivist Perspectives

The Vienna Circle was led by Moritz Schlick, and included figures such as Otto Neurath, Rudolph Carnap, Hans Hahn, and Herbert Feigel. These and several other intellectuals shared a particular professional characteristic: while they ended up as philosophers (or at any rate, as 'social thinkers'), each started out in mathematics or in the positive sciences. For example, Schlick, Carnap, and Feigl trained as physicists, and Hahn was a mathematician. While Neurath started out as a sociologist and an economist, his early research was very much in the positivist tradition.

The logical positivists' discussions might not have amounted to much (at least not much beyond academia) had they kept their discussions to the lecture halls of the University of Vienna. However, in 1929, as politics in Austria became more and more violent, the Ernst Mach Association decided to go public, and to start what was, in effect, an academic public relations campaign. The campaign included the publication of a manifesto; editing a journal (*Erkenntnis*, which translates roughly as 'knowledge recognition') and some book series. Crucially, it also entailed producing themselves influential conferences about logical positivism. The conferences were held across a number of European and the North American cities, including Cambridge, Copenhagen, Prague, and Cambridge, Massachusetts. In one sense at least, the campaign was remarkably successful; as suggested earlier, what today we describe as positivism probably owes at least as much to the writing of Schlick and his fellow scientists-cum-philosophers as it does to Comtian positivism.

So what did the members of the Vienna Circle propose? Comtian positivism was explicitly political, however much it tried to replace existing political structures with a scientific intelligentsia. By contrast, the logical positivists set out to draw an absolute line between scientific and any other kinds of discourse, and to reserve for scientific discourse, and scientific discourse alone, the right to make any truth claims. To this end, they suggested that the analysis of any matters involving truth claims ought to be addressed exclusively via a 'scientific conception of the world', as presented in a manifesto with that very title: *The Scientific Conception of the World. The Vienna Circle.*[7]

In the manifesto, the logical positivists suggest that the scientific conception of the world 'is characterised not so much by theses of its own, but rather by its basic attitude, its points of view and direction of research', viz., '... [it] knows only empirical statements about things of all kinds, and analytic statements of logic and mathematics':

---

[7] Hans Hahn, Otto Neurath, and Rudolf Carnap, *Wissenschaftliche Weltauffassung. Der Wiener Kreis.* English translation, *The Scientific Conception of the World. The Vienna Circle.* Vienna, Austria, August 1929. Pamphlet republished in Marie Neurath and Robert S. Cohen (Eds.) *Otto Neurath: Empiricism and Sociology* (Boston: D. Reidel, 1973), 299-318.

If someone asserts "there is a God", "the primary basis of the world is the unconscious", "there is an entelechy which is the leading principle in the living organism"; we do not say to him: "what you say is false"; but we ask him: "what do you mean by these statements?" Then it appears that there is a sharp boundary between two kinds of statements. To one belong statements as they are made by empirical science; their meaning can be determined by logical analysis or, more precisely, through reduction to the simplest statements about the empirically given. The other statements, to which belong those cited above, reveal themselves as empty of meaning if one takes them in the way that metaphysicians intend. One can, of course, often re-interpret them as empirical statements; but then they lose the content of feeling which is usually essential to the metaphysician. The metaphysician and the theologian believe, thereby misunderstanding themselves, that their statements say something, or that they denote a state of affairs. Analysis, however, shows that these statements say nothing but merely express a certain mood and spirit. To express such feelings for life can be a significant task. But the proper medium for doing so is art, for instance lyric poetry or music.[8]

This text has been quoted in some length to show the extent to which logical positivists at once privileged a certain scientific discourse, and denied any real ontological value in metaphysical discourse. As far as the logical positivists were concerned, if the aforementioned discursive separation was strictly observed, and with it the methods proposed by the Vienna Circle, then science should cut through the confusions of theological and metaphysical discourse, and so undermine the rationales that underpinned the actions of fascist groups in Austria, Germany, and elsewhere in the world.

In Austria itself, the logical positivists' project came to nought after fascist forces dissolved the parliament in 1934 and instituted a dictatorship that ultimately resulted in the country's annexation by Nazi Germany (the so-called *Anschluss* of 1938). In the run-up to the Anschluss, the threat of violence was such for liberal, socialist and pacifist intellectuals that most members of the Vienna Circle chose to exile themselves elsewhere in Europe, and later in Britain and the United States. One notable and tragic exception was Schlick. Schlick decided to stay in Vienna, and was murdered in 1936 by Johan Nelböck, a former student. In the subsequent trial, Nelböck was portrayed as a psychopath, but after the *Anschluss*, the Nazi authorities effectively pardoned Nelböck's crime and released him from jail.

---

[8] Otto Neurath, Hans Hahn, and Rudolf Carnap, *The Scientific Conception of the World.*

Part 1 Positivist Perspectives

At a time when there are increasingly obvious parallels to be drawn between events in contemporary US and Europe and the Europe of the 1930s, it is tempting to sympathise with any effort to establish seemingly ironclad guarantees of truth and objectivity. And indeed, faced as we are now with very deliberate attempts to blur the boundary between truth and abject propaganda, it might well seem that one solution is to put science on a discursive pedestal, in the way that the logical positivists did. In fact, now as in the 1930s, there are good reasons for rejecting any such strategy.

A first set of reasons involves the epistemological limitations of positivism (logical or other), and in the remainder of this chapter, I will consider these limitations via the critique offered by the historian of ideas Leszek Kolakowski (1927-2009).[9] A second set of reasons involve an irony. The logical positivists were liberal, or in some cases socialist intellectuals who attempted to use scientific discourse to combat right-wing discourse. However, after World War II, key aspects of the logical positivist discourse went mainstream, and did so in ways that arguably strengthened some of the very forces that logical positivists opposed. Between the mid 1940s and 1970s, positivism became a key ideological prop for the dominant institutions of both capitalist and communist nations throughout much of the world. As such, it played, and even now still plays an important role in supporting very much the kind of politics that the Vienna Circle opposed. I will discuss this irony as part of the account of the work of Wilson and Marr in Chapters 2 and 3, respectively.

First, let's turn to Leszek Kolakowski's critique of positivist philosophy.[10] Many positivists would deny that they adhere to a *philosophy*, arguing instead that they subscribe to what are simply tried and tested scientific methods. With Kolakowski it is possible to better understand how positivism *is* based on a philosophy, and with it, a certain ontology and metaphysics, i.e. a notion of 'what there is', what *being* is all about, and with these, a particular discourse about the nature of knowledge (an epistemology). According to Kolakowski, positivism may be defined as a 'a collection of prohibitions concerning human knowledge, intended

---

[9] Leszek Kolakowski, *Positivist Philosophy: From Hume to the Vienna Circle,* translated by Norman Guterman (Harmondsworth: Penguin Books, 1969 [1966]).

[10] Kolakowski's biography involves an irony of its own. Having started out in Poland as a new left critic of Stalinism, Kolakowski was forced into exile in Britain, where he was eventually accused of betrayal by a leading member of the British New Left, and went on to become something of a hero of the right. See Edward P. Thompson, 'Open Letter', *Socialist Register,* 10 [1973], 1-100). See also Kolakowski's reply in the following issue of the *Socialist Register* ('My Correct Views on Everything', in *Socialist Register,* 11 [1974], 1-20). Thanks to this exchange, but also to Kolakowski's support for the *Solidarity* movement in Poland, the obituary published by *The Economist* on 20 July 2009 described Kolakowski as 'Marxism's most perceptive opponent' [<https://www.economist.com/node/14120114>[Accessed 5 April 2018].

to confine the name "knowledge" or "science" to the results of those operations that are observable in the evolution of the modern sciences of nature'. Positivism, Kolakowski explains, opposes metaphysical speculation of every kind, and so is against 'all reflection that either cannot found its conclusions on empirical data or formulates its judgements in such a way that they can never be contradicted by empirical data'.[11]

Kolakowski suggests that four assumptions underpin this approach, each of which can and must be problematised. The first involves what Kolakowski describes as the *rule of phenomenalism* – the idea there is no difference between 'essence' and 'phenomenon'.[12] Accordingly, scientists are entitled to record only that which is actually manifested in experience;[13] when, for example, a chemical ecologist studies the effect of *P. clavata* stings, it is probably possible to determine quite unequivocally when envenomation has occurred. From a positivist perspective, the essence of envenomation – and positivism entails that, an ostensibly non-metaphysical quest for essences – is to be found not so much in the pain expressed via exclamations or words by the victim, but in the quantifiable chemical and neurological processes provoked by the sting.

The rule of phenomenalism might seem to be a matter of scientific common sense. However, if this kind of approach is taken to human practices a problem may arise that was famously illustrated by the anthropologist Clifford Geertz.[14] In an analogy that Geertz borrows from the philosopher Gilbert Ryle, he describes two boys who are 'rapidly contracting the eyelids of their right eyes'.[15] In one of the boys, this is an involuntary twitch, while in the other, it is a conspiratorial signal to a friend. In as much as the two movements are, as physical movements, identical – with Kolakowski, we might say identically *manifest* – the analogy serves to show why the rule of phenomenalism, and its concomitant essentialism may be misleading: as Geertz puts it, 'the difference, however unphotographable, between a twitch and a wink is vast; as anyone unfortunate enough to have had the first taken for the second knows. The winker is communicating, and indeed communicating in quite a precise and special way'[16] – a way that, to be understood, requires a knowledge, however implicit or unconscious, of a social code. Geertz underscores this point with reference to a third boy, who 'parodies the first

---

[11] Leszek Kolakowski, *Positivist Philosophy*, p. 18.

[12] Leszek Kolakowski *Positivist Philosophy*, p. 11.

[13] Leszek Kolakowski *Positivist Philosophy*, p. 11.

[14] Clifford Geertz, 'Thick Description: Toward an Interpretive Theory of Culture', in *The Interpretation of Cultures* (New York: Basic Books, 1973), 3-30.

[15] Clifford Geertz, 'Thick Description', p. 6.

[16] Clifford Geertz, 'Thick Description', p. 6.

boy's wink as amateurish, clumsy, obvious, and so on... Here, too, a socially established code exists... and so does a message'.[17]

A second positivist assumption is what Kolakowski describes as the *rule of nominalism*. This rule 'states that we may not assume that any insight formulated in general terms can have any real referents other than individual concrete objects.'[18] As Kolakowski notes, this ostensibly self-evident rule is clearly contradicted by geometry: when we say that the sum of the angles of any triangle is equal to two right angles, what does this statement refer to? Clearly, not to one or another actual, as in 'physical' triangular body. But of course, the statement clearly does still refer to *something* – something which, strictly speaking, is not recognised by positivist philosophy thanks to its advocates' insistence that we only have 'a right to acknowledge the existence of a thing ... when experience obliges us to do so'.[19] Note that *this same argument would offer precisely a reason for contradicting the rule of nominalism*; however, from a radically positivist perspective, constructs of the kind produced by geometry, mathematics, and indeed physics are purely ideal in the sense that they are no more than a convenient way of describing a reality that is in and of itself *utterly empirical*; as Kolakowski puts it, paraphrasing the argument of positivists, '[t]here is no reason to suppose, because we assume such situations for the convenience of our calculations, that they must actually exist anywhere else in reality. The world as we know it is a collection of individual observable facts.'[20]

I will return to the problems with nominalism in Part 2. With the philosopher Gonzalo Rodríguez-Pereyra,[21] I will explain that there are two different varieties of nominalism, each with what is potentially a very slippery ontology.[22] Here it suffices to point out that it would appear that the kind of nominalism that Kolakowski attributes to positivism is one that in some sense doubts the reality of abstract concepts. In this study I will oppose this, and indeed one other kind of nominalism (which rejects universals) by way of an ideal-realist approach that foregrounds the semeiotic dimensions of wildlife observation. From the perspective of such an approach, the triangles of geometry are *signs* that are at least as real as the objects they re/present. Signs may be employed as part of complex as-

---

[17] Clifford Geertz, 'Thick Description', p. 6.

[18] Leszek Kolakowski *Positivist Philosophy*, p. 13.

[19] Leszek Kolakowski *Positivist Philosophy*, p. 14.

[20] Leszek Kolakowski *Positivist Philosophy*, p. 15.

[21] Gonzalo Rodríguez-Pereyra, 'Nominalism in Metaphysics', *The Stanford Encyclopedia of Philosophy* (Winter 2016 Edition), Edward N. Zalta (Ed.) <https://plato.stanford.edu/archives/win2016/entries/nominalism-metaphysics/>[Accessed 21 July 2018].

[22] Kolakowski is attributing to positivism the kind of nominalism that rejects abstract objects. In forthcoming chapters, I will refer to a slightly different variety of nominalism, namely, the kind that rejects universals (see Parts 2 and 3).

semblages (cf. Part 3) to transform even the overtly empirical objects that logical positivists would prefer to stick with (perhaps I should say, with another wink, to *triangulate*). If the idea that signs can be and *are* used to change the world seems strange, it may be noted that if a person asks somebody politely to open a door, it is quite likely, *ceteris paribus*, that they will do so. Then again, it may also be noted that engineers use geometry (and other mathematical systems, themselves based on signs) to design dams, build roads and other structures – the very structures that have often been held responsible for untold environmental destruction.

A third rule, which may be described as the *rule of objectivism* (Kolakowski does not supply a name of his own), suggests that value judgements and normative statements do not constitute legitimate knowledge.[23] This rule follows on from the mistakes of the first and second rules: phenomenalism suggests that values cannot be discovered in the way that empirically verifiable things can, and nominalism suggests that there cannot be objective 'values in themselves', only idealised or subjective constructs; it must thus be true that, while we are of course entitled to express value judgements about the world, 'we are not entitled to assume that our grounds for making them [value judgements] are scientific; in other words, the only grounds for making them are our own arbitrary choices.'[24] This is, one might say, a very precise way of describing that everyday opposition between that which is thought to be subjective (one's 'values' or value judgements), and that which is thought to be objective (the 'hard facts' revealed via properly scientific inquiry, which is, or from a positivist perspective *ought* to be, wholly independent of value judgements). If, however, the premises of this approach are mistaken, so is the argument itself; ironically, a good example of a value judgement is precisely one that suggests that *values* should not be a part of a scientific explanation. If it is true that science is premised on values even on this fundamental level, then it cannot be valid to argue that science is, or ought to be, 'value-free'.

Saying this does not mean that 'anything goes' (as per the caricatural rejoinder offered by some advocates of positivism). It *does* mean that one cannot produce knowledge without some kind of 'prejudgement'. However, far from being an impediment to objective research practice, such judgement is part of what actually *enables* the scientist to propose hypotheses, and to test them in a way that might then be scrutinised by others.[25] The problem is not thereby to try to *avoid* any kind of prejudgement, but rather to try to elucidate the character, and the consequences of different kinds of tacit evaluations. However, at least from the perspective of critical theory, no amount of scepticism can ever yield a *tabula rasa* from

---

[23] Leszek Kolakowski *Positivist Philosophy*, p. 16.

[24] Leszek Kolakowski *Positivist Philosophy*, p. 17.

[25] This is a hermeneutic perspective famously developed by the philosopher Hans-George Gadamer. See *Truth and Method* 2nd revised edition, translation revised by Joel Weinshemier and Donald G. Marshall (London: Sheed and Ward 1989).

which to begin the research; the particular dispositions, the political discourses, the social ideologies and the cultural habits that researchers adhere to are never entirely transparent to themselves or anyone else, however much the participants in a research project may attempt to elucidate them.

The fourth rule, which may be described as the *rule of consilience* (here again, Kolakowski provides no term), is a belief in 'the essential unity of the scientific method'.[26] Accordingly, the principles, if not the methods of natural and social sciences should be identical. This may be regarded as a lasting legacy of Comte, and has long been one of the moral cudgels used by some positivists to attack the humanities and the critical social sciences: the latter sciences are not really *sciences* because they fail to adhere to the positivist standards of the physical sciences.[27] The folly of this stance can be illustrated inversely: while the critiques of phenomenalism, nominalism and objectivism certainly apply to research in the physical sciences, it does not necessarily follow that such critiques prove the need for, say, a *humanities-led* consilience, e.g. one based on the kind of constructivism that I will problematise at a later stage in this volume. Instead, it points to the importance of what I began to describe in the Preface as the principle of *methodological pluralism*, i.e. the principle that no *one* way of knowing can provide all the answers for all questions (or quite often, all the answers to any one *good* question). Stated more positively, methodological pluralism is the principle that, whatever else it may entail, knowledge production should be guided by the acknowledgement of the heteroglossic, or at least the fundamentally heterogeneous nature of the cosmos.

To conclude this chapter, I would like to discuss one more tendency that is found throughout much positivist research, but which is not described as a rule by Kolakowski. The tendency involves the role accorded to causation. Strictly speaking, logical positivists propose to replace what they regard as an old-fashioned quest for causal links with research into functional interdependence. However, at least across what might be described as 'popular positivism' in the second half of the twentieth century, one finds numerous public statements by scientists

---

[26] Leszek Kolakowski *Positivist Philosophy*, p. 17.

[27] This logic remains very much alive and kicking in the late 2010s. In 2018, the Guardian Online had a headline that effectively took a positivist potshot at the social sciences: 'Attempt to Replicate Major Social Scientific Findings of Past Decade Fails'. The article itself clarified that 'Some of the most high profile findings in social sciences of the past decade do not stand up to replication, a major investigation has found. [...] The project, which aimed to repeat 21 experiments that had been published in Science or Nature – science's two preeminent journals – found that only 13 of the original findings could be reproduced.' See Hannah Devlin, *Guardian Online*, 27 Aug 2018 < https://www.theguardian.com/science/2018/aug/27/attempt-to-replicate-major-social-scientific-findings-of-past-decade-fails>[Accessed 27 August 2018].

(or those who make it their business to comment on science) which seem to assume that all relations – physical or social – can be explained in terms of a linear, or mechanical conception of causation (see for example, my account of the sociobiological principles of Edward O. Wilson, in Chapter 2). What might be described as the *rule of causalism* is one of the reasons why the critical (as in non-positivist) social sciences are often dismissed by positivists. After all, what is the purpose of scientific research, if not to reveal the underlying causes of all manner of phenomena, and to be able to use that knowledge to *predict* phenomena? Simplifying somewhat, from this perspective to explain something is to discover, in an empirically verifiable manner, how one factor (the so-called independent variable) produces a certain effect (the dependent variable) with law-like regularity. In some scientific circles, any form of research that falls short of this standard may be dismissed as being at best descriptive, and at worst purely *introspective*, i.e. based on little more than subjective rumination.

Questions regarding the nature of causation are complex, and so not really amenable to a synoptical account. Here it suffices to explain that, contrary to what might be expected, in this study I will *not* join those in the critical social sciences who effectively dismiss the explanation of causation as a matter of scientism, *tout court*. The explanation of causation certainly *does* have a place in the critical social sciences, albeit with at least two substantial caveats. First, not all social relations are best described in terms of causation; a case can be made, for example, that the concept of *motivation* provides a more meaningful account of at least some kinds of social relations, or aspects of those relations.[28] Second, even those relations that *are* best conceived as a matter of causation may entail something rather different from the common sense notion of mechanical cause and effect, and with it, a one-way determination of *things* or even *properties*.

I agree in this sense with the philosopher Mario Bunge when he suggests that causation entails a relation among events, as opposed to a relation among properties, states or ideas[29] (though I will argue in Part 2 that events have, at the very least, a semeiotic dimension, and indeed ideas, regarded as signs, may also constitute events: 'I had an idea...'). The last caveats notwithstanding, as Bunge puts it, 'When we say that thing A caused thing B to do C, we mean that a certain event (or set of events) in A generated a change C in the state of B'; '[u]nlike other relations among events, the causal relation is not external to them, as are the relations of conjunction, or coincidence, or succession. *Every effect is somehow produced (generated) by its cause(s). In other words, causation is a mode of energy transfer.'*[30]

---

[28] See for example Paul Ricoeur, *Lectures on Ideology and Utopia* edited by George H. Taylor (New York: Columbia University Press, 1986).

[29] Mario Bunge, *Causality and Modern Science*, 3rd revised edition (Mineola, NY: Dover Publications, 1979) p. xix.

[30] Mario Bunge, *Causality and Modern Science*, p. xx.

Contrary to what many in the critical social sciences and humanities have argued, even semiotic relations can entail causality. For example, the media representations that influenced many visitors' interpretations of Barro Colorado (cf. my account of my ethnographic research in the introduction to this volume) may indeed be regarded as *events* (A) that generated a change (C) in the state of audiences (B), and so might qualify as an instance of a causal relation, by Bunge's criteria. If, however, there *is* causality in such processes, then it entails a sociological *intríngulis* that is easily overlooked. And indeed, Bunge distinguishes amongst a variety of forms of determination (in his words, a 'spectrum of categories of determination'), including causal determination or causation, *sensu stricto* (determination as per the earlier definition of a causal nexus); quantitative self-determination (the determination of the consequent by the antecedent); interaction (reciprocal causation, or functional interdependence); mechanical determination (determination of the consequent by the antecedent, usually with the addition of efficient causes and mutual actions); structural or wholistic determination (determination of the parts by the whole); teleological determination (determination of the means by the ends, or goals); and dialectical determination (or qualitative self-determination, whereby the whole process is determined by inner 'strife' and eventual subsequent synthesis of its essential opposite parts).[31] Bunge employs this spectrum to suggest firstly that only one qualifies as causation, *sensu stricto* (the first), but also, that the above are not mutually exclusive categories, and that on the contrary, 'no type of determination is found to operate in all purity, to the exclusion of all the others, save in ideal cases'.[32]

Of these, what Bunge describes as interaction, i.e. determination involving reciprocal causation, seems to come closest to what happens in the case of media messages in so far as such messages are not like a bullet that shatters a glass (as per Bunge's own example of causal determination). For the meaning to have a certain effect on an audience member, that individual must interpret it in a certain way, and such an interpretation is itself a kind of determination – one which is typically the result of at once structural, teleological and dialectical forms of determination (if we are to stick to Bunge's list, which Bunge acknowledges to be incomplete).

Critical and detailed as Bunge's account is, at least in my reading two aspects are either missing, or not given the importance that they deserve. The first involves what I will describe in Part 2 as 'Thirdness'. The second is purposiveness or motivation. Where the first aspect is concerned, is it not possible, indeed *necessary* to conceive of relations of causality that are 'three-way', in a semeiotic sense of 'three-way', i.e. entailing objects, representamens and their interpretations?(cf.

---

[31] Mario Bunge, *Causality and Modern Science*, pp. 17-19.

[32] Mario Bunge, *Causality and Modern Science*, pp. 20.

Part 2).[33] This issue can be illustrated by going back to the formidable sting of the bullet ant. If one adopts an abjectly phenomenalist and causalist stance, then epistemologically or ontologically speaking there is no reason for approaching a wink differently from a sting. Whatever the specificities of the physiological response to bullet ant venom, the *orbicularis oculi* muscle only closes the eyelids when a certain chemical action causes a certain reaction. From this perspective, just as the kinin contained in one of the fractions of the bullet ant venom causes the contraction of smooth muscle, an eyelid will only close if there is a release of acetylcholine into the synaptic cleft between the motor neuron and any skeletal muscle. So causation must be queen even in the sociological sciences.

What such a crude account cannot explain is how social codes of the kind alluded to by Geertz may *mediate* the chemical reaction itself (inhibiting, or generating it), let alone the actions that may follow on from said reaction. In a context where human interaction and symbolic forms play a key role, recourse to a linear or mechanical concept of causation is woefully simplistic.

And then there is the issue of purposiveness. Even if we accept that there is no such thing as 'free will', and that *of course* chemical reactions of the kind just described *do* and *must* take place when someone winks, the interpretation *of* a wink, like the decision *to* wink, requires rather more than the one-way transfer of chemical signals to sarcomeres. Someone must want to wink at someone else, and for that wink to have the requisite effect, the other person must not only be able to, but be *willing* to respond in a certain way. There is, however, unlikely to be much that is automatic in such a response – at least, automatic in the way that certain chemical reactions are.

---

[33] Although I take it that Bunge is critical of causalism, he also seems extraordinarily hostile to Peircian philosophy; in one passage, he seems to assume, completely mistakenly, that Peirce is a propounder of 'accidentalism', *tout court*. (See Mario Bunge, *Causality and Modern Science*, p. 12.) Put simply, Bunge suggests that for Peirce, all is chance (Bunge refers to what Peirce himself describes as *Tychism*, the doctrine that one form of evolution involves absolute chance). As will become clear in Part 2, this attribution says more about Bunge than it does about Peirce; ironically, across the critical social sciences, Peirce has more often been attacked for his insistence on the importance of causal determination, and the lawfulness even of certain types and classes of signs.

2
Sociobiology and Biophilia

Let's return to the bullet ants. And let's suppose, purely for the sake of argument, that all human beings have lived around such ants for hundreds of thousands of years, experiencing their formidable stings time and again. In the fullness of evolutionary time, would we not expect a tendency to avoid *P. clavata* – or for that matter, other hymenopterans with similarly bad stings – to be passed on genetically to successive generations of humans? If so, would we not also expect to find an innate fear amongst all humans with respect to the members of the order (*Hymenoptera*)? This is, in a nutshell, the kind of argument that I will consider in this chapter, which is devoted to a sociobiological account of human perceptions of beyond-human animals.

Before considering the precise arguments, it is pertinent to say something about the cultural context in which sociobiology emerged. As I noted in Chapter 1, the Vienna Circle was forced to disband in the 1930s. However, in the United States as in many other industrialised nations, positivist philosophy made an extraordinary comeback after World War II. An analysis of the reasons for this return is far beyond the scope of this chapter.[1] Here it suffices to note that, in the postwar period, the resurgence of positivism was particularly evident in some academic fields (e.g. sociology and history), and in the rise of public bureaucracies charged with administering state institutions – not least, the institutions responsible for the emergent *welfare* state. In capitalist, socialist and communist states where one or another form of social welfarism was embraced, government agencies were often given the task of finding ostensibly fact-based solutions to countless social issues arising in the management of economies, health systems, and the environment (to mention just three spheres). In nations with a private sector, much the same logic prevailed especially in the larger corporations, which often employed scientists to develop new products, to control their products' manufacturing processes, and eventually even to research and to try to control the *use* of those products by consumers in everyday life.

Of course, the emergence of policy science was no guarantee that the theories generated by positivist researchers accurately explained what was actually going on in everyday life. On the one hand, in large organisations as in nation states, what Comte would describe as metaphysical (if not theological) 'mind' continued to prevail despite numerous politicians' and CEOs' proclamations to the contrary, and this of course influenced the research. On the other hand, the many blind spots of especially the cruder forms of positivist social research overlooked, and

---

[1] See for example the analysis of George Steinmetz, 'Positivism and Its Others in the Social Sciences', in George Steinmetz (Ed.) *The Politics of Method in the Human Sciences: Positivism and its Epistemological Others* (Durham, NC: Duke University Press, 2005), pp. 1-58.

in some sense facilitated the proliferation of tactical responses of the kind described in the last quarter of the twentieth century by the more critical anthropologists and sociologists.[2] Even so, Comte would no doubt have been very impressed by the extent to which, across much of the world, order and progress really did seem to be marching to the tunes of a positivist hymnal in the decades immediately after the end of World War II.

By contrast, it is to be suspected that Schlick, Neurath, Carnap and the rest of the Vienna Circle would have been appalled. Even as state and corporate bureaucracies appeared to be putting their decision-making on solidly scientific footing, positivist research principles were also used to develop new forms of social control, and to perfect new weapons of mass destruction. One famous example was the use of Albert Einstein's and other physicists' research to develop nuclear weapons as part of the Manhattan Project. A somewhat less well-known example is Secretary of Defence Robert McNamara's use of systems analysis, developed in the private sector, to quantify, and to analyse virtually all aspects of the US war effort during the Vietnam War – the first foreign war that the US lost, and which played a major role in the rise during the 1960s of a counter-cultural movement. One of the byproducts, if it can be called that, of this movement was that a growing number of actors began to question the identification of positivism, order, and progress.

So far my examples have been drawn from the US, but it is easy to forget that before the US took over, it was France that fought a colonial war in Vietnam. And it was also France that, between 1954 and 1962, fought a similarly brutal war to stop the independence of Algeria. Then again, any lingering doubts amongst leftwing thinkers regarding the benevolence of Soviet communism – and state-sponsored positivism – came to an end after the ruthless suppression of popular uprisings in Hungary (1956) and Czechoslovakia (1968). A world that thought it had put an end to fascism by beating Hitler and Mussolini discovered fascism's hydra-like nature in precisely the societies that had suffered the most terrible losses fighting the Axis. As scholars began to investigate how authoritarian forms of governance could persist in such societies and even in advanced capitalist democracies, many began to suspect that positivism, and with it perhaps even scientific realism, might well play a key role.

In this context, after an initial resurgence immediately after WW II, positivism increasingly came under attack both within and beyond academia. For critical scholars as for many feminist, anti-racist, anti-capitalist, and so-called 'Third World' activists, it was apparent that scientific practice based on positivist principles was *not* above culture and politics; on the contrary, many viewed it as another instrument of oppression, a way of securing order *over* social progress. The

---

[2] See for example, Michel de Certeau, *The Practice of Everyday Life,* translated by Steven Rendall (Berkeley, CA: California University Press, 1984).

search was thus on for alternative ways of engaging in more progressive forms of critical reasoning.

In Parts 3 and 4, I will describe several of the proposals offered especially by Continental philosophers and cultural theorists. Before that, I would like to consider two forms of positivist research that not only bucked the critical trend, but went on to become highly influential paradigms within their own fields, and across much broader cultural formations. Somewhat surprisingly, the forms in question arose not in sociology or history or even philosophy, but from the work of scientists doing research about subjects that would seem to be a world away from politics. In this chapter, I will consider some of the proposals of a biologist with a research specialism in myrmecology, the study of ants. The scholar in question is Edward O. Wilson, who in 1975 published *Sociobiology: A New Synthesis*,[3] a book that became a rallying cry for something like a positivist counter-reformation.

As those familiar with Wilson's *oeuvre* will know, it includes a very large volume of scientific papers, as well as books and articles published in the popular media for so-called 'lay' audiences. In this chapter I will be primarily concerned with Wilson's theory, or what he later describes as his *hypothesis* regarding 'biophilia'.[4] This being the case, I will begin by presenting this hypothesis, and briefly considering its implications for wildlife observation. Thereafter I will offer a critique of the biophilia hypothesis as part of a more general critique[5] of Wilson's adaptationist approach to the matter. Simplifying somewhat, adaptationism is the doctrine that natural selection amongst the members of a population constitutes the only really effective cause for the modification of a trait. Adaptationists are thereby inclined to either exclude or diminish, *a priori*, the role of any other kinds of explanation, any other forms of modification – not least, those that acknowl-

---

[3] Edward O. Wilson, *Sociobiology: The New Synthesis* (Cambridge, MA: Harvard University Press, 1975).

[4] See Edward O. Wilson, *Biophilia: The Human Bond with Other Species* (Cambridge, MA: Harvard University Press, 1984). See also Stephen R. Kellert and Edward O. Wilson (Eds.), *The Biophilia Hypothesis* (Washington D.C.: Island Press, 1993).

[5] I am aware that, following the writings of Bruno Latour and others, 'critique' has become something of a bad word in the critical social sciences. This book is not the place for a 'critique of the critique of critique'; as the last only half-humorous phrase is meant to make clear, in my view of there is no getting around critique as a key element of critical theorising. One may debate how best to engage in critique, but I am not persuaded by suggestions that critique is an unnecessarily confrontational, let alone 'masculinist' way of engaging in critical theorising.

edge cultural-historical contingency.[6] In so far as everything boils down, in one way or another, to evolutionary processes with durations in the hundreds of thousands of years, then the 'clock' of human nature must have been set in the Pleistocene. From this perspective, everything that follows is, in one way or another, no more than a derivation of the changes that our species or genus had achieved by then.

Back, then, to the biophilia hypothesis itself. According to Wilson, biophilia is 'the innate tendency to focus on life and lifelike processes'.[7] 'From infancy', Wilson argues, 'we concentrate happily on ourselves and other organisms. We learn to distinguish life from the inanimate and move toward it like moths to a porch light. Novelty and diversity are particularly esteemed; the mere mention of the word *extraterrestrial* evokes reveries about still unexplored life, displacing the old and once potent *exotic* that drew earlier generations to remote islands and jungled interiors. That much is immediately clear...'[8] As this introductory description begins to suggest, Wilson's biophilia is not an account of the perception of wildlife in any narrowly physiological sense. It is, however, certainly an account of the perception of wildlife in so far as it offers an explanation of the ways in which people attend to, and interpret nonhuman, or as as I describe them, *beyond-human* animals.

Wilson's paradigmatic example of biophilia involves humans' ostensibly universal reaction to snakes. At the start of the sixth chapter of *Biophilia*, Wilson argues that 'Science and the humanities, biology and culture, are bridged in a dramatic manner by the phenomenon of the serpent'.[9] 'The snake's image', Wilson suggests, 'enters the conscious mind with ease during dreams and reverie, fabricated from symbols and bearing portents of magic. It appears without warning and departs abruptly, leaving behind not the perception of any real snake but the vague memory of a more powerful creature, the serpent...'.[10] Snakes are, in Wilson's account, the archetypal instance of organisms that have 'a special impact on mental development'.[11] The mind, he says, 'is primed to react emotionally to the sight of snakes, not just to fear them but to be aroused and absorbed in their de-

---

[6] For an introductory account to adaptationism, see Steven H. Orzack and Patrick Forber, 'Adaptationism', Edward N. Zalta (Ed.) *The Stanford Encyclopedia of Philosophy*, Spring 2017 edition <https://plato.stanford.edu/archives/spr2017/entries/adaptationism>[Accessed 2 June 2017].

[7] Edward O. Wilson, *Biophilia*, p. 1.

[8] Edward O. Wilson, *Biophilia*, p. 1.

[9] Edward O. Wilson, *Biophilia*, p. 83. This sentence provides an inkling of the role that Wilson's account of biophilia plays in his larger sociobiological account.

[10] Edward O. Wilson, *Biophilia*, p. 83.

[11] Edward O. Wilson, *Biophilia*, p. 85.

tails, to weave stories about them'.[12] While children under five 'feel no special anxiety' over snakes, thereafter, and in contrast to other fears that Wilson argues wane as people grow older (for example, fear of the dark, of strangers or of loud noises), 'the tendency to avoid snakes grows stronger with time'.[13]

What is it, he asks, that makes snakes at once repellant and fascinating? The deceptively simple answer, Wilson suggests, lies in snakes' ability to 'remain hidden, the power of their sinuous limbless bodies, and the threat from venom injected hypodermically through sharp hollow teeth'; given such threats, Wilson argues that '[i]t pays in elementary survival to be interested in snakes and to respond emotionally to their generalized image, to go beyond ordinary caution and fear. The rule built into the brain in the form of a learning bias is: become alert quickly to any object with the serpentine gestalt. *Overlearn* this particular response in order to keep safe.'[14]

So more than any innately *positive* affiliation to other organisms – the idea that is implied by the very notion of *biophilia* (bio/life + philia/a fondness or love for something) – biophilia may best be understood as an attempt to explain this 'learning bias' as it relates to human evolution. And indeed, in an essay[15] presented nearly a decade after he published *Biophilia* in 1984, Wilson revisits his initial account in an attempt to correct the impression given by the name he chose for the phenomenon. It is in *The Biophilia Hypothesis,* which Wilson co-edited with the evolutionary psychologist Stephen Kellert, that biophilia is described as a hypothesis. As Kellert puts it, 'We treat the biophilia notion as a hypothesis to underscore the need for systematic inquiry ... The idea of a hypothesis, moreover, emphasizes the scientific convention that a proposition does not "exist" until proven otherwise'.[16] Having started out suggesting that biophilia is 'the innate tendency to focus on life and lifelike processes', in his later account Wilson suggests that biophilia is actually not a single instinct but 'a *complex of learning rules* that can be teased apart and analyzed individually. The feelings molded by the learning rules fall along several emotional spectra: from attraction to aversion, from awe to indifference, from peacefulness to fear-driven anxiety'.[17] Wilson further suggests that biophilia is better understood as a matter of 'gene-culture co-

---

[12] Edward O. Wilson, *Biophilia*, p. 86.

[13] Edward O. Wilson, *Biophilia*, p. 95.

[14] Edward O. Wilson, *Biophilia*, p. 92-93. Italics in the original.

[15] Edward O. Wilson, 'Biophilia and the Conservation Ethic', in Stephen R. Kellert and Edward O. Wilson (Eds.), *The Biophilia Hypothesis*, pp. 31-41.

[16] Stephen R. Kellert, 'Introduction', in Stephen R. Kellert and Edward O. Wilson (Eds.) *The Biophilia Hypothesis*, 20-27, p. 21.

[17] Edward O. Wilson, 'Biophilia and the Conservation Ethic', p. 31. Italics added to the original.

evolution'[18] – a reference to a theory that Wilson developed in the 1980s with the biologist Charles J. Lumsden,[19] and which argues that biological and cultural evolution go hand-hand.

I will return to the theory of co-evolution below. First it may be noted that, in order to support his claims regarding biophilia, Wilson invokes much the same kind of evidence that he employs across his evolutionary psychological *ouevre*. First, he turns to the anthropological evidence, or at any rate what he regards as such evidence, and suggests that it overwhelmingly confirms a cross-cultural 'biophilic tendency'. Here as in other parts of Wilson's work, we must take his word for it in so far as he does not offer a comprehensive review of the relevant literature. Doubtless this failure is prompted at least in part by the ambition of Wilson's approach; it would probably be impossible for one researcher, or even a team of researchers to produce the requisite review given the sheer number of anthropological studies that might be pertinent. So we are left instead with comments like 'the biophilic tendency ... is so clearly evinced in daily life and widely distributed as to deserve serious attention. It unfolds in the predictable fantasies and responses of individuals from early childhood onward. It cascades into repetitive patterns of culture across most or all societies, a consistency often noted in the literature of anthropology'.[20]

A second kind of evidence involves looking for similarities with other primates, the logic presumably being that if similar tendencies are found in the taxa most closely related to the human genus, then that too, provides an indication of the likelihood that a certain behaviour (or what is *treated* as 'a behaviour') is a product of an evolutionary process. So for example, Wilson notes that two common monkeys in Africa have evolved a similar disposition with respect to snakes: 'When guenons and vervets, the common monkeys of the African forest, see a python, cobra or puff adder, they emit a distinctive chuttering call that rouses other members in the group', a call that is different from those used to signal other kinds of danger; '[t]he monkeys in effect broadcast a dangerous-snake alert, which serves to protect the entire group and not solely the individual who encountered the danger. The most remarkable fact is that the alarm is evoked most strongly by the kinds of snakes that can harm them. Somehow, apparently through the routes of instinct, the guenons and vervets have become competent herpetologists'.[21]

---

[18] Edward O. Wilson, 'Biophilia and the Conservation Ethic', p. 33.

[19] See for example, Charles J. Lumsden and Edward O. Wilson, 'The Relation between Biological and Cultural Evolution', *Journal of Social and Biological Structure* 8(October 1985), 343-359.

[20] Edward O. Wilson, *Biophilia*, p. 85.

[21] Edward O. Wilson, *Biophilia*, p. 93.

Note the use of 'apparently' in this account. But note also that, once again, Wilson does not say whether he has investigated if this is the case with all other primates that share a habitat with poisonous snakes, let alone if a similar argument can be made about primate responses to the threats posed by other kinds of wild animals; presumably here too, the claim would be so broad that it would be difficult to research. Then again, at the risk of stating the obvious, it is by no means straightforward that all species within one taxon evolve in identical ways. So at best, this is arguably little more than circumstantial evidence.

I've now said enough about the biophilia hypothesis to give the reader some sense of what it entails. What would be the implications of such an approach for a theoretical approach to wildlife observation? If one were to take Wilson's approach at face value, then wildlife observation, and in particular the desire to observe certain creatures – what I described in the introductory chapter as the 'charismatic fauna' – could be interpreted as an instance, or better yet an expression of biophilia. In particular, many tourists' undoubted interest in poisonous snakes and apex predators such as crocodiles and big cats could be regarded as strong examples of much the same 'biophilic tendency' that Wilson claims for 'our' relation to snakes.

More generally, in so far Wilson's biophilia hypothesis is part of a broader adaptationist approach to all aspects of human behaviour, then any attempt to interpret wildlife observation as biophilia would need to adopt a similarly adaptationist perspective with respect to the contextual elements for such observation: for example, the very existence of tourism (wildlife or other) could be regarded as being fundamentally a matter of biological determination, however much cultural 'filters' (as Wilson describes them in *Biophilia*) might also play a role.

As I will explain in Part 2, this study *does* incorporate evolutionary theory, albeit via the cosmology and the phenomenology of the philosopher, mathematician and physical scientist Charles Sanders Peirce. There can be little doubt, for example, that aspects of what the ecological psychologist James Gibson describes as our human perceptual systems (cf. Chapter 4) certainly *have* been shaped by evolutionary processes. The last point notwithstanding, there are a number of problems with the biophilia hypothesis, as well as fundamental theoretical, methodological, philosophical and ethical problems with Wilson's (as other sociobiologists') adaptationist perspective. In the remainder of this chapter, I would like to consider these problems in some detail.

Let's begin with the nuts and bolts of the biophilia hypothesis, considered from the perspective of the positivist framework that it is based on. Much of what passes for evidence in Wilson's account is circumstantial at best, and poor common sense at worst. The reader will recall, for example, that Wilson claims that from infancy 'we concentrate happily on ourselves and other organisms. We learn to distinguish life from the inanimate and move toward it like moths to a porch light. Novelty and diversity are particularly esteemed; the mere mention of the word *extraterrestrial* evokes reveries about still unexplored life, displacing the old and once potent *exotic* that drew earlier generations to remote islands and jungled interiors. That much is immediately clear...' In fact, many parents might argue that it is actually toys that are known, and familiar (in every sense of this last term) that are often the ones that are most cherished by children; at the very least, the researcher would have to clarify crucial matters of meaning and context before stating one or another point as easily, and as categorically as Wilson does. Then again, if it is true that we move towards life 'like moths to a porch light', why is it that some in the US and elsewhere have deemed it necessary to start a 'back to nature' movement to try to encourage children to spend more time in the 'great outdoors'?[22]

A similar point can be made about interest in extraterrestrial life forms – an interest which is by no means universally shared even amongst people living in late capitalist societies. Indeed, the admission that, at least in Wilson's mind, the extraterrestrial has displaced the 'exotic' would itself point to the importance of cultural-historical contingency. In each of the 'universal' examples offered by Wilson, a strong case can be made that the determinants of *attention* (a term that seems more useful than *concentration*) must also involve social, cultural and indeed ecosystematic *contexts*. This being the case, attention must itself surely be at least partly a matter of social and cultural, as well as ecological contingency. If this is so, then at least a crudely adaptationist take on human/beyond-human cannot be valid.

To reiterate a point made in the introductory chapter, and lest there be any misunderstanding, I am *not* arguing that children's or adults' attention is a matter of 'free will'; far from it. The point instead is that erasing contextual motivations in favour of supposedly universal 'underlying emotions' is to skew the argument in a direction that facilitates an adaptationist interpretation. As noted by the biologist Stephen Jay Gould, sociobiologists must at least *try to claim* universality; as Gould explains, 'If certain behaviors are invariably found ... amongst humans ... a circumstantial case for common, inherited genetic control may be advanced';[23] conversely, in the absence of such universality, the edifice of adaptationism as applied to 'human behaviour' comes a tumbling down.

It might further be noted that, despite efforts to tighten up the approach in *The Biophilia Hypothesis*, the very conceptualisation of biophilia is riddled with polysemy and ambiguity – two bogeys of positivist research. The notion of an 'innate tendency to focus on life and lifelike processes'[24] is vague on at least two counts: 'life and lifelike processes' could refer to just about any organic process; images of

---

[22] See for example, Richard Louv's bestselling – and arguably crypto-Christian *The Last Child in the Woods: Saving Our Children from Nature Deficit Disorder*, revised edition (London: Atlantic Books, 2009).

[23] Stephen Jay Gould, 'Biological Potentiality vs. Biological Determinism', in *Ever Since Darwin*, 251-259, p. 254.

[24] Edward O. Wilson, *Biophilia*, p. 1.

bacteria, threadworms, ticks and other parasites come to mind, but so do images of plants, and fungi. Do we have universal, and *predictable* fantasies about these organisms as well?

Then again, notions like 'focus', 'feelings' and 'emotional spectra' are certainly evocative, but would require, at least by logical positivist standards, much further specification if Wilson is to make the kinds of claims that he does. This issue is important not least because, as Yannick Joye and Andreas de Block note in a closely argued critique of biophilia, if Wilson and Kellert are indeed doing no more than formulating a hypothesis (which is, of course, fair enough), then 'The minimal prerequisite for any hypothesis ... is that it is open to falsification, perhaps in conjunction with the larger theory in which it is embedded'. But as this Popperian[25] perspective suggests, falsification cannot occur if there is 'terminological sloppiness' (as Joye and de Block describe it).[26]

Thus far I have focussed on problems with Wilson's biophilia hypothesis, but there are even more fundamental problems with Wilson's evolutionary psychology, which provides the theoretical scaffolding for biophilia. Evolutionary psychology is the term used to refer to sociobiology applied to humans. Wilson first proposed this extension of his work as a biologist and myrmecologist in *Sociobiology: The New Synthesis*. Anyone who leafs through this massive work (a large format book, 599 pages in length) could be forgiven for believing the work is all about the biology of beyond-human species. And indeed, as Wilson puts it in the first chapter, his book 'makes an attempt to codify sociobiology into a branch of evolutionary biology and particularly of modern population biology'.[27] Alas, in that first chapter as in the last, Wilson announces to the world that he has rather greater ambitions for his sociobiological theory. The first sentences in the book are devoted to refuting Albert Camus' claim that the only serious philosophical question is suicide. I doubt that many philosophers would agree with Camus, but Camus' provocative claim pales into insignificance when one reads what Wilson goes on to say. According to Wilson, Camus was wrong 'even in the strict sense intended. The biologist, who is concerned with questions of physiology and evolutionary history' – itself a provocative claim – 'realizes that self-knowledge is constrained and shaped by the emotional control centers in the hypothalamus and limbic systems of the brain. These centers flood our consciousness with all the emotions – hate, love, guilt, fear, and others – that are consulted by ethical philosophers who wish to intuit the standards of good and evil.'[28] In other words,

---

25 Karl Popper, *The Logic of Scientific Discovery* (London: Routledge, 2002 [1959]).

26 Yannick Joye and Andreas de Blocke, '"Nature and I are Two": A Critical Examination of the Biophilia Hypothesis', *Environmental Values*, 20:2 (2011), 189-215, p. 191.

27 Edward O. Wilson, *Sociobiology*, p. 4.

28 Edward O. Wilson, *Sociobiology*, p. 3.

the problem is not to study the great 'moral' questions; it is to study the physiological 'control centers'.

This is, to put it mildly, a rather large claim. It is, however, in the last chapter of *Sociobiology* – Chapter 27 – that Wilson most clearly proclaims his academic ambitions. The chapter in question begins with the following statement: 'Let us now consider man [sic] in the free spirit of natural history, as though we were zoologists from another planet completing a catalog of social species on Earth. In this macroscopic view the humanities and the social sciences shrink to specialized branches of biology; history, biography, and fiction are the research protocols of human ethology; and anthropology and sociology together constitute the sociobiology of a single primate species'.[29]

Three years after publishing *Sociobiology*, Wilson upped the ante by devoting a whole book to this subject. In his Pulitzer Prize-winning *On Human Nature*,[30] Wilson reiterates his neo-positivist project, but this time with a somewhat softer – the more accurate term is *hegemonic* – appeal to the need to bridge the 'Two Cultures', a reference to the physical chemist Charles Percy Snow's popular, if somewhat crudely dualistic account of the ostensive failure of scholars in the physical sciences and the humanities to communicate with each other.[31] 'The biological principles', Wilson argues, 'which now appear to be working reasonably well for animals in general can be extended profitably to the social sciences'; 'the time has at last arrived to close the famous gap between the two cultures ... general sociobiology, which is simply the extension of population biology and evolutionary theory to social organization, is the appropriate instrument for the effort'.[32]

I suspect that most non-sociobiological biologists would balk at the very idea of making pronouncements of this kind. It might nonetheless by assumed that most would share in an adaptationist perspective such as Wilson's, and to agree with its associated ontology. In fact, at least the more extreme versions of adaptationism have long been contested *within* the biological sciences. This not least because Darwin himself rejected such a stance when he noted in *The Origin of the Species* that he was 'convinced that Natural Selection has been the most important, *but not the exclusive*, means of modification'.[33] For these and other reasons that I will explain in a moment, in the two decades that followed the publication

---

[29] Edward O. Wilson, *Sociobiology: The New Synthesis*, p. 547.

[30] Edward O. Wilson, *On Human Nature* (Cambridge, MA: Harvard University Press, 1978).

[31] an account famously presented in a Rede Lecture delivered in Cambridge in 1959. See Charles P. Snow, *The Two Cultures and a Second Look* (Cambridge: Cambridge University Press, 1964).

[32] Edward O. Wilson, *On Human Nature*, p. ix-x.

[33] Charles Darwin, *The Origin of the Species By Means of Natural Selection*, 6th edition, (London: Senate, 1994 [1872]), p. 4. Italics added to the original.

of *Sociobiology,* Wilson was forced to fight what one journalist has described as the 'Darwin Wars'.[34] Between the 1980s and the early 2000s, Wilson's work attracted highly critical attention from scholars in both the social and physical sciences.[35] Although Wilson and fellow sociobiologists like Steve Pinker and Richard Dawkins eventually claimed to have won the arguments, the very notions of sociobiology and evolutionary psychology became, by Wilson's own account, tainted.[36]

A comprehensive review of the numerous objections raised vis-a-vis sociobiological theory is beyond the scope of this chapter. Here I would like to engage in some detail with two intimately interrelated problems with Wilson's sociobiological account of human sociality: its biologically reductionist conception of culture, as developed via what Wilson conceives as a theory of 'gene-culture coevolution';[37] and what has to be described as a deeply conservative, to not say reactionary metaphysics.

It is pertinent to start the problematisation by noting that Wilson is not, on the face of it, a biological determinist, let alone a nineteenth century-style Social Darwinist. On the contrary, Wilson himself would probably disagree quite vehemently with any suggestion that his work is biologically determinist, not least because in his later writings he often makes statements that appear to accept that culture *does* play a key role in social practice. For example, in *Consilience: The Unity of Knowledge,*[38] Wilson states, apparently unequivocally, that 'We know that virtually all of human behavior is transmitted by culture'.[39] To return to an earlier theme, Wilson would probably argue that, far from being biologically determinist, his theory is premised on an approach that is designed to *bring together* biology and culture, science and the humanities as part of his take on C. P. Snow's lament regarding the 'Two Cultures.'

---

[34] Andrew Brown, *The Darwin Wars: The Battle for the Scientific Soul of Man* (London: Simon & Schuster, 1999).

[35] For a critique of evolutionary psychology that includes perspectives from across the critical social sciences, see Hilary Rose and Steven Rose (Eds.), *Alas, Poor Darwin: Arguments Against Evolutionary Psychology* (London: Vintage, 2001). For critiques by scholars based in the physical sciences, see for example Stephen Jay Gould, *Ever Since Darwin: Reflections in Natural History* (New York: W.W. Norton, 1979) and Richard C. Lewontin, Steve Rose, and Leon J. Kamin, *Not In Our Genes: Biology, Ideology and Human Nature* (Harmondsworth: Penguin, 1984).

[36] See David Sloan Wilson and Edward O. Wilson, 'Rethinking the Theoretical Foundation of Sociobiology', *Quarterly Review of Biology,* 82:4 (2007), 327-348, p. 343.

[37] Charles J. Lumsden and Edward O. Wilson, 'The Relation between Biological and Cultural Evolution'.

[38] Edward O. Wilson, *Consilience: The Unity of Knowledge* (London: Abacus, 1998).

[39] Edward O. Wilson, *Consilience*, p. 138.

Close analysis nevertheless reveals that Wilson's theory of gene-culture coevolution, like his evolutionary psychology more generally, *is* biologically determinist and reductionist in so far as it *evolutionises culture*. As the historian Joseph Fracchia and the geneticist Richard C. Lewontin note, sociobiologists try to produce a theory of cultural evolution that is *isomorphic* to the Darwinian variational scheme.[40] This being the case, cultural analogs must be found for that scheme's fundamental principles, viz., the Principle of Random Variation; the Principle of Heredity; and the Principle of Differential Reproduction.[41] While there has been significant debate amongst cultural evolutionists as to how best to do this, Fracchia and Lewontin point out that the otherwise scholarly debate has no regard for the consequences of Darwinian evolutionism for the very conception of culture. Yet the 'translation' *does* radically transform how one understands culture thanks to at least six discursive procedures: what I describe as simplification, hierarchisation, linearisation, reduction, atomization, and population(ism).

Simplification. Amongst critical social scientists, it has long been a maxim that the role of the researcher is to identify, explain, and thereby make more intelligible the real complexity of social practice. By contrast, Wilson adheres, as positivists more generally do, to a perspective based on an aesthetics of simplicity. The term *aesthetics* is used advisedly; according to Wilson, the scientists 'most esteemed by their colleagues' have as 'their principle aim' to 'discover natural law marked by *elegance*, the right mix of simplicity and latent power'.[42] Elegance, Wilson recognises, 'is more a product of the human mind than of external reality'; however, and astonishingly, Wilson claims that elegance is nonetheless 'best understood as a product of organic evolution. The brain depends on elegance to compensate for its smallish size and short lifetime'[43] (here again we have an example of the kind of unverifiable assertion discussed earlier).

One consequence of the 'elegance of simplicity and latent power' approach is that the researcher adopts, *a priori*, a reductive outlook: 'less is more'. In some cases, that may well produce useful, and as Wilson puts it, 'elegant' results; in other cases, it may simplify to the point of distortion what are themselves complex pro-

---

[40] Joseph Fracchia and Richard C. Lewontin, 'Does Culture Evolve?', *History and Theory*, 38:4(1999), 52-74, p. 68.

[41] Fracchia and Lewontin summarise the three principles as follows: '1. Individual organisms within populations vary from one another in their characteristics. This variation arises from causes within organisms that are orthogonal to their effects on the life of the organism' (Principle of Random Variation). '2. Offspring resemble their parents (and other relatives) on the average more than they resemble unrelated organisms' (Principle of Heredity). '3. Some organisms leave more offspring than others' (Principle of Differential Reproduction). Joseph Fracchia and Richard C. Lewontin, 'Does Culture Evolve?', p. 68.

[42] Edward O. Wilson, *Biophilia*, p. 60. Italics in the original.

[43] Edward O. Wilson, *Biophilia*, p. 60.

cesses. A case in point, a process such as wildlife observation, which involves an ensemble, or as I will describe at a later stage, an *assemblage* of elements which are not best conceived organically, i.e. as a matter of essentially interdependent determinations.

Hierarchisation: As noted earlier, Wilson ostensibly sets out to bridge the 'Two Cultures', but what he builds is not so much a bridge, as a *bridgehead* for a veritable assault on the principle of methodological pluralism. Wilson's sociobiology is premised on a very clear hierarchy: it is biology at the top (or rather, at the fundamental 'base'), and culture on the bottom (or rather, the superstructural 'product'). Perhaps the passage that most clearly and succinctly expresses this is the following, which appears in *Consilience*:

> *Culture is created by the communal mind, and each mind in turn is the product of the genetically structured human brain. Genes and culture are therefore inseverably linked. But the linkage is flexible, to a degree still mostly unmeasured. The linkage is also tortuous: Genes prescribe epigenetic rules, which are the neural pathways and regularities in cognitive development by which the individual mind assembles itself. The mind grows from birth to death by absorbing parts of the existing culture available to it, with selections guided through epigenetic rules inherited by the individual brain.*[44]

This passage is a particularly revealing one in so far as it shows just how skilfully Wilson manages to make biology the fundamental determination in a culture-nature relation that is otherwise made to seem 'flexible' or even 'tortuous'. Anyone accusing Wilson of biological determinism might thus be easily 'refuted' with reference to those parts of Wilson's writing that appear to make the opposite case.

As part of the above summary, Wilson offers the link, as he sees it, between biophilia (or at least his narrative about the snake and the serpent), and an adaptationist account of culture:

> To visualize gene-culture coevolution more concretely, consider the example of snakes and dream serpents, which I used earlier to argue the plausibility of complete consilience [read: the necessity of a positivist, and sociobiological approach to 'human behaviour']. The innate tendency to react with both fear and fascination toward snakes is the epigenetic rule. The culture draws on the fear and fascination to create metaphors and narratives. The process is thus: […] *As part of gene-culture coevolution, culture is reconstructed each generation collectively in the minds of individuals. When oral tradition is supplemented by writing and art, culture can grow indefinitely large and it can even skip gen-*

---

[44] Edward O. Wilson, *Consilience*, p. 139. Italics in the original.

erations. But the fundamental biasing influence of the epigenetic rules, being genetic and ineradicable, stays constant.[45]

As Wilson explains in response to the question, 'What is human nature?', it is neither the genes 'which prescribe it', nor culture, 'its ultimate product'; instead, it is 'the epigenetic rules, the hereditary regularities of mental development that bias cultural evolution in one direction as opposed to another, and thus connect the genes to culture'.[46] Again, the casual reader might interpret this feint as a sign of a kind of culture-nature neutrality; in fact, as far as Wilson is concerned, there is one set of such rules that is shared right across the human species. So Wilson is still saying, albeit in a more subtle manner, that it really is *biology* all the way down (or up). This stance makes virtually unthinkable the notion that the cultural process might work to determine human biology; or that what Wilson describes as epigenetic rules must really be the rules of different cultures, full stop. More fundamentally, Wilson's approach makes it difficult to conceive nature-culture relations that are not about a top-down (or bottom-up) *transmission.*

Linearisation: This last term (transmission), requires scrutiny in its own right. Contrary to what might be assumed, to treat culture as a matter of cultural *transmission* is by no means the neutral, or value-free terminological choice that it might seem to be. The term and its underlying connotations, as allied to the rest of the theoretical and methodological choices highlighted thus far, work to render culture amenable to sociobiological theory. Indeed, the use of the concept of 'transmission' is perhaps the key way of discursively transforming cultural process into a mechanical, or quasi-mechanical process. To say that culture is a matter of 'horizontal transmission' of knowledge is to begin to suggest that, as a process, it is not that different from the process whereby genes are transmitted to offspring (so-called *vertical* transmission), or indeed, the process whereby a chemical signal is transmitted by a motor neuron to a muscle across a synapse. Again, this might seem like a perfectly *natural* way of describing cultural process. But as Fracchia and Lewontin ask, 'Is culture "transmitted" at all? An alternative model, one that accords better with the actual experience of acculturation, is that culture is not "transmitted" but "acquired." Acculturation occurs through a process of constant immersion of each person in a sea of cultural phenomena, smells, tastes, postures, the appearance of buildings, the rise and fall of spoken utterances'.[47]

Unfortunately, and as Fracchia and Lewontin note, theories of cultural evolution only pay lip service to such complexities in so far as they 'persist in treating culture as merely the sum total of individual cultural units at a given stage in the selection process'; doing so has the effect of denying culture any system-specific

---

[45] Edward O. Wilson, *Consilience*, p. 139. Italics in the original.

[46] Edward O. Wilson, *Consilience*, p. 139. Italics in the original.

[47] Joseph Fracchia and Richard C. Lewontin, p. 73.

characteristics, and this in turn 'allows all cultures to be explained according to the same (transhistorical and therefore ahistorical) selectionist logic'.[48] '... [I]f the passage of culture cannot be contained in a simple model of transmission, but requires a complex mode of acquisition from family, social class, institutions, communications media, the work place, the streets, then all hope of a coherent theory of cultural evolution seems to disappear.'[49]

Reduction, Atomisation and Population(ism): Given the quest for what might be called 'elegant simplicity', it is hardly surprising that evolutionary psychology is itself inherently reductionist in its theory and ethic. There are different forms of reductionism, but the kind of reductionism that sociobiology partakes in is the one described by Richard Lewontin, Steven Rose and Leon Kamin in their classic critique, *Not in Our Genes* (the authors are, respectively, a neuroscientist, a geneticist, and a psychologist who has conducted extensive research with beyond-human animals). Lewontin *et al.* suggest that '... reductionists try to explain the properties of complex wholes – molecules, say, or societies – in terms of the units of which those molecules or societies are composed. They would argue, for example, that the properties of a protein molecule could be uniquely determined and predicted in terms of the properties of the electrons, protons, etc., of which its atoms are composed'; this way of thinking then leads sociobiologists (and any other similarly inclined scholars) to argue that 'the properties of a human society are similarly no more than the sums of the individual behaviors and tendencies of the individual humans of which that society is composed. Societies are "aggressive" because the individuals who compose them are "aggressive," for instance'; put more generally and abstractly, 'reductionism is the claim that the compositional units of a whole are ontologically prior to the whole that the units comprise. That is, the units and their properties exist before the whole, and there is a chain of causation that runs from the units to the whole'.[50]

As Lewontin *et al.* also note, sociobiology is *biologically* reductionist in so far as its advocates assume that 'human lives and actions are inevitable consequences of the biochemical properties of the cells that make up the individual; and these characteristics are in turn uniquely determined by the constituents of the genes possessed by each individual', with all human behaviour being 'ultimately determined by a chain of determinants that runs from the gene to the individual to the sum of the behaviours of all the individuals'.[51]

One consequence of this approach is that it makes individuals the 'units' of analysis, but at one and the same time, entities that are treated theoretically and methodologically as *atomistic* monads – i.e. as so many atoms susceptible to com-

---

[48] Joseph Fracchia and Richard C. Lewontin, p. 71.

[49] Joseph Fracchia and Richard C. Lewontin, p. 73.

[50] Richard C. Lewontin, Steven Rose, and Leon J. Kamin, *Not In Our Genes*, pp. 5-6.

[51] Richard C. Lewontin, Steven Rose, and Leon J. Kamin, *Not In Our Genes*, p. 6.

bination and recombination, ostensibly without changing the fundamental properties of the unitised entities. In *Biophilia*, Wilson himself puts it thus: 'molecules compose cells, cells tissues, tissues organisms, organisms populations, and populations ecosystems. To understand any given species and its evolution requires a knowledge of each of the levels of organization sufficient to account for the one directly above it.'[52]

That is precisely the rationale that Wilson proposes to apply to culture and to societies. As Wilson explains in *On Human Nature*, 'The psychology of individuals will form a key part of this [kind] of analysis. Despite the imposing holistic traditions of Durkheim in sociology and Radcliffe-Brown in anthropology, cultures are not superorganisms that evolve by their own dynamics. Rather, cultural change is the statistical product of the separate behavioral responses of large numbers of human beings who cope as best they can with social existence. [...] *When societies are viewed strictly as populations, the relationship between culture and heredity can be defined more precisely.*[53] In later texts, Wilson appears to change his mind, and suggests that cultures are, after all, 'superorganisms'. That certainly is what is implied in his 'eusocial' account of the evolutionary 'function' of homosexuality (more on this point below).

In *Consilience*, Wilson presents the reductionist orientation even more explicitly: 'The great success of the natural sciences has been achieved substantially by the reduction of each physical phenomenon to its constituent elements, followed by the use of the elements to reconstitute the holistic properties of the phenomenon.'[54] It follows, as far as Wilson is concerned, that this same method should be applied to the study of culture. This 'viewing strictly as populations' is what I describe as *populationism*, a procedure that has significant theoretical, methodological and ethical implications in its own right.

Take, for example, Wilson's attempt to find a 'unit of culture' – a requirement of a reductionist approach of the kind just described. What, Wilson asks, 'is the basic unit of culture?' He suggests that 'the natural elements of culture can be reasonably supposed to be the hierarchically arranged components of semantic memory, encoded by discrete neural circuits awaiting identification',[55] or what fellow sociobiologist Richard Dawkins describes as a 'meme'.[56] Wilson defines the meme as a 'node of semantic memory and its correlates in brain activity'[57] – also a reference to his earlier research with Charles J. Lumsden on gene-culture coevolution, with whom Wilson defines nodes as 'units of semantic memory', and distin-

---

[52] Edward O. Wilson, *Biophilia*, p. 44.

[53] Edward O. Wilson, *On Human Nature*, p. 78. Italics added to the original.

[54] Edward O. Wilson, *Consilience*, p. 147.

[55] Edward O. Wilson, Consilience, 149.

[56] Richard Dawkins, *The Selfish Gene* (Oxford: Oxford University Press, 1976).

[57] Edward O. Wilson, *Consilience*, p. 149.

guishes between three levels: concepts, propositions, and schemata.[58] Wilson notes that, like words, nodes are meaningful in so far as they are interconnected with each other. 'This linkage', Wilson suggests, 'with all the emotional coloring pulled up with it, is the essence of what we refer to as meaning.'[59]

Anyone familiar with semiotic theory will see a parallel between this account, and that of the sign. And indeed, the Science and Techologies Studies (STS) scholar and feminist philosopher Donna Haraway has noted the extent to which, between the 1930s and 1950s, biological research was influenced by the kind of semiotic theory developed by Charles Morris[60] and other scholars with pragmatist, positivist, and cybernetic tendencies.[61] In the following chapter I will say more about cybernetic conceptions of communication; here it suffices to say that, if one can characterise Wilson's sociobiology as being semiotically inclined, it is an inclination that is itself steeped in biologism. Wilson explains that he 'realize[s] ... that the assignment of the unit of culture to neuroscience might seem at first to attempt to short-circuit semiotics, the formal study of all forms of communication', but actually that objection is unjustified, or so Wilson claims, because his purpose is to 'establish the plausibility of the central program of consilience ... the causal connections between semiotics and biology'.[62] While Wilson puts semiotics first in the above sentence, it is clear that what he really means is that it is *biology* that causes a certain matrix of memes-nodes, and indeed, his research makes an appeal to the supposedly universality of certain nodes.[63]

Evidently, Wilson sees no contradiction between a reductionist approach on the one hand, and his mechanistic version of what might be described as a *biosemiotics* on the other. However, as Fracchia and Lewontin note, not all semiotics is about mechanism or even functionalism; at least the semiotician Ferdinand de Saussure's[64] fundamental insight is that meaning is system-specific, and

---

58 Charles J. Lumsden and Edward O. Wilson, 'The Relation between Biological and Cultural Evolution', p. 344.

59 Edward O. Wilson, *Consilience*, p. 148.

60 Charles William Morris, *Foundations of the Theory of Signs* (Chicago: University of Chicago Press, 1938).

61 See for example Donna Haraway, 'A Semiotics of the Naturalistic Field, From C.R. Carpenter to S.A. Altman, 1930-55', in *Primate Visions: Gender, Race, and Nature in the World of Modern Science* (New York: Routledge, 1989), 84-114. See also Donna Haraway, 'The Biological Enterprise: Sex, Mind, and Profit from Human Engineering to Sociobiology', in *Simians, Cyborgs, and Women: The Reinvention of Nature* (London: Free Association Books, 1991) 43-68.

62 Edward O. Wilson, *Consilience*, p. 150.

63 Charles J. Lumsden and Edward O. Wilson, 'The Relation between Biological and Cultural Evolution', p. 344-45.

64 Ferdinand de Saussure, *Course in General Linguistics*, Charles Bally and Albert Sechehaye (eds.), translated by R. Harris (London: Duckworth 1983 [1916]).

that each term, each *sign* acquires its meanings by virtue of its place within a discrete set of differential relations.[65] I would add that, from Saussure's perspective – a perspective that I will question in Part 2 for quite different reasons – such relations are not primarily a matter of memes-nodes that reside in the 'circuitry' of individual brains, however much aspects of that circuitry are shared across the species; they are a matter of *instituted* relations which are emphatically *not* psychological in the individualist, and atomistic sense of the expression (if they can be said to be psychological, they must then be a mtter of a *social* psychology). Fracchia and Lewontin rightly suggest that, by focussing entirely on the aggregate rather than the systemic in an attempt to produce cultural molecules out of cultural atoms, coevolutionists are unable to move beyond an *ontological* individualism.[66] Simplifying somewhat, this is the notion that an atomistically conceived individuality is itself entirely natural.

It might be inferred from Fracchia and Lewontin's critique that it is only a 'social' semiotics such as Saussure's that adopts a non-mechanicistic stance. In fact, in the field of what is today formally described as biosemiotics, it is possible to find perspectives which take strong exception to the kind of adaptationism espoused by Wilson and other sociobiologists. Part 4 will refer, for example, to the work of the biologist Jakob von Uexküll,[67] who questioned not only any attempt to reduce all matters biological to a matter of natural selection, but also efforts to ground the biological science in the kind of physics- and chemistry-based scaffolding that is taken for granted by Wilson (cf. Chapter 14).

Earlier, I referred to what I described as Wilson's conservative metaphysics, and I would like to conclude this chapter by considering this aspect of Wilson's evolutionary psychology, which I regard as an ideological one, through and through. To some readers, the issues that I've raised thus far may seem like hopelessly abstract technicalities – the conceit, as it were, of scholars battling it out over the minutiae of this or that arcane theoretical position. In fact, when societies and their cultural processes are reduced to so many units that are thought to be more or less hard-wired to genes, or even to 'epigenetic rules' of the universal kind conceived by Wilson, then there is a risk that those 'units' end up being treated as little more than the marbles in the jars used by some textbooks to illustrate genetic drift. Despite the kind of ontological individualism that haunts the reductionist approach, the end game is emphatically *not* to consider the fate of each marble, or of certain groups of marbles, as they may be affected by those around them. Fracchia and Lewontin make this point sharply when they note that

---

[65] Joseph Fracchia & Richard C. Lewontin, 'Does Culture Evolve?', p. 71.

[66] Joseph Fracchia & Richard C. Lewontin, 'Does Culture Evolve?', p. 71.

[67] See for example Jakob von Uexküll, *Theoretical Biology*, translated by D. L. MacKinnon (New York: Harcourt Brace, 1926).

... the dissolution of societies into populations ... precludes the possibility that social systems might have properties unique to them as organized systems, that is, that social relations might be characterized by structures of unequal power that affect individual social behaviour and the fitness of cultural traits. This dissolution means, in turn, that social hierarchy and inequality are explained as just the consequence of the differential cultural fitness of individuals or the cultural traits they bear, rather than, say, as a consequence of antagonistic and exploitative social relations.[68]

In a footnote to this statement, Fracchia and Lewontin refer to a passage from Karl Marx's *Grundrisse*: '*Society does not consist of individuals, but expresses the sum of interrelations, the relations within which these individuals stand.*' The sentences in *Grundrisse* that follow, which Fracchia and Lewontin leave out, are also worth quoting in so far as they lay bare the ethos, and indeed the politics of the apparently neutral turn to populationism. As Marx puts it, it's 'As if someone were to say: Seen from the perspective of society, there are no slaves and no citizens: both are human beings. Rather, they are that outside society. To be a slave, to be a citizen, are social characteristics, relations between human beings A and B. Human being A, as such, is not a slave. He is a slave in and through society'.[69]

It might seem to some readers that I am engaging in a rather big conceptual leap in so far as I, like Fracchia and Lewontin, go from the intricacies of evolutionary theory to social issues of the kind just mentioned. But so did Wilson. Today it may have been forgotten that, right from the start, Wilson accompanied his attempts to promote sociobiology/evolutionary psychology with extraordinarily controversial claims involving patriarchy, homophobia, classism, and racism. For example, the same year that Wilson published *Sociobiology* – and at a time that second-wave feminists were engaging with conservatives in battles over gender inequalities, expressed then as now not least via sheer 'domestic' violence – Wilson penned an article in the *New York Times Magazine* that suggested that, amongst other 'basic human patterns', 'adult males are more aggressive and are dominant over females'. The article further suggested that, amongst the qualities 'that are so distinctively, ineluctably human that they can be safely classified as genetically based' is 'the weaker but still strong tendency for sexually bonded women and men to divide their labor into specialized tasks'.[70]

Equally if not more troubling is a statement in *Sociobiology* about the nature of so-called 'perfect societies', in which Wilson proposes to define as 'societies that

---

[68] Joseph Fracchia & Richard C. Lewontin, 'Does Culture Evolve?', p. 70.

[69] Karl Marx, *Grundrisse (Fundamentals of Political Economy Criticism)*, 'Notebook II – The Chapter on Capital', translated by Martin Nicolaus (Harmondsworth: Penguin Books, 1973). Italics added to the original.

[70] Edward O. Wilson, 'Human Decency is Animal', in *New York Times Magazine*, 12 October 1975.

lack conflict and possess the highest degrees of altruism and coordination', and which he suggests 'are most likely to evolve where all of the members are genetically identical'.[71] As if this were not problematic enough, in the aforementioned *New York Times* article, Wilson speculates that 'homosexuals regularly served as *a partly sterile caste, enhancing the lives and reproductive success of their relatives* by a more dedicated form of support than would have been possible if they produced children of their own. If such combinations of interrelated heterosexuals and homosexuals regularly left more descendants than similar groups of pure heterosexuals, the capacity for homosexual development would remain prominent in the population as a whole'.[72]

However unintentionally, the comments on gender relations amount to a crude naturalisation of patriarchy, and those that are presented as a liberal take on homosexuality are clearly an example of homophobia. It is, moreover, difficult not to read the equation of perfect society with genetic identity as a not-so-subtle echo of the explicitly racist political philosophies that have haunted the United States for centuries. Whether this was his intention or not, these and numerous other passages in Wilson's work arguably lent, and even now still lend a pseudo-biological credence to some of the most aggressively reactionary ideologies of our times.

To be clear, if Wilson's arguments were based on critical evidence, then it would of course be wrong to suppress them in the interests of feminism, LGBT rights, or indeed any other kind of counter-ideological activism. But as I have explained throughout this chapter, much of the evidence that Wilson does present is either circumstantial, or a matter of poor common sense. Perhaps the strongest example of this can be found in a passage of *Consilience*, in which Wilson tries to bolster his claims regarding the biological determination of cultural practice by referring to a compendium of 'universals of culture' proposed in 1945 by the anthropologist George P. Murdock. Wilson presents these alleged universals as an ostensibly self-explanatory list of 67 terms, organised alphabetically. The list includes 'cleanliness training', 'cooking', 'division of labor', 'games', 'gestures', 'hair styles', 'inheritance rules', 'language', 'magic', 'marriage', 'personal names', 'surgery' and 'weather control' (to cite just 13 terms found in the list).[73] What Wilson clearly hopes is to persuade the reader that, if there is such a long list of 'universals' covering so many everyday practices then there must be what he describes as a 'human nature' (with emphasis on the singular), and that 'human nature' must be, as he puts it in *Consilience*, on a 'genetic leash'. But of course, to emphasise the universality of, say, cooking over the still vast differences that exist in cultural cuisines is absurd (would we say that all ants are essentially similar because they ... feed?). A similar point can be made about the rest of the cate-

---

[71] Edward O. Wilson, *Sociobiology: The New Synthesis*, p. 314.

[72] Edward O. Wilson, 'Human Decency is Animal'. Italics added to the original.

[73] Edward O. Wilson, *Consilience*, p. 162.

gories. Take for example 'cleanliness training' and 'hygiene' – both areas for which one can find big differences in outlook and everyday practice even within modern cultures. To use a list of 'universals' *made* such by one anthropologist's evidently ethnocentric approach is an example of the poor common sense that pervades sociobiology at least in its evolutionary psychological guise.

In making this point I should acknowledge that Wilson is surely right to remind us that there are many domains of human experience which *are* engaged, in one way or another, by cultures across the world. Wilson argues that if the discursive emphasis is entirely on identifying differences, then one loses sight of all that we humans have in common – a commonality that is important not least as an antidote to any quasi-biological protestations of racists. In Part 3 I myself will consider this kind of issue in philosophical detail. But of course, in so far as one veers in the opposite direction, and uses a common sensical commonality to define a univocal 'human nature', then the obverse ideological dynamic prevails: real difference is denied in the service of what is bound to be an ethnocentric perspective, i.e. *one's own socially situated* definition of human nature.

Some scholars have attempted to excuse these and other rather manifestly politicised aspects of Wilson's sociobiology / evolutionary psychology by suggesting that they reflect the discursive tendencies of the 1970s, a time when a conservative backlash to the counter-cultural movement had started that would culminate in the election first of Margaret Thatcher as Prime Minister in the UK, and then Ronald Reagan as president of the US – both arguably racist leaders that made it their mission to curtail not just the welfare state, but what they regarded as liberal excesses across all social domains. In fact, Wilson continued to espouse very much the same discourse by the late 1990s and beyond. Indeed, this is what Wilson had to say about social class as late as 1998, when he published *Consilience*: 'People do not merely select roles suited to their native talents and personalities. They also gravitate to environments that reward their hereditary inclinations. Their parents, who posses similar inborn traits, are also likely to create a family atmosphere nurturing development in the same direction. The genes, in other words, help to create a particular environment in which they will find greater expression than would otherwise occur'.[74] A page later, having just discussed the 'scores for heritability in IQ and measurable personality traits in white Americans', Wilson asks if 'we wish to change these numbers'. The answer, he suggests, is negative, 'at least not as a primary goal. Imagine the result if a society became truly egalitarian, so that all children were raised in nearly identical circumstances and encouraged to enter any occupation they chose within reach of their abilities. Variation in the environment would thus be drastically reduced, while the original innate abilities and personality traits endured. Heritability in such a society

---

[74] Edward O. Wilson, *Consilience*, p. 154.

would drastically increase. Any socioeconomic class divisions that persisted would come to reflect heredity as never before'.[75]

It should be noted that, in the mid-2000s, Wilson signalled what might be taken as a major shift vis-a-vis this stance when he embraced group, and multilevel selection theory.[76] While this shift certainly has implications for the more traditional sociobiological stance – indeed it generated a controversy with figures such as Richard Dawkins and Steven Pinker, who clearly regarded it as something of a betrayal of sociobiological first principles – its main implication from a critical social scientific perspective is to shift adaptationism to the level of groups, which David Wilson and Edward O. Wilson define as 'something other than a family and to be composed of individuals that need not be closely related'.[77] Despite this shift, the problem is still to identify much the same mechanisms described above, albeit for the social behaviour of groups, thus defined. 'From an evolutionary perspective, a behavior can be regarded as social whenever it influences the fitness of other individuals in addition to the actor'.[78]

Aside from potentially opening a new door for racism ('Group selection is an important force in human evolution in part because cultural processes have a way of creating phenotypic variation among groups, even when they are composed of large numbers of unrelated individuals'[79]), this shift allows Edward O. Wilson to make even more ambitious claims about the explanatory role of sociobiology, and evolutionary psychology more particularly: as he and David Wilson put it,

> If a new behavior arises by a genetic mutation, it remains at a low frequency within its group in the absence of clustering mechanisms such as associations among kin. If a new behavior arises by a cultural mutation, it can quickly become the most common behavior within the group and provide the decisive edge in between-group competition ... The importance of genetic and cultural group selection in human evolution enables our groupish nature to be explained at face value. Of course, within-group selection has only been suppressed, not entirely eliminated. Thus *multi*level selection, not group selection alone, provides a comprehensive framework for under-

---

[75] Edward O. Wilson, *Consilience*, p. 155.

[76] See David Sloan Wilson and Edward O. Wilson, 'Rethinking the Theoretical Foundation of Sociobiology'.

[77] David Sloan Wilson and Edward O. Wilson, 'Rethinking the Theoretical Foundation of Sociobiology', p. 328.

[78] David Sloan Wilson and Edward O. Wilson, 'Rethinking the Theoretical Foundation of Sociobiology', p. 329.

[79] David Sloan Wilson and Edward O. Wilson, 'Rethinking the Theoretical Foundation of Sociobiology', p. 343.

standing human sociality. [...] These ideas can potentially explain the broad sweep of recorded history in addition to the remote past'.[80]

So there is a further concession, and with it another attempt to incorporate the arguments of the 'opposition', this time by recognising, or half-recognising that cultural processes can, after all, modify the 'natural' ones. Note, though, that the fundamental conception continues to be completely evolutionist, and as part of that, *transmission-ist*. Whatever one's view of these aspects, I hope this chapter has made it clear that the sociologists cannot avoid a metaphysics, i.e. an appeal to values, and a conception of the cosmos that goes beyond what is allowed, however er implicitly or explicitly, by the rules of phenomenalism, nominalism, objectivism and consilience.

---

[80] David Sloan Wilson and Edward O. Wilson, 'Rethinking the Theoretical Foundation of Sociobiology', p. 343.

3

The Computation of Vision

During the 1980s, sociobiology arguably became the queen of *media* science, with Wilson, and several like-minded scholars publishing popular science books, giving interviews, writing articles for influential newspapers and magazines, and generally going about the business of telling the world that human sociality (or lack thereof) was down to the genes, and to natural selection – indeed one sociobiologist went so far as to conjure the image of 'selfish' genes.[1] As I began to suggest in Chapter 2, critics of sociobiology would argue that it was no coincidence that this period also saw the rise of the New Right, a political movement devoted to promoting precisely the kind of Hobbesian discourse naturalised by the sociobiologists; and this, however ironically, on the basis of precisely the kind of explicitly metaphysical and theological grounds that arch-positivists like Wilson might be expected to reject.

Alas, it says something about the times that even as a sociobiological reductionism dominated the popular scientific airwaves, another even more radical reductionism was in the making. The reductionism in question would eventually come to command far more media attention as part of a shift to a second wave of neoliberalism – the one symbolised not by Reagan or Thatcher, but by the compassionate Bill and Hillary Clinton, and by the liberal Tony Blair. The reductionism in question was the one proposed by the *neurosciences* – a set of fields premised on privileging the study of neurones (or neurons, as the term is now spelled throughout most English language writing) and neural networks.[2]

Perhaps the most categorical statement concerning the promise of this approach is the one produced in 1995 by Francis Crick, better known for his role as the co-discoverer of the DNA double helix. Crick achieved a new claim to fame by suggesting in a popular science book that, in the end, we humans are nothing but a 'pack of neurons': '"You", your joys and your sorrows, your memories and your ambitions, your sense of identity and free will, are in fact no more than the behaviour of a vast assembly of nerve cells and their associated molecules. As Lewis Carroll's Alice might have phrased it: "You're nothing but a pack of neurons"'.[3]

This chapter will consider aspects of the work of one scientist who would almost certainly have rejected such abject neurological reductionism, but who ironically played a significant role in developing a key element in its theoretical scaffolding: the suggestion that neurological function can be likened to a computa-

---

[1] Richard Dawkins, *The Selfish Gene*.

[2] For an account of this shift, see Hilary Rose and Steven Rose, *Can Neuroscience Change Our Minds?* (Cambridge: Polity, 2016).

[3] Francis Crick, *The Astonishing Hypothesis: The Search for the Soul* (New York: Scribner, 1995) p. 3.

tional logic, and as a kind of correlate, that human perception may be conceived as computer-like information processing. The scholar in question is the mathematician, neuroscientist, and psychologist of perception David Marr (1945-1980). Marr's *Vision*,[4] published posthumously following his untimely death in 1980,[5] is still widely regarded as one of the classic texts on the subject of visual neuroscience, as well as a treatise on key methodological issues for neuroscientific research more generally.

Before discussing Marr's contribution, it may be helpful to engage in a brief historical excursus on the rise of the neurosciences. The shift in question is clearly linked to, but cannot be simply explained by reference to scientists' growing awareness of the role that neurons play in the central and peripheral nervous systems of humans. At the very least, the shift has to be explained as a remarkable *paradigm* shift; but even the concept of paradigm falls short in so far as it leads the researcher to overlook the social and cultural politics involved then as now.

Although scientists had discovered cells by the 1830s, limitations in microscopy prevented them from observing precisely and characterising the structure of neurons until the late 1880s. At that point, new staining techniques perfected by Camillo Golgi and by Santiago Ramón y Cajal led to the discovery that neurons are very particular kind of cells – cells that have, amongst other features, branched dendrites and an axon. Even then, Golgi was convinced that neurons were all connected to each other as part of a single network (the so-called reticular theory), a belief that Ramón y Cajal contested via research with cells taken from the brain of a bird. Despite this fundamental difference in perspective, the two scientists were awarded a joint Nobel Prize in Physiology or Medicine in 1906.

The matter was not conclusively resolved, at least from a conventional scientific perspective, until the development of electron microscopy in the 1950s. At that point it was confirmed that, as Ramón y Cajal had argued, each neuron *was* a distinct cell connected to others via synapses. Electron microscopy, as coupled to a positivist logic, eventually facilitated research based on much the same kind of reductionism that I described in Chapter 2 – if each cell was indeed distinct, then here was an apparently non-arbitrary biological 'unit', a fundamental 'building block' with which to utterly focus scientific attention.

Significant as this development was, it is necessary to turn to the period between the 1930s and early 1950s to explain the particular political (with a small 'p') way in which neurons have come to be conceived and researched. At this time, fundamental steps were taken to bring together research in mathematics, electrical engineering, computing, and human physiology. The field where this conjunction developed most powerfully was arguably the one that came to be

---

[4] David Marr, *Vision: A Computational Investigation into the Human Representation and Processing of Visual Information* (San Francisco: W. H. Freeman, 1982).

[5] In 1978, Marr was diagnosed with acute leukemia, and died two years later, at the age of 35.

known as *cybernetics*. And the figure that arguably played the most important role in the emergence of said field was Norbert Wiener (1894-1964), a mathematician and eventually also a keen social critic.

In 1948, Wiener published *Cybernetics: Or Control and Communication in the Animal and the Machine*.[6] The subtitle of the book gives a sense of the concerns of the new field. To better understand those concerns, it may be helpful to briefly consider the research that Wiener conducted as part of the US war effort during World War II. By Wiener's own account, in 1940 he sent Vannevar Bush, the engineer in charge of the US Office of Scientific Research and Development (the OSRD), some ideas for a machine that might be useful in war. The OSRD was the US's wartime clearing house for all research and development that might have military applications, and Wiener's idea was essentially for a computer – a machine that would dramatically speed up and increase the accuracy of machine-based arithmetic calculations. Wiener proposed to use an electronic tube-based mechanism with a numerical system based on a scale of two, and to replace various human interventions with an automated calculating process internal to the machine. The machine would work by producing sequences of operations with data that could not only be stored in the same machine, but might be recorded, read, and if necessary quickly erased.[7] That Wiener describes this proposal in *Cybernetics* must be interpreted as an attempt to lay a claim for the invention of computers. Of more interest to this study is that Wiener goes on to suggest that, whether or not his ideas had 'some effect in popularizing [his model for a computer] among engineers', these ideas are 'of interest in connection with the study of the nervous system'.[8]

To explain why Wiener would suggest such a link, it is necessary to turn to another surprising development: Wiener's main research project during World War II, which involved finding ways of improving the accuracy of anti-aircraft guns. Wiener made important breakthroughs in the development of new methods of targeting by applying new methods of predictive analysis. By Wiener's own account, the work of the mathematician Willard Gibbs, who originated statistical mechanics, was an important inspiration for this research.[9] Gibbs, who was also an engineer, contended that in so far as no physical measurements can ever be absolutely precise, then the calculation of changes involving machines and other dynamic phenomena must be approached as a matter of their most likely *distribution*. From this perspective, the problem is not simply to arrive at the explanation

---

[6] See Norbert Wiener, *Cybernetics: Or Control and Communication in the Animal and the Machine*, 2nd edition (Cambridge, MA: M.I.T. Press, 1961 [1948]). See also Norbert Wiener *Cybernetics and Society: The Human Use of Human Beings*, 2nd edition (London: Free Association Books, 1989[1954]).

[7] Norbert Wiener, *Cybernetics*, p. 4.

[8] Norbert Wiener, *Cybernetics*, p. 4.

[9] As explained by Wiener in *Cybernetics and Society*, pp. 7-11.

of phenomena on the basis of ostensibly immutable laws, as per the Newtonian (and indeed Comtian) perspective, but to use maths to try to quantify and thereby arrive at some control over uncertainty and contingency, or what Wiener characterises as a problem of *entropy*.

Returning to anti-aircraft guns, by the beginning of World War II the guns were still aimed with the aid of mechanisms that calculated the expected position of enemy aircraft by extrapolating the present course along a straight line. Building in part on statistical mechanics, Wiener now suggested that the systems should also include a statistical calculation of how factors such as acceleration, delays in regimes of flow over aircraft surfaces (e.g. over the rudder or the ailerons) and other similarly complex dynamics would be most likely to constrain otherwise unpredictable evasive manoeuvres. At the same time, Wiener realised that the firing system should also try to calculate the response of the pilots themselves. Crucially, he became aware that 'an aviator under the strain of combat conditions is scarcely in a mood to engage in a very complicated and untrammeled voluntary behavior, and is quite likely to follow out the pattern of activity in which he has been trained'; this being the case, additional variables based on such training might be factored in to calculate and predict 'the performance of certain human functions'.[10] Henceforth, mathematical analysis would be employed to predict not just the mechanical determinants of aircraft's trajectories, but also *pilots'* responses to anti-aircraft fire as part of a more or less integrated system. As Wiener puts it, in this research as in the research about computing systems more generally, a key aspect involves *usurping* a specifically human function.[11] If the new firing system worked, then anti-aircraft gunners would not have to second-guess the pilots.

Even as Wiener incorporated the logic of statistical mechanics, he also became aware of, and began to theorise *feedback*, and more generally, *communication* as part of the predictive analysis. As Wiener explains, 'I [class] communication and control together.' 'When I communicate with another person, I impart a message to him[sic], and when he communicates back with me he returns a related message which contains information primarily accessible to him and not to me. When I control the actions of another person, I communicate a message to him...'.[12] So as far as Wiener was concerned, at least from the point of view of control, people are not fundamentally different from machines: 'When I give an order to a machine, the situation is not essentially different from that which arises when I give an order to a person. I am aware of the order that has gone out and of the signal of compliance that has come back'. 'To me, personally, the fact that the signal in its intermediate stages has gone through a machine rather than through a person is

---

[10] Norbert Wiener, *Cybernetics*, p. 6.

[11] Norbert Wiener, *Cybernetics*, p. 6.

[12] Norbert Wiener, *Cybernetics and Society*, p. 16.

irrelevant and does not in any case greatly change my relation to the signal. Thus the theory of control engineering, whether human or animal or mechanical, is a chapter in the theory of messages.'[13] It is, in this sense, entirely appropriate that Wiener decided to describe his new approach as *cybernetics* – a neologism that he derived from the Greek term for a ship's steersman.[14]

Returning to Wiener's work on computers, whilst working with a medical doctor on human physiology, Wiener realised that he could liken neurological activity to the workings of the computer prototype described earlier. One passage in *Cybernetics* is particularly evocative of this shift: in the course of welcoming what today we would describe as a neurophysiologist to the interdisciplinary team that Wiener and the doctor were assembling, Wiener took it upon himself to explain to the new member the workings of modern vacuum tubes. In *Cybernetics*, he describes his account as follows: the tubes are

> ... ideal means for realizing in the metal the equivalents of his [the new scientist's] neuronic circuits and systems. From that time, it became clear to us that the ultra-rapid computing machine, depending as it does on consecutive switching devices, must represent almost an ideal model of the problems arising in the nervous system. *The all-or-none character of the discharge of the neurons is precisely analogous to the single choice made in determining a digit on the binary scale*, which more than one of us had already contemplated as the most satisfactory basis of computing-machine design. *The synapse is nothing but a mechanism for determining whether a certain combination of outputs from other selected elements will or will not act as an adequate stimulus for the discharge of the next element, and must have its precise analogue in the computing machine.* The problem of interpreting the nature and varieties of memory in the animal has its parallel in the problem of constructing artificial memories for the machine.[15]

As Wiener himself recognises in his work, he was by no means alone in developing a cybernetic approach to 'control and communication in the animal and the machine'. And indeed, even as Wiener published his influential *Cybernetics*, in 1948 Claude Shannon (1916-2001), a mathematician and cryptanalyst with whom Wiener had numerous exchanges, proposed a theory of communication that built on Wiener's insights to formalise the cybernetic approach to communication even

---

[13] Norbert Wiener, *Cybernetics and Society*, pp. 16-17.

[14] Norbert Wiener, *Cybernetics*, p. 11.

[15] Norbert Wiener, *Cybernetics*, p. 14. Italics added to the original.

as it provided a way of linking it to computers.[16] Writing in the technical journal of the Bell Telephone Laboratories in which he worked, Shannon suggests that 'The fundamental problem of communication is that of reproducing at one point either exactly or approximately a message selected at another point'[17] (see Figure 2, below).

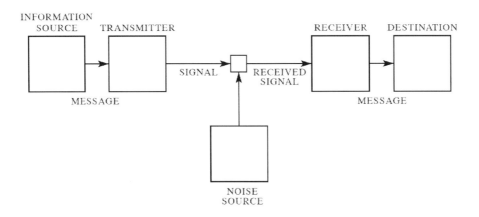

Figure 2: Shannon's Linear Model of Communication[18]

Shannon further suggests that the 'semantic aspects' of communication, i.e. the vicissitudes of discourse, 'are irrelevant to the engineering problem'; what matters 'is that the actual message is one selected from a set of possible messages', and so the system 'must be designed to operate for each possible selection, not just the one which will actually be chosen since this is unknown at the time of design. If the number of messages in the set is finite then this number or any monotonic function of this number can be regarded as a measure of the information produced when one message is chosen from the set, all choices being equally likely'.[19]

Shannon proposes that this approach should employ a logarithmic function and measure: 'The choice of a logarithmic base corresponds to the choice of a unit for measuring information. If the base 2 is used the resulting units may be called

---

[16] See Claude Shannon 'A Mathematical Theory of Communication', *Bell System Technical Journal* 27 (July and October 1948), 379-423 & 623-656. See also the later work published with Warren Weaver, *The Mathematical Theory of Communication* (Champagne, IL: Illinois University Press, 1963).

[17] Claude Shannon 'A Mathematical Theory of Communication', p. 379.

[18] Claude Shannon 'A Mathematical Theory of Communication', p. 380.

[19] Claude Shannon 'A Mathematical Theory of Communication', p. 379.

binary digits, or more briefly *bits*, a word suggested by J. W. Tukey. A device with two stable positions, such as a relay or a flip-flop circuit, can store one bit of information. N such devices can store N bits, since the total number of possible states is 2^N and $\log_2 2^N = N$.'[20] In what is another instance of brilliant synthesis, Shannon thereby provides a means not only of characterising information mathematically, but of doing so in a way that would render it useful, i.e. calculable and manipulable, for the purposes of electrical engineering, and more specifically for the emerging computing technologies.

The significance of this last aspect is difficult to overestimate. As suggested earlier, already by the early 1940s the possible structure and function of modern computers had begun to be proposed, thanks not only to Wiener, but more famously to the work of another mathematician and cryptanalyst, Alan Turing. Turing's conceptual invention, the 'Turing Machine', which he proposed in 1936,[21] described a system for the coding of a potentially infinite set of instructions via alphabetical characters inscribed on tape (an itself brilliantly practical solution for the storage aspect of the system that Wiener described). Shannon's model of communication, as allied to a cybernetic rationality of the kind described above, was a key step in the development of machines that could be used to read characters, and to then perform increasingly complex sets of calculations as per the instructions of themselves increasingly complex algorithms. Even before World War II ended, primitive versions of such machines had become indispensable tools for cryptanalysis; after the war, they would become fundamental to virtually all institutionalised aspects of modern life. This not only as so many high-speed *calculating* machines, but as a key part of an apparatus increasingly designed to attain the kind of control – technical, social, and perhaps eventually also biological – theorised by Wiener, and now in a growing number of domains.

Before returning to Marr's research, it is pertinent to mention one other development whose significance is easy to overlook so long as one's discourse focusses on mathematics, technical innovation, or the genius of individuals (as per the romantic discourse of invention). During and after World War II, the interface between science and the production of new technologies became increasingly mediated by the bureaucracies of large institutions – private, public, and hybrid – and this gave the research at once a certain impetus, and a rather different set of motivations from those that had once driven at least so-called 'pure' scientific investigation. Where once it could be argued – however naively – that at least those scientists devoted to 'pure theory' stood above the 'contamination' of political and economic discourse (we return by this route to logical positivism), after the Manhattan Project scientists would no longer be able to draw so neat a line between pure and applied science, between scientific discourse and the world. It might be

---

[20] Claude Shannon 'A Mathematical Theory of Communication', p. 381.

[21] Alan M. Turing 'On Computable Numbers, with an Application to the Entscheidungsproblem', *Proceedings of the London Mathematical Society*, 2, 42:1 (1936), pp. 230–65.

noted as an aside that, having played a key role in opening the Pandora's box of cybernetics, Wiener distanced himself from the corporate-military apparatus that started to take over his field and then the closely related field of Artificial Intelligence; he not only refused to accept funding from sources linked to the mentioned apparatus, but publicly denounced them.[22]

It may seem that I have digressed from the rise of the neurosciences. In fact, during the second half of the twentieth century, neuroscientific research emerged in close relation to the developments outlined above. Even as scientists acquired a greater understanding of neuronal function, the cells and the networks that they formed came to be regarded as the organismic equivalent of cybernetic systems. In particular, in so far as neurons could be said to 'transmit' electrical and chemical signals across synapses, then they too, might be at once characterised, and studied by way of the linear and mathematising models of communication proposed by Turing, Wiener, Shannon and their many successors.

Equally if not more importantly, at a time when many regarded the emergence of true Artificial Intelligence as being imminent, a computational logic became the analogy of choice to describe the workings of neurological networks. In this context, the mind came to be regarded as an organic machine capable of processing 'inputs' via complex algorithms which might then result in certain 'outputs', viz. an 'information processing' machine. If in the seventeenth and eighteenth centuries philosophers of nature employed first the clock and then the *camera obscura* as the key mechanical metaphors for the nature of the cosmos and the working of mind,[23] from the 1960s and 70s onwards it was computer-based information processing that dominated the collective imaginations of scientists devoted to research about subjects linked to neurological function – not least, perception.

Let's go back then, to the work of David Marr. In the aforementioned period, Marr worked to develop a theory of the function of the archicortex, cerebellum, and neocortex in mammalian brains. He then decided to specialise in the study of vision, and from 1973 onwards did so at the Artificial Intelligence Laboratory at the Massachusetts Institute of Technology (MIT). By his own account, this was

---

[22] As Wiener put it in a now famous letter he sent to the director of Boeing, who requested information about Wiener's WWII research about anti-aircraft control systems, '... the bombing of Hiroshima and Nagasaki ... has made it clear that to provide scientific information is not necessarily an innocent act, and may entail the gravest consequences. One therefore cannot escape reconsidering the established custom of the scientist to give information to every person who may enquire of him'. Norbert Wiener, letter published in the *Atlantic Monthly*, January 1947, and reprinted in Norbert Wiener, *Cybernetics and Society*, pp. xxvi-xxvii.

[23] For an account of the importance of the clock metaphor, see Carolyn Merchant, *The Death of Nature: Women, Ecology and Scientific Revolution* (San Francisco, CA: HarperSan Francisco, 1980). For an account of the significance of the *camera obscura*, see Jonathan Crary, *Techniques of the Observer* (Cambridge, MA: MIT Press, 1990).

thanks in no small part to that university's superior computing facilities (until then, Marr had worked at Cambridge University). Marr's *Vision* describes the result of that research: an approach that develops what is primarily a computational theory of mind as applied to particular aspects of visual perception.

In the introductory chapters of *Vision*, Marr offers a historical contextualisation of his own. It seems that in the early stages, many budding neuroscientists were deeply impressed by the research conducted between the 1950s and 70s by the physiologist Horace Barlow, who discovered that if a certain stimulus was presented to 'intact frogs' (presumably frogs not dissected), they would 'turn towards the target and make repeated feeding responses consisting of a jump and a snap'.[24] According to Barlow, this suggests that 'a large part of the sensory machinery involved in a frog's feeding responses may actually reside in the retina rather than in mysterious "centres" that would be too difficult to understand by physiological methods. The essential lock-like property resides in each member of a whole class of neurons and allows the cell to discharge only the appropriate key pattern of sensory stimulation'.[25]

Barlow's discovery led him to engage in something of a leap of his own: he argued that, in general, 'A description of that activity of a single nerve cell which is transmitted to and influences other nerve cells and of a nerve cell's responses to such influences...is a complete enough description for functional understanding of the nervous system'; '[t]here is nothing else "looking at" or controlling this activity, which must therefore provide a basis for understanding how the brain controls behaviour'.[26] '[P]hysiological experiments might thus be able to answer questions of "psychological interest"'.[27]

In Barlow's research, we find perhaps the clearest example of the kind of reductionism that I described in Chapters 1 and 2, when I noted that, from a positivist, and *biologically* reductionist perspective, there is, in principle, no good epistemological reason for approaching the observation of an ant and the effects of an ant sting differently: both may be treated as purely physiological events, however much the events may involve different parts of the human physiology.

It will thus come as no surprise that, according to Marr, in the wake of this discovery there was a period when aspiring neuroscientists interested in the neurophysiology of perception raced each other to discover the equivalent of so-called 'grandmother cells' – a half humorous reference to cells that fire only when one's grandmother comes into view.[28] If such cells could be discovered, then scientists

---

[24] Horace B. Barlow 'Single Units and Sensation: A Neuron Doctrine For Perceptual Psychology?', *Perception* 1 (1972), 371-394, p. 373.

[25] Horace B. Barlow, 'Single Units and Sensation', p. 373.

[26] Horace B. Barlow, 'Single Units and Sensation', p. 380.

[27] Horace B. Barlow, 'Single Units and Sensation', p. 372.

[28] David Marr, *Vision*, p. 15.

would be well on the road to attaining a cybernetic ideal: the possibility not just of explaining, but eventually of controlling at least some aspects of the perceptions, if not more generally the behaviour, of intact *humans*.

Alas, as far as Marr was concerned, Barlow's research proved to be something of a dead end. In a critique that nonetheless attests to the continuing importance of cybernetic discourse, Marr suggests that none of the studies that immediately followed Barlow's 'succeeded in elucidating the *function* of the visual cortex. [...] It is difficult to say precisely why this happened, because the reasoning was never made explicit and was probably largely unconscious. However, various factors are identifiable. In my own case ... even if the theory was correct, it did not much enlighten one about the motor system – it did not, for example, tell one how to go about programming a mechanical arm.'[29]

To be sure, and as Marr puts it in *Vision*,

> Suppose ... that one actually found the apocryphal grandmother cell. Would that really tell us anything much at all? It would tell us that it existed ... but not *why* or even *how* such a thing may be construed from the outputs of previously discovered cells. Do ... the simple and complex cells ... tell us much about how to detect edges or why one would want to, except in a rather general way through arguments based on economy and redundancy? If we really knew the answers, for example, we should be able to program them on a computer.[30]

As these quotes begin to suggest, even if some scholars today describe Marr's proposals as a kind of antidote to a technologically reductionist neuroscience (more on this, below), a cybernetic rationality is very much in evidence in his work, and in keeping with this it will hardly come as a surprise that Marr recommends, as an alternative, a study of vision that should be centred on what he describes as *information processing*: 'Vision is ... first and foremost, an information-processing task...' – a *computing-based* information processing: 'The need to understand information-processing and machines has arisen only quite recently. Until people began to dream of and then to build such machines, there was no very pressing need to think deeply about them. Once people did begin to speculate about such tasks and machines, however, it soon became clear that many aspects of the world around us could benefit from an information-processing point of view.'[31] This relatively reflexive account is followed directly by a statement that makes evident the extent to which Marr conflates not just information processing and machines, but the human organism and a computational logic: 'Most of the

---

[29] David Marr, *Vision*, p. 15.

[30] David Marr, *Vision*, p. 15. Italics in the original.

[31] David Marr, *Vision*, p. 3.

phenomena that are central to us as human beings – the mysteries of life and evolution, of perception and feeling and thought – are primarily phenomena of information processing, and if we are ever to understand them fully, our thinking about them must include this perspective'.[32]

Marr qualifies this view with two important caveats: First, '… if we are capable of knowing what is where in the world, our brains must somehow be capable of *representing* this information – in all its profusion of color and form, beauty, motion, and detail. The study of vision must therefore include not only the study of how to extract from images the various aspects of the world that are useful to us, but also an inquiry into the nature of the internal representations by which we capture this information…'.[33]

Second, information-processing and representation must be part of a complex, modular, and multi-stage process that goes beyond the mere activation of neurons, or neurological networks. Marr *is* critical, in this sense, of the kind of reductionism suggested by Barlow (according to Barlow, 'the activities of neurons, quite simply, *are* thought processes'[34]); today he would most likely also be critical of the so-called *connectionist* perspective that predominates in the neurosciences – and for which the model of explanation involves little more than describing the activities of neural *networks* in the hopes of establishing links between patterns of such activities, and particular kinds of 'outputs', i.e. a kind of neuronal behaviourism.

By contrast, and building on his own and other researchers' investigations into the anatomy of the human brain, Marr suggests that 'There must exist an additional level of understanding at which the character of the information-processing tasks carried out during perception are analyzed and understood in a way that is independent of the particular mechanisms and structures that implement them in our heads'.[35] For an account of such a level to be rigorous, Marr argues that it is necessary to distinguish between different aspects of the process. By adopting this stance, '[i]t becomes possible … to make explicit statements about what is being computed and why and to construct theories stating that what is being computed is optimal in some sense or is guaranteed to function correctly. The ad hoc element is removed, and the heuristic computer programs are replaced by solid foundations on which a real subject can be built'.[36]

So what are those different aspects? Marr distinguishes between three levels: what he describes as 'computational theory' ('What is the goal of the computation, why is it appropriate, and what is the logic of the strategy by which it can be

---

[32] David Marr, *Vision*, pp. 3-4.

[33] David Marr, *Vision*, p. 3.

[34] Horace B. Barlow, 'Single Units and Sensation', p. 380. Italics in the original.

[35] David Marr, *Vision*, p. 19.

[36] David Marr, *Vision*, p. 19.

carried out'); 'representation and algorithm' ('How can this computational theory be implemented? In particular, what is the representation for the input and output, and what is the algorithm for the transformation?'); and finally, 'hardware implementation' ('How can the representation and algorithm be realized physically'?).[37] Significantly, Marr introduces this scheme by way of an analogy involving a cash register. The computational theory is the equivalent of what the device does, and why it does it, processes that Marr characterises entirely in terms of the mathematical operations that the machine must be able to engage. The representation and algorithms aspect is the equivalent of the kinds of numbers (e.g. Arabic, as opposed to Roman) that are employed by the cash register, as linked to the kinds of calculations that are made possible by the chosen representational system. Finally, the hardware implementation aspect of the process involves the kind of machine that is used to carry out the operations. Marr highlights that, at each stage, there are many possible ways of engaging with the constraints posed by each aspect of the process; so, for example, addition is commutative and associative, but also, someone at the checkout could do the maths on a piece of paper, or using the latest cash register, etc.

This analogy is a useful one from the point of view of a propaedeutics of Marr's computational account. The reader may nonetheless notice that this theory says absolutely nothing about the reasons why cash registers, let alone the organisations that they are part of, exist. This may seem like a trivialising, or even facetious point to make about a theory that is, after all, concerned with explaining human *vision*; in fact, and as I will explain in more detail in Chapter 4 and in the conclusions to Part 1, it is meant to highlight the extent to which Marr's theory of perception utterly decontextualises perception from social, cultural, and environmental determinations.

I would now like to turn to the representational aspects of Marr's theory – aspects which would appear to contradict at least the most neurologically reductionist approaches to visual perception (if not to *mind* more generally). According to the latter approaches, even if we humans might not jump and snap at black discs in the way that intact frogs did for Barlow, more complex versions of such mechanisms, and purely neurological descriptions of them ought to serve as the basis for explanation; anything else is superfluous.

Marr rejects this view by arguing that representation constitutes a fundamentally enabling (and limiting) aspect of the process. However, in keeping with his computational approach, he defines representation as 'a formal system for making explicit certain entities or types of information, together with a specification of how the system does this.'[38] In other words, he conceives semiosis, or the semiotic process (terms which Marr himself does not use) as a kind of mechanical *encoding*

---

[37] David Marr, *Vision*, p. 25.

[38] David Marr, *Vision*, p. 20.

process, e.g. one of the kind that might be used for a Morse-like code, or, more pertinently, for the programming of computers.

Here again, this approach leaves out questions about ambiguity, polysemy, contradiction, or the unselfconsciously coded dimensions of representational forms – arguably themselves fundamental aspects of communication. Marr is nonetheless aware that how any particular representation makes certain information explicit is important 'because how information is represented can greatly affect how easy it is to do different things with it ... It is easy to add, to subtract, and even to multiply if the Arabic or binary representations are used, but it is not at all easy to do these things – especially multiplication – with Roman numerals.'[39] While Marr uses numbers in this example, he suggests that a representation for a shape also qualifies as a 'formal scheme', as does a musical score, or indeed any other symbolic form. 'The phrase "formal scheme"', he explains, 'is critical to the definition, but the reader should not be frightened by it. The reason is simply that *we are dealing with information-processing machines, and the way such machines work is by using symbols to stand for things* – to represent things, in our terminology'.[40]

Building on physiological research, Marr proposes to marry computational logic with this conception of representation to produce what he describes as a 'representational framework for vision'. Key to this aspect of the research is what I have begun to characterise as a functionalist perspective on perception. In a manner that is evocative of sociobiological research, Marr approaches vision by considering the biological function that it must fulfill. By this account, human vision is far less specialised than that of, say, a fly (an organism whose visual system Marr considers in some detail thanks to the research of two investigators, Werner Reichardt and Tomaso Poggio, the latter with whom Marr has conducted joint research). Human vision may be less specialised than that of the fly, but Marr is convinced that it too, has a function that is universal to the species. In order to determine what that function might be, Marr turns to the findings of research into brain lesions, and in particular to research that shows the differing effects on vision of lesions on the left or right parietal lobe. Translating this research, Marr arrives at the conclusion that human vision 'tells about shape and space and spatial arrangement'.[41] While he acknowledges that aspects such as discerning illumination and motion are important, they are in some sense secondary to what Marr describes as the 'main job of vision': 'to derive a representation of a shape'.[42]

How, then, do humans derive a representation of a shape? Marr argues that the process must involve more than one step. He proposes that 'shape information' is at once produced by, and derived from images across three steps. First, human

---

[39] David Marr, *Vision*, p. 21.

[40] David Marr, *Vision*, pp. 20-21. Italics added to the original.

[41] David Marr, *Vision*, pp. 35-36.

[42] David Marr, *Vision*, pp. 35-36.

vision must start with the production of a 'retinocentric' two-dimensional *primal sketch* that is achieved primarily via the detection of light intensity changes, the representation and analysis of local geometrical structure, and the detection of illumination effects such as light sources, highlights and transparency. This is followed by a *2¹/₂ -dimensional sketch*, itself still retinocentric, which produces a representation of visible *surfaces*. This step makes explicit the orientation and rough depth of such surfaces, as well as contour discontinuities in these aspects relative to the viewer (e.g. distance from viewer, discontinuities in depth or surface orientation, etc.) The final step, which Marr characterises as the *3-D model representation*, describes shapes and their spatial organisation in an object-centred coordinate frame that uses a modular hierarchical representation that includes information on the volume of space that an object occupies, as well as its shape as a surface.[43]

In the rest of *Vision*, Marr fleshes out the details of his model. Those details are certainly not unimportant, but in this chapter I will omit their consideration in favour of a discussion of the implications of this kind of research for this study. More than Marr's actual findings or the specific mechanism that he proposes, I am interested in his epistemological framework, and in the methodology of his research. Marr is not only telling us how 'vision works', but how it may be studied, and in the process, he is also explaining what vision *is*. Just as anyone adopting the theory (or hypothesis) of biophilia would need to adopt an adaptationist perspective overall, anyone adopting Marr's framework for representation would have to adopt a *computationalist*, and also a *functionalist* perspective to wildlife observation as a process. The latter perspective is at least as problematic as the former.

Let's begin by considering the implications of a functionalist approach. As I noted earlier, Marr approaches visual perception by asking what function it achieves as compared to the visual systems of other creatures. On the face of it, this might seem like a perfectly valid approach; if vision fulfills certain purposes in other animals, then surely it does the same thing for humans, and the problem is thus to figure out what that purpose, what that *function* is in the case of our own species. As I have already explained, Marr's answer is that, whatever else human vision does, it produces representations of shapes.

A first problem with this approach is that it negates the immense implications of cultural mediations. Marr adopts, at least on this level of his analysis, a stance which is little different from the biological reductionism of Wilson, Lumsden, and the evolutionary psychologists more generally. That said, one need not go back to the Darwin Wars to find problems with Marr's functionalism. By Marr's own account, vision does 'other things' as well, just not things that are as important as representing shapes. An obvious issue here is whether one can isolate the representation of shapes from everything else that vision can do. In Chapter 4, I will agree with the psychologist James J. Gibson when he offers an ecological rationale

---

[43] David Marr, *Vision*, p. 42ff.

for opposing this reduction. Simplifying somewhat, to focus on representation in the way that Marr does is to abstract the visual sense not just from some of the very cues that it purportedly reproduces, but from the overlap across different *perceptual systems* (to borrow Gibson's concept). For these reasons too, Marr's approach is a radically reductionist one, however much Marr rightly opposes the *abject* neurological reductionism proposed by Barlow et al. – this not least by invoking a principle of complexity which includes a role for representation.

Yet here too, there are problems. I would certainly agree that representation – or as I will prefer to describe it, *semeiosis*, which is to say a semeiotic process – *does* play a central role in observation in general, and in wildlife observation in particular. However, Marr's debt to a linear and unidirectional conception of information, itself a part of an attempt to construe vision as a matter of information processing along computationalist lines, is, paradoxically, at once uncritically *realist*, and uncritically *idealist* when it comes to explaining the role of representation in vision.

Uncritically realist: in so far as the receptors on human retinas clearly *do* respond to phenomena that originate beyond the individual observer, and do so in ways that the observer has little or no control over, then on some level it *must* be true that a realist conception is required to explain at least aspects of visual perception: put crudely, something that happens 'out there', that is at least partly independent of the observer, has consequences 'in here' (on the receptors, and in the brain itself). The question is nonetheless, at what point, or better yet, in what *ways* does the 'in here' affect what the observer perceives, and this in a way that is not simply mechanically reproductive (as per Marr's 'scanning' model). We know intuitively that there are countless examples of situations in which something appears to be 'staring a person in the face', but remains unnoticed because the mind is working to look for something else. In so far as that is the case, it must also be true that an unqualifiedly realist perspective fails to account for what is, at the very least, a dynamic of reciprocal causation: it is not only the perceived world that *reflects*.

Uncritically idealist: Given its realist approach to the problem, Marr's conception of information processing might well be expected to be equally steeped in realism. Instead, we find a parallel with Wilson in so far as Marr's computational turn is an exercise in the highest idealism, in the sense of a blind faith (no pun intended) in a computational conception of mind. While algorithms can of course be remarkably complex and even 'life-like', their fundamental coding procedure owes everything to a mathematical *idea*: that in the end as in the beginning, reality can be reduced to 0's or 1's, or any string of such binary alternatives. A purer, and indeed more absolute idealisation is hard to imagine. Marr himself notes that 'how information is represented can greatly affect how easy it is to do different things with it'.[44] But of course, this functionalist conception of representation fails

---

[44] David Marr, *Vision*, p. 21.

to recognise that it is not just a matter of *ease of doing*; even if one adopts, as I will do, a pragmatic approach to meaning, the researcher ignores at her/his peril the importance of a logic of *mediation* that does not itself obey a binary or dualistic logic (cf. Part 2).

A final problem returns us to the role of reflection or thought in representation. Put simply, there is no place for subjectivity and purposiveness in Marr's theory. A preliminary way of approaching this problem is by way of the philosopher John Searle's famous Chinese Room analogy, which he first proposed in 1980:[45]

> Imagine a native English speaker [sic] who knows no Chinese locked in a room full of boxes of Chinese symbols (a data base) together with a book of instructions for manipulating the symbols (the program). Imagine that people outside the room send in other Chinese symbols which, unknown to the person in the room, are questions in Chinese (the input). And imagine that by following the instructions in the program the man in the room is able to pass out Chinese symbols which are correct answers to the questions (the output). The program enables the person in the room to pass the Turing Test for understanding Chinese but he does not understand a word of Chinese.[46]

The 'Turing Test' that Searle refers to is the famous test of Artificial Intelligence proposed in 1950 by Alan Turing.[47] The Turing Test suggests that a computer can be said to exhibit intelligent behaviour, or actions which closely imitate human intelligence, if a human observer poses questions to two hidden respondents – one a person, and the other a computer with human-like responses – and after the exchange cannot distinguish which of the respondents is the human, and which is the computer. In such a situation, a computer can be said to 'think'. But according to Searle, the Chinese Room situation shows that

> the implementation of the computer program is not by itself sufficient for consciousness or intentionality ... Computation is defined purely formally or syntactically, whereas minds have actual mental or semantic contents, and we cannot get from syntactical to the semantic just by having the syntactical operations and nothing else. To put this point slightly more technically, the notion "same implemented program" defines an equivalence class that is specified independently of any specific physical realization. But such

---

[45] John R. Searle, 'Minds, Brains, and Programs', *The Behavioral and Brain Sciences*, 3 (1980), 417-457.

[46] This description is a summary offered by Searle in 1999. John R. Searle, 'The Chinese Room', in Robert A. Wilson and Frank C. Keil (Eds.) *The MIT Encyclopedia of the Cognitive Sciences* (Cambridge, MA: MIT Press, 1999).

[47] Alan Turing, 'Computing Machinery and Intelligence', *Mind*, 59:236 (October 1950), 433-460.

a specification necessarily leaves out the biologically specific powers of the brain to cause cognitive processes. A system, me, for example, would not acquire an understanding of Chinese just by going through the steps of a computer program that simulated the behavior of a Chinese speaker.[48]

Returning to Marr, we can say that this kind of critique applies to his theory in so far as Marr himself arguably fails to consider the biologically specific powers of the human brain to observe the world; discerning 'shapes' is but one aspect of those powers. This is ironic given that, as noted earlier, Marr conducted significant research early in his career into the physiology of the function of the archicortex, cerebellum, and neocortex in mammalian brains. By the time of his death, however, the computational metaphor has so completely taken over the research field that it has acquired, as discourses so often do, a life of its own.

If, however, it was suggested earlier that the Chinese Room argument offers a *preliminary* way of problematising a computational approach, it was because Marr's conception also fails to consider human *social* powers – not least, the powers attained via institutions and cultural formations, and which may themselves cause individuals and their brains to perceive certain aspects of the world, and to do so in some ways and not others. And indeed, as the last point begins to make clear, what is at stake is not a positivist understanding of the subject(ive); instead, it is a conception of subjectivity of the kind developed by critical forms of sociological inquiry especially in the last quarter of the twentieth century (cf. Part 3): subjectivity, viz., subjectivity as a matter of instituted, and instituting modes of consciousness, and as part of these, collective ways of imagining the self, and the self as it relates to others (human or beyond-human). Modern subjects are often encouraged to believe that they have a 'self of all selves' (cf. Chapter 4) – a self that is entirely one's own. But from the critical perspective mentioned above, we humans are by no means the sovereign subjects, the subjects of a purely personal construction that some enlightenment thinking makes us out to be. As well as being biological entities, we are at once the agents and products of socio-cultural processes which work, for the most part implicitly, to direct our attentions towards some objects and not others, and to shape our *interpretations* of those objects in some ways and not others – a process that will be theorised in some detail especially in Chapter 8.

This kind of problematisation raises the question: if a computational theory of mind – and of perception – is wrong, how is it that computers can, for example, play chess and win over the world's best masters? Or how is that face recognition software now allows digital surveillance systems to locate individuals in crowds? (to name but two examples).

---

[48] John R. Searle, 'Why Dualism (and Materialism) Fail to Account for Consciousness' in Richard E. Lee (Ed.) *Questioning Nineteenth Century Assumptions about Knowledge III: Dualism.* (New York: SUNY Press, 2010), p. 17.

Part 1 Positivist Perspectives

Anyone who knows the ins and outs of the 'Deep Blue' story will realise that the famous competition between IBM's engineers and Gary Kasparov in 1996/1997 was not what IBM made it out to be. But even if one overlooks the politics of such macho tournaments, a first answer is that it may well be that neuroscience has become a social force that may actually *achieve* a kind of working conflation between mind and machine by way of what John Searle and other speech act theorists describe as the illocutionary and perlocutionary dimensions of discourse – in this case, the illocutionary and perlocutionary dimensions of the research itself. Contrary to what logical positivists might assume, even the most rigorously logical and analytical research findings are not merely statements grounded in empirically verifiable facts; they *do* something in the process of representing (what Speech Act Theory describes as the illocutionary dimension), and crucially, such *doing* has *consequences* (the perlocutionary dimension).[49] In the context of the rise and rise of 'neuro-', we may end up with a world in which thinking and computing are so utterly entangled *by the researchers and the applications of their research* that it is either no longer possible, or at any rate, no longer *legitimate* to distinguish the one from the other. Truth becomes an algorithm, or better yet, a function of an algorithm.

A second answer addresses the questions more directly. Face recognition software undoubtedly reveals the power of computers to compare like with like. In so far as the main, indeed the only criterion is one of discrimination, then in principle the only thing holding back AI is the capacity of technicians to develop more and more subtle technologies of discrimination (I use the last term advisedly; as anyone who has followed developments in AI will know, one of its primary uses is the segregation of social, ethnic and 'racial' groups by political, security, and other types of organisations). But of course, the whole point is that human minds, and with them the perceptual systems, are not only machines that specialise in discrimination and comparison. Our minds have a capacity not just to analyse, but to synthesise, and crucially, to do so in ways that are not context-dependent in mechanical ways. While it would be romantic – and frankly naive – to deny that AI will increasingly take over the vast domains whose practices *can* be reduced to the binary logic in question, it is doubtful that they will go much further than that any time soon; this not least thanks to the poverty of the models still used by many neuroscientists to try to explain the workings of mind.

It is, in this sense, symptomatic of the state of the field that today some in neuroscience see Marr's work as a kind of antidote to what has become not only neurological, but abject *technological* reductionism across the field. In 2017 a group of researchers led by John A. Krakauer suggested in a paper that 'the emphasis in neuroscience has transitioned from ... larger scope questions to the development

---

[49] See John L. Austin, *How To Do Things With Words* (Oxford: Oxford University Press, 1975 [1962]), and John R. Searle, *Speech Acts: An Essay in the Philosophy of Language*, rev. edition (Cambridge: Cambridge University Press, 1969).

of technologies, model systems, and the approaches needed to analyse the deluge of data they produce'.[50] Yet just one year earlier, in what was widely described as a landmark paper, Eric Jonas and Konrad Kording showed that using such techniques to analyse a relatively simple microprocessor – the MOS 6502 microchip used to run classic video games such as *Donkey Kong* and *Space Invaders*, which the scientists treated as a 'model organism' – failed to meaningfully describe its hierarchy of information processing. If this was true for a simple processor, the researchers asked, then what hope of using such methods to explain the human brain with its 100 billion or so neurons, each with the potential for thousands of links to other neurons.[51]

A final set of problems with Marr's approach has to do with what I describe as the discourse of *retinalism*. The advocates of retinalism suggest that, when it comes to explaining vision, most if not all of what we need to know may be researched by examining the workings of those organs that are most directly involved in the perceptual process. In the case of vision, and of organisms with single-chambered eyes such as we humans have, this entails examining the workings of the *retina* as it relates to the central nervous system (hence my choice of name for the discourse). The underlying logic is a version of the following: if we can explain how light reflected from an optic array is mapped onto receptor cells; and if we can further explain how such mapping is transformed into a pattern which is transmitted to, and further processed by the central nervous system (e.g. the sequence retina, optic nerve, brain), then we will have managed to understand just how it is that we are able to detect relevant information, and represent it in patterns of neural activity that translate what we see, or think we see.

I should clarify that the name I've chosen for this tendency may suggest that it only applies for research into visual perception, but exactly the same tendency accrues to much research conducted in relation to other senses as well; as I conceive it, retinalism is a way of referring to this kind of reductionism in any kind of research about human perception. The last point notwithstanding, it is also possible to find important variations on this theme, not least regarding the extent to which researchers conform to more or less drastic forms of neurological reductionism. Marr, for example, was aware of the importance of avoiding the more extreme forms of reductionism even if his theory focussed almost entirely on those aspects of visual perception that involve the retina as linked to the central nervous system.

---

[50] John W. Krakauer, Asif A. Ghanzanfar, Alex Gomez-Marin, Malcolm A. MacIver, David Poeppel, 'Neuroscience Needs Behavior: Correcting a Reductionist Bias', *Neuron* 93:3 (2017), 480-490.

[51] Eric Jonas and Konrad Kording, 'Could a Neuroscientist Understand a Microprocessor?', *PLOS Computational Biology*, 13:1(2017) <http://journals.plos.org/ploscompbiol/article?id=10.1371/journal.pcbi.1005268>[Accessed 19 April 2018].

It might be thought that retinalism is all about the body, to the exclusion of mind. But in at least one sense, Marr's variety of retinalism *decorporealises* vision by cutting it off from the rest of the senses, the rest of the *body*, and from the environment with which that body interacts. From this perspective, Marr's theory of visual perception reproduces a tendency that goes at least as far back as Descartes' infamous mind-body dualism (cf. Chapter 10). This is ironic, to say the least; a visual neuroscientist like Marr, and indeed neuroscientists more generally would probably argue that, if anything, they have settled the mind-body dualism by focussing entirely on the workings of the body – a body which generates, via the neurons, what Francis Crick would probably describe as a simulacrum of mind. However, one way of thinking of the 'rise and rise' of the neurosciences is precisely as a kind of return of the only barely repressed: an ontology that clearly favours mind over body, but now conceives mind as a computer.

4

Ecological Psychology and Direct Perception

Even as evolutionary psychology and neuroscience gathered followers in academia and beyond, another field was rather more quietly developing – one that, despite adhering to several key tenets of positivist philosophy, approached perception in a very different way. I refer to ecological psychology, a field that was founded by James J. Gibson[1] (1904-1979), the third and last of the researchers whose work I will consider in this part of the study. Gibson's ecological approach to visual perception fundamentally subverts the notion that one can explain perception simply by focussing on the workings of the organs of perception – especially but not only the stimulation of receptors in the retina. In so doing, it offers a particularly sharp critique of retinalism (though Gibson does not call it that).

To better understand how and why Gibson critiques retinalism, it may be useful to begin by providing a social and philosophical context for his theory. Gibson is today usually described as a psychologist, but he majored in philosophy at Princeton University in 1925. While psychology did become the focus of his research during his postgraduate studies at the same university, Gibson's work continued to be influenced by philosophical debate. One of the founders of psychology was the pragmatist philosopher William James (1842-1910), and James' writing continued to shape the field long after he died. Indeed, one of Gibson's mentors was the philosopher and psychologist Edwin B. Holt, who himself had William James as a teacher during his undergraduate education at Harvard.

I will describe aspects of William James' philosophy, and of pragmatism more generally in Part 2. Here it suffices to note that James subscribed to a radically nominalist, and empiricist perspective. And Holt continued this line of thought as a leading member of a post-Jamesian intellectual movement known as the New Realism. Beginning the 1910s, the new realists developed an arch-positivist perspective on perception, and on epistemological matters more generally. In a manifesto and a series of texts which they published in 1910,[2] six leading philosophers in the movement invoked the need for unity across the sciences, and indeed within philosophy itself – what I have described as the positivist rule of consilience. In their manifesto, the philosophers go so far as to suggest that 'Philosophy is famous for its disagreements, which have contributed not a little towards bringing

---

[1] James J. Gibson, *The Perception of the Visual World* (Boston: Houghton Mifflin, 1950); James J. Gibson, *The Senses Considered as Perceptual Systems* (Boston: Houghton Mifflin, 1966), and James J. Gibson, *The Ecological Approach to Visual Perception* (London: Psychology Press, 2015 [1979]).

[2] Edwin B. Holt, Walter T. Marvin, W.P. Montague, Ralph Barton Perry, Walter B. Pitkin, and E. G. Spaulding, 'The Program and First Platform of Six Realists', *The Journal of Philosophy, Psychology and Scientific Methods*, 7:15 (2010), 393-401.

it into disrepute as being unscientific, subjective, or temperamental. These dis-agreements are due in part, no doubt, to the subject matter of philosophy, but chiefly to the lack of precision and uniformity in the use of words and to the lack of deliberate cooperation in research. In having these failings philosophy still dif-fers widely from such sciences as physics and chemistry.'[3]

In his own entry, Holt espouses much the same kind of empiricism as his men-tor William James, asserting the following principles: that '[t]he entities (objects, facts, etc.) under study in logic, mathematics, and the physical sciences are not mental in any usual or proper meaning of the word "mental"'; that '[t]he being and nature of these entities are in no sense conditioned by their being known'; and that '[t]he degree of unity, consistency, or connection subsisting among enti-ties is a matter to be empirically ascertained'.[4]

As we shall see, Gibson adopts what is itself an arch empiricist perspective when it comes to visual perception, arguing that direct perception does not in-volve memory or representation. What somewhat tempers this empiricism is a Jamesian recognition that subjects are active interpreters of sense data. As part of his analysis in the *Principles of Psychology* regarding the nature of the innermost self – what he called the 'self of all the other selves' – William James suggests that this self may be characterised as 'the *active* element in all consciousness; saying that whatever qualities a man's [sic] feelings may possess, or whatever content his writing may include, there is a spiritual something in him which seems to *go out* to meet these qualities and contents, whilst they seem to *come in* to be received by it'.[5] As the philosopher Josiah Royce, another mentor of Holt, as well as a friendly critic of William James puts it, for James '[e]xperience is never yours merely as it comes to you. Facts are never mere data. They are data to which you respond. Your experience is constantly transformed by your deeds. That this should be the case is determined by the most essential characteristics of your consciousness'. 'The simplest perception, the most elaborate scientific theory, illustrate how man never merely finds, but also always cooperates in creating his world'.[6]

I will explain below how fundamental this perspective is for Gibson's theory. First it should be noted that, important as the new realist philosophy was to be for Gibson's research, a rather more practical set of experiences also played a key role in the development of Gibson's theory of visual perception, which he first fully presented in 1950 in his first book, *The Perception of the Visual World*. Like Wiener and Shannon, Gibson also contributed to the US war effort in World War II. In

---

[3] Edwin B. Holt, Walter T. Marvin, W.P. Montague, Ralph Barton Perry, Walter B. Pitkin, and E. G. Spaulding, 'The Program and First Platform of Six Realists', p. 393.

[4] Edwin B. Holt, 'The Programme and Platform of Six Realists', p. 394.

[5] William James, *Principles of Psychology*, Vol. I (New York: Henry Holt, 1890), p. 297.

[6] Josiah Royce, 'William James and the Philosophy of Life', in *William James and Other Essays on the Philosophy of Life* (New York: Macmillan, 1911), 3-48, p. 37.

Gibson's case, he joined the Army Air Force, where he became the director of the Motion Picture Research Unit in the Aviation Psychology Program. The purpose of this unit was to develop visual aptitude tests for the screening of prospective pilots.[7]

By Gibson's own account, this research led directly to a critique of a key part of the scaffolding of the discourse of retinalism: the tendency to approach perception from the perspective of a Newtonian conception of 'empty' space. As Gibson explains

> The space in which the pilot flies is not the abstract space of theories, nor the lines and figures of the stereoscope, nor the space of the usual laboratory apparatus for studying depth perception. It does not consist of objects at varying empty distances. It consists chiefly of one basic object, a continuous surface of fundamental importance – the ground. ... The spatial situation which needs to be analysed, therefore, must involve the ground and everything that it implies. Instead of calling it a space it would be better to call it a world. [...] The conception of an empty space of three dimensions was a conception of philosophers and physicists. It was appropriate for the analysis of the abstract world of events defined by Newton. It was and still is of enormous value for analysis in the physical sciences. But the fact that it simplifies such problems does not make it the best starting point for the problem of visual perception. Space, time, points and instants are useful terms, but not the terms with which to start the analysis of how we see, for no one has ever seen them. [...] The world with a ground under it – the visual world of surfaces and edges – is not only the kind of world in which the pilot flies; it is the prototype of the world in which we all live. *In it, one can stand and move about.*[8]

The last point begins to explain another aspect of Gibson's critique: the criticism of what may be described as retinalism's *immobility bias*. Although Gibson had not yet worked out a precise theory to explain this problem in 1950, his experiments with pilots revealed the flaws of adopting an approach that assumed an immobile observer. As Gibson puts it in his later work, 'Observation implies movement, that is, locomotion with reference to the rigid environment, because all observers are animals and all animals are mobile. Plants to not observe but animals do, and plants do not move about but animals do.'[9] We might note in passing that Gibson switches seamlessly between accounts of human, and beyond-human organisms in a manner that is redolent of a positivist stance; the more im-

---

[7] Julian Hochberg, 'James Jerome Gibson: A Biographical Memoir' (Washington, D.C.: National Academy of Sciences, 1994), p. 154.

[8] James J. Gibson, *The Perception of the Visual World*, pp. 59-60. Italics added to the original.

[9] James J. Gibson, *The Ecological Approach to Visual Perception*, p. 65.

portant point for now is that Gibson was determined to develop an alternative conception that not only moved away from Newtonian conceptions of space, but also from immobile, or perhaps one should say *immobilising* conceptions of the observer.

Following the publication of *The Perception of the Visual World,* Gibson spent the better part of the next three decades refining what he eventually described as an ecological psychology of perception, an approach that received its fullest treatment in *The Ecological Approach to Visual Perception* (published in 1979, the year that Gibson died). The reference to ecology signals Gibson's determination to link psychology to a burgeoning ecological turn across the biological sciences.

In keeping with these developments, Gibson argued that it is not possible to understand perception without understanding the environment in which perception occurs. As he puts it in *The Ecological Approach,* '... it is often neglected that the words animal and environment make an inseparable pair. Each term implies the other. No animal could exist without an environment surrounding it. Equally, although not so obvious, an environment implies an animal (or at least an organism) to be surrounded'.[10] By this account, the environment is not the same as the physical world, if by that one means the kind of space, time, matter, and energy conceived by physics. As Gibson puts it, 'We do not live in "space"'.[11] While the physical world as conceived by the physical sciences goes from the extremes of atoms and their ultimate particles to light-years and beyond, 'neither of these extremes is an environment. The size-level at which the environment exists is the intermediate one that is measured in millimeters and meters',[12] and it is this band that the ecological theory of perception must be concerned with, albeit in a way that is itself different from that assumed by the traditional physical sciences.

This is true in at least two ways. First, the relation between the units involves not the discrete units conceived by reductionist approaches, but what Gibson describes as *nesting.* So, for example, 'canyons are nested within mountains; trees are nested within canyons; and cells are nested within leaves. There are forms within forms both up and down the scale of size ... Things are components of other things. They would constitute a hierarchy except that this hierarchy is not categorical but full of transitions and overlaps.'[13] As allied to the notion of the mutuality between organism and environment, this means that for the terrestrial environment, 'there is no special proper unit in terms of which it can be analyzed for once and for all'; the reductionist inclination to find a fundamental 'unit' for this

---

[10] James J. Gibson, *The Ecological Approach to Visual Perception*, p. 4.

[11] James J. Gibson, *The Ecological Approach to Visual Perception*, p. 27.

[12] James J. Gibson, *The Ecological Approach to Visual Perception*, p. 5.

[13] James J. Gibson, *The Ecological Approach to Visual Perception*, p. 5.

or that is misguided in so far as it leads the researcher to overlook the fact that there are subordinate and superordinate units.[14]

Second, it is also the case that, from an ecological perspective, 'the unit you choose for describing the environment depends on the level of the environment you choose to describe'.[15] From this perspective, the problem is to describe things at the ecological level 'because we all behave with respect to things we can look at and feel, or smell and taste, and events we can listen to'.[16] It follows that the unit one chooses for describing the environment depends on the level of the environment – and of course the organismic *relation* – one chooses to describe.[17]

Gibson's critique of the motionlessness of retinalism is particularly pertinent in an ecological context. Gibson notes that the motions of things in the environment are of a different order from the motions of bodies in space. 'So different, in fact, are environmental motions from those studied by Isaac Newton that it is best to think of them as changes of structure rather changes of position of elementary bodies, changes of form rather than of point locations, or changes in layout rather than motions in the usual meaning of the term.'[18] Then again, in so far as animals move about, 'the structure of an optic array at a stationary point of observation is only a special case of the structure of an optic array at a moving point of observation'; normally, the 'point of observation ... proceeds along a path of locomotion, and the "forms" of the array change as locomotion proceeds.'[19] One key consequence is that there is a play of change and non-change; while some features of the optic array persist, others don't. Where retinalism conflates the two on the assumption of a static perspective (or static perspectivalism), Gibson distinguishes between the invariant, or relatively invariant structure of an environment, and the changing perspective structure.[20]

This is a key theoretical step for at least two reasons: on the one hand, Gibson conceives perception as a process involving motions for which the cues are provided by the dynamic relation between the invariant aspects and the changing perspective structure. A key, if not *the* key to perception, Gibson suggests, is the ability to navigate, and to explore objects on the basis of an understanding of what aspects or features are invariant, or relatively invariant.

On the other hand, Gibson's theory of direct perception is contingent on the discovery of what he describes as 'affordances' of objects and surfaces, which are themselves premised, as affordances, on the interplay of continuity and change.

---

[14] James J. Gibson, *The Ecological Approach to Visual Perception*, p. 5.

[15] James J. Gibson, *The Ecological Approach to Visual Perception*, p. 5.

[16] James J. Gibson, *The Ecological Approach to Visual Perception*, p. 5.

[17] James J. Gibson, *The Ecological Approach to Visual Perception*, p. 5.

[18] James J. Gibson, *The Ecological Approach to Visual Perception*, p. 10.

[19] James J. Gibson, *The Ecological Approach to Visual Perception*, p. 65.

[20] James J. Gibson, *The Ecological Approach to Visual Perception*, p. 66.

We can say provisionally that affordance has to do with what an object or event can 'offer' the individual perceiver. However, in this conception, an observer is not merely one who sees *à la* Marr, but one who *moves*, and actively *explores* spaces on the basis of their known and relatively invariant qualities – yet this with the purpose of discovering the unknown qualities of new or changing objects and spaces. We return by this route to a Jamesian conception of the active subject.

In keeping with this approach, Gibson develops a theory of what he describes as the *ambient* optic array viz., structured or 'arranged' light that surrounds a position (and by implication, any observer at that point), and which affords information about the surrounding environment. On the basis of his ecological critique of Newtonian space, Gibson suggests that the environment is not simply a matter of so many objects 'vectoring' in empty space. In so far as there are objects on the earth or in the sky, and in so far as those objects may be nested, as opposed to detached or segregated. '... the environment is *all* of these various things – places, surfaces, layouts, motions, events, animals, people, and artifacts that structure the light at points of observation'.[21]

If the concept of the ambient optic array problematises the application of Newtonian space to perception, it is also premised on a rejection of the typically retinalist notion that perception is primarily a matter of the *stimulation* of receptors in the eye. In one telling analogy, Gibson notes that in a situation in which an observer finds itself in a fog-filled medium, there will still be stimulation of the retina's receptors; however, in so far as the light entering the pupil is the same in all directions, the observer will not be able to focus on any one thing: there will be no retinal image 'because the light on the retina would be just as homogeneous as the ambient light outside the eye', and so the possessor of the eye will not be able to fix the eye on anything.[22] By contrast, when there is what Gibson describes as *structured ambient light*, the environment specifies information thanks to the way in which light is reflected by opaque surfaces, or emitted or transmitted by luminous and semitransparent surfaces.

According to Gibson, the hypothetical case of a fog-filled medium demonstrates the difference 'between the retina and the eye, that is, the difference between receptors and a perceptual organ. Receptors are *stimulated*, whereas the organ is *activated*. There can be stimulation of a retina by light without any activation of the eye by stimulus information. Actually, the eye is part of a dual organ, one of a pair of mobile eyes, and they are set in a head that can turn, attached to a

---

[21] James J. Gibson, *The Ecological Approach to Visual Perception*, p. 59. Italics added to the original.

[22] James J. Gibson, *The Ecological Approach to Visual Perception*, p. 47.

body that can move from place to place'.[23] Those organs, Gibson suggests, make up a hierarchy that constitutes what he describes as a *perceptual system*.[24]

I will return to Gibson's theory of perceptual systems below. First it is pertinent to explain that the concept of the ambient optic array also rejects the retinalist – and it might be added, characteristically positivist – notion that vision is *either* 'objective' or 'subjective': *objective* in so far as it involves aspects such as retinal images, that are said to be completely independent of the observer's intentionality or purposiveness; or *subjective* in so far as it entails no more than a purely individual and psychological response to the signals sent to the brain by the senses. Instead, Gibson proposes that an optic array *is ambient at a point*, meaning that the optic array is observed from a point, or rather a *place* or *set of places*, in *ecological* space, 'in a medium instead of in a void'. 'Whereas abstract space consists of points, ecological space consists of places – locations or positions'.[25] However, and crucially, by this account observation must be *at once subjective and objective* in so far as it starts from a particular place, but perceives structures that are not themselves subjective, and this in *ways* that are not purely subjective.

To explain this last point, it is necessary to understand in more detail what is perhaps the most interesting, but also the most controversial aspect of Gibson's theory: his theory of affordances, which I mentioned earlier, and with it his suggestion that perception is in fact direct perception. As Gibson puts it, 'I argue that the seeing of an environment by an observer existing in that environment is direct in that it is not mediated by visual sensations or sense data. The phenomenal visual world of surfaces, objects, and the ground under one's feet is quite different from the phenomenal visual field of color–patches ... I assert that the latter experience, the array of visual sensations, is not entailed in the former. Direct perception is not based on the having of sensations. The suggestion [is] that it is based on the pickup of information.'[26] What is 'picked up' is the *affordance* of whatever is observed.

Affordance is a neologism that Gibson coins in order to explain perception as a process that discerns what an environment can offer the observer. 'The affordances of the environment are what it offers the animal, what it provides or furnishes, either for good or ill.'[27] As an example, Gibson refers to a terrestrial surface that affords support for some animals but not others. So, for example, a hard surface able to carry the weight of an adult human will allow the adults to walk

---

[23] James J. Gibson, *The Ecological Approach to Visual Perception*, p. 47. Italics in the original.

[24] James J. Gibson, *The Ecological Approach to Visual Perception*, p. 47.

[25] James J. Gibson, *The Ecological Approach to Visual Perception*, pp. 58-59.

[26] James J. Gibson, 'A Theory of Direct Visual Perception', in A. Noe & E. Thompson (Eds.), *Vision and Mind: Selected Readings in the Philosophy of Perception* (Cambridge: MIT Press, 2002), 77–89, p. 77.

[27] James J. Gibson, *The Ecological Approach to Visual Perception*, p. 119.

or run along it, whereas the surface of at least a deep body of water will not. The strong, hard surface is 'stand-on-able', permitting an upright posture for bipeds, and so 'walk-on-able' and 'run-over-able'[28] (as opposed to, say, 'sink-into-able' like water; note though, that this last affordance may be very different for, say, a pond skater [*Gerridae* spp.]).

Affordance applies not just to such macro-structural aspects of the environment, but to any and all objects in the environment, be they attached or detached. An elongated object of a moderate size and weight affords wielding, while a rigid object with a sharp dihedral angle, an edge, affords cutting and scraping, and so is a knife.[29] According to Gibson, some *events* also have affordances; '[a] fire', Gibson suggests, affords warmth on a cold night; it also affords being burnt. An approaching object affords either contact without collision or contact with collision; a tossed apple is one thing, but a missile is another'.[30]

From Gibson's point of view, perception does not just *involve* the appraisal of affordance; in some sense it *is* the appraisal of affordance. His theory flatly rejects the conventional notion that we perceive those objects that we can represent, or as Gibson puts it, those objects whose qualities or properties we can discriminate. Where traditional psychology assumes that objects are composed of their qualities, Gibson argued that 'what we perceive when we look at objects are their affordances, not their qualities. We can discriminate the dimensions of difference if required to do so in an experiment, but what the object affords us is what we normally pay attention to. The special combination of qualities into which an object can be analyzed is ordinarily not noticed'.[31] Or as he also puts it, 'An affordance is an invariant combination of variables'.[32]

Gibson clarifies that objects can have different affordances. A stone, for example, may serve as a missile, a paperweight, a bookend, or a hammer. However, in his view, this does not contradict the theory in so far as such affordances are all 'consistent with one another'.[33] '[T]he arbitrary names by which they are called do not count for perception. If you know what can be done with a graspable detached object, what it can be used for, you can call it whatever you please'. 'The theory of affordances rescues us from the philosophical muddle of assuming fixed classes of objects, each defined by its common features and then given a name'.[34]

As this account makes clear, Gibson's approach to perception is very different from Marr's information processing. Marr's information processing approach is

---

[28] James J. Gibson, *The Ecological Approach to Visual Perception*, p. 119.

[29] James J. Gibson, *The Ecological Approach to Visual Perception*, p. 125.

[30] James J. Gibson, *The Ecological Approach to Visual Perception*, p. 94.

[31] James J. Gibson, *The Ecological Approach to Visual Perception*, p. 126.

[32] James J. Gibson, *The Ecological Approach to Visual Perception*, p. 126.

[33] James J. Gibson, *The Ecological Approach to Visual Perception*, p. 126.

[34] James J. Gibson, *The Ecological Approach to Visual Perception*, p. 126.

what is known as an 'indirect' theory of perception in so far as it assumes that representation plays a key role. By contrast, for Gibson

> *Information* ... refers to specification of the observer's environment, not to specification of the observer's receptors or sense organs. The qualities of objects are specified by information; the qualities of the receptors and nerves are specified by sensations. Information about the world cuts right across the qualities of sense. [...] The term information cannot have its familiar dictionary meaning of *knowledge communicated to a receiver* ... picking up information is not to be thought of as a case of communicating. The world does not speak to the observer. Words and pictures convey information, carry it, or transmit it, but the information in the sea of energy around each of us, luminous or mechanical or chemical energy, is not conveyed. It is simply there. [...] Shannon's concept of information applies to telephone hookups and radio broadcasting in elegant ways but not, I think, to the firsthand perception of being-in-the-world, to what the baby gets when first it opens its eyes.[35]

So what precisely is Gibson's alternative? Gibson describes it as the theory of information *pickup,* and it involves the play of the invariant and variable features of the ambient optic array, and of the environment more generally: '...*perceiving* is a registering of certain definite dimensions of invariance in the stimulus flux together with definite parameters of disturbance. The invariants are invariants of structure, and the disturbances are disturbances of structure.'[36]

In the case of vision, the structure in question is that of the ambient optic array. For example, as one's head turns and surveys a scene, the occluding edge of the nose is an invariant which helps to specify the self even as the sweeping motion over the ambient array specifies that one's head is turning, and as the acquisition and loss of texture at the edge of the field of view specifies that there is motion of an object over the ground.[37] In the context of perceptual events like this, the *process of pickup* 'is postulated to depend on the input-output loop of a perceptual system', but where such a loop does not depend on communication or on memory; 'according to pickup theory, information does not have to be stored in memory because it is always available'. Returning to the earlier examples, one does not have to remember that the ground outside one's home is hard, and is 'walk-able'; a similar point might be made about the affordance of everything from toilet seats to kitchen knives to one's front door keys.

---

35 James Gibson, *The Ecological Approach to Visual Perception*, pp. 231-232. Italics in the original.

36 James Gibson, *The Ecological Approach to Visual Perception*, p. 238.

37 James Gibson, *The Ecological Approach to Visual Perception*, p. 239.

Part 1 Positivist Perspectives

I would now like to turn to what may be described as Gibson's *recorporealisation* of vision. As I began to suggest in Chapter 3, approaches like Marr's have the paradoxical quality of *decorporealising* vision. Their advocates may claim that they have moved beyond the mind-body dualism, when in fact their theory excludes from perception all those parts of the body that are not to do with the central nervous system, and with the organs deemed to be most directly involved with one or another perceptual system. If this is true for relations within the body of the observer, it is even more true for the relation between a perceiving body and its environment.

Gibson, by contrast, argues that if we cannot neatly separate off perception from the environment in which perception takes place, we cannot neatly separate any one sense from the rest of the organs, the rest of the body. As Gibson puts it, 'We are told that vision depends on the eye, which is connected to the brain. I ... suggest that natural vision depends on the eyes in the head on a body supported by the ground, the brain being only the central organ of a complete visual system.'[38]

In *The Senses Considered as Perceptual Systems* (which Gibson published in 1966 and constitutes something like a halfway point in the development of his theory), Gibson clarifies that it is not only a complete *visual* system that is at stake; there are other senses, or as Gibson describes them, other perceptual systems. Once again, these systems are not to be confused with those aspects of perception that merely receive sensory inputs; when Gibson refers to *systems* he means the sensory nerves and the brain, but also the perceptual organs acting in tandem with muscles and other aspects of the body to enable an active detection and investigation of the information provided by an environment. As Gibson explains, 'It has always been assumed that the senses were channels of sensation ... but the fact is that there are two different meanings of the verb *to sense*, first, *to detect something*, and second, *to have a sensation*. When the senses are considered as perceptual systems the first meaning of the term is being used.'[39] The second meaning tends to be associated with a relatively passive state: 'When the senses are considered as channels of sensation (and this is how the physiologist, the psychologist, and the philosopher have considered them), one is thinking of the passive receptors and the energies that stimulate them, the sensitive elements in the eyes, ears, nose, mouth, and skin'[40] – energies that are, by definition, in a constant state of flux, and which in approaches such as Marr's are thought to impinge on the retina and to produce signals that determine a certain representation in the brain. By contrast, Gibson's approach focusses on the detection of what he describes as the 'higher-order variables of stimulus energy' – for example, ratios and proportions

[38] James J. Gibson, *The Ecological Approach to Visual Perception*, p. xiii.

[39] James J. Gibson, *The Senses Considered as Perceptual Systems*, p. 1.

[40] James J. Gibson, *The Senses Considered as Perceptual Systems*, p. 3.

– that do *not* change.[41] It is these invariant perceptions that the *active* observer gets despite varying sensations.[42]

On the basis of this approach, Gibson insists that the senses, or rather the perceptual systems, are active and mutually imbricating, as opposed to being completely separate, and so ought to be classified not as modes of conscious quality (as per more traditional psychological approaches) but as modes of *activity*.[43] The modes in question are 1) the basic orienting system, with a mode of attention geared to general orientation; 2) the auditory system, geared to listening;   3) the haptic system, geared to touching; 4) the taste-smell system, geared to smelling and tasting; and 5) the visual system, geared to looking. A key conceptual displacement is that, while the channels of sensation are not subject to modification by learning – as Gibson puts it, the sense data are by definition given – the perceptual systems *are* amenable to learning.[44] 'It would be expected that an individual, after practice, could orient more exactly, listen more carefully, touch more acutely, smell and taste more precisely, and look more perceptively than he could before practice',[45] and indeed Gibson notes that this insight is born out by abundant psychological research.

Having provided an overview of Gibson's approach, what might its implications be for this study? Unlike the proposals developed by Edward O. Wilson and David Marr, Gibson has much to offer a theoretical approach to wildlife observation. Here I would highlight three aspects: First, unlike Marr, Gibson describes, rightly in my view, the need to regard observation (or what he more often describes as perception) as a process that entails at once a physiology, an intentional or purposive subject, and an environment. Anyone studying wildlife observation in a tropical forest will recognise the importance of this seemingly obvious displacement vis-a-vis retinalist perspectives; it *must* be true that observation is at least as much a matter of what is observed, as it is of a certain physiology, as it is an active observational process. While Marr would presumably accept this point in general, his theory focusses entirely on visual/neurological activities considered from a computational perspective, and as a matter of an interface between channels of sensation (the eyes) and the central nervous system.

Second, where both Marr and Wilson adopt perspectives that take for granted the centrality of representation and memory, Gibson suggests in sharp contrast that, at least in the case of perception *sensu stricto*, neither 'pictures' nor 'storage' really matter. There is, according to Gibson, something like a pre-representational, or better yet an *a*-representational process of the kind that is best illustrated by

---

[41] James J. Gibson, *The Senses Considered as Perceptual Systems*, p. 3.

[42] James J. Gibson, *The Senses Considered as Perceptual Systems*, p. 3.

[43] James J. Gibson, *The Senses Considered as Perceptual Systems*, p. 49.

[44] James J. Gibson, *The Senses Considered as Perceptual Systems*, p. 51.

[45] James J. Gibson, *The Senses Considered as Perceptual Systems*, p. 51.

our relation the ground each time we take a step. Unless there is something wrong with the basic orienting system, or the nature of the ground changes noticeably, our perception of the ground is a matter of a learned, but thereafter a more or less automatic presumption of affordance. This is the aspect of Gibson's approach that I will be most interested in problematising, but not before recognising that, at the very least, some aspects of the way in which people observe wildlife (or the environments in which the wildlife are observed) *must* entail something very much like a relation based on affordance, if not direct perception. If nothing else, so many of the activities that we engage are based on a mixture of bodily and more-than-bodily *habit* that it must be true that an entirely rationalist perspective – e.g. playing on Descartes' famous formulation, 'I think, therefore I perceive' – must be mistaken.

James Gibson's proposals regarding an ecological approach to the psychology of learning bring me to a third implication: while Gibson would certainly agree with Wilson (and any other adaptationist biologist) that the human physiology has evolved, and poses species-specific constraints particularly on the level of the structure of the senses and the perceptual organs, his approach does not have to appeal to anything like universal epigenetic rules. On the contrary, and very much in keeping with a Jamesian empiricism (or what we might also describe as *active* empiricism), a learning process occurs anew with each individual. Each individual acquires a knowledge of affordances via the trial and error that begins from the earliest childhood, and continues throughout that individual's lifetime. It is to be deduced that this kind of dynamic also applies to the perception of beyond-human animals; there is, in this sense, no need to appeal to a 'mysterious' (Barlow, *passim*), and by positivism's own standards, *unresearchable* 'epigenetic rule' to explain why we learn to avoid or otherwise relate to certain creatures. Eleanor Gibson, who worked closely with James Gibson in the development of ecological psychology, describes this kind of learning process as a matter of *perceptual learning.*[46]

There are a number of other more specific aspects of Gibson's theory that might well be useful in a study about wildlife observation and I will refer to some of these in the course of later chapters. For now, I would like to problematise two aspects of Gibson's approach: his theory of direct perception, and his naturalistic conception of environment.

Earlier, I noted that Gibson was influenced by the New Realism, a philosophical movement that took up William James' empiricist philosophy. This is evident across several aspects of Gibson's theory, but it is perhaps most obvious in his attempt to move away from an indirect theory of perception. Gibson is clear that

---

[46] See Eleanor J. Gibson, *Principles of Perceptual Learning and Development* (New York: Century Psychology Series, 1969). See also Eleanor J. Gibson and Anne D. Pick, *An Ecological Approach to Perceptual Learning and Development* (Oxford: Oxford University Press, 2000).

representation has its purposes, but not when it comes to perception 'itself'. If this is the case, Gibson must formulate an alternative, and he does so via the theory of affordances, and the related notion of the 'pickup' of information. While the former aspect is clearly described and formulated, the pickup of information is not as clearly formulated, beyond the suggestion that, as Gibson puts it, *'perceiving is a registering of certain definite dimensions of invariance in the stimulus flux together with definite parameters of disturbance'*.[47] To be fair, there is much to be said about Gibson's critique of linear, and mathematised conceptions of communication (e.g. Shannon's, and by extension, Marr's and Wilson's), but in my view there is no equally precise, and well-formulated alternative with respect to information pickup. It is no coincidence that, by the time of *The Ecological Approach to Visual Perception*, Gibson's last work, Gibson is increasingly referring not to *information* pickup (or the pickup of information), but simply to *pickup*.

Again, it might be argued that affordance, for which there *is* an explanation, tells us all that we need to know about direct perception. But the theory of affordance is not without its own issues, and indeed one of its main problems can be illustrated via the repurposing of objects. While Gibson recognises that one same object can have different affordances, he argues that this does not contradict the theory in so far as differing affordances will nevertheless all be 'consistent with one another'.[48] However, it seems to me that the argument of consistency negates or minimises the enormity of the challenge posed to Gibson's approach by, say, the repurposing of familiar objects.

Consider, for example, the following scenario. Someone needs to move several large boxes out of their house, but it's a windy day and the front door keeps slamming shut. They decide to use as a doorstop an old desktop computer that is sitting next to the front door (the old computer was supposed to be taken to the local recycling centre, but a busy schedule has kept the owner from doing so). A part of this process, which clearly involves a multimodal (multichannel) form of perception, fits well with the kind of logic described by Gibson: having learned from previous experience that the computer is quite heavy, and that it has features that keep it from sliding about (e.g. rubber feet), the person will know intuitively that it will 'do the trick' and hold the door open, despite the gusts of wind. In Gibson's terms, the computer has the requisite affordance, an affordance that is, again in his terms, 'consistent' at least with the machine's physical characteristics.

What Gibson's theory leaves out – or rather, tries to separate off – is the role that an 'indirect' calculation, itself impromptu, plays in this kind of everyday activity. That calculation, as calculation, *must* entail a combination of memory and representation. The person only knows from past experience that the computer is heavy, and so must remember this. This entails, in Gibson's own terms, 'storage'.

---

[47] James J. Gibson, *The Ecological Approach to Visual Perception*, p. 238. Italics in the original.

[48] James J. Gibson, *The Ecological Approach to Visual Perception*, p. 126.

And if it entails 'storage', which is really to say memory, then it must surely also involve some kind, some degree of representation 'in' the mind, however much the object and its affordance are also *beyond* the mind. Can we really separate off, as Gibson suggests, the 'direct perception' of an object such as the computer, from contextual calculations of the kind I have just illustrated?

As part of his notion of pickup, Gibson is keen to emphasise the flow of energy, and certainly he is right to do so over and against the reductionist and immobilising analogies favoured by retinalist approaches. But aside from the possibility that references to 'flows of energy' pose similar issues to the notion of 'empty space',[49] the very idea of a *flow* suggests a more general way of posing the question I asked above: can Gibson really deny that recollection and representation must also be an integral aspect of that selfsame flow of energy?

To be clear, I am not arguing that a 'thought process' or 'representation', conceived in the abstract, must necessarily *guide* perception – if there *is* some kind of guidance, it is certainly not inevitably a voluntary or conscious one. That would be a return to the kind of rationalism that Gibson rightly critiques. Rather, I am questioning the obverse movement, which either shuts out, or simply *adds* representation as a kind of afterthought. To say that we can only really perceive what we can imagine is as problematic as suggesting that what we perceive has nothing to do with what we can imagine. Doubtless Gibson is right to point out that in some situations, representation – certainly representation conceived along Saussurean lines (cf. Chapter 5) – has little to do with affordance. But that is very different from effectively opposing perception to semiotic activity, *tout court*; doing so reveals the extent to which, despite numerous differences vis-a-vis Wilson and Marr, Gibson himself remains very much in the thrall of the rules of phenomenalism and nominalism (cf. Part 1). In so far as he also is an advocate of something like consilience, then Gibson too, may described as a positivist.

In the perspective that I will develop in this study, there must always be a *sense* of affordance, in the semiotic connotation of sense as meaning, and not just in the connotation of detection. But of course, on occasion that sense may be proven wrong by phenomena which are either not semiotic (at least not, *symbolically* semiotic, cf. Part 2), or not immediately *experienced* as being semiotic. This is, we shall see, an important distinction which may ultimately spare the researcher from having to choose between direct and indirect theories of perception.

I would like to conclude this chapter by saying something about Gibson's naturalistic conception of environment. Before doing so I would like to reiterate how welcome Gibson's foregrounding of the environment is; given this study's focus on wildlife observation, Gibson's development of an *ecological* approach is a particularly felicitous one. Amongst many other aspects, the critique of 'empty' space

---

[49] Gibson critiques, rightly in my view, any effort to simply transfer the conceptual baggage of physics to the study of perception; yet this is arguably what he does when he refers to flows of 'energy', some of whose invariant aspects are 'picked up' by observers.

chimes with the social theorist Henri Lefebvre's critique of the double illusion[50] which I will present in Part 4.

There is, nevertheless, at least one problem with Gibson's conceptualisation of the nexus between perception and environment: while Gibson certainly acknowledges that a space can be, and often is in some sense constructed by human activity, he makes little or no allowance for the possibility that that selfsame constructive activity might fundamentally alter perception. Both James and Eleanor Gibson would agree that what they describe as perceptual learning can and will occur in any new environment. However, this kind of learning can only take into account what might be described as the *proximal* environment, and indeed, those aspects of that environment that have physical presentation: surfaces, edges, objects with a certain texture, events with what are themselves relatively invariant aspects, and the like.

By contrast, when it comes to remembered environments, or environments generated by virtual means – be they actual physical spaces, or environments represented by pictures or via the moving image – Gibson goes so far as to state that a drawing (a term that he uses in a very general sense that includes pictures) 'does not have ecological validity'.[51] While Gibson is referring to the use of drawings in psychological experiments, the implication is clear: drawings are inferior, perceptually speaking, to the real McCoy of perception. This is a stance that is echoed in the final chapter of *An Ecological Approach to Visual Perception*, which considers motion pictures. In it Gibson suggests that 'the art of film-editing should be guided by knowledge of how events and the progress of events are naturally perceived'.[52] Here again, the aesthetic and ethical implications are clear: even though motion pictures – or as Gibson prefers to call them, the *progressive* pictures – yield 'something closer to natural visual perception',[53] they are still inferior, perceptually speaking, to actual, as in immediate perceptual experience.

In the introductory chapter, I noted that the visitors with whom I conducted ethnographic research on Barro Colorado seemed surprised by the gap between the represented space constructed by wildlife TV programmes (and other nature media), and the space of *in situ* wildlife observation in an actual tropical forest. Gibson's theory would certainly help to explain some of the differences across the two contexts, and so provide a critical perspective on what I have described as dynamics of transmediation. The problem is that Gibson cannot account theoretically for the equally important fact that, if there *was* disillusion amongst some tourists over the differences, this was precisely because the nature media repre-

---

[50] Henri Lefebvre, *The Production of Space*, translated by Donald Nicholson-Smith (Oxford: Blackwell, 1991 [1974]), pp. 27-29.

[51] James J. Gibson, *The Ecological Approach to Visual Perception*, p. 256.

[52] James J. Gibson, *The Ecological Approach to Visual Perception*, p. 288.

[53] James J. Gibson, *The Ecological Approach to Visual Perception*, p. 288.

sentations helped to generate expectations. Not just 'general' expectations in the sense of 'ideas' or vague impressions, but very concrete *observational* expectations at once borne by, and taken *from* the techniques of observation deployed by both photographic and cinematographic media. The conclusion has to be that there was, in the Gibson's own terms, perceptual learning – *but via the media*. If this is the case, then surely one cannot claim so categorically that pictures, be they drawings or progressive pictures, 'lack ecological validity'. More importantly, this issue suggests another reason for assuming that James Gibson's decision to exclude memory and representation from perception is mistaken. If representations can affect how one perceives the world, it is difficult to argue that representations have nothing to do with the perception of the world.

Conclusions to Part 1

Let's return to the bullet ants, and let's say that a graduate research student who is taking a course in a biological field station in Central America has three encounters with said species on three different days, in environmental circumstances that are, for all practical purposes, identical: the encounters all occur in the same forest, along the same trail, and indeed next to the same large tree, at the base of which bullet ants have built a nest. It is the dry season, and this being a *seasonal* moist forest (cf. Part 4), there is no rain or significant cloud cover on any of the three days. In so far as the encounters all occur at the same *time* of day, there is identical illumination, and finally, each time that the researcher reaches the tree, there is the same level of activity amongst the ants (i.e. each time, several ants can be observed climbing along the same observable part of the tree trunk).

In the first encounter – let's call it E1 – the researcher is looking out for blue morpho butterflies (*Morpho menelaus*). Although the researcher walks right past the aforementioned tree, with several of the large ants crawling up its base at that precise moment, the researcher does not notice them. In one sense, this is not really an *encounter* in so far as at least the human observer does not attend to the ants. However, the graduate student will nonetheless have had the tree very clearly within view, and will have been less than two metres away from its base for at least a few seconds.

In the second encounter – E2 – the researcher is walking along the same trail, but this time is looking for army ants (*Eciton* spp.), and so is looking carefully at the ground. As the graduate student travels along the trail, the researcher detects a slight movement at the base of a large tree. A glance reveals that it is only bullet ants, and so the researcher does not stop.

In the third encounter – E3 – the researcher is leading a group of tourists who are taking a tour of the forest (the field station encourages graduate research students to hone their communicational skills with 'lay' members of the public). The tourists expect to have numerous encounters with charismatic fauna, but thus far, the group has only encountered an agouti (*Dasyprocta punctata*). As the group approaches a large tree along the trail, the researcher-turned-tour-guide remembers the bullet ants at the base of the tree, and luckily the ants are still there. So the researcher motions the group to stop, looks intently at the immediate surroundings to make sure that there are no ants beyond the bole, and then points to the ants. As the tour members gather round and peer at the huge ants moving along the trunk, the researcher explains that these are bullet ants (*Paraponera clavata*), also known in the Spanish language as the *Hormiga Veinticuatro* – the 24 (Hour) Ant. The researcher further explains that the vernacular names of the ants refer to the excruciating, and long-lasting pain caused by their venom...

The first and most obvious point that these three scenarios reveal is that what an individual sees, or more generally what the individual *observes* may be at least partly determined by the contingencies of what they are looking out for, what they *expect* to observe. Suggesting this is not to regress to a romantic conception of wholly voluntary perception; on the contrary, as I began to suggest in earlier chapters, on occasion individuals fail to see the very object they are looking for, even when it is clearly, indeed *centrally* in their field of view. At least a part of the reason involves involuntary and unselfconscious dynamics. If the grad student did not notice the bullet ants in E1, it was because the researcher was looking for *flying* organisms, as opposed to creatures crawling up trees. This being the case, the researcher's perceptual system may well have 'picked up', in Gibson's vocabulary, the ants on the tree, but the mind was so focussed on finding butterflies that the ants did not, in a manner of speaking, *register*, and so were not observed – they were not attended to. We may, in this sense, establish a parallel here with Gibson's account of a fog-filled environment, and his critique of 'stimulationist' accounts of perception. But wildlife *observation* is not just a matter of *perception* – at least not perception of the kind theorised by Marr; whatever else it entails, I will argue that it is perception plus attention, and more often than not, conception and interpretation.

The second point is somewhat more subtle, and involves the difference between E2 and E3. In both of those encounters, the researcher *did* attend to the ants, did consciously perceive them. Notice, however, that the researcher did so quite differently across the two encounters; in one case (E2) it was a glance, in the second (E3) it was a far more sustained *gaze*. This is a difference that Marr's approach does not theorise at all, and it is one that Gibson's approach fails to characterise in sufficient detail. While Gibson's approach certainly acknowledges the importance of exploratory perceptual activity, his concept of information pickup, tied as it is to *affordance*, does not, indeed *cannot* adequately explain what might be described as *observational contingency as it relates to meaning*. It may well be true that, under 'normal' circumstances, the ground need not be questioned, just as bullet ants and venomous snakes should be avoided; but of course, under what may well be normal circumstances for certain *scientists*, they might well go out of their way to encounter a venomous snake, or indeed to scrutinise the ground as if they had never walked on it before, looking for signs of particular geological phenomena. Then again, we know thanks to the example of Justin O. Schmidt, and thanks also to the research conducted by anthropologists, that some people go out of their way to provoke the stings of bullet ants. For example, the Sateré-Mawé, an indigenous group of roughly ten thousand members who live in the Amazonia, are said to have an initiation ceremony that is designed to test the mettle of boys as prospective warriors. Apparently the Sateré-Mawé need warriors capable of protecting the group from hostile neighbours and from the Amazon's wild animals, as well as to hunt for food in what are often adverse conditions. To this end, boys are reportedly subjected to an initiation ritual that involves putting both hands into specially made 'mittens' that have numerous live bullet ants woven into them; it seems the ants are initially sedated with unspecified, but presumably locally available substances, and then sown into the mittens with the stingers pointing inwards. In order to be considered as a prospective warrior, boys who in some cases are only just reaching their teens are made to thrust their hands into the mittens for five minutes or more, being stung the entire time, even as other members of the group sing and dance.[1]

The more general point I am making is that, contrary to what is suggested (however differently) by both Wilson and Gibson, when it comes to the perception of animals, the normal is not necessarily the natural, or at any rate, the *naturalistic*. Depending on a variety of aspects of one's social background, but also one's most immediate context, there may be at least partly different forms of perception, or what I will prefer to describe as partly different *techniques of observation*. From this perspective, Gibson's suggestion that 'The theory of affordances rescues us from the philosophical muddle of assuming fixed classes of objects, each defined by its common features and then given a name'[2] may inadvertently be a discursive inversion: the only way that the theory of affordance can actually work is precisely by assuming more or less fixed classes, if not of objects, then certainly of affordances – the argument of 'consistency' across variables.

Now Gibson would probably argue that the issues I have just raised have nothing to do with perception 'itself'; perception, in itself, as opposed to perception mistaken for something remembered or something represented, is all about affordance. But ironically, given Gibson's otherwise sharp critique of retinalism, such a stance is only really tenable if one adopts *Marr's* at once reductionist and indirect conception of perception (i.e. perception as a wholly autonomic process that is prior to purposiveness, embodiment, or environment, but which does involve representation). If, by contrast, one really *does* want to argue that perception is an active and purposive process, then surely it has to be accepted that memory and representation are integral to the process – albeit not necessarily in the way suggested by voluntarist, rationalist, or indeed decorporealised theories of representation.

The last issue that I would like to raise, this time via the third encounter, is the one that perhaps most clearly reveals the problematic character of a positivist approach to perception. In the case of E3, observation entails at once a collaborative and instituted process. The process is collaborative in so far as several people come together to engage in observational practices that are not only shared, but which may be initiated by *someone else* – in this context, attention depends not so much on William James' 'self of all other selves', but rather of *selves among other selves*. But those *selves* are neither the monads of populationism, nor the perceptu-

---

[1] See for example, Anand N. Bosnia, R. Shane Tubbs, and Christoph J. Griessenauer, 'Ritualistic Envenomation by Bullet Ants Amongst the Sateré-Mawé Indians in the Brazilian Amazon', *Wilderness and Environmental Medicine*, 26(2015), 271-273. See also, Steve Backshall, 'Bitten by the Amazon', *Sunday Times*, 6 January 2008 <https//www.thetimes.co.uk/article/bitten-by-the-amazon-ptmjffcg8ff>[Accessed 20 April 2018].

[2] James J. Gibson, *The Ecological Approach to Visual Perception*, p. 126.

al sovereigns described by ecological psychology. They are members of a *group* whose practices are *instituted* in the obvious sense that an institution provides a guide who leads the tour according to certain organisational criteria, and in the not-so-obvious sense that all the members of the group have what are more or less subtly instituted *dispositions* – i.e. a tendency to respond, to act, to engage in practice in a manner that is guided, however implicitly and unselfconsciously, by shared codes. This being true, it is not sense *or* institution, as implied by positivist approaches, but sense *and* institution, or better yet, the institution of sense.

And yet... in the so far the researcher takes this perspective too far, then they may well end up reproducing the very dualism that they seek to contest. To argue that it is culture all the way down is no better than to argue that it is biology all the way up. Equally if not more fundamentally, to engage in a dispute over the identity of any 'ultimate determinant' is to fail to acknowledge the importance of the haecceity – the '*thisness*' – of the different participants in any encounter with wildlife. For these and for many other reasons that will be considered in the following chapters, it will be necessary to question both positivism, and the kind of constructivism that has long prevailed in the critical social sciences and humanities.

PART 2

Signs of Wildlife Observation
Semeiotic-Phenomenological Perspectives

By late April Barro Colorado's dry season is coming to an end. Torrential down-pours have become increasingly frequent, and for a few days, bright yellow and pink flowers appear on the island's *Guayucán* trees (*Tabebuia* spp.). There is, more-over, a renewed faunal activity in the island's tropical moist forest; even as some of the trees blossom, more animals can be heard calling and seen foraging as Bar-ro Colorado's forest becomes, once again, a forest with rain.

I have joined another group of visitors on a tour of the island, and as we make our way down a ravine, the group's guide stops us and asks, 'Can you hear that?'

Many of the tour members look up quizzically; the forest is full of sounds, and at least to the inexpert ear, nothing appears to stand out. After the guide provides some indications, an intermitting, relatively high pitched, indeed almost insect-like 'chirping' sound becomes apparent. It is coming from one of the trees next to the trail. Is it a cricket, or some other insect? The guide explains that it is in fact a small frog – a green and black poison dart frog (*Dendrobates auratus*). After a brief search, the guide finds the amphibian perched on the trunk of a large tree, and we gather round it while the guide provides information about the frog's ecology.

Even as we listen to the account, several members of the group take pho-tographs of the frog. One man with a digital SLR camera seems particularly keen, taking pictures from one angle and then another. When he is done, he uses the screen on the back of the camera to show the pictures to his partner, who remarks that the frog looks much larger in some of the close-ups than it does in real life. Another tourist comments that she's seen a similar frog in her local zoo. Eventual-ly the frog, which has remained motionless throughout our discussion, starts to hop away, and the group moves on.

If the guide was able to perceive the calls of the frog amid a multitude of other sounds of the forest; if once she heard the call, she was able to recognise its source; if, thereafter, she was able to locate the amphibian, recognise its shape, and then point to it; and if visitors, for their part, could comprehend the words uttered by the guide, follow her indications, themselves see the frog, take pho-tographs of it, and compare this to other experiences – all these events were made possible, in one way or another, by *signs*.

We can say provisionally that signs are anything that conveys meaning about something to someone. The call of *D. auratus*; the name 'poison dart frog' or '*Den-drobates auratus*'; the guide's pointing action; any comments such as 'Look, it's here'; the photographs taken by the visitors; any *thoughts* the guide or tourists may have entertained; and even the form and movements of the frog itself, as perceived by guide and visitors – and of course by other frogs, and other wildlife – all these may be regarded as signs.

As this list begins to make clear, signs allow humans to conceive and interpret their surroundings. But signs are also what enable us to *perceive* beyond-human animals and any other objects or processes (and vice-versa, at least in the case of animate objects). Had the frog not called out, our group would doubtless have walked right past it; equally, had the *form* of its call not alerted the guide that the

118

sound was the one produced by *D. auratus*, she might well have ignored it. Conversely, had the frog not itself sensed the presence of potentially threatening forms, then it might have remained on the tree trunk. Signs may thereby be regarded as an integral, indeed an indispensable aspect of encounters with wildlife, and by implication, of wildlife observation. More generally, signs can be said to be an integral, and indispensable aspect of life itself. In so far as this is the case, the theory and analysis of signs, known as *semiotics*, constitutes a fundamental aspect of the theoretical approach that I will develop across this study. This being so, I will devote Part 2 to a detailed introduction to semiotic theory.

Semiotics has been described as a science,[1] but the kind of semiotics that I will be employing does not engage in the kind of inquiry prescribed by positivist philosophy. Instead, it entails what is at once a more modest, and a more ambitious investigative endeavour. It is more modest in so far as semioticians do not typically seek to use research about signs for predictive purposes, let alone for the purposes of policy science. But it is more ambitious in so far as the pursuit of semiotic truths is not constrained by the narrowest of positivist forms of inquiry, i.e. those that are fixated on the rules of phenomenalism, nominalism, objectivism, consilience, or causalism (cf. Chapter 1). Put very simply, at least the kind of semiotics that this part of the study will be concerned with seeks to elucidate how it is that signs at once constitute, present, and mediate all manner of phenomena.

As this account begins to suggest, there are different ways of going about semiotic inquiry. At least amongst social scientists and scholars in the humanities, the best known of these build on the insights of the founder of modern linguistics, Ferdinand de Saussure. I myself will not use Saussure's theory, and indeed, Chapter 5 will begin by explaining why I have opted instead for an approach that is based on the writings of the philosopher and scientist Charles Sanders Peirce. In Chapter 6, I will then consider in some detail the fundamental elements of Peirce's semiotic theory, and also his phenomenology. Thereafter, Chapter 7 will consider Peirce's complex typology and classification of signs as these might be employed to analyse aspects of wildlife observation.

---

[1] Charles Morris famously described semiotics as the 'science of signs'. See Charles William Morris, *Foundations of the Theory of Signs* (Chicago: University of Chicago Press, 1938).

5
Peircian Pragmatism

For many readers, any reference to semiotics probably evokes notions of signs regarded as dyadic (two-part) constructs involving signifieds and signifiers. This shows the extent to which the theory of Ferdinand de Saussure[1] (1857-1913) continues to have a profound influence on the field of semiotics. For academics with a more specialised knowledge, semiotics may also evoke the work of post-Saussurean luminaries such as Roland Barthes[2] and Jacques Derrida,[3] who continued to employ the dyadic or bilateral conception of the sign even as they raised questions about other aspects of Saussure's theory and methodology. In Part 3 I will explain that the legacy of Saussure is even apparent in the writing of the philosophers Gilles Deleuze and Félix Guattari, who attack semiotic structuralism but nevertheless use the theory of Louis Hjelmslev.[4]

As I began to explain in this volume's introduction, I myself will not be adopting a Saussurean or even a post-Saussurean approach. However, for both theoretical and propaedeutic reasons it is pertinent to explain why I will not be doing so. So I will begin this chapter with a very brief introduction to the Saussurean semiotics, or what is perhaps better described as *semiology*.

Saussure's contribution to semiotic theory, and indeed to modern linguistics more generally is articulated in his *Course in General Linguistics*, which was originally published in French in 1916. I say 'his *Course*', but actually, while the book is attributed to him, Saussure didn't actually write it. After Saussure died without publishing a work on his life-long investigations, three colleagues at the University of Geneva used notes that students produced of Saussure's lectures between 1906 and 1911 to compile a written version of his theory. This means that the *Course* is probably best regarded as a book written by at least four authors (one, apparently the most important, *in absentia*). The *Course* stands, in this sense, as a kind of silent tribute to one of the hallmarks of a Saussurean, and more generally a structuralist conception of semiotics: however important individuals are to

---

[1] Ferdinand de Saussure, *Course in General Linguistics*, Charles Bally and Albert Sechehaye (eds.), translated by R. Harris (London: Duckworth 1983 [1916]).

[2] See for example, Roland Barthes, *Mythologies* translated by Annette Lavers (London: Paladin, 1972[1957]).

[3] See for example, Jacques Derrida's 'Structure, Sign, and Play in the Discourse of the Human Sciences' in *Writing and Difference*, translated by Alan Bass (London: Routledge, 1981[1966]), 278-294.

[4] See for example, Louis Hjelmslev, *Prolegomena to a Theory of Language*, translated by Francis J. Whitfield, revised English edition (Madison: University of Wisconsin Press, 1961 [1943]).

meaning-making, the semiotic process is ultimately a matter of signs which go beyond the individual act of signification, the individual *use* of signs.

The *Course*'s account of the science that Saussure set out to develop gives some sense of the kind of enquiry that Saussure envisioned: 'It is ... possible to conceive of a science *which studies the role of signs as part of social life*. It would form part of social psychology, and hence of general psychology. We shall call it *semiology* (from the Greek *sēmeion*, 'sign'). It would investigate the nature of signs and the laws governing them.'[5]

According to Saussure, signs are dyadic entities which conjoin concepts – or what Saussure terms *signifieds* – with the *forms* employed to express the concepts – what he calls *signifiers* or signals. An example of a sign is any word, which at once expresses a concept (or the signified) and gives that concept a particular form, e.g. w-o-r-d (the signifier). Meaning is produced by way of such a conjunction, and in everyday discourse, speakers draw on a system of such conjunctions, or such 'units' of conjunction. Saussure makes the case that it is this mutually imbricating and inherently relational *system* that enables its users to generate meaning, and so it is the system, as opposed to, say, the history of each of its units, that the semiotician must analyse. It follows that the semiotician should adopt not what Saussure describes as a *diachronic* perspective of the kind developed by philologists (studying the changing meanings of terms across time), but instead a *synchronic* perspective – a kind of snapshot of the dictionary-like system of meanings as it exists at one point in time.

Key to this approach is Saussure's insight that each sign entails a general articulation, a *structure* made up of signified/signifier that can and ought to be analysed in its own right. As I began to explain above, such a structure transcends any particular instance of use, and any individual intentionality. This being true, Saussure argues that semiology should concern itself with explaining the general structures of signs as opposed to their 'application' by particular users – hence, the structura*lism* of his approach. As noted earlier, Saussure saw semiology as a part of social psychology, but in the *Course* he rejects efforts to study 'the mechanism of the sign in the individual', or in its 'individual execution'.[6] The sign, Saussure suggests, 'always to some extent eludes control by the will, whether of the individual or of society: that is its essential nature'.[7]

Saussure further argues that the structure of signs is not only a matter of social convention, but is entirely *arbitrary*. As Saussure puts it, 'There is no internal connexion, for example, between the idea 'sister' and the French sequence of sounds s-ö-r which acts as its signal. The same idea might as well be represented by any other sequence of sounds. This is demonstrated by differences between lan-

---

[5] Ferdinand de Saussure, *Course in General Linguistics*, p. 15. Italics in the original.

[6] Ferdinand de Saussure, *Course in General Linguistics*, p. 16.

[7] Ferdinand de Saussure, *Course in General Linguistics*, p. 16.

guages, and even by the existence of different languages. The signification 'ox' has as its signal *b-ö-f* on one side of the frontier, but *o-k-s (Ochs)* on the other side'.[8]

Now the *Course* was published in 1916, but Saussure's theory was not widely understood or applied in the decades immediately after it appeared. In time, it nonetheless came to be admired for its powerful critique of psychologism and naturalism in the study of language. During the second half of the twentieth century, structuralist conceptions of language became a conceptual foundation not just for modern linguistics, but for research in fields such as social anthropology, as well as cultural, literary, film, media and communication studies. In their diverse ways, all of these fields sought to explain the socially and culturally contingent character of practice, and for decades Saussure's brilliant insights offered a model not just for the reconceptualisation of language, but for communication and cultural practices more generally. If linguistic utterances could be shown to be a matter of social convention and arbitrary construction, and if they could be analysed via a structuralist methodology, then a similar approach, or so many scholars believed, could and should be applied to beyond-linguistic systems.[9] As I began to note earlier, even after aspects of Saussure's theory began to be critiqued, many poststructuralist accounts continued to employ Saussure's dyadic conception of the sign; there might be a number of issues with Saussure's *structuralism*, but more often than not, the dyadic conception of the sign was itself left unquestioned.

In this study, I reject a Saussurean approach to semiotics (semiology) for a number of reasons. First, Saussure severs the analysis of signs from any objects beyond the semiological dyad. In Saussure's own words, 'The linguistic sign is … a two-sided psychological entity […] [whose] two elements are intimately linked and each triggers the other. Whether we are seeking the meaning of the Latin world *arbor* or the word by which Latin designates the concept 'tree', it is clear that only the connexions institutionalised *in the language* appear to us as relevant. *Any other connexions there may be we set on one side.*'[10] Philosophers of language have explained why this is a problematic methodological turn from the point of view of the referential function of language.[11] While signs do have an internal ar-

---

[8] Ferdinand de Saussure, *Course in General Linguistics*, pp. 67-68. Saussure, who was himself Swiss, was presumably referring to the boundary between a Francophone part of Switzerland and Germany.

[9] A good example may be found in the work of Claude Levi-Strauss. See Claude Levi-Strauss, *Structural Anthropology*, translated by Claire Jacobson (New York: Basic Books, 1974[1963]). See also Roland Barthes' famous analysis in 'Myth Today', *Mythologies*, 109-159.

[10] Ferdinand de Saussure, *Course in General Linguistics,* 66-67, italics added.

[11] See for example Paul Ricoeur, *Interpretation Theory: Discourse and the Surplus of Meaning* (Fort Worth: Texas Christian Press, 1976).

ticulation, people use words – or any other kinds signs – to *refer* to the world, its objects and processes. A semiotic analysis must thus recognise *both* the internal structures, or what have come to be known as the *syntactic* aspects, *and* language as it is actually employed to refer to the world in particular social contexts – what have come to be known as the *pragmatic* dimensions of any act of signification.[12] An apparently simple example drawn from my research on Barro Colorado may help to explain why this is important. Amongst guides and tourists observing wildlife along a trail, the production of signs was certainly a matter of the instantiation of general systems of meaning, but it was also about pointing to, and more generally responding to objects in the *hic et nunc* of wildlife observation: 'There it is!'; 'What's it doing now?'; or 'Where's it gone?!'. Any suggestion that the question of reference, and of actual use is somehow external to such communications is clearly mistaken.

A second, related criticism applies to the question of the relation between the production of meaning and social motivations: if guides, tourists, and indeed STRI's Visitors Programme sought to observe wildlife, it was arguably because each group was motivated to do so, however unselfconsciously, by a variety of more-than-semiotic forces, including those associated with institutions, cultural formations, and particular geographies. Far from being secondary to, or indeed external to the perception, conception and interpretation of wildlife, such forces are what motivated the different actors to engage in the quest for wildlife encounters in the first place. The forces in question also played a key role in determining each group's responses (linguistic and other) to the different species. Yet ironically, even as Saussure claims that his approach entails a form of *social* psychology, and even as his approach focusses on what he himself describes as institutionalised connexions, Saussure puts precisely the institutional and cultural connections to one side, and in so doing effectively severs the semiotic analysis from any kind of sociological or anthropological inter-relation. Of course, it is possible to add the consideration of the latter relations to a semiotic analysis, and this is what many scholars have attempted to do whilst leaving the fundamental semiotic

---

[12] The distinction between the syntactic, semantic, and pragmatic aspects of language was first proposed by Charles W. Morris. For Morris' definitive account, see his *Writings on the Theory of Signs* (The Hague: Mouton de Gruyter, 1971). Morris was another of the founding figures of semiotic theory, albeit one who embraced positivism and behaviourism.

conception intact. A critique of this kind of approach is the starting point of what has become known as Social Semiotics,[13] and Critical Discourse Analysis.[14]

Thus far my critique has focussed on the absence of anything like a 'social' or beyond-human referential aspect in Saussure's account. However, in a tropical forest, the objects of signification are not only *referred to* and *interpreted* by humans; the objects may themselves play a key role in *producing* signs. Returning to the opening example in this part of the volume, if the guide and then the visitors could detect the presence of a poison dart frog, it was because it called out, and the call was certainly a sign – a sign to a guide keen to find poison dart frogs for the benefit of the tourists, but of course also a sign to other members of its species. A similar point can be made with respect to one particularly salient aspect of poison dart frogs' form: scientists believe that the frog's remarkably bright colours are meant to act as an aposematic warning to would-be predators; as such, the very morphology of frogs and of other creatures may serve beyond-human semeiotic purposes. But here again, Saussure, like many semiotic, hermeneutic and linguistic scholars, makes no allowance for this phenomenon. The *arbor* in the earlier quotation matters not because it refers to something that is alive and able to partake actively in the semiotic process, but because it is a *concept* of tree. There is thus, at the very heart of the structuralist semiological mode of inquiry, at once a deep anthropocentrism, and a deeply paradoxical void or absence: the very *thisness*, which is to say the haecceity of what is represented, is either ignored or bracketed in favour of an analysis of the representational entities, i.e. the things in themselves are ignored in favour of their semiological identities. If this is problematic in general, it seems like a particularly questionable stance if one is interested in explaining wildlife observation in an actual forest.

The last problem with Saussure's approach that I will consider concerns what he describes as the 'signifiers', or the 'signals' of signs. While Saussure appears to recognise that semiology might have a much broader remit than just the study of oral or written language, he makes the linguistic sign the model for all other sign systems. This logocentrism (or as the semiotician Thomas A. Sebeok calls it, 'glottocentrism'[15]) leads Saussure to develop a theory that effectively ignores the differences between linguistic and non-linguistic signs, and indeed the differences across non-linguistic signs. The difficulty with this approach can be illustrated by returning to the poison dart frog: even if one sticks to a representationalist frame,

---

[13] See for example, Robert Hodge and Gunther Kress, *Social Semiotics* (Cambridge: Polity Press, 1988).

[14] See Gunther Kress, 'Against Arbitrariness: The Social Production of the Sign as a Foundational Issue in Critical Discourse Analysis', *Discourse and Society*, 4:2 (1993), 169-191. For a more general account of Critical Discourse Analysis, see for example, Michael Toolan (Ed.) in *Critical Discourse Analysis: Critical Concepts in Linguistics, Volumes I-IV* (London: Routledge, 2002).

[15] Thomas A. Sebeok, *Global Semiotics* (Bloomington: Indiana University Press, 2001).

the words '*Dendrobates auratus*' do not represent Dendrobatid frogs in the same way that a tourist's photograph of such creatures does, and the same can be said of the *call* of the amphibian when compared to a photograph. This is actually not just a matter of differences in the 'signifiers'; it also has to do with fundamentally different relations between *all* the elements of the sign; a call is produced by the frog, whereas a photo of the frog is produced by a camera and a photographer. This being true, there are also different objects/subjects entailed.

If a Saussurean approach to semiotic theory constitutes a dead end for someone keen to develop a theoretical approach to wildlife observation, what is the alternative? In this study I will adopt the semiotic theory developed by Charles Sanders Peirce (1839-1914),[16] one of North America's most important philosophers, and arguably the first to propose a truly modern form of semiotic inquiry.

Peirce developed and started to publish his own approach to semiotics decades before Saussure's *Course* appeared. His approach reflects not just his interest in logic, but the fact that his original training, and longest employment were both in the physical sciences. During his undergraduate studies at Harvard, Peirce obtained a degree in chemistry, and then he devoted much of his life to working as a geodesist for the U.S. Coast Survey (later Coast and Geodetic Survey). The work for the Coast Survey involved making painstaking measurements with pendulums in order to try to ascertain the force of gravity at different locations. The purpose of this research was to determine the precise shape of the planet, a question that had fascinated philosophers of nature from ancient times, and which by Charles Peirce's era was the subject of a transnational scientific effort. Peirce's ability to conduct research in this field reflected not just his considerable technical skills, but his extraordinary knowledge of advanced mathematics, physics, and astronomy, all fields in which he had become a leading scholar by the beginning of the twentieth century.

---

[16] See Charles Sanders Peirce, *Collected Papers of Charles Sanders Peirce*, 8 Vols., edited by Charles Hartshorne, Paul Weiss, and Arthur Burks (Cambridge: Harvard University Press, 1931-1958). Throughout this study, I adopt the bibliographical conventions used by Peircian scholars. Accordingly, '*CP*' will refer to Charles Sanders Peirce, *Collected Papers*, where the first number refers to the volume while the second refers to the paragraph number. In turn, '*EP1*' will refer to Nathan Houser & Christian Kloesel (eds.), *The Essential Peirce Volume 1* (Bloomington: Indiana University Press, 1992). '*EP2*' will refer to The Peirce Edition Project (eds.), *The Essential Peirce, Volume 2* (Bloomington: Indiana University Press, 1998), and '*SS*' will refer to Charles Sanders Peirce, *Semiotic and Significs: The Correspondence Between Charles S. Peirce and Victoria Lady Welby* (Bloomington: Indiana University Press, 1977). In the last three references, the second number refers to the page number, so, for example, Charles Sanders Peirce, *EP* 2.123 is a reference to *The Essential Peirce* Volume 2, p. 123. Finally, *MS* refers to Peirce's Manuscripts, for which the numbering refers to Richard S. Robin's *Annotated Catalogue of the Papers of Charles S. Peirce* (Amherst: University of Massachusetts Press, 1967).

Chapter 5 Peircian Pragmatism

In an appendix to this part of the volume (*Appendix 1: Biography of Charles Sanders Peirce*), I offer a brief account of Peirce's trajectory as a scientist and philosopher. If I mention aspects of that biography here, it is to counter any suggestion that Peirce's semiotic theory can be characterised as a typically 'postmodern' endeavour, or as an example of the kind of relativism that many perceive, rightly or wrongly, to inform research across the critical social sciences, and semiotics in particular. As we shall see, Peirce's semiotic theory is, if anything, very much a realist one, and Peirce's varied interests come together in what might today be described as a philosophy of science. As the Peirce scholar Nathan Houser puts it, Peirce was concerned to 'build an adequate theory of science and an objective theory of rationality'.[17] For these and other reasons which I will present in due course, I distinguish Peircian semiotics from Saussurean/post-Saussurean semiotics by adding an extra 'e' to semiotics (semeiotics) – a spelling which Peirce himself used, and which to be sure is closer, by Saussure's own account, to the Greek *sēmeion*.

In so far as Peirce's approach to semeiotic theory is closely intertwined with his philosophy, I would like to begin with a brief account of key aspects of that philosophy. This poses some challenges, for Peirce's philosophy spanned across an astonishing range of philosophical issues. I will nevertheless begin with one set of issues that Peirce himself might well have regarded as being central to his work: that of the contest between realist and nominalist perspectives.

In Chapter 1, I offered a critique via the historian of ideas Leszek Kolakowski of the positivist rule of nominalism – the notion, in the terms used by Kolakowski, 'that we may not assume that any insight formulated in general terms can have any real referents other than individual concrete objects.'[18] However, in that chapter, I also began to explain that there is more than one variety of nominalism, and here it is pertinent to preface Peirce's views with a more detailed account of the difference between a nominalism that rejects abstract objects and a nominalism that rejects universals (the kind that I will explain that Peirce opposes). According to the philosopher Gonzalo Rodríguez-Pereyra, the former kind of nominalism 'asserts that there are concrete objects and that everything is concrete', while the latter asserts that 'there are particular objects and that everything is particular'.[19] The two forms of nominalism are independent of each other; hence, someone can assert that everything is particular without denying the reality of

---

[17] Nathan Houser, 'Introduction', in Nathan Houser and Christian Kloesel (eds.)*The Essential Peirce: Selected Philosophical Writings, Volume 1 (1867-1893)* (Bloomington: Indiana University Press, 1992), xix-xli, p. xxx.

[18] Leszek Kolakowski, *Positivist Philosophy: From Hume to the Vienna Circle*, translated by Norman Guterman (Harmondsworth: Penguin Books, 1969 [1966]), p. 13.

[19] Gonzalo Rodríguez-Pereyra, 'Nominalism in Metaphysics', *The Stanford Encyclopedia of Philosophy* (Winter 2016 Edition), Edward N. Zalta (Ed.) <https://plato.stanford.edu/archives/win2016/entries/nominalism-metaphysics/>[Accessed 21 July 2018].

abstract objects, and someone can assert that everything is concrete without denying the reality of the particular.[20] According to Rodríguez-Pereyra, both forms of nominalism entail an anti-realism. Where one kind of nominalism denies the existence, and therefore the reality, of universals (also known as generals), the other denies the existence and so the reality of abstract objects. Nominalism either denies the existence of the entities involved, or accepts their existence but argues that they are particular or concrete.[21] As Rodríguez-Pereyra understands it, nominalism has nothing against properties, numbers, propositions, possible worlds, etc., as such. Instead, it rejects that such entities should be accepted as universals or abstract objects: '... the mere rejection of properties, numbers, possible worlds, propositions, etc., does not make one a nominalist – to be a nominalist one needs to reject them because they are supposed to be universals or abstract objects.'[22]

This kind of stance raises slippery questions about how one defines abstract objects, and universals. A detailed discussion of such questions is beyond the scope of this chapter. Here I will simply echo Rodríguez-Pereyra when he suggests that abstract objects are those that lack an existence in space and time, and which are causally inert. For their part, universals can be conceived by way of relations of realisation or instantiation: 'something is a universal if and only if it can be instantiated (whether it can be instantiated by particulars or universals) — otherwise it is a particular. Thus while both particulars and universals can instantiate entities, only universals can be instantiated. If whiteness is a universal then every white thing is an instance of it.'[23]

Back, then, to Peirce: one way of interpreting Peirce's philosophy is in terms of a kind of anti-nominalism – of both kinds of nominalism, but especially the kind that rejects universals. Peirce would presumably reject the kind of nominalism that denies abstract objects if only because he is convinced of the reality of signs. And while Peirce actually started off as a nominalist of the second kind, by the time his philosophical writing reached maturity, he was quite vehemently opposed to that kind of nominalism (the kind that rejects universals). Peirce defined (this kind of) nominalism as 'the doctrine that nothing is general but names; more specifically, the doctrine that common nouns, *man*, *horse*, represent in their generality nothing in the real things.'[24]

---

20 Gonzalo Rodríguez-Pereyra, 'Nominalism in Metaphysics'.

21 Gonzalo Rodríguez-Pereyra, 'Nominalism in Metaphysics'.

22 Gonzalo Rodríguez-Pereyra, 'Nominalism in Metaphysics'.

23 Gonzalo Rodríguez-Pereyra, 'Nominalism in Metaphysics'.

24 Charles Sanders Peirce, in William D. Whitney (ed.), *The Century Dictionary and Cyclopedia* (New York: The Century Company, 1895), Volume 5, p. 4009. This kind of nominalism is associated with the scholastic philosophy of the medieval philosopher William of Ockham. See Paul Forster, *Peirce and the Threat of Nominalism* (Cambridge: Cambridge University Press, 2011).

As I noted above, Rodríguez-Pereyra suggests that both kinds of nominalism involve an anti-realism. While strictly speaking this must be true, if one stands back and examines who adopts what kind of nominalism in what context, a case can be made that the nominalism that rejects abstract objects is associated with the realism of positivist philosophy and empiricist science. By contrast, the nominalism that rejects universals tends to go hand in hand with the kind of constructivism (itself an anti-realism) that has long dominated the critical social sciences and humanities. The realist nominalism is one that may lead scientists – and politicians – to treat abstractions as little more than a semiotic convenience in a world that is really about concrete things, i.e. things that exist in space and time, and which entail causal relations. The constructivist nominalism is one that may lead philosophers, and critical social scientists to acknowledge the reality of abstractions and of representations more generally, but to draw a very strong line between these and the world that they represent. Saussure's doctrine of the arbitrariness of the sign makes him a nominalist of the second kind.

As I began to suggest earlier, Peirce started out as a nominalist of the second kind but ended up as a committed realist,[25] albeit a *scholastic* realist. I will say more about this kind of realism in a moment. First it may be noted that Peirce apparently grew to believe that the voyage from nominalism to realism was a good one. The Peirce scholar Max Fisch quotes Peirce as suggesting that '"Everybody ought to be a nominalist at first, and to continue in that opinion until he [sic] is driven out of it by the *force majeure* of irreconcilable facts"'.[26] Indeed, '"What distinguishes the nominalist is that he does *not* admit certain elements. The realist, if he is a sound thinker, must once have occupied the same position"'.[27]

So how does Peirce understand realism? According to Peirce, from a general philosophical perspective, realism is the belief 'in the real existence of the external world as independent of all thought about it'.[28] By contrast, *scholastic* realism is 'the conclusion that general principles are really operative in nature'.[29] The scholastic specification is important not least because it allows Peirce to reconcile his conviction that logic is ultimately a matter of semeiotic relations, and his belief that scientific practice constitutes an apt vehicle with which to arrive at empirically verifiable truths. All ideas may be regarded as signs, but there is continuity be-

---

[25] Max H. Fisch, 'Peirce's Progress from Nominalism toward Realism' in Kenneth Laine Ketner and Christian J. W. Kloesel (eds.), *Peirce, Semeiotic and Pragmatism: Essays by Max H. Fisch*, 184-200.

[26] Peirce in Max H. Fisch, 'Peirce's Progress from Nominalism toward Realism', p. 184. Italics in the original.

[27] Peirce in Max H. Fisch, 'Peirce's Progress from Nominalism toward Realism', p. 184. Italics in the original.

[28] Charles Sanders Peirce, *The Century Dictionary and Cyclopedia*, Volume 6, p. 4986

[29] Charles Sanders Peirce, *CP* 5.101.

tween ideas, thus conceived, and the cosmos, which Peirce believes to be 'perfused' with signs, 'if it is not composed exclusively of signs'[30] as part of what may be described not only as a 'pansemeiotic',[31] but as a *panpsychist* cosmology.

We can say in this sense that Peirce's realism is tempered by a certain *idealism*, which he defines as 'the metaphysical doctrine that the real is of the nature of thought; the doctrine that all reality is in its nature psychical'.[32] Numerous critiques of psychologism in the twentieth century would suggest the folly of this perspective;[33] however, when Peirce refers to the psychical, he does not only mean the individual, let alone the purely personal, mind; he refers to semeiotic relations. Moreover, especially his later conception of semeiosis (semeiotic process) is itself one that foregrounds not so much consciousness, as the importance of triadic, or three-way relations of mediation.

This initial characterisation raises the prospect of a contradiction: are idealism and realism not antithetical? By the end of his long scientific and philosophical career, Peirce himself sees no contradiction in his articulation of these tendencies. On the contrary, he describes his philosophical approach as a form of 'ideal-realism'[34] – one that, in effect, transcends the traditional opposition between the two perspectives.[35] To understand how this is the case, I will now introduce three key aspects of Peirce's writing: his *pragmatist* philosophy; his *evolutionary cosmology*; and what he describes as the doctrine of *synechism*. Doing so is pertinent to this

---

[30] Charles Sanders Peirce, *CP* 5.448, fn.

[31] Winfried Nöth, *Handbook of Semiotics* (Bloomington: Indiana University Press, 1990), p. 41.

[32] Charles Sanders Peirce, *The Century Dictionary and Cyclopedia,* Volume 4, p. 2974

[33] In philosophy, psychologism is the tendency to explain events in ways that emphasise the importance of the psychological to the detriment of the social, or 'beyond-psychological' aspects. This is what Peirce himself had to say about the matter when he realised that his early version of the pragmatic maxim (discussion to follow) showed a tendency to psychologism: 'But how do we know that belief is nothing but the deliberate preparedness to act according to the formula believed? [...] My original article carried this back to a psychological principle. The conception of truth, according to me, was developed out of an original impulse to act consistently, to have a definite intention. But in the first place, this was not very clearly made out, and in the second place, I do not think it satisfactory to reduce such fundamental things to facts of psychology. For man could alter his nature, or his environment would alter it if he did not voluntarily do so, if the impulse were not what was advantageous or fitting. Why has evolution made man's mind to be so constructed? That is the question we must nowadays ask, and all attempts to ground the fundamentals of logic on psychology are seen to be essentially shallow.' Charles Sanders Peirce, *CP* 5.28.

[34] Nathan Houser, 'Introduction', p. xxxv.

[35] For a provocative discussion of this interpretation, see Carl R. Hausman, *Charles Sanders Peirce's Evolutionary Philosophy* (Cambridge: Cambridge University Press, 1993).

study in so far as my own geosemeiotics of wildlife observation owes much to these aspects.

Let's start with Peirce's pragmatist philosophy. Pragmatism (the more general term for the philosophy) was a philosophical movement that emerged in New England during the second half of the nineteenth century as a response to the challenge posed to traditional forms of philosophy by the growing cultural authority of modern science – in particular, the challenge presented by evolutionary theory. Josiah Royce, one of Peirce's students at Johns Hopkins, and himself a leading exponent of pragmatism, gives a sense of the times when he speaks of the 'storm-and-stress' period of Darwinism.[36]

Peirce is widely regarded as one of founding figures of the movement. But even today, pragmatism is often more closely associated with the writing of the philosopher and founder of modern psychology, William James (1842-1910). James provides an inkling of the way in which pragmatist philosophers tried to address the aforementioned 'storm-and-stress' when he characterises pragmatism as a way of reconciling the claims of a 'tough-minded' empiricism (i.e. 'science') and a 'tender-minded' rationalism (i.e. 'philosophy').[37] James suggests that such a reconciliation can be achieved by engaging in a methodical form of reasoning which reveals the spurious nature of ostensibly deep disagreements between the advocates of empiricism and rationalism. This reasoning should examine not the forms of the arguments, as per traditional approaches to logic, but the practical consequences of each position. Subjected to the test of actual practice, argumentative differences can often be shown to be no more than formalities.

James credits his good friend Charles Peirce with penning the founding statement for this stance – what came to be known as the 'pragmatic maxim'. In an article titled 'How To Make Our Ideas Clear' that was published in 1878 in the *Popular Science Monthly*, Peirce suggests 'Consider what effects, that might conceivably have practical bearings, we conceive the object of our conception to have. Then, our conception of these effects is the whole of our conception of the object.'[38] A lecture which James presented in Berkeley in 1898 provides a useful, even if ultimately misleading way of introducing what Peirce was getting at in his curiously worded maxim. A part of that lecture is worth quoting at length:

> The soul and meaning of thought, he [Peirce] says, can never be made to direct itself towards anything but the production of belief, belief being the demicadence which closes a musical phrase in the symphony of our intellectual life. Thought in movement has thus for its only possible motive the

---

[36] Josiah Royce, 'William James and the Philosophy of Life', in *William James and Other Essays on the Philosophy of Life* (New York: Macmillan, 1911), 3-48, p. 12.

[37] William James, *Pragmatism: A New Name for Some Old Ways of Thinking* (London: Dover Publications, 1995 [1907]).

[38] Charles Sanders Peirce, *CP* 5.402.

attainment of thought at rest. But when our thought about an object has found its rest in belief, then our action on the subject can firmly and safely begin. Beliefs, in short, are really rules for action; and the whole function of thinking is but one step in the production of habits of action. If there were any part of a thought that made no difference in the thought's practical consequences, then that part would be no proper element of the thought's significance. Thus the same thought may be clad in different words; but if the different words suggest no different conduct, they are mere outer accretions, and have no part in the thought's meaning. If, however, they determine conduct differently, they are essential elements of the significance. "Please open the door," and, "*Veuillez ouvrir la porte,*" in French, mean just the same thing; but "D–n you, open the door," although in English, *means* something very different. Thus to develop a thought's meaning we need only determine what conduct it is fitted to produce; that conduct is for us its sole significance. And the tangible fact at the root of all our thought-distinctions, however subtle, is that there is no one of them so fine as to consist in anything but a possible difference of practice. To attain perfect clearness in our thoughts of an object, then, we need only consider what effects of a conceivably practical kind the object may involve – what sensations we are to expect from it, and what reactions we must prepare. Our conception of these effects, then, is for us the whole of our conception of the object, so far as that conception has positive significance at all.[39]

By way of an aside, readers may note a remarkable contrast between this passage, and one of the ones I quoted earlier from Saussure's *Course in General Linguistics*. Where Saussure points to differences in the signifiers across languages as one of the cornerstones of his theory and methodology, Peirce, at least as translated by James, appears to minimise the importance of such differences.

In fact, the qualification 'at least as translated by James' is an important one; while James identifies his own pragmatist philosophy with Peirce's, by the late 1890s Peirce has revised his own approach to pragmatism in important ways. According to John Dewey (1859-1952), another of the founding pragmatists, where James adopts a nominalist, and quite radically empiricist *approach*, Peirce develops a realist *theory* of pragmatism (what Peirce himself describes as *pragmaticism* to differentiate it from Jamesian pragmatism). This theory offers, amongst other things, a way of testing the value of hypotheses.[40] A brief account of the difference

---

[39] William James, 'Philosophical Conceptions and Practical Results', *University Chronicle* [University of California at Berkeley], 1(1898), pp. 290-291.

[40] John Dewey, 'The Pragmatism of Peirce', in Charles Sanders Peirce, *Chance, Love and Logic: Philosophical Essays*, Morris R. Cohen (ed.) (London: Kegan Paul, Trench, Trubner and Co., 1923) 301-308 pp. 303.

between the two stances may help to clarify what is entailed by Peirce's ideal-realist philosophy.

James argues that '...the effective meaning of any philosophic proposition can always be brought down to some particular consequence, in our future practical experience, whether active or passive; the point lying rather in the fact that the experience must be particular, than in the fact that it must be active.'[41] Peirce agrees, in his terms, that '[t]he rational meaning of every proposition lies in the future'.[42] But he disagrees strongly with James in so far as Peirce believes that such a future also involves further propositions, and indeed, *signs*; and that the consequences are not just *particular* – a tacit invocation of nominalism – but *general* ones. As Peirce puts it in an entry to the Baldwin *Dictionary of Philosophy and Psychology*,

> If it be admitted ... that action wants an end, and that that end must be something of a general description, then the spirit of the [pragmatic] maxim itself, which is that we must look to the upshot of our concepts in order rightly to apprehend them, would direct us towards something different from practical facts, namely, to general ideas, as the true interpreters of our thought.[43]

Dewey offers a helpful gloss on this difference vis-a-vis James' writing when he notes that Peirce's pragmatism

> identifies meaning with formation of a habit, or way of acting having the greatest generality possible, or the widest range of application to particulars. Since habits or ways of acting are just as real as particulars, it is committed to a belief in the reality of "universals" [or what Dewey also described as 'generals'] ... Moreover, not only are generals real, but they are physically efficient. The meanings "the air is stuffy" and "stuffy air is unwholesome" may determine, for example, the opening of the window.[44]

Now it might be inferred that Peirce conceives of pragmaticism along foundationalist lines, i.e. that even if philosophers must look forward to the ends of actions, they should do so on the basis of a firm foundation of beliefs already con-

---

41 William James, 'Philosophical Conceptions and Practical Results', p. 291.

42 Charles Sanders Peirce, *CP* 5.427.

43 Charles Sanders Peirce, 'Pragmatic and Pragmatism' in James Mark Baldwin (Ed.), *Dictionary of Philosophy and Psychology*, Volume 2 (New York: Macmillan, 1901-1905) p. 321.

44 John Dewey, 'The Pragmatism of Peirce', pp. 303-304.

firmed by science.[45] In fact, the opposite is true, and is more consistent with Peirce's philosophy: Peirce adopts what was an explicitly *fallibilist* stance, which is to say one that emphasises the provisionality of all scientific belief. As he puts it, 'fallibilism is the doctrine that that our knowledge is never absolute but always swims, as it were, in a continuum of uncertainty and of indeterminacy'.[46] Moreover, Peirce is keenly aware of the importance of inter-subjectivity in the production of knowledge, and so his philosophy is the better part of a century ahead of that of figures such as Karl Popper and Thomas Kuhn. In Peirce, a fallibilist perspective is at once expressed in, and based on his pragmaticist philosophy that Truth (with a capital 'T') is a matter of the ongoing production, but also the endlessly deferred validation, of meaning.

This last aspect might seem like a recipe for relativism, but as I will explain across the rest of Part 2, Peirce is first and foremost a realist, not least in so far as he recognises that certain objects may be connected indexically to signs, and are thereby capable of *causing* certain meanings. Equally if not more importantly, Peirce embraces an evolutionary cosmology, an aspect of his writing that I would now like to turn to.

As Peirce himself describes it, the cosmology in question involves a beginning from utter nothingness, a

> chaos of unpersonalized feeling, which being without connection or regularity would properly be without existence. This feeling, sporting [i.e. mutating spontaneously] here and there in pure arbitrariness, would have started the germ of a generalizing tendency. Its other sportings would be evanescent, but this would have a growing virtue. Thus, the tendency to habit would be started; and from this with the other principles of evolution all the regularities of the universe would be evolved. At any time, however, an element of pure chance survives and will remain until the world becomes an absolutely perfect, rational, and symmetrical system, in which mind is at last crystallized in the infinitely distant future.[47]

This cosmology is evolutionary because it suggests that *everything* in the universe – 'all the regularities', and so presumably not just those of biotic phenomena – evolves, and continues to evolve by chance (or at any rate with an element of chance) beginning from that 'chaotic unpersonalised feeling'. Here the notion of evolution is not used casually or metaphorically; Peirce was an early, but remark-

---

45 Perhaps the founding foundationalist was René Descartes, who argued that he could not doubt that he existed if he was thinking, or capable of doubting (hence the '*cogito ergo sum*', I think, therefore I am) (cf. Chapter 10). Peirce believed that Descartes was wrong on this point; see Charles Sanders Peirce *CP* 5.388-410.

46 Charles Sanders Peirce, *CP* 1.171.

47 Charles Sanders Peirce, *CP* 6.33.

ably critical adopter of the Darwinian theory of evolution.[48] While he acknowledges Darwin's insight that chance is constrained by environmental factors, Peirce not only rejects the social Darwinism that prevailed by the late nineteenth century, but turns to competing theories of evolution to correct what he rightly sees as the limitations of an overly deterministic conception of natural selection. So it is, for example, that via Jean Lamarck, he conceives of the possibility that some 'sportings' might be a consequence of effort and exercise by individuals within a lifetime[49] – a point that is especially important when it comes to critiquing biological determinism in the context of human practices (cf. Chapter 2). Moreover, by way of the geologist Clarence King, the founder of the American Geology Society, Peirce also conceives the now widely accepted possibility that, while species might remain scarcely modified under ordinary circumstances, they might also be 'rapidly altered after cataclysms or rapid geological changes'.[50]

In keeping with his radically evolutionary perspective, Peirce's cosmology rejects the Cartesian opposition of mind and matter (an opposition that I will consider in some detail in Part 3). But his stance is the antithesis of the kind of reductionism that would be adopted in the twentieth century by sociobiologists and neuroscientists. Most of those scientists tend to assume that there is either no mind, or that mind is a figment of a neurological imagination. In Chapter 3, I referred to the now famous quote by Francis Crick, to the effect that we humans are 'nothing but a pack of neurons'.[51] By contrast, Peirce adopts what is in some respects exactly the opposite approach: he assumes that everything in the cosmos is a matter of mind.

This may seem outlandish, especially if mind is conceived as consciousness, or as a capacity to engage in more or less elaborate thoughts of the kinds entertained by humans. In fact, Peirce conceives of mind rather differently: he suggests that mind is, in effect, *semeiosis* – a process where objects, representamens (or as Peirce also calls them, *grounds*) and interpretants engage in an ongoing, and indeed potentially infinite process of mediation. Any one relation of this kind is what Peirce describes as a sign. I will explain in some detail just what is entailed by semeiosis in the following chapters. Here it suffices to begin to explain that for Peirce, the universe is mind in so far as it is perfused, or perhaps even composed exclusively of signs[52] – but signs understood not as 'representations', but as *three-way interrelations*, with one thing affecting another in relation to a third, *ad infinitum*. This explains why, in a letter to an English interlocutor, the philosopher of language

---

[48] Charles Darwin, *The Origin of the Species* (London: Senate 1995 [1872]).

[49] Charles Sanders Peirce, *CP* 6.16.

[50] Charles Sanders Peirce, *CP* 6.17.

[51] Francis Crick, *The Astonishing Hypothesis: The Search for the Soul* (New York: Scribner, 1995).

[52] Charles Sanders Peirce, *CP* 5.448, fn.

Lady Victoria Welby-Gregory, Peirce famously suggests that it has never been in his power to study anything – and he offers a list that includes such diverse subjects as mathematics, ethics, metaphysics, chemistry, comparative anatomy, astronomy, wine, and 'men and women' – 'except as a study of semeiotic'.[53]

Given everyday understandings of signs, it might seem that this cosmology constitutes an example of psychologism. But as I began to explain earlier, Peirce is very clear that the mindfulness of the universe is not merely a matter of thought or consciousness. On the contrary, much of it entails what Peirce describes as 'effete mind' – by which he means mind that has lost so much of its spontaneity through habit that it seems entirely set in a law-like regularity. This regularity is what scientists attempt to explain, though of course most would not coach their research in semeiotic terms. That Peirce can conceive of the cosmos in this way explains how he can reconcile his philosophical and semeiotic inclinations with a life that was also devoted to advanced scientific research.

Now it might be objected that Peirce's semeiotic reduction is no better than the genetic reduction of sociobiology, or the neuronal reduction of neuroscience. However, unlike the biological reductionists, Peirce develops a cosmology which makes a place for everything from chance, to the facts of action and reaction, to symbols. Peirce rejects, in this sense, any attempt to reduce the world to the kind of mechanical determinism associated with Newtonian physics, and so to dyadically conceived relations of cause and effect. That said, Peirce is by no means averse to the notion of causation, or indeed to the idea that two objects might relate in 'brute' manner to each other as part of what he describes as relations of secondness (cf. Chapter 6). Rather, he suggests that the three-way interrelations that I referred to above, or what he describes as thirdness, must not *themselves* be reduced to 'brute' action and reaction, even if there is, of necessity, always an element of action and reaction 'within' thirdness. Moreover, Peirce argues, as per his evolutionary cosmology, that the element of chance, of indetermination, of positive possibility – what he calls firstness – itself remains 'in', or 'across', secondness and thirdness.

I will return to Peirce's categories of firstness, secondness and thirdness at a later stage. For now I would like to turn to Peirce's doctrine of *synechism*. In a remarkable essay which he published in 1893 ('Immortality in the Light of Synechism'), Peirce explains that this neologism (one of many in Peirce's work) is derived from the English form of the Greek *synechismos*, and from *syneches*, meaning continuous. If '*materialism* is the doctrine that matter is everything, *idealism* the doctrine that ideas are everything, *dualism* the philosophy which splits everything in two ... *synechism* mean[s] the tendency to regard everything as continuous'.[54] In the same essay, he suggests that

---

[53] Charles Sanders Peirce, *SS* 85-86.

[54] Charles Sanders Peirce, *CP* 7.565.

Synechism, even in its less stalwart forms, can never abide dualism, properly so called. It does not wish to exterminate the conception of twoness ... But dualism in its broadest legitimate meaning as the philosophy which performs its analyses with an axe, leaving, as the ultimate elements, unrelated chunks of being, this is most hostile to synechism. In particular, the synechist will not admit that physical and psychical phenomena are entirely distinct, – whether as belonging to different categories of substance, or as entirely separate sides of one shield, – but will insist that all phenomena are of one character, *though some are more mental and spontaneous, others more material and regular*.[55]

Earlier I referred to Peirce's fallibilism, and according to Peirce, '[t]he principle of continuity is the idea of fallibilism objectified', 'the doctrine of continuity is that *all things* ... swim in continua'.[56] As the statement on physical and psychical phenomena begins to explain, Peirce's synechism is based on what is itself a synechistic theory of mind. The philosopher Joseph Esposito makes this point succinctly when he notes that for Peirce, 'Philosophy cannot start with a *cogito* or with sense impressions. It starts with a recognition that sensation is judgment; judgment is generalization, and generalization requires generality. The next step is to link generality with significance',[57] at which point he quotes Peirce as saying that

Just as a continuous line is one which affords room for any multitude of points, no matter how great, so all regularity affords scope for any multitude of variant particulars; so that the idea [of] continuity is an extension of the idea of regularity. Regularity implies generality; and generality is an intellectual relation essentially the same as significance, as is shown by the contention of the nominalists that all generals are names. Even if generals have a being independent of actual thought, their being consists in their being possible objects of thought whereby particulars can be thought. Now that which brings another thing before the mind is a representation; so that generality and regularity are essentially the same as

---

[55] Charles Sanders Peirce, *CP* 7.570. Italics added to the original.

[56] Charles Sanders Peirce, *CP* 1.171.

[57] Joseph Esposito, 'Synechism: the Keystone of Peirce's Metaphysics', in M. Bergman & J. Queiroz (Eds.), *The Commens Encyclopedia: The Digital Encyclopedia of Peirce Studies. New Edition*, Pub. 130510-1417a, 2005, <http://www.commens.org/encyclopedia/article/esposito-joseph-synechism-keystone-peirce%E2%80%99s-metaphysics> [Last accessed 13 March 2017].

significance. Thus, continuity, regularity, and significance are essentially the same idea with merely subsidiary differences.[58]

There are several implications of this stance for this study which will become clear as I flesh out the details of Peirce's semeiotic theory, and his phenomenology. Here I would simply like to mention three.

First, as I began to suggest earlier, Peirce's pansemeiotic and panpsychist cosmology allows us to avoid the mind-body dualism by suggesting that everything is or has 'mind', everything entails semeiosis, or at least, potential semeiosis (albeit with the provisos presented earlier). In Part 3 of this volume, I will note that over the last two decades or so, Baruch Spinoza's metaphysics, as interpreted via the philosophy of difference of Gilles Deleuze, has come to the forefront of the critical social sciences. Peirce's panpsychism and synechism – themselves influenced by Spinozan philosophy – may be read as forerunners of this shift. Consider for example the continuities between Peirce's previously quoted statement that '… all regularity affords scope for any multitude of variant particulars' and Deleuze's interest in Spinozan modes (cf. Chapter 10). Or the strongly Spinozan 'the synechist will not admit that physical and psychical phenomena are entirely distinct, – whether as belonging to different categories of substance, or as entirely separate sides of one shield, – but will insist that all phenomena are of one character, though some are more mental and spontaneous, others more material and regular'.

In keeping with this perspective, and second, Peirce's philosophy certainly allows for the possibility, indeed the *necessity* of a certain semeiotic capacity amongst beyond-human animals. According to Peirce biographer Joseph Brent, it seems that Peirce first began to think that other animals possessed reasoning powers when he realised that the family parrot had learned to imitate his brother Jem in order to tease Jem's dog, a spitz. When he returned home, the brother would call out to his dog, 'Spitz! Spitz! Spitz!' In response, the parrot invented, in effect, a game which involved imitating the calls – 'Spitz! Spitz! Spitz!' – so as to make the dog run downstairs, its tail wagging in anticipation, only to be met by the jeers of his winged tormentor.[59]

In later life, Peirce probably regarded this remarkable example not just as evidence of powers of reasoning – powers that have been more than confirmed by the small but growing group of ornithologists devoted to research with

---

[58] Charles Sanders Peirce, *CP* 7.535 in Joseph Esposito, 'Synechism'. I have added the first sentence for greater clarity. Peirce goes on to say that 'That this element is found in experience is shown by the fact that all experience involves time. Now the flow of time is conceived as continuous' – an issue I will return to in Part 4, Chapter 14.

[59] Joseph Brent, *Peirce: A Life*, 2nd edition (Bloomington: Indiana University Press, 1998), p. 47.

psittacines[60] – but of course of a remarkable *semeiotic* power. The biosemiotician Jesper Hoffmeyer notes that in one passage written in 1905, Peirce suggests that 'All thinking is by signs; and the brutes [beyond-human animals, sic] use signs. But they perhaps rarely think of them as signs. To do so is manifestly a second step in the use of language. Brutes use language, and seem to exercise some little control over it. But they certainly do not carry this control to anything like the same grade that we do. They do not criticize their thought logically.'[61] Whatever the validity of this perspective in light of the last decades' findings regarding the communications of, say, cetaceans and dolphins, there can be no doubt that semiotics as a discipline should not be exclusively concerned with the signs made by humans; on the contrary, there is clearly a need for what is now known as a 'biosemiotics'.[62]

A third implication of Peirce's semeiotic theory, his phenomenology, and his later work's realism is that all three allow for phenomena whose primary expression is less-than-semeiotic in the sense that they are about 'brute' action and reaction. In so far as Peirce acknowledges the importance of such relations, and goes so far as to suggest that some signs signify primarily by virtue an indexical relation between object and sign, then he cannot be simply accused of being a relativist.

---

[60] See for example Catherine Ann Toft & Timothy F. Wright, *Parrots of the Wild: A Natural History of the World's Most Captivating Birds* (Oakland, California: University of California Press, 2015).

[61] Charles Sanders Peirce, *CP* 5.534, in Jesper Hoffmeyer, 'Animals Use Signs. They Just Don't Know It', in Torkild Thellefsen and Beren Sorensen (Eds.) *Charles Sanders Peirce In His Own Words: 100 Years of Semiotics, Communication and Cognition* (The Hague: De Gruyter Mouton 2014), 411-414, p. 411. Peirce's unfortunate use of the term 'brutes' clearly reflects the discourse on non-human animals that prevailed at the time.

[62] See for example, Claus Emmeche, 'Biosemiotics', in J. Wentzel Vrede van Huyssteen (Ed.) *Encyclopedia of Science and Religion* (New York: Macmillan Reference 2003) 63-64. For a more recent account of biosemiotics, see Klaus Emmeche and Kalevi Kull (Eds.) *Towards a Semiotic Biology: Life is the Action of Signs* (London: Imperial College Press, 2011). For an account of the related *zoo*semiotics, see the special issue of *Semiotica* edited by Timo Maran, 'Dimensions of Zoosemiotics', *Semiotica*, 198:1/4 (2014).

6
Semeiotic Logic and Phenomenology

Having provided a broad introduction to Peirce's philosophy, I would now like to examine in greater detail Peirce's semeiotic theory, as well as his phenomenology. As we shall see, both of these aspects are intimately inter-related, but for the sake of clarity, I will introduce each of them in turn, beginning with the semeiotic theory.

Earlier, I noted that a number of scholars have defined semiotics as a science. Peirce offers what is perhaps a more useful definition when he suggests that semeiotics is 'the analytic study of the essential conditions to which all signs are subject'.[1] A key aim for semeiotics, thus conceived, is the characterisation of a universal logic of signs. Today many critical scholars would balk at any such characterisation, or indeed the very possibility of separating off the semiotic from other aspects of any phenomenon.[2] However, this is not a problem that is unique to semiotics; might we not query, for example, whether it is possible to separate off the instituted from the cultural (as per at least the more traditional forms of sociology), or indeed the cultural from the semiotic (as per the more traditional forms of anthropology)? The researcher must of course proceed as if it *were* possible, albeit hopefully in such a way as to make it clear that such distinctions are analytical ones, and that in fact what are described as different domains (the semiotic, the cultural, the social, etc.) are part of a continuum.

How, then, does Peirce characterise the fundamental semeiotic logic? Peirce offers accounts of such a logic in at least 76 texts or passages.[3] The following definition, published in the Baldwin Dictionary, shows just how broadly Peirce came to understand the concept: a sign, Peirce suggests, is '[a]nything which determines something else (its *interpretant*) to refer to an object to which itself refers (its *object*) in the same way, the interpretant becoming in turn a sign, and so on *ad infinitum*.'[4] A sign is a sign, and *only* a sign by dint of interrelating these three elements; this interrelation constitutes the semeiotic logic, and in so far as any phe-

---

[1] Charles Sanders Peirce, *MS* 774.6. For a more comprehensive account of this and other aspects of Peirce's semeiotic, see James Liszka, *A General Introduction to the Semeiotic of Charles Sanders Peirce* (Bloomington, IN: Indiana University Press, 1996).

[2] See for example, Gilles Deleuze and Félix Guattari, *A Thousand Plateaus: Capitalism and Schizophrenia*, translated by Brian Massumi (London: Continuum, 1988[1980]), pp. 111 and ff.

[3] See Robert Marty, '76 Definitions of The Sign by C.S. Peirce' at <http://www.cspeirce.com/rsources/76DEFS/76defs.htm>[Last accessed 3 September 2014].

[4] Definition republished in James Hoopes (Ed.) *Peirce on Signs: Writings on Semiotic By Charles Sanders Peirce* (Chapel Hill, NC: University of North Carolina Press, 1991), p. 239.

nomenon whatsoever can be said to involve this logic, then it may be regarded as – and from Peirce's perspective *will work* as – a sign.

It should be noted that there is a remarkable ontological ambiguity here: Peirce acknowledges that, strictly speaking, nothing is in itself a sign; however, in so far as the cosmos is all about three-way relationships of the kind just referred to, then *everything* is a sign, or perhaps we should say, a part of a semeiotic relation – potential, actual, past, or in the making. To explain how and why this is the case, let's examine in greater detail each of the aspects of the sign, beginning with one that is deceptively obvious: the *object*. Peirce himself provides a comprehensive account in the following passage:

> The Objects – for a Sign may have any number of them – may each be a single known existing thing or thing believed formerly to have existed or expected to exist, or a collection of such things, or a known quality or relation or fact, which single Object may be a collection, or whole of parts, or it may have some other mode of being, such as some act permitted whose being does not prevent its negation from being equally permitted, or something of a general nature desired, required, or invariably found under certain general circumstances.[5]

In another text, a letter to his good friend the philosopher William James, Peirce puts it as follows: 'A sign is a Cognizable that, on the one hand, is so determined (i.e. specialized, *bestimmt*) by something *other than itself*, called its Object'.[6] Peirce nonetheless notes that here the word 'determined' should not be understood in an overly narrow way: 'It may be asked, for example, how a lying or erroneous Sign is determined by its Object, or how if, as not infrequently happens, the Object is brought into existence by the Sign.'[7] Peirce provides the answer with the following example: 'A person who says Napoleon was a lethargic creature has evidently [had] his mind determined by Napoleon. For otherwise he could not attend to him at all. But here is a paradoxical circumstance. The person who interprets that sentence (or any other Sign whatsoever) must be determined by the Object of it through collateral observation quite independent of the action of the Sign.'[8] The person must know, by way of other signs, who Napoleon is; or, put more generally, the interpretation at once given to, and caused by an object will typically depend on other signs.

The last point can be further illustrated by going back to the call of the poison-dart frog: if the call is the ground, the object of the call can be said to be the frog

---

[5] Charles Sanders Peirce, *CP* 2.232.

[6] Charles Sanders Peirce, *CP* 8.177.

[7] Charles Sanders Peirce, *CP* 8.178.

[8] Charles Sanders Peirce, *CP* 8.178.

itself in so far as the frog produces the call. But of course, to someone who does not know the call or the frog in question, it may mean little or nothing; in that person's mind, the call will simply be an unidentified noise, if it is attended to at all. There is clearly a difference in kind between the first and the second object, a fundamental difference even, which I will consider in a moment. First it is necessary to consider the other two aspects of the sign: the representamen, or what Peirce also describes as the '*ground*'[9]; and the interpretant.

In at least one definition, Peirce describes the *ground* as an idea – in the everyday sense of 'one man catch[ing] another man's idea'.[10] But as noted earlier, in his later work Peirce defines the sign more generally as *anything* that determines something else as per the triadic relation, and indeed by the early 1900s he also describes the ground, and signs more generally as *mediums*: 'For the purposes of this inquiry a *Sign* may be defined as a Medium for the communication of a Form.'[11] Or as he explains in his letters to Lady Welby (herself what today we would describe as a semiotician), Peirce uses the word 'Sign' 'in the widest sense for any medium for the communication or *extension* of a Form (or feature). Being medium, it is determined by something else, called its Object, and determines something, called its interpretatnd [sic] or interpretand'.[12] This approach, as opposed to the ideational, is arguably the more powerful one in so far as it leaves behind the psychologism of earlier accounts, and increasingly foregrounds the importance of *form*; whatever else it is, the ground is certainly a form that serves to convey meaning.

But why the term 'ground'? Doubtless Peirce employs this somewhat peculiar name to highlight a key feature of any sign: that far from being a neutral copy of an aspect of the world, a sign always re/presents its object in some particular respect, and so its form establishes the ground, in the sense of basis, for re/presentation. The Peirce scholar James Liszka makes this point helpfully when he notes that 'The sign always presents the object *as* that object in some regard or respect, and so serves to present its object aspectively and partially'.[13] The call of the frog is not the frog itself; it is only a partial sign (of presence, of reproductive intent, etc.) in so far as it is temporary, in some sense disembodied, and merely acoustic in character.

---

[9] Until recently, I used the term 'representamen' for this aspect; I am, however, persuaded that the convenience of this term may cause confusion to those familiar with Peirce's work. Peirce at times used representamen to refer to signs as a whole, and so from this publication onwards I plan to stick to 'ground', which is somewhat awkward but should avoid any confusion.

[10] Charles Sanders Peirce, *CP* 2.228.

[11] Charles Sanders Peirce, *MS* 793.

[12] Charles Sanders Peirce, *SS* 196. Italics added to the original.

[13] James Liszka, *A General Introduction*, p. 20

To say that it is 'merely acoustic' is not to suggest that this partiality would not apply to more-than-acoustic signs; the same point can be made with respect to *any* sign produced by any creature, be it a specimen of *D. auratus*, or a producer who makes a 'full' audio-visual representation of the kind found on TV. The TV image is partial not least because it is two-dimensional, and an image-movement taken from a particular angle, at a particular point in time (cf. Volume 2, Part 2). But even the mental image – according to Peirce, itself a sign – formed by a tourist in the very presence of an actual frog, is also ineluctably partial in so far as someone can only perceive, conceive and interpret some aspects of the frog, and not others at any one point in space and time.

This brings us to the third aspect of the sign: the *interpretant,* which may be defined as a *translation* of one sign into additional signs. As this definition suggests, the interpretant is not to be confused with the *interpreter,* which is to say the interpreting subject or actor. In the example of the green and black poison dart frog calling, the 'whole' sign, if one can ever speak of a whole sign, is not just the call and its object(s), but these aspects plus whatever interpretation, indeed whatever *consequence* they give rise to *via additional signs*. The interpretant is, in this sense, the *interpretation-consequence*: the consequence regarded as an interpretation, or the interpretation regarded as a consequence of the semeiotic process (semeiosis). In the context of a tour of a tropical forest, an obvious instance of an interpretant is the production of linguistic signs by guides or tourists. For example, the frog's call might lead a guide to say, 'Oh, I can hear a poison dart frog calling'; alternatively, a visitor might also hear it and ask, 'What was that?'. In both cases, the words, regarded as a response, constitute the interpretant.

If, however, I also speak of a *consequence,* it is because for Peirce, *any* response, be it a linguistic utterance or a physical displacement, qualifies as interpretant. A jaw dropping in surprise, an exclamation of pain, a detailed analysis based on one or another herpetological theory, or indeed a tourist nearly running away in a panic at the roar of Howler Monkeys (as happened with one of the groups I accompanied) are all examples of interpretants. And lest we forget, the responses of beyond-human creatures to any ground also qualify as interpretants; so if, for example, a female *D. auratus* responds to the call of the male by hopping towards it, or by wrestling away any competing females, then that action too, may be regarded as an interpretant.

Each additional interpretant is a sign in its own right, viz., a three-way relation of object, ground and interpretant. Each sign, as Peirce explained via the example of Napoleon, can only be interpreted via additional signs: one sign must be interpreted by another, and that one by another again *ad infinitum*. This is why I raised a question earlier as to whether it is ever possible to speak of a 'whole' sign, or indeed a 'real' sign in any crudely empiricist manner; from a Peircian perspective, every sign is always at least partly virtual, or to put it differently, no sign is ever truly complete if only because it must form part of another sign to be meaningful – hence, no doubt, Peirce's fallibilism, but also, his doctrine of synechism: if every part is 'completed' by another, if the cosmos is made up of interconnected signs, then it is not possible to wield an epistemological axe saying, in effect, 'The sign stops here!'

The last point may seem to be a pretty abstraction until one thinks of any concrete situation involving objects, in Peirce's very broad sense of the term which includes actions or processes. When, for example, a female *D. auratus* hops towards a male that is calling out, that action may be, and doubtless *will* be treated as a *de facto* sign by any other specimen close enough to notice. Indeed, this action may generate an interpretant on the part of a third frog, which may also hop towards the first frog. Then again, a broad-billed motmot (*Electron platyrhynchum*) might hear the semeiotic commotion, and upon discovering signs of such culinary delight (Motmots are able to eat *D. auratus*, despite the poison), it might well 'interpret' one or two frogs by snapping them up – and so on down (or up) the food chain. From an biosemeiotic perspective, one way of interpreting an ecosystem is precisely in terms of a chain of such interpretants involving not just frogs, motmots or any other brightly coloured creatures, but all of the living organisms that make up the 'semeiosphere' – the semiotic equivalent of the biosphere. Here I refer to the concept coined by the semiotician Juri Lotman, who defines semeiosphere (he actually spelled it as semiosphere) as a 'specific semiotic continuum, which is filled with multi-variant semiotic models situated at a range of hierarchical levels'.[14] Lotman's reference to multi-variant semiotic models seems particularly appropriate for the remarkably complex biosemeiotics of any tropical forest, which is replete not just with a variety of beyond-human sign systems, but with signs operating with different levels of generality and consequence: compare, for example, the significance for a seasonal tropical forest of the first heavy rains, and a single dendrobatid frog calling out in the manner illustrated at the beginning of this part of the study.

As this account makes clear, for Peirce semeiosis is inevitably 'interpretive' if only because it always entails an interpretant, viz. a translation into additional signs. It is also ineluctably pragmatic in so far as meaning-making is not just a matter of synchronic relations between signs across a system (*à la* Saussure); it is also a matter of future *events* understood as interpretants. The last points notwithstanding, one of the many remarkable features of Peirce's semeiotic is that, even as it suggests that the world can only be *interpreted* via the concatenation of elements that makes up each sign, it also acknowledges a world with elements that are in some way not in themselves semeiotic, but are quite capable of producing, or at least partly determining, a semeiotic relation.

The last point brings me to a distinction that is key to Peircian semeiotics: that of dynamical and immediate objects. Umberto Eco provides a useful preamble for a discussion of this aspect of Peirce's theory when he notes that, while much has been written in semiotics about whether and how we use signs to *refer* to some-

---

[14] Juri Lotman, 'On the Semiosphere', in *Sign Systems Studies*, 33 (2005 [1984]), p. 206.

thing,[15] very little has been said about '[w]hat is that something that induces us to *produce* signs'. 'Beyond a doubt the only person who made this problem the very foundation of his theory – semiotic, cognitive, and metaphysical all at once – was Peirce'.[16] 'To use an expression that is efficacious albeit not very philosophical, the Dynamical Object is Something-that-sets-to-kicking-us and says "Talk!" to us – or "Talk about me!" or again, "Take me into consideration!".[17] The Immediate Object is, in turn, the object that is in some sense generated by the semeiotic process. To return to the example of the frog calling out, if the frog itself is the dynamical object, a ground in the form of a call, or indeed the frog's shape will produce an interpretant in an observer, and that interpretant will lead the observer to perceive and interpret that frog in a certain way. That perception/interpretation of the dynamical object is the immediate object.

Peirce himself characterises the difference as follows:

> ... we have to distinguish the *Immediate Object*, which is the Object as the Sign itself represents it, and whose Being is thus dependent upon the Representation of it in the Sign, from the *Dynamical Object*, which is the Reality which by some means contrives to determine the Sign to its Representation.[18]

In his correspondence with Lady Welby, he puts the matter in this way:

> The Form, (and the Form is the Object of the Sign), as it really determines the ... Subject, is quite independent of the Sign; yet we may and indeed must say that the object of a sign can be nothing but what that sign represents it to be. Therefore, in order to reconcile these apparently conflicting Truths, it is indispensible [sic] to distinguish the *immediate* object from the *dynamical* object.[19]

This distinction is at once a fundamental and a slippery one. Put simply, if a semiotic theory (Peircian or Saussurean, Hjemslevian or Barthian) is unable to take into account dynamical objects, then there can be no attribution of truthful-

---

[15] Consider for example the critique of semiotic structuralism offered by the philosopher Paul Ricoeur, in 'What is a Text? Explanation and Understanding' in John B. Thompson (ed. and translator) *Paul Ricoeur: Hermeneutics and the Social Sciences* (Cambridge: Polity Press, 1981), 145-164.

[16] Umberto Eco, 'On Being', in *Kant and the Platypus: Essays on Language and Cognition*, translated by Alastair McEwen, (London: Secker and Warburg, 1999 [1997]), 12-56, pp. 12-13. Italics added to the original.

[17] Umberto Eco, 'On Being', p. 14.

[18] Charles Sanders Peirce, *CP* 4.536, italics added.

[19] Charles Sanders Peirce, *SS* 196.

ness, and it becomes almost impossible to avoid a creeping relativism, and for related reasons, what I described in the introductory chapter as *culturalism*: the doctrine that everything in human affairs (and one might add, in *semiotic* affairs), is a matter of culture.[20] A version of this is, arguably, the *de facto* doctrine of Saussurean, and not a few post-Saussurean theories of semiotics.[21] For Saussure and many of his followers, all that seems to matter is the already instituted relation between a signified and a signifier; this effectively eliminates from the analysis dynamical objects, i.e. all those objects that have a real capacity to partly or wholly determine the ground and even the interpretant. This is a problematic approach in so far as objects may have a power of interpellation that in some sense preempts *reference*. In a tropical forest, the wild animals are not simply the *referents* of human signification; they are, on the one hand, beings with a certain capacity to *produce* their own signs. But on the other hand, they are also creatures with a certain capacity to *provoke* signs, or as suggested earlier, to *cause* signs in or by other creatures.

Now it might be assumed that the distinction between an immediate and a dynamical object is as neat as that of an 'internal' mental object and an 'external' empirically verifiable object. But this is not how Peirce understands the distinction. As he puts it, 'the dynamical object does not mean something out of the mind. It means something forced upon the mind in perception, but including more than the perception reveals. It is an object of actual Experience.'[22] By contrast, immediate objects are ineluctably *partial* in the sense that they can *only* convey an *aspect* of a dynamical object – whatever aspect is actually represented by the ground of a sign, and by its interpretant. This does not mean that immediate objects are 'subjective' in the positivist sense of the term, let alone untruthful (although of course they may be just that). It *does* mean that no immediate object can ever convey, for once and for all, the Truth, *tout court*, of a dynamical object.

It might also be assumed that another way of distinguishing between immediate and dynamical objects is that immediate objects cannot themselves produce additional signs. However, over time, in so far as immediate objects sediment and come to be taken for granted by one or more individuals, those objects may acquire objective dynamical qualities in their own right. For instance, it might be thought that nobody in their right mind would confuse a TV image of a tropical forest with the real McCoy; but as the introductory chapter began to explain, some tourists were so persuaded by the realism of the images, that it led them to

---

[20] Terry Eagleton, 'Culture and Nature', in *The Idea of Culture* (Oxford: Blackwell 2000), p. 91.

[21] In Saussure's case we might have to speak not of culturalism, but of 'socialism'. Examples of culturalism in the critical social sciences are rife; consider for example some of the Baudrillard-influenced analyses of nature and the hyper-real, e.g. Timothy Luke, 'Southwestern Environments as Hyperreality: the Arizona-Sonora Desert Museum' in *Museum Politics: Power Plays at the Exhibition* (Minneapolis: University of Minnesota Press, 2002).

[22] Charles Sanders Peirce, *SS* 197.

arrive on Barro Colorado with a very definite set of expectations. Those expectations must themselves be regarded as interpretants, and so we can say in this sense that the TV programmes' immediate objects, as mediated by the interpretants produced by the audiences-turned-tourists, generated new signs.

Now many accounts of Peirce's semeiotic theory focus entirely on its semeiotic categories. However, the semeiotic categories are interrelated with those of Peirce's *phenomenology*, and indeed, the phenomenology is key to understanding the logic that underlies Peirce's typology and classification of signs, which I will introduce in the following chapter. As Peirce himself puts it, '... I found Logic [semeiotic] largely on a study which I call Phaneroscopy [Peirce's later name for phenomenology], which is the keen observation of and generalization from the direct Perception of what we are immediately aware of'.[23] This being the case, I would now like to introduce Peirce's phenomenological categories.

It may be helpful to note from the outset that these categories, and the kind of account involved, both antecede, and are quite different from those proposed by the phenomenologists associated with so-called Continental philosophy (e.g. Edmund Husserl or Maurice Merleau-Ponty, cf. Chapter 9). That said, Peirce describes phenomenology in a manner that suggests clear continuities across the academic contexts:

> The initial great department of philosophy is phenomenology whose task it is to make out what are the elements of appearance that present themselves to us every hour and every minute whether we are pursuing earnest investigations, or are undergoing the strangest vicissitudes of experience, or are dreamily listening to the tales of Scheherazade...[24]

And he adds:

> ...what we have to do, as students of phenomenology, is simply to open our mental eyes and look well at the phenomenon and say what are the characteristics that are never wanting in it, whether that phenomenon be something that outward experience forces upon our attention, or whether it be the wildest dreams, or whether it be the most abstract and general conclusions of science.[25]

As will perhaps be clear from these accounts, Peirce's understanding of phenomenology goes far beyond consciousness understood in any narrowly 'subjective' or even psychological sense. In keeping with this, Peirce opens his, and our

---

[23] Charles Sanders Peirce, *EP* 2.501.

[24] Charles Sanders Peirce, *EP* 2.147

[25] Charles Sanders Peirce, *CP* 5.41.

eyes in a characteristically logical, abstract, and universalising, to not say *totalising* manner: he suggests that the elements in question can be explained by way of just three categories, one of which might be more prominent in any one phenomenon, 'but all of them belonging to every phenomenon'[26]: a First, a Second, and a Third, or what he eventually calls firstness, secondness, and thirdness.

> Category the First is the Idea of that which is such as it is regardless of anything else. That is to say, it is a *Quality* of Feeling.
>
> Category the Second is the Idea of that which is such as it is as being Second to some First, regardless of anything else and in particular regardless of any *law*, although it may conform to a law. That is to say, it is *Reaction* as an element of the Phenomenon.
>
> Category the Third is the Idea of that which is such as it is as being a Third, or Medium, between a Second and its First. That is to say, it is Representation as an element of the Phenomenon.[27]

If one adds the suffix 'ness' to each of these categories (First becomes Firstness, Second becomes Secondness, and Third Thirdness), then one may arrive at a first understanding of the categories considered as 'conditions', and as constructed from the point of view of Peirce's phenomenology. This initial understanding is nonetheless still a very abstract one, so I will now explain each category in more concrete terms, and with examples. In so doing I will reverse Peirce's order of presentation and begin with *thirdness*.

Thirdness is the *semeiotic* dimension of consciousness (though as we have seen, it also goes far beyond *consciousness* to embody any and all forms of *mediation*). Peirce clearly calls it 'thirdness' precisely because he conceives it as a matter of a triadic relation of the kind described with respect to object, ground, and interpretant. As may be clear by now, this does not mean that thirdness is purely a matter of *inscribed*, let alone *linguistic* signs; in so far as Peirce adopts the pragmaticist perspective that I described in Chapter 5, then thirdness may also involve the consequences, even the physical consequences, of a semeiotic relation.

Thirdness can be exemplified by returning to the encounter with the green and black poison dart frog. If the guide, and then the visitors became conscious of the frog's presence, it was because the frog called out, and the call became a sign of its presence. Such a sign had several consequences, or at least enabled certain events: the guide and visitors not only became aware of the Dendrobatid's presence, but after finding the specimen, employed its location as the site for an informal mode of environmental education that produced additional signs (interpretants) about

---

[26] Charles Sanders Peirce, *CP 5.43*.

[27] Charles Sanders Peirce, *CP 5.66*.

the species. Several visitors took photographs, there was excited commentary, and so forth. All of these are examples of thirdness, and each reminds us of what I said at the start of this part of the volume: signs are what allow individuals to perceive, conceive, and interpret each other, and the world more generally. It follows that, without thirdness, there can be no wildlife observation, no *observation* full stop.

According to Peirce, the logic of thirdness, *as* thirdness, does not admit further reduction. This point may be explained by way of an admittedly crude analogy: if two people are fighting and a third intervenes to mediate between the first two, that mediation, *as mediation*, cannot be explained merely by describing the fight between the two antagonists. The last point notwithstanding, a similar point might be made in the opposite direction: the blows, regarded purely as blows (as distinct from signs of anger or discord) are themselves not of the same order as semeiosis. This is why Peirce includes *secondness* as part of his phenomenology. Secondness is 'the element of *Struggle*',[28] 'brute' action and reaction. The following is a vivid example drawn from Peirce himself:

> If while you are walking quietly along the sidewalk a man carrying a ladder suddenly pokes you violently with it in the head and walks on without noticing what he has done, your impression probably will be that he struck you with great violence and that you made not the slightest resistance; although in fact you must have resisted with a force equal to that of the blow. Of course, it will be understood that I am not using force in the modern sense of a moving force but in the sense of Newton's *actio*; but I must warn you that I have not time to notice such trifles.[29]

Such 'trifles' do not matter, *phenomenologically* speaking, because as noted above, secondness is a matter of 'brute' force, which is to say a force without reason or law (be it Newtonian, or any other law):

> Generally speaking genuine Secondness consists in one thing acting upon another, – brute action. I say brute, because so far as the idea of any *law* or *reason* comes in, Thirdness comes in. When a stone falls to the ground, the law of gravitation does not act to make it fall. The law of gravitation is the judge upon the bench who may pronounce the law till doomsday, but unless the strong arm of the law, the brutal sheriff, gives effect to the law, it amounts to nothing. True, the judge can create a sheriff if need be; but he

---

[28] Charles Sanders Peirce, *CP* 5.45.

[29] Charles Sanders Peirce, *CP* 5.45. Peirce was delivering a lecture at Harvard for which he had been given strict time limits; however, at the same time he was making the point that follows.

must have one. The stone's actually falling is purely the affair of the stone and the earth at the time. This is a case of *reaction*. So is *existence* which is the mode of being of that which reacts with other things.[30]

How, then, does secondness relate to observation? Returning to the encounter with *D. auratus*, before a guide can say 'That's a green and black poison dart frog calling' (thirdness), it is not only necessary for a specimen of the species to call out, but to do so thanks to, and by way of processes which are themselves in some sense at least partly a matter of brute action and reaction. On Barro Colorado Island, *D. auratus* males call out for mates during the raining season that usually takes place between late April and December. However much rain has semeiotic consequences (as experienced by humans, and no doubt, by frogs too), considered purely as drops of water hitting surfaces, rain is a matter of secondness. Moreover, rain may be regarded as a result of an itself 'brute' movement of moisture across contrasting temperatures, such that the change produces precipitation. We can say, in this sense, that the category of secondness allows the phenomenologist (and in this study's case, the researcher) to acknowledge that the cosmos is also about the kinds of relations that have long interested physical scientists, and which certainly involve dyadic relations (whatever else they involve).

Returning to the poison dart frog, regarded as a trigger for the new mating season, rain becomes a sign, which is to say primarily a matter of thirdness: the first rains presumably produce a response, which is to say an interpretant which itself then generates further interpretants, and so forth. This perspective sheds light on the research of the neurophysiologist Horace Barlow, which I referred to in Chapter 3. Barlow discovered that if a certain stimulus is presented to 'intact frogs', they 'turn towards the target and make repeated feeding responses consisting of a jump and a snap'.[31] Contrary to what Barlow suggested, such responses clearly entail thirdness, in so far as the jump and the snap are responses to a sign. All that said, we may still grant that the responses must entail an element of secondness in the form of the kind of 'brute' response that led Barlow to assume (mistakenly, as it happens), that the response was *entirely* 'neurological', and so from his perspective, not just entirely physical but also entirely *mechanical*.

It is, of course, not just the frogs' neurological systems that are affected by one or another 'stimulus'. When a frog calls, it will produce a vibration of the tympanum in any *human* with a working version of the organ; when the vibration occurs, it too, entails what is in some respects a 'raw' action and reaction in the human sensory apparatus (although the vibration, like the bodily organ itself, has, of course, a form with consequences). Whatever the individual's interpretation of such vibration, it is clear that a physical process is required before the interpreta-

---

[30] Charles Sanders Peirce, *CP* 8.330.

[31] Horace B. Barlow 'Single Units and Sensation: A Neuron Doctrine For Perceptual Psychology?', *Perception* 1, 371-394, p. 373.

tion of at least that particular call can occur. Understood in this manner, some form, or some *degree* of secondness clearly plays a role in the perceptual process, and so must itself be an aspect of consciousness, and indeed of the *observational* process. But this point must be made very cautiously because as Peirce reminds us again and again, firstness, secondness and thirdness are always *all* present to some degree; any attempt to neatly separate out secondness may lead one to forget that there are always also semeiotic forms involved.

I take it that a focus on secondness is what explains how many neuroscientists approach perception (perception as a matter of chains of secondnesses) – but also what James Gibson is referring to when he suggests that 'When the senses are considered as channels of sensation (and this is how the physiologist, the psychologist, and the philosopher have considered them), one is thinking of the passive receptors and the energies that stimulate them, the sensitive elements in the eyes, ears, nose, mouth, and skin'.[32] A problem occurs when someone adopts a *retinalist* discourse (cf. Chapters 3 and 4), and in so doing tries to suggest that neurological action and reaction also explain thirdness – a point that Gibson makes in his own critical manner when he suggests the need to distinguish between channels of sensation and perceptual systems.

I've now characterised secondness and thirdness, but Peirce proposes another category, firstness, which is at once the most intriguing, and the most difficult one to describe. It may be helpful to begin by explaining that, on one level, Peirce's logic is clearly mathematical, or at any rate, numerical: where there are threes and twos, there must also be ones, and so where there is thirdness and secondness, there must be firstness. Put differently, if thirdness is a matter of a third 'element' coming between two others, and if secondness is a matter of a relation between those two others considered without the mediation of the third, then logically there must also be individual elements, the elements considered in themselves, and only themselves. That is firstness.

There is, however, rather more than a purely numerical logic (ordinal or cardinal) to Peirce's understanding of the category – one that involves a fundamental question of ontology and metaphysics. Peirce offers a particularly helpful account in the following passage, which is thereby worth quoting at length:

> There is a point of view from which the whole universe of phenomena appears to be made up of nothing but sensible qualities. What is that point of view? It is that in which we attend to each part as it appears in itself, in its own suchness, while we disregard the connections. Red, sour, toothache are each *sui generis* and indescribable. In themselves, that is all there is to be said about them. Imagine at once a toothache, a splitting headache, a jammed finger, a corn on the foot, a burn, and a colic, not necessarily as ex-

---

[32] James J. Gibson, *The Senses Considered as Perceptual Systems* (Boston: Houghton Mifflin, 1966), p. 3.

isting at once – leave that vague – and attend not to the parts of the imagination but to the resultant impression. That will give an idea of a general
quality of pain. We see that the idea of a quality is the idea of a phenomenon or partial phenomenon considered as a monad, without reference
to its parts or components and without reference to anything else. We must
not consider whether it exists, or is only imaginary, because existence depends on its subject having a place in the general system of the universe. An
element separated from everything else and in no world but itself, may be
said, when we come to reflect upon its isolation, to be merely potential. But
we must not even attend to any determinate absence of other things; we are
to consider the total as a unit. We may term this aspect of a phenomenon the
monadic aspect of it. The quality is what presents itself in the *monadic* aspect.

Peirce adds that

> The phenomenon may be ever so complex and heterogeneous. That circum-
> stance will make no particular difference in the quality. It will make it more
> general. But one quality is in itself, in its monadic aspect, no more general
> than another. The resultant effect has no parts. The quality in itself is inde-
> composable and *sui generis*. When we say that qualities are general, are par-
> tial determinations, are mere potentialities, etc., all that is true of qualities
> reflected upon; but these things do not belong to the quality-element of ex-
> perience.[...] Experience is the course of life. The world is that which experi-
> ence inculcates. Quality is the monadic element of the world. Anything
> whatever, however complex and heterogeneous, has its quality *sui generis*,
> its possibility of sensation, would our senses only respond to it.[33]

So firstness can be explained in terms of a monadism, an indivisibility, and by
implication, 'i-mediacy' – the absence not just of three-way mediation, but even of
two-way action and reaction. Of course, it might be argued that nothing in the
universe is 'i-mediate'. But Peirce might well reply that we cannot conceive of
mediation without presupposing 'parts' that are mediated, which is not to say
that such 'parts' *add up* to make the 'whole'. From a phenomenological perspective focussed on consciousness, humans cannot dispense of monads – we attend
to one thing, however heterogeneous in itself, at a time. If this is true of the way
in which our attention works in general, it seems to be particularly true for the
way in which we attend to things with our visual sense, which is able to focus on
a very narrow segment of the optical array – just a couple of degrees' worth – at
any point in time. This ability may itself be taken as a metaphor for the way in
which our attention may engage in a certain filtering of data; something may be
'staring us in the face' but if the mind is focussed on something different, the ob-

---

[33] Charles Sanders Peirce, *CP* 1.424-426.

ject may well be overlooked. We return by this route to the kind of phenomenon that I illustrated in the conclusions to Part 1, when I sketched the three scenarios involving bullet ants; so long as the researcher was looking for butterflies, the grad student did not 'see' the bullet ants even though they were right in the line of sight.

Now Peirce suggests that, in some cases, it is precisely the sheer *presentness*[34] of *one* thing that dominates consciousness – one might say, there is both a sheer *feeling* of presentness and a sheer *presentness* of feeling. Of course, the i-mediacy of such a feeling, its undefined and undefinable quality, must give way to *actual* sensation (secondness), and to some *interpretant* of sensation (thirdness). As noted earlier, Peirce is clear that, ultimately, all of the categories (firstness, secondness, and thirdness) belong to every phenomenon. However, in every case, one of the phenomenological elements is likely to be the more prominent, however fleetingly. And in every case, the firstness of something will also be a matter of more or less *potential*.

The last point requires elaboration. For Peirce, at least as important as indivisibility or presentness is the potential of the monad. As I began to explain in my account of Peirce's evolutionary cosmology, Peirce conceives a universe that starts from absolute potential – the nothingness before Big Bang – and ends, in the infinite future, in death: the cosmos acquires such regularity, is so law-bound, that nothing changes any more, and so life, which is premised at least as much on change as it is on continuity, no longer exists. Until that moment, however, everything continues to have at least some potential for change; as Peirce puts it, 'At any time, however, an element of pure chance survives and will remain until the world becomes an absolutely perfect, rational, and symmetrical system, in which mind is at last crystallised in the infinitely distant future'.[35] I am suggesting that, instead of interpreting this aspect of the cosmology as an example of rationalist teleology, one may interpret it as an acknowledgement that in the infinitely distant future the universe itself will die. In the meantime, this 'element of chance' is part of what gives things a potential to change in unexpected ways. If this is important for the theory of evolution, it is also crucially important for all manner of everyday events – not least, whether tourists actually encounter wild animals, or not.

---

[34] Charles Sanders Peirce, *CP* 5.44.

[35] Charles Sanders Peirce, *CP* 6.33.

7
Types and Classes of Signs

Let's return to the encounter with the poison dart frog. As I began to note earlier, a photograph of the frog, the word 'frog', or someone's perception of a frog *in situ* are of course all quite different from each other. The signs might all share a semeiotic logic – each entails a relation between object, ground, and interpretant – but the precise character of this relation varies across the mentioned examples. On one end of the semeiotic spectrum, words such as '*D. auratus*' would appear to have little or nothing in common with the actual frog, and so might well seem to be arbitrary in much the manner that Saussure described. There are, in fact, grounds on which to dispute this approach even in the case of linguistic signs, but we will not concern ourselves with that issue here.[1] On the other end of the spectrum, the same cannot be said about an *actual* frog, which, if anything, would appear to do no more than *present* (as opposed to *re*present) itself. Indeed, and as suggested in Chapter 6, from a certain perspective an actual frog, regarded in and of itself, is not a sign at all; it is a dynamical object. The last point notwithstanding, the frog becomes a sign as soon as it becomes a part of a three-way relation, including situations in which it comes to be treated, however unconsciously, as a sign of something by another creature. In so far as this is, in practice, bound to always be the case, then from a Peircian perspective a frog is always *also* a sign, whatever else it is.

Peirce was keenly aware of the differences between different kinds of signs, and so sought to develop a typology of signs capable of reflecting differences such as I have just described. This typology changed as his work matured; in this study I will employ the version that appeared in 1903.[2] According to this typology, a sign's type may be articulated by way of a scheme that cross-references two sets of criteria: first, if the sign is considered from the point of view of its different semeiotic elements (i.e. object, ground or interpretant); or second, if the sign is considered from the perspective of its phenomenological aspects (i.e. firstness, secondness, or thirdness). The resulting matrix divides all possible signs into three trichotomies. Table 1 (below) represents the matrix diagrammatically:

---

[1] See Gunther Kress, 'Against Arbitrariness: The Social Production of the Sign as a Foundational Issue in Critical Discourse Analysis'.

[2] James Liszka distinguishes between what he calls the *original*, the *interim*, the *expanded*, and the *final* typologies, and suggests that of these, the interim (1903) offers the most fruitful avenue for analysis. James Liszka, *A General Introduction*, p. 34. Peirce offers an account of this version of the typology in 'Nomenclature and Divisions of Triadic Relations, as Far as They Are Determined', which appears in *CP* 2.233-72, and also in *EP* 2. 289-299. The account that follows supersedes the one I offered in the appendix of Nils Lindahl Elliot, *Mediating Nature*, pp. 241-254.

Phenomenology	Semeiotic		
	Object	*Ground*	Interpretant
Firstness	Icon	Qualisign	Rheme
Secondness	Index	Sinsign	Dicent
Thirdness	Symbol	Legisign	Delome

Table 1: Peirce's Typology of Signs[3]

If one approaches signs from the phenomenological perspective, then *Icon, Qualisign,* and *Rheme* are types for which firstness may be more prominent; *Index, Sinsign,* and *Dicent* are types for which secondness may be more prominent; and *Symbol, Legisign,* and *Delome* are signs for which thirdness may be more prominent. If, on the contrary, one starts from the semeiotic perspective, then *icon, index* and *symbol* are signs in/for which the *representational* dimension may be the more important; *qualisign, sinsign,* and *legisign* are signs in/for which the *presentational* dimension may be the more important; and *rheme, dicent* and *delome* are signs in/for which the *interpretive* dimension may be the more important.[4]

Let's consider then, each type in turn, beginning with the first trichotomy, icon/index/symbol, the trichotomy for which Peirce is best known, and the one that he apparently considered to be the most important. In one of his earlier accounts, Peirce describes an *icon* as a 'likeness,' a type of sign whose relation to its object is 'a mere community in some quality'.[5] In later accounts Peirce renders the type more complex by suggesting that it 'refers to the Object that it denotes merely by virtue of characters of its own and which it possesses, just the same, whether any such Object actually exists or not'. Peirce adds that 'It is true that unless there really is such an Object, the icon does not act [as] a sign; but this has nothing to do with its character as a sign. Anything whatever, be it quality, existent individual, or law, is an icon of anything, in so far as it is like that thing and used as a sign of it.'[6] So an icon may be a sign that *resembles* an object, as happens

---

[3] This table is an interpretation of one proposed by Winfried Nöth, *Handbook of Semiotics,* p. 45.

[4] I borrow the notions of representational, presentational and interpretive from James Liszka, *A General Introduction,* p. 20-31.

[5] Charles Sanders Peirce, *CP* 1.558.

[6] Charles Sanders Peirce, *CP* 2.247.

in, say, the case of the photographs of the specimen of *D. auratus* taken by the tourist in Part 2's opening example. However, it may also be a likeness of the kind found in metaphors or in abstract art.

For its part, an *index* is a sign that refers to an object 'by virtue of being really affected by that Object'; it involves, Peirce suggests, an icon 'of a peculiar kind'; however, 'it is not the mere resemblance of its Object ... which makes it a sign, but it is the actual modification of it by the Object'.[7] Examples of indexes are devices such as thermometers, weathervanes, or in the context of the opening example in this chapter, the photographs taken by tourists. The reading on the thermometer will be actually modified by a change in the degree or intensity of energy that is present in an object or an environment. The weathervane will point in a new direction if the wind itself changes direction. And while we may agree that photographs are icons, the precise icon will be contingent on the nature of any reflected light (at least in the case of analogue photography). In each case, the resulting grounds, if not the interpretants, will be linked *indexically* to dynamical objects.

Both icons and indexes may be contrasted with a third type, the *symbol*, which Peirce defines as a sign 'which refers to the Object that it denotes by virtue of a law, usually an association of general ideas, which operates to cause the Symbol to be interpreted as referring to that Object'.[8] Here 'law' may be taken to mean a law in an experimental scientific sense, but also a binding system of rules, as in a social convention, e.g. a social *code* or what I will describe in Part 3 as a *coding orientation*. So it is, for example, that a guide's discourse about the ecology of a particular forest will be constructed with words, which are symbols in Peirce's sense, viz., a way of representing something on the basis of an association of ideas. The obvious example of words notwithstanding, non-linguistic signs may signify in a similar way; we have only to think of the association between certain animals, or indeed certain landscapes, and particular *symbolisms*.

Where icons can be said to be closer to firstness, and indexes closer to secondness, symbols are, as a type, all about thirdness. However, even a symbolic sign's instantiation in a particular context may endow it with more or less firstness, more or less secondness: an exclamation such as 'Ouch' may be a symbol, but at the time it is uttered, it may be indexically linked to a source of pain, and may quite perfectly resemble, affectively speaking, the immediateness of that pain, thereby making it an icon. This example shows not only the potentially polyvalent character of any sign, but also its more or less complex character as a semeiotic entity that involves more than one type of sign, an issue that I will return to below.

Now the first trichotomy focuses on the *representational* qualities of signs. But as the above examples begin to suggest, some types of signs signify not so much by representing objects, but by conveying more or less directly their presence, by

---

[7] Charles Sanders Peirce, *CP* 2.248.

[8] Charles Sanders Peirce, *CP* 2.249.

presenting them. For example, the call of a frog, or a sudden rainstorm, might be significant for their 'eventness', their very occurrence, in which case the *ground* of the sign, as opposed to its object, would be the most prominent aspect. Here Peirce distinguishes between the *qualisign*, the *sinsign*, and the *legisign*.

A *qualisign* is, as the name suggests, 'a quality which is a Sign',[9] where the notion of quality takes us back to firstness. Peirce suggests that, 'since a quality is whatever it is positively in itself, a quality can only denote an Object by virtue of some common ingredient or similarity; so that a Qualisign is necessarily an Icon'.[10] However, in the case of a qualisign, a quality such as 'redness' is more than a *representation* of redness; it is a quality that signifies in its own right, *as quality*.

For its part, a *sinsign* is the *very presentation*, the 'present-ness' of a sign. Put differently, sinsigns are signs regarded from the perspective of the actual *evenement* or 'performance' that is required to produce a meaning. By this account all signs, including qualisigns, must become sinsigns in order to be perceived, even if that does not mean that they will necessarily be treated as being *primarily* sinsigns. However, in some cases, signs *will* be mainly sinsigns in the sense that they will be signs which are signs by virtue of *becoming* signs, as in the case of an alarm suddenly going off, or a creature suddenly calling out – or indeed appearing – in the undergrowth. The term *sinsign*, Peirce suggests, is derived from the syllable *sin*, which is taken as meaning '"being only once," as in *single*, *simple*, Latin *semel*, etc.' It is 'an actual existent thing or event which is a sign. It can only be so through its qualities; so that it involves a qualisign, or rather, several qualisigns. But these qualisigns are of a peculiar kind and only form a sign through being actually embodied.'[11] The call of the specimen of *D. auratus*, or rather *the very occurrence* of the call, is a good example of a sinsign. But so would the sudden emergence of any fauna, *as* sudden appearance.

In the context of this study, the sign types that I have presented thus far raise an issue which is particularly important to wildlife observation, and which helps to explain why Peirce's semeiotic may, paradoxically, be regarded as a 'non-representational' theory, or at least as one with non-representational dimensions. In all those forms of semiotic theory that focus on *re*presentation, the object of the sign, if the semiotic theory allows for an object, is necessarily absent: to represent is, as the term suggests, to present again something that is, by implication, absent in itself. While some passages in Peirce would appear to suggest a similar approach, as I interpret his work, in some cases the ground may be, one might say, more or less perfectly 'attached'. In some of his writing, Peirce discusses a concept that sheds light on this phenomenon: that of *entelechy*. According to Peirce,

---

[9] Charles Sanders Peirce, *CP* 2.244.

[10] Charles Sanders Peirce, *CP* 2.254.

[11] Charles Sanders Peirce, *CP* 2.245.

> Aristotle gropes for a conception of perfection, or *entelechy*, which he never succeeds in making clear. We may adopt the word to mean the very fact, that is, the ideal sign which should be quite perfect, and so identical,—in such identity as a sign may have,—with the very matter denoted united with the very form signified by it. The entelechy of the Universe of being, then, the Universe *qua* fact, will be that Universe in its aspect as a sign, the "Truth" of being. The "Truth," the fact that is not abstracted but complete, is the ultimate interpretant of every sign.[12]

In so far as any one sign is, *qua* sign, always partial or aspectival, then in one sense no sign can ever be perfect, no sign can ever truly achieve entelechy. Or rather, only the universe 'itself' can in so far as one regards the universe as the sign made up of all possible signs. But doubtless Peirce would be the first to recognise that some signs come closer at least to *a* truth, and so a continuity with 'their' objects. This is certainly the case for indexical signs, and for sinsigns produced by the dynamical objects 'themselves'. When a tourist (or for that matter, any prospective predator) sees an actual poison dart frog, clearly the amphibian's morphology, including its overall shape and colour, together with its ethology (including movements, calls, etc.) are, or become, signs. But those signs are not (only) signs of something that is absent (though of course for some observers they may invoke a species or type, which as species or type, *is* necessarily absent). Instead, they *are* the 'thing', or at any rate a part of the thing, a part of the object, which at the very least, co-occurs with ground, with the interpreter. In such a case, the observed creature will be at once an object of direct experience, a dynamical object, and a sign that affords information concerning its object – what with Peirce I will describe below as a dicent indexical sinsign.

Returning to the typology, the third type in this trichotomy, the *legisign,* 'is a law that is a sign'[13], one which Peirce notes is usually established by humans. The qualification is necessary because Peirce suggests that while every conventional sign is a legisign, even signs that are not a matter of convention in and of themselves (at least not human convention or code) may become legisigns. So it is, for example, that the howl of a wolf has come to be recognised as a conventional representation of the haunted, despite being an icon of one of the sounds made by a real creature; the occurrence of such a howl as part of the conventional ambient noise in the haunted house of an amusement park would be a good example of a legisign. But that would of course change were an actual wolf to appear to a doubtless very surprised visitor.

---

[12] Charles Sanders Peirce, *EP* 2.304.

[13] Charles Sanders Peirce, *CP* 2.246.

Every legisign, Peirce suggests, 'signifies through an instance of its application, which may be termed a *Replica* of it'.[14] If, for example, a guide says 'The skin of this poison dart frog is capable of secreting toxins', 'The' is a replica of the definite article 'the', which is a legisign, and the same is true for the use of every word or term. An analogous point might be made about propositions and the instantiation of any other conventional form.

As noted earlier, each of the above types can be said to be closer to one or another of the phenomenological elements. So it is that qualisigns are all about the monadism of firstness; Signsigns suggest the kind of 'brute' eventness of secondness; and legisigns are all about the lawful mediation that is the hallmark of thirdness.

I would now like to consider the third trichotomy, which centres on the sign regarded from the perspective of the interpretant – the response, itself semeiotic, that the object and ground at once give rise to, and receive. Here Peirce distinguishes between the *rheme*, the *dicent*, and the *argument* – or what he sometimes also describes as the *delome*, a term that I will prefer in order to avoid confusion with the commonplace understanding of argument.

A *rheme* is 'a sign which, for its Interpretant, is a sign of qualitative possibility, that is, is understood as representing such and such a kind of possible Object. Any rheme, perhaps, will afford some information; but it is not interpreted as doing so.'[15] In the terms of traditional logic, a rheme is the *term* in the triad of term/ proposition/argument. As Peirce also puts it, a rheme 'is said to represent its object in its characters merely';[16] it is 'any sign that is not true nor false', like almost any single word except terms such true or false.[17] The rheme entails, in this sense, what appears to be a kind of degree-zero interpretant, and like the icon, can thereby be said to be closer to firstness. Peirce uses the example of a sentence in which various of the elements have been blanked out:

> Let a heavy dot or dash be used in place of a noun which has been erased from a proposition. A blank form of proposition produced by such erasures as can be filled, each with a proper name, to make a proposition again, is called a *rhema*, or, relatively to the proposition of which it is conceived to be a part, the *predicate* of that proposition. The following are examples of rhemata:

---

[14] Charles Sanders Peirce, *CP* 2.246.

[15] Charles Sanders Peirce, *CP* 2.250.

[16] Charles Sanders Peirce, *CP* 2.252.

[17] Charles Sanders Peirce, *SS* 34.

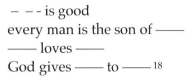

– – - is good
every man is the son of ——
—— loves ——
God gives —— to ——[18]

Perhaps the easiest way to illustrate a rheme is by way of any single term such as 'human', so long as it is employed in such a way that it does not indicate any particular human being, or suggest a particular predicate of the form of, say, 'humans are inherently...'. In a tour of Barro Colorado, a guide might produce a rheme simply by uttering a term such as 'poison dart frog'.[19] The all-important caveat is that any such utterance would be *bound* to be accompanied by a gesture, or indeed by a comment such as 'Here is a...', in which case the rheme would point to something, and become a *dicent*.

A dicent is a sign 'which, for its Interpretant, is a Sign of actual existence', a sign that something 'does exist'.[20] Returning to traditional logic, if a rheme is a term, a dicent is a proposition. The manner of the indication is regarded, as in the case of the index, as being closer to secondness. 'Here is a green and black poison dart frog' indicates the *existence* of a specimen (and by implication, the species), but at least in one sense, does little more than that – it is, or appears to be, *no more than an indication*.

The last type in this trichotomy, that of an *argument* or *delome*, is a sign 'which, for its Interpretant, is a Sign of law'; or, a sign which 'is said to represent its Object in its character as sign',[21] and so, like the symbol, a matter of relatively 'pure' thirdness. As Peirce also puts it, 'An Argument [Delome] is a sign whose interpretant represents its Object as being an ulterior sign through a law, namely, the law that the passage from all such premises to such conclusions tends to the truth. Manifestly, then, its Object must be general; that is, the Argument must be a Symbol. As a Symbol it must, further, be a Legisign'.[22]

The paradigmatic example of a delome is an argument in the sense given to the term in logic, such that something is inductively, deductively or abductively ('hypothetically') explained: in effect, a *meta-sign* in the sense that *signs are used to explain signs*. So it is, for example, that any account given by a guide of the characteristics of *D. auratus* would be bound to involve one or more delomes, e.g. 'The bright colours of the frog are a sign to potential predators that it's not good to eat,

---

[18] Charles Sanders Peirce, *CP* 4.438.

[19] Of course, critical discourse analysts would note the social and discursive charge of any indication *qua* classification (cf. Chapter 8).

[20] Charles Sanders Peirce, *CP* 2.251.

[21] Charles Sanders Peirce, *CP* 2.252.

[22] Charles Sanders Peirce, *CP* 2.263.

so most predators will avoid eating it'. What distinguishes the delome from the legisign is that the former constitutes a response, with signs, that interprets other signs' 'signedness', which is to say their character as a semeiotic entity.

As the previous pages have begun to show, Peirce often suggests that one type of sign includes another, and that another, and so forth. This is actually consistent with his semeiotic theory; if all signs in a system are eventually inter-related by way of interpretants, then it would be expected that all *types* of signs would eventually be inter-related. We return, by this route, to Peirce's synechism. Contrary to what might be assumed, this is not a recipe for relativism, any more than it is relativistic to say that we humans share a surprising amount of DNA with amphibians. What *is* relative, which is to say *relative to context*, is the precise meaning given to particular signs.

For example, someone not familiar with the call of the howler monkeys (*Alouatta palliata*) might respond, as more than one visitor did during my ethnographic research on Barro Colorado, with fear at the sheer occurrence of the loud sound (the calls acting then primarily as unidentified *sinsigns* generating *rhemes* of fear...). By contrast, for someone very familiar with such calls, the primary response might well be recognition, whereby the calls might well be treated as *legisigns* of the *A. palliata's* famously sonorous chorus. These examples reveal the importance of the pragmatics of semeiosis, in the sense of the importance of the role that context plays in interpreting signs – a point that I will explore in more detail when I consider the cultural sociology of wildlife observation in Part 3.

Now Peirce was clear that, while it might be possible to isolate certain types of signs for analytical purposes – for example, icons, qualisigns, and rhemes – in practice semeiosis involves ensembles of what are themselves compound, and so more or less complex signs. What is true for signs in general is certainly true for the signs generated during encounters with wildlife.

To illustrate this point, we may return to the encounter with the specimen of *D. auratus*. As the reader may recall, it was the call of the frog that first alerted the guide, and then the group of tourists to its presence. From the point of view of at least those tourists who were unfamiliar with the call, the call was a sinsign in so far as it was an occurrence, an event. But it was also a rheme in so far as it represented its object – to begin with, no more than the call itself – in its characters merely. And yet both aspects, working together, would have served to in effect 'point' to the call, if not to the frog itself, and so the call would also have acted as an index. This combination is, precisely, one of ten classes of signs that Peirce proposed, and which he called a *rhematic indexical sinsign*. In what remains of this chapter, I would like to consider the different classes, which I will then use in the conclusions of Part 2 to develop a typology of different kinds, or *modes* of wildlife observation.

I will start by listing and exemplifying the classes, and will then say something about the manner in which Peirce arrived at them. In addition to naming all of the classes, the list below provides an abbreviated account of each type (paraphrased from Peirce[23]), and offers two examples: one given by James Liszka (interpreting Peirce's own examples)[24] and another drawn from my research about wildlife observation in tropical forests.

1. Rhematic Iconic **Qualisign** (bold lettering shows the dominant aspects, and the abbreviated names that Peirce himself chose for each class[25]): Any quality in so far as it is a sign. This sign is necessarily an icon, and a rheme. Example A: Any undefined feeling (as opposed to a clearly conceived sense directed towards a particular Object). Example B: The shock of someone being struck by something and not knowing what has happened (see conclusions to Part 2 for an example involving a crocodile attack).

2. Rhematic **Iconic Sinsign**: Any object of experience in so far as some quality of it makes it determine the idea of an Object. It embodies a qualisign and can only be interpreted as a rheme. Example A: The only existing sculpture of Socrates, in so far as we take it to be identical to the philosopher. Example B: An actual specimen of *D. auratus*, encountered and actually seen by someone for the very first time.

3. **Rhematic Indexical Sinsign**: Any object of direct experience so far as it directs attention to an Object by which its presence is caused. It brings the attention of the interpreter to the very Object denoted. Example A: A spontaneous cry. Example B: The call of a specimen of *D. auratus*, as perceived by a tourist for the first time w/o prior knowledge of what causes the sound.

4. **Dicent** Indexical **Sinsign**: Any object of direct experience in so far as it is a sign, and, as such, affords information concerning its Object. It can only do this by being really affected by its Object and so necessarily an index. But the only information it can afford is of actual fact. It involves an iconic sinsign to embody information and a rhematic indexical sinsign to indicate the Object to which the information refers. Example A: A weathervane being pushed by the wind in a certain direction. Example B: The *Guayacán* trees flowering on Barro Colorado thanks to the onset of the raining season.

5. Rhematic **Iconic Legisign:** Any general law or type, in so far as it requires each instance of it to embody a definite quality which renders it fit to call up in the mind the idea of a like Object. Being an icon, it must be a rheme. Being a legisign, its mode of being is that of governing single replicas, each of which will be an iconic sinsign of a peculiar kind. Example A: A diagram. Example B: An actual sign in a park with a diagrammatic rendering of information (e.g. a diagrammatic map).

---

[23] In Charles Sanders Peirce *CP* 2.254-263.

[24] James Liszka, *A General Introduction*, pp. 48-52.

[25] Peirce is not consistent in this respect; as the list below shows, he did keep the full three names for some classes.

6. **Rhematic Indexical Legisign**: Any general type or law, however established, which requires each instance of it to be really affected by its Object in such a manner as merely to draw attention to that Object. Each replica will be a rhematic indexical sinsign. The Interpretant of a rhematic indexical legisign represents it as an iconic legisign. Example A: A demonstrative pronoun (this, that, those). Example B: An arrow on a sign along a trail.

7. **Dicent Indexical Legisign**: Any general type or law, however established, which requires each instance of it to be really affected by its Object in such a manner as to furnish definite information concerning that Object. It involves an iconic legisign to signify the information and a rhematic indexical legisign to denote the subject of that information. Each replica is a dicent sinsign. Example A: The street cry of someone selling a product such as food or a newspaper. Example B: The 'dawn chorus' of Howler Monkeys, recognised as such by someone who hears their roars.

8. **Rhematic Symbol** Legisign: A sign connected with its Object by an association of general ideas in such a way that its replica calls up an image in the mind which image, owing to certain habits or dispositions of that mind, tends to produce a general concept, and the replica is interpreted as a sign of an Object that is an instance of that concept. Example A: Any common noun e.g. 'dog'. Example B: Any noun used by a guide or visitor in the course of a tour of a tropical forest, abstracted from the rest of the discourse.

9. **Dicent Symbol** Legisign: An ordinary Proposition, a sign connected with its Object by an association of general ideas, and acting like a rhematic Symbol, except that its intended Interpretant represents the dicent Symbol as being, in respect to what it signifies, really affected by its Object, so that the existence or law which it calls to mind must be actually connected with the indicated Object. Example A: Proposition, e.g. 'This is a big dog'. Example B: A guide saying, 'This is a green and black poison dart frog'; or the conventional symbols for the presence of toilets in a park building, located just outside the facilities (and so saying, in effect, 'Here are the toilets').

10. **Argument** Symbolic Legisign (shortened to Argument, or as explained earlier, what Peirce and this study will describe as a **Delome**): A sign whose Interpretant represents its Object as being an ulterior sign through a law, namely, the law that the passage from all such premisses to such conclusions tend to the truth. It must be a symbol, and as a symbol, it must further be a legisign, its replica being a dicent sinsign. Example A: A syllogism. Example B: A guide explaining, 'Bright markings on animals often warn predators that an animal is not good to eat. Poison dart frogs are poisonous; their bright colours signal that that they're not good to eat'.

I would now like to say something about the rationale behind this classification of signs. As explained by James Liszka,[26] this classification evokes the par-

---

[26] James Liszka, *A General Introduction*, p. 44.

adigm of *mathematical* classification, e.g. for something to count as a regular convex polygon, it must have more than four sides, it must have the same number of angles as of sides, etc. In a similar way, any class of signs must satisfy the formal conditions that determine the grammar of any sign, i.e. that it should re/present an object, do so via a ground that is of necessity partial, and in so doing generate an interpretant that translates the object with reference to that partiality. However, and as the typology of signs discussed earlier began to illustrate, not all signs will conjugate these three conditions in the same way, and that is why a system of classification is proposed.

The manner in which Peirce articulates this problem is to establish a grid of signs for which each class includes one of the elements of each of the three trichotomies presented above (icon/index/symbol, qualisign/sinsign/legisign, rheme/dicent/delome). However, in each class, some of the elements are deemed to be more important for a given combination (hence the terms highlighted in bold lettering).

This approach could result in many more classes of signs than Peirce allows for (3x3x3=27), but Peirce reduces the number to ten thanks to a series of classificatory filters, or what James Liszka describes as the following 'leading principles':

- *Composition Rule*: Signs are, by definition, constituted by triadic relations; it follows that every sign must partake in each of the three 'dimensions' or aspects just described. It also follows that all signs must be classified and indeed be classifiable in relation to each of the three trichotomies.

- *Qualification Rule*: Although Peirce privileges semeiotics, he considers semeiotics to be subordinate to phenomenology. His phenomenology, which itself has a triad as its basis – the categories of firstness, secondness, and thirdness – establishes a hierarchy whereby a first can only determine a second, and a second a third. According to Peirce, qualisigns, icons, and rhemes are phenomenologically typed as Firsts; sinsigns, indexes and dicents as Seconds; while legisigns, symbols and delomes are typed as Thirds. Consequently, the total number of permutations suggested by the first rule – 27 – is in fact reduced to 10 in so far as
    - Qualisigns will always represent their objects iconically;
    - Qualisigns will always be interpreted rhematically;
    - Icons will always be interpreted rhematically;
    - Arguments can only represent their objects symbolically.

- *Dominance Rule*: every sign requires an element from each of the three trichotomies, but some subset of these will predominate over the others. For example the fact that a sign occurs suddenly (sinsign) may be more important than the fact that it is produced by a certain object (index).

- *Instantiation Rule*: all signs, to be signs, must be instantiated, and therefore must be a sinsign. This, however, does not mean that a sign will necessarily be interpreted as being *primarily* a sinsign.
- *Inclusion Rule*: Finally, given the principles derived from the phenomenological hierarchy of signs, it is the case that thirdness always incorporates secondness, and secondness, firstness. Symbols are always to some extent iconic and indexical; delomes contain rhemes and dicents; legisigns are to some extent qualisigns and sinsigns.[27]

Now anyone who tries to apply these principles soon finds that it is not a simple matter to employ the categories for the description of semeiotic phenomena. More than one class may often be employed to describe one same form, and indeed Peirce himself does not offer many examples, raising the suspicion of a kind of heuristic. But Peirce is very candid, to not say casual, on this point: 'It is a nice problem to say to what class a given sign belongs; since *all the circumstances* of the case have to be considered. But it is seldom requisite to be very accurate; for if one does not locate the sign precisely, one will easily come near enough to its character for any ordinary purpose of logic.'[28] In so far as Peirce was a mathematician and a scientist devoted over decades to making very precise measurements for the purposes of geodesy, I assume that Peirce defined 'ordinary purpose' according to a very high standard.

---

[27] James Liszka, *A General Introduction*, p. 44-46. I first presented a summary of Liszka's interpretation in Nils Lindahl Elliot, *Mediating Nature*, p. 250.

[28] Charles Sanders Peirce, *CP* 2.265. Italics added to the original.

Conclusions to Part 2

From a Peircian perspective, wildlife observation is all about the detection and the interpretation of signs. Signs are what enable visitors to detect the presence of wild animals, and indeed of any other objects in a tropical forest. Conversely, signs are what allow beyond-human animals to 'make themselves known' (or in some cases 'unknown', as in crypsis) and to detect the presence of *humans*. Signs both enable and require interpretation, in the sense of translation via interpretants, viz., further signs.

To say that wildlife observation is all about signs is nonetheless misleading in at least two respects. On the one hand, Peirce conceives of signs as triadic entities that comprise dynamical objects, grounds, and interpretants. In so doing, Peirce builds into his approach a deliberate semeiotic, phenomenological, and indeed one might say *ontological* ambiguity. From a perspective such as Peirce's, the poison dart frog in the opening example, like all of the other dynamical objects found in the forest, has a quality which confounds the ancient philosophical maxim that something cannot be, and *not* be, all at once. The frog *is what it is* in so far as it is a *dynamical* object, or what I will describe in Part 3 as a dynamical *body* with a real existence of its own. The last point notwithstanding, when a frog, poison dart or other, is treated, however unselfconsciously, as a sign – as it must be, if it is to be observed – it becomes something else: it becomes an *immediate* object, i.e. the object as produced by the sign or the semeiotic process. This dimension, this medium, is the frog as frog presented or represented by something other than itself: a call heard by something else, or indeed the image formed by an *in situ* observer, etc.

Now it would be tempting to leave things there, and to adopt the kind of nominalist stance which has long been the philosophical mainstay of the critical social sciences and humanities. But Peirce's phenomenology, as linked to his semeiotic, takes an important step to avoid precisely such a stance. It is true, Peirce suggests, that some signs, which he calls symbols (but also legisigns, and delomes), entail what today might be described as a 'constructed' or 'conventional' relation to any object of representation, or interpretant of interpretation. But it is also true that some signs, which he calls indexes, entail a real link between object, ground, and interpretant. In so far as this is the case, in some circumstances – a really hardnosed realist might say in *all* circumstances – signs may be linked, quite directly, to at least some aspects of a dynamical object. This is the kind of relation (and realism) that Marr in some sense takes for granted when he speaks of the three stages of perception (*primal sketch, $2^1/_2$ -dimensional sketch, 3 D model representation*) (cf. Chapter 3). What differentiates Peirce's approach is that Peirce adopts, by his own account, an ideal-realist perspective, and does so explicitly. As part of such an approach, he theorises, in great detail, signs which entail *combinations* of iconicity, indexicality, and symbolism. This is not only in keeping with his semeiotic types and classes, but with his phenomenological proposal that *all* signs entail a degree of firstness, secondness, and thirdness.

Building on this remarkable articulation of semeiotics and phenomenology, it is possible to distinguish between what might be characterised very generally as three semeiotic-phenomenological modes (as in short for modalities) of wildlife observation (and indeed, of observation more generally) – what I will describe as the immediate, the dynamical and the mediate.[1] In order to explain these modes more clearly, it may be helpful to refer to a somewhat extreme encounter with wildlife – one which enables a clear exposition of the modes.

Two researchers conducting botanical research decided to take a swim in the Gatún Lake, just off the shoreline of one of the peninsulas that make up Barro Colorado Nature Monument. Temperatures in the region typically exceed 30 C during daytime, so a dip in the Gatún's inviting waters would have seemed like a good way of cooling down. Soon after they jumped in, one of the researchers received a huge blow. By that researcher's account, the attack was so sudden, and so unexpected that there was a split second when it was not clear what had happened. But incomprehension gave way almost instantly to the realisation that the researcher been given a blow in the chest by *something*, and then that the something was a crocodile – an American crocodile (*Crocodylus acutus*), a species that can grow over four meters in length, and weigh almost a tonne.

The crocodile surfaced in front of the researcher, opened its mouth and arched its back and tail before sinking ominously beneath the surface. The researcher expected the worst, but there was no second attack, let alone the notorious 'death rolls'. It seems that the crocodile had swum up from below, and used its snout to strike her, perhaps to drive the researcher away from its territory. Whatever the case, the researcher had the presence of mind to call for help whilst trying to remain as still as possible – no easy feat in a reservoir whose deep waters fill what was once a valley along the Chagres River. Help was forthcoming when the second researcher managed to swim back to shore, and then brought over their boat – but reportedly not before several cinematographically failed attempts to start its temperamental outboard motor. The researcher who was struck not only lived to tell the tale (which I heard first-hand some months later) but did not suffer any life-changing injuries.

From a semeiotic phenomenological perspective, it is possible to characterise the attack in the following manner. In the first instant, there was a blow that was so unexpected that for a split second there was, by the researcher's own account, utter incomprehension, or I would say *sheer blowness*: there was nothing but the feeling of a tremendous knock. 'Sheer blowness' is, from the perspective of Peircian phenomenology, a good example of firstness, and firstness is precisely what

---

[1] I first proposed a scheme of this kind in Nils Lindahl Elliot, *Mediating Nature*, pp. 35-36. In that account, I spoke of immediate, dynamic, and habitual modes of observation. The current terminology and theoretical account supersede the earlier one.

predominates in what I term the *immediate* mode of wildlife observation. In such situations there is at least a moment, or what we might even call a 'noment' (no *moment*, strictly speaking) when there is no comprehension, or even a conscious reaction. There is just a 'feeling'.

Thus far I have emphasised the firstness of the attack, yet as soon as the crocodile struck, the researcher would have suffered a 'brute' action with a corresponding set of reactions; her body would have been knocked in the water, and almost instantly, a series of autonomic responses would have resulted (e.g. surge of epinephrine, faster heart rate etc.). In so far as aspects of this kind prevailed, then there was what I describe as the *dynamical* mode of wildlife observation – one in which, and for which secondness dominates. In practice, this is the mode that most likely predominated during the first seconds of the attack, and this not least because it had such manifestly corporeal consequences.

The last point notwithstanding, as soon as the researcher began to wonder what had produced the blow, signs of a primarily symbolic type and class would have produced. In so far as the researcher's response became increasingly governed by such signs (the individual had, for example, the 'presence of mind' to try to keep as still as possible), then a more or less explicit thirdness would have begun to dominate at least on the level of consciousness. This is what I describe as the *mediate* mode of wildlife observation. Any form of wildlife observation for which symbolic signs play a leading role in generating interpretants entails this mode; it is thereby to be expected that during the vast majority of wildlife encounters experienced by tourists, it is this mode that predominates, albeit with some important caveats that I will now turn to.

The first is that it might be assumed that the modes that I have just introduced are mutually exclusive. This is emphatically not the case. The reader will recall that Peirce suggests that even if one or another phenomenological aspect may be more pronounced, *all three belong to every phenomenon*. This is also true for the different modes of observation as I conceive them. In any instance, indeed in any moment of wildlife observation, there is likely to be a certain *preeminence* of one of the three categories (firstness, secondness, and thirdness); however, the other two categories will nonetheless be co-present. All instances of wildlife observation must involve signs, and so thirdness. But all instances of wildlife observation must also entail a more or less 'brute' physics (or more generally, action and reaction), and so secondness. And then again, all wildlife observation involves an element of chance and of objects (in the broadest semiotic sense of the expression) that, however inter-related, entail a monadic quality, a certain haecceity.

Now it may also be assumed that the immediate mode only occurs in the most extreme, in the sense of life-threatening, encounters. Again, this is not the case. Consider, for example, the encounters that visitors and I experienced on Barro Colorado with the crested guans (*Penelope purpurascens*), large turkey- or pheasant-like birds found in Central America and northern South America. Crested guans are often found roosting in pairs in the understory; when surprised by humans or any other prospective predators, the birds will suddenly erupt in flight with loud calls, and an itself surprisingly loud beating of their large wings. The encounters tend to generate, if only for a split second, the kind of 'noment' I referred to earlier, and with it a certain *stupefaction* in the original Latin sense of *stupere*, 'to be struck senseless'. This first instant of i-mediacy turns into a dynamic of action and reaction, and once visitors recover from the initial surprise, they may and will begin to objectify the creatures in the primordial sense of becoming aware of them as objects (always in the semeiotic sense), and in the sense of beginning to question what has just happened, what species was involved, etc. – an example of the *dynamical* mode of observation turning into the *mediate*.

Even this kind of encounter might seem somewhat extreme when compared to the almost delicate qualities of the encounter with which Part 2 began: an encounter that, as the reader will recall, involved at first a sound which the visitors not only did not recognise, but did not even pay attention to until the guide invited them to do so. In this encounter too, one might speak of a transition from an immediate, to a dynamical, to a mediate: at first there was no conscious perception (immediate), even though the calls would have caused the visitors tympanums to vibrate (dynamical); when the guide indicated the call, the dynamical would have given way to the mediate mode in so far as the call was identified, located, and associated with a particular creature, and a particular name.

From this and the previous examples, it might be assumed that the different modes always occur in the same sequence. In some sense this must be true, if only because logic suggests that when observing wildlife, the perception of any sign requires the firstness of an object to trigger the secondness of a reaction that results in the thirdness of an interpretant. However, even if this is a valid account in the most general and theoretical sense, at least as experienced by tourists across a certain duration, different sequences may occur. For example, in many cases, silent specimens were only observed after a guide pointed out their presence; in a certain sense, such encounters began in the mediate mode in so far as the guide used explicit signs to alert the visitors.

Additional differences in the sequence may arise in so far as some encounters develop in such a manner as to involve sudden changes in a mode dominated by mediacy and thirdness. For example, in the case of a group that is observing a docile, or at any rate subdued specimen at close range, even as a guide provides information about its ecology, that self-same creature may suddenly turn on its observers, producing, if not a wound, then certainly extreme surprise amongst the members of the group. In such a case, observation may quickly go from the mediate to the immediate to the dynamical mode, and back to the mediate: from discourse, to a wholly unexpected attack and stupefaction, followed by recoil, or even horror and pain, back to a rueful awareness of the folly of getting so close, etc. Then again, even highly knowledgeable individuals (for example, a scientist with a specialism in the ecology of a particular species) may come across a specimen which is instantly recognised, and which poses no threat (mediate mode), but which is so unusual, owing perhaps to its size, its behaviour, etc. that the scientist experiences a very strong feeling (e.g. amazement, elation, etc.). This feeling may so pervade the encounter that it is more accurate to describe the encounter as being dominated by firstness, and so by the immediate mode. A similar phenomenon might occur in all those cases where very rare species are sighted.

It is possible to associate particular classes of signs with each mode of observation, and I would now like to provide an account of such associations. Table 2 offers a schematic representation of the associations:

Mode of Observation	Phenomenology	Sign Classes
Immediate	Firstness	Qualisigns/Iconic Qualisigns (Classes 1 and 2, as per the classes presented in Chapter 7, above.)
Dynamical	Secondness	Iconic Sinsigns, Rhematic Indexical Sinsigns, Dicent Sinsigns (Classes 3 to 5)
Mediate	Thirdness	Rhematic Indexical Legisigns, Dicent Indexical Legisigns, Rhematic Symbol, Dicent Symbol, and Delomes (Classes 6 to 10)

Table 2: Semeiotic-Phenomenological of Modes of Wildlife Observation

Let's consider these associations in greater detail. Starting with the immediate mode, the firstness of the *possibility* of an encounter, and then again the sheer presence or feeling of an unexpected encounter, are undoubtedly a matter of *qualisigns*. As Peirce notes, qualisigns are 'any quality in so far as it is a sign', and which involve not just icons (in so far as the quality must denote an object by way of some form of similarity), but rhemes (in so far as 'a mere logical possibility... can only be interpreted as a sign of essence').[2] So what I describe as the immediate mode of observation is all about *qualisigns* and *iconic qualisigns*.

Moving on to the dynamical mode, which is about brute action and reaction (secondness), strictly speaking this mode begins when there starts to be any kind of reaction to some thing, and with it a response to some *object*. This being so, there will be what Peirce terms an *iconic sinsign*, 'any object of experience in so far as some quality of it makes it determine the idea of an Object.'[3] But for there to be this kind of reaction, there must be some kind of *indication*, i.e. indexicality, and so also a *rhematic indexical sinsign*, a sign whose definition might almost have been

2 Charles Sanders Peirce, *CP* 2.254.

3 Charles Sanders Peirce, *CP* 2.255.

written for the purposes of this study: as Peirce puts it, 'any object of direct experience so far as it directs attention to an Object by which its presence is caused. It necessarily involves an Iconic sinsign of a peculiar kind, yet is quite different since it brings the attention of the interpreter to the very Object denoted'.[4]

The transition from the dynamical to the mediate mode of wildlife observation will typically involve the production of *dicent sinsigns*. The reader will recall that dicents are signs which, for their Interpretants, are signs of actual existence, signs that something 'does exist'. As Peirce defines it, a dicent sinsign is 'any object of direct experience in so far as it is a sign, and, as such, affords information concerning its Object. This it can only do by being really affected by its Object; so that it is necessarily an index. The only information it can afford is of actual fact'.[5] In the case of an encounter with wildlife, *there is something to be observed, and so by implication, it does exist.*

Once a presence is confirmed – once some creature is detected – then typically the guide and/or the visitors will seek to alert each other of the creature's presence via pointing actions, expressions such as 'Look!' or 'There!' or indeed 'Here!' – in effect, what Peirce terms *rhematic indexical legisigns*, which are general types or laws which require each instance – we might say each 'performance', or in Peircian terms, each *Replica* – to be really affected by its object, albeit in such a manner as to 'merely draw attention to that Object'.[6] When that happens, we have the beginning of the *mediate* mode of wildlife observation.

From that point onwards, rhematic indexical legisigns are likely to turn into *dicent indexical legisigns*. Peirce defines such signs as 'any general type or law, however established, which requires each instance of it to be really affected by its Object in such a manner as to furnish definite information concerning that Object. ... Each Replica of it will be a *dicent sinsign* of a peculiar kind'.[7] The dicent indexical legisign occurs, for example, at the moment that a guide (or indeed a visitor) at once alerts others about the presence of a creature, but in so doing identifies it: '*Collared Peccary!*' or '*Keel-Billed Toucan!*'

If a guide then offers more or less detailed information about the wildlife encountered – or indeed a visitor entertains various thoughts about what s/he is observing that build on habitual ways of indexing, classifying, and framing objects (what I will describe as coding orientations in Part 3), then the predominant signs may become *dicent symbols* (propositions of the kind 'X is y') strung together, more or less complexly, to produce *delomes,* which is to say arguments regarding the nature of the object just encountered.

---

4 Charles Sanders Peirce, *CP* 2.256.

5 Charles Sanders Peirce, *CP* 2.257.

6 Charles Sanders Peirce, *CP* 2.259.

7 Charles Sanders Peirce, *CP* 2.260.

Conclusions to Part 2

It should be noted finally that, at least with the most able guides, the delomes in question go beyond the field guide-like recitation of information, to offering insights into the specifics of the *specimen*, and its behaviour 'in real time'. When this happens, the production of delomes is made more complex by the fact that, especially in the case of encounters involving *fauna*, the creatures often produce additional iconic sinsigns which afford the opportunity, or on occasion generate a visitor *demand*, for further explanation: 'What's it doing now?' In practice, this means that the delomes produced in the field need to be permeable to the appearance of new signs in a way that those found in actual texts (e.g. field guides) are not. This point highlights the importance of taking into account dynamical objects, as *dynamical* objects and with them, the agency of the creatures themselves. But it also serves as another example of the persistence of firstness and secondness, or immediate and dynamical modes, 'within' a frame otherwise dominated by thirdness, or mediate modes of observation.

Appendix 1: On the Biography of Charles Sanders Peirce

Semiotic theory may seem especially esoteric to readers not familiar with the critical social sciences. Some readers located in the physical sciences may also assume that semiotics entails a kind of postmodernism, and so a discipline that is ostensibly predicated on relativism. Then again, as I explained at the beginning of Part 2, many of those who *are* familiar at least with Saussurean semiology may believe that signs are arbitrary constructs which conjoin concepts (signifieds) and their forms of expression (signifiers). This being the case, they may assume that, whatever else it entails, semiotics is devoted to explaining the structuring of signifying entities (or indeed systems) conceived along structuralist, or at any rate poststructuralist lines.

I have explained across Part 2 that both sets of assumptions overlook the possibility of an alternative: a form of semiotics – what I describe as Peirce's *semeiotics* – which is neither Saussurean, nor predicated on social constructionism, let alone the much maligned, if often misconstrued postmodern philosophies. Aware that for some readers this may be a bit of a surprise (a form of semiotic theory that is not simply relativist?), in this appendix I would like to offer a brief account of aspects of Peirce's remarkable biography that will help to explain why Peirce would have been highly critical of some forms of postmodernism. It is also my hope that this account will provide a sense of why, to this day, Peirce's work is not as widely known as it might have been.

Peirce was born in 1839 in Cambridge, Massachusetts, and spent most of his life living in the northeastern United States. He has often been described as North America's leading philosopher. However, his longest career path was in the physical sciences, and indeed in what was then a minutely empirical form of scientific practice: geodesy. Peirce spent decades patiently swinging pendulums at various locations and jotting down countless measurements as part of a transnational effort to determine the precise shape of the Earth. During many of those same years, he also wrote scientific treatises that were often far ahead of their time. This earned Peirce a place in the US National Academy of Sciences, as well as high regard amongst a number of European scientists, whose scientific activities were perhaps less encumbered by puritanical morals than were many of Peirce's North American associates.

Even as he engaged in the mentioned scientific pursuits, Peirce developed his own philosophy and with it, a semeiotic theory. Peirce identified semeiotics with the study of logic, his life-long passion. Indeed, Max Fisch suggests that Peirce's major but unfinished work was to be called *A System of Logic, Considered as Semei-*

Appendix 1 On the Biography of Charles Sanders Peirce

otic.[1] It seems that from a very early age, Peirce's father, Benjamin Peirce, himself a leading North American mathematician and astronomer at Harvard University, encouraged the intellectually precocious Charles to solve increasingly difficult logical problems, and eventually Charles became a voracious reader of treatises on logic.

As a teenager, Charles obtained an undergraduate degree in chemistry from Harvard, and thanks initially to Benjamin's position on the board of the United States Coast Survey (later renamed as the United States Coast and Geodetic Survey), Charles came to be employed as a geodesist for that institution, a post that he held from 1861 to 1891. This work enabled Charles Peirce to apply his growing knowledge as an astronomer, physicist, and eventually as one of the nineteenth century's leading mathematicians. However, both Charles and Benjamin always regarded the geodetic job as a kind of prelude to the main event: a long and doubtless very promising career as a professor at Harvard.

Before discussing the fate of that expectation, I would like to provide a sense of the shear breadth of the scientific knowledge that Charles Peirce had acquired by the early 1900s. By then, Peirce had become such a polymath that he was able to write all of the definitions on logic, metaphysics, mathematics, mechanics, astronomy, and weights and measures for the six-volume *Century Dictionary*, one of the largest and most prestigious encyclopaedic dictionaries of its time.[2] Doubtless he could have written highly authoritative entries for several other subjects as well. For example, Peirce was an early, and remarkably critical adopter of Darwin's theory of evolution; he was also well-versed in the principles of the then-new field of psychology; and towards the end of his life, he evidently still knew enough about chemistry to be able to engage work as a consultant chemical engineer.

Alas, during his life, the significance of Peirce's scientific and philosophical contribution was not widely recognised, and as late as the 1980s, Peirce remained relatively unknown to many scholars in the humanities and the social sciences. There are several reasons for this obscurity. First, while philosophical pragmatism was an important movement in the US especially during the last quarter of the nineteenth century, throughout much of the twentieth century it was eclipsed by behaviourism and by positivism. And even though pragmatism did have a revival on both sides of the Atlantic beginning the 1970s, in an ironic twist, by then the realism of Peirce's philosophy, and Peirce's many invocations of experimental science may have put off some scholars opposed to positivism, and more enter-

---

[1] Max H. Fisch, 'Peirce's Place in American Thought', in Kenneth Laine Ketner and Christian J. W. Kloesel, *Peirce, Semeiotic and Pragmatism: Essays by Max H. Fisch*, 305-355 (307).

[2] Max H. Fisch, 'Charles Sanders Peirce', in Kenneth Laine Ketner and Christian J. W. Kloesel (eds.) *Peirce, Semeiotic and Pragmatism: Essays by Max H. Fisch* (Bloomington: Indiana University Press, 1986), 1-5 (1).

tained first by the rise of Saussurean semiology, and then by poststructuralist and postmodern philosophy.[3]

It didn't help that, by his own account, Peirce did not have the most agreeable of writing styles. According to Peirce biographer Joseph Brent, late in his life Peirce blamed this on the fact that he was left-handed.[4] It would, however, arguably be contrary to Peirce's own philosophy to invoke biological determinism to explain his penchant for neologism, or indeed those aspects of Peirce's writing that are the result of a vast, and complex architectonic structure with many thematic wings.[5] Peirce was rightly contemptuous of what today we might describe as the notion of 'plain English', viz. the idea that texts in the humanities and the social sciences should not only avoid jargon, but should be written in an 'absolutely untechnical' way.[6] However, at times it almost seems that Peirce went out of his way to coin not one, but *several* neologisms to explain phenomena that might equally be represented by one new term, if not by existing terms. Whatever the intention, it seems clear that the terminology, as allied to the architectonic, played a role in generating a particularly exclusive hermeneutic circle: in Peirce sizeable portions of his work do not make sense without an understanding of the whole, and vice-versa.

Of course, a similar issue may be raised with respect to many great philosophers' writings. The key difference is perhaps that, in the case of Peirce's work, the whole is characterised by subtly changing formulations which are spread out across a vast body of more or less fragmentary texts. There is no single book written by Peirce that offers a detailed and accessible synthesis of his philosophy, let alone the interrelationships across all his research in what we now know as the humanities, the social and the physical sciences. Perhaps there cannot be, given the sheer breadth of Peirce's knowledge. By one account, there are some 100,000 pages of Peircian manuscripts, approximately one half of which are likely to be of philosophical interest.[7] Again, a similar point might be made about many philosophers' work, but in Peirce case there is not even a set of books that present

---

[3] It might nonetheless be noted that some postmodernists see in Peirce's conception of the interpretant the very epitome of the relativism that they hold dear – a stance that Peirce himself would doubtless have opposed.

[4] Joseph Brent, *Peirce: A Life*, 2nd edition (Bloomington: Indiana University Press, 1998), p. 43.

[5] For an account of the architectonic, as architectonic, see Murray G. Murphey, *The Development of Peirce's Philosophy* (Cambridge: Hackett, 1993[1961]).

[6] I borrow this expression from William James, 'Philosophical Conceptions and Practical Results', *University Chronicle* [University of California at Berkeley], 1(1898), 287-310, p. 287.

[7] Joseph Ransdell, 'Some Leading Ideas of Peirce's Semiotic', Version 2.0, 20 November 1997, online publication <http://www.iupui.edu/~arisbe/menu/library/aboutcsp/ransdell/LEADING.HTM> [Last accessed 26 September 2014].

different parts of his thought at different stages of his life (as *is* typically the case for the more famous philosophers).

It was only after Peirce's death that his extraordinary body of work began to be edited and published, and this leads me to a third reason for Peirce's very belated recognition as a major thinker and scientist. Peirce's *Collected Papers*, long the most comprehensive edition of his writing, did not begin to appear until 1931 (nearly two decades after Peirce died), and that edition was not completed until 1958. However, as anybody who has used the *Collected Papers* will know, its structure does not facilitate a strategic understanding of Peirce's work (though that was its purpose). On the contrary, anyone trying to make sense of Peirce's overall corpus will come away bewildered as to its sense and progression, and this not least because the papers are not presented in anything like a chronological order. To be sure, the *Collected Papers* do not collect all of Peirce's papers; even now, a century or so after his death, Peirce's manuscripts are still being edited, re-edited, and published.[8]

Given these issues, if Peirce has become much better known since the 1980s, it is thanks to the sympathetic work of a group of scholars who, in the early 1990s, began to publish fresh introductions to Peirce's work, and to produce new compilations of his most important papers.[9] These developments beg two questions: why was Peirce's work relatively fragmentary, and why was so much of it not published during his lifetime?

Beyond the sheer scope and diversity of his writing, a significant, if not determining factor was that Peirce procured his first and last properly academic post relatively late in his life (in 1879, when he was 40), and then lost it just five years later (in 1884). To be sure the post, a lectureship in logic at the then-new Johns Hopkins University, was only a part-time and non-tenured position which forced him to continue engaging in the exacting empirical research for the Coast Survey.

According to Brent, Peirce ostensibly lost the post at Johns Hopkins thanks to a sexual scandal generated by academics driven by the Victorian morality of the New England elites. The scandal started after Peirce's first wife, Melusina ('Zina')

---

[8] See the *Peirce Edition Project* website at <http://www.iupui.edu/~peirce/> [Last accessed 1 October 2014].

[9] See for example, Charles Sanders Peirce, *The Essential Peirce Vol. I*, and Charles Sanders Peirce, *The Essential Peirce, Vol. II*. More recently the internet has enabled a veritable explosion of Peirce resources. See for example the Peirce Edition Project's own website at http://www.iupui.edu~peirce//; or any of the now numerous online introductions to Peirce's life and work (e.g. Robert Burch, 'Charles Sanders Peirce', The Stanford Encyclopedia of Philosophy (Winter 2014 edition), Edward N. Zalta (ed.), <http://plato.stanford.edu/entries/peirce> [last accessed 18 November 2014]. A particularly helpful resource with respect to Peirce's terminology may be found at Mats Bergman and Sami Paavola's (eds.), *The Commens Dictionary: Peirce's Terms in His Own Words, New Edition.* <http://www.commens.org/dictionary> [Last accessed 23 October 2014].

Appendix 1 On the Biography of Charles Sanders Peirce

Fay Peirce, herself an author and feminist thinker, left Peirce in 1875, apparently hoping that Peirce would put an end to his abusive ways and profligate lifestyle. Peirce was devastated, but after the separation took place, he commenced a relationship with a new partner, Juliette Froissy.[10] Once it became known that the relation had begun before the formal divorce from Zina Fay (which took place in 1883, with Peirce marrying Froissy soon afterwards), the affair became one of the reasons for Johns Hopkins University's dismissal of Peirce – though it seems that Peirce was never informed that this was the case.

The importance of this scandal notwithstanding, there are other, possibly more important reasons why Peirce failed to acquire the status and position that might have been expected from such a talented, and well-connected thinker. Perhaps the single most important reason why Peirce failed to obtain a tenured academic post is that he was extraordinarily incompetent when it came to managing the day-to-day politics required to keep jobs and to protect his own reputation. Brent documents how, time and again, Peirce made remarkably foolhardy social and financial choices, not least when dealing with individuals who were in a position to damage his career prospects. A case in point, Peirce's relations with Charles William Eliot, the formidable first non-clerical president of Harvard, Peirce's *alma mater* and the university in which he hoped to obtain a post. At the time, Charles Peirce's evident brilliance, and the fact that his father Benjamin Peirce was the director of the Harvard mathematics department should have guaranteed a tenured position at the institution. Instead, Eliot, who was one of Peirce's undergraduate instructors, came to so strongly dislike Charles Peirce that during his 40-year tenure as Harvard's president (1869-1909), he repeatedly stepped in to block Peirce's appointment to a variety of posts at the university.

Alas, Eliot was perhaps not Peirce's most determined foe. Brent suggests that another, perhaps even more determined enemy was Simon Newcomb, a professor of mathematics at Johns Hopkins who started his own career as something of a *protégé* of the Peirce family, but who appears to have turned on Charles Peirce. Brent suggests that, for reasons that may never be well understood, Newcomb so hated Peirce that he went out of his way to destroy him. Unbeknownst to Peirce, it was Newcomb that whispered in the ear of Daniel Coit Gilman, the president of Johns Hopkins, about what were perceived as Peirce's marital improprieties. Thereafter, it seems that Newcomb repeatedly (but secretly) sabotaged Peirce's efforts to attain employment, or to obtain support amongst a variety of entities for the publication of his work.

It might also be noted that Peirce's intellectual style generated further animosity even amongst those who did not have a personal relationship with him. He was, on the one hand, a highly independent thinker who was not only intellectu-

---

[10] Juliette's true maiden name remains unknown to this day, as does her country of origin; it seems that when she emigrated to the U.S. she gave a false name. Joseph Brent, *Peirce: A Life*, pp. 141.

ally far ahead of his time, but did not suffer gladly those whom he regarded as fools. On the other hand, Peirce had beliefs regarding religion and politics that were controversial, to say the least. For example, he was rumoured to be an atheist. And like his father, he defended the enslavement of Afro-Americans – a stance that, beyond entailing a murderous racism, presumably reflected his class position as the US equivalent of a British Tory, viz., a so-called 'Boston Brahmin'. But of course, Peirce lived in a part of the United States that otherwise opposed slavery, and so this stance may well have had costs for him in a way that it would not have had for a 'southern gentleman'.

As if this combination of ideology and failed everyday politics were not enough, from his late teens onwards Peirce suffered from an excruciatingly painful illness known today as Trigeminal neuralgia. As a result, he was forced to self-medicate with morphine and other highly addictive drugs, and these may well have contributed to a string of nervous breakdowns. Brent goes so far as to suggest that Peirce's alternation between such breakdowns and periods of extraordinary productivity are evidence of a bipolar personality disorder.[11] I myself wonder if Peirce's evident difficulties in relating to other people, except on strongly intellectual grounds, are not evidence of an Aspergers-like expression of the Autistic Spectrum Disorder.

Such speculations to one side, even as Peirce was dismissed from Johns Hopkins in 1884, his 30-year career at the Coast Survey began to collapse. One reason for this may be that in the early 1880s Peirce lost the key people that helped him to look after his professional interests. In 1880 Benjamin Peirce died, and was followed in 1881 by Superintendent (director) Carlile Patterson, who was Peirce's champion at the Coast Survey. In 1883, Peirce finalised the divorce with Melusina Fay Peirce, and so severed another vital relationship.

From this perspective, it is probably no coincidence that soon after the end of the mentioned relationships, Peirce's job at the Coast Survey itself came under attack as part of a controversy involving financial and other scandals at the institution. In 1885 a congressional report publicly censured Peirce even as it forced leading figures at the Survey to resign from their posts. Here again, Peirce did not help matters by repeatedly delaying the completion of technical reports that he owed the Survey – reports which in some cases he had undertaken to finish several years earlier. In 1891, at the age of 52, Peirce was himself finally forced to resign from the Survey, and he was never able to gain employment at any other institution, though certainly not for want of trying.

The combination of a lack of employment and Peirce's profligate lifestyle led to growing poverty. Indeed, Brent notes that by 1897, Charles and Juliette Peirce were near starvation and were forced to steal food. Throughout the 1890s and 1900s, Peirce tried his hand at a series of get-rich-quick schemes even as he continued to publish short pieces for dictionaries and for magazines, and worked oc-

---

[11] Joseph Brent, *Peirce: A Life*, pp. 40-41.

casionally as a consultant chemical engineer. Thanks to the support of the philosopher William James, Peirce's long-term friend and benefactor – but ironically perhaps the scholar whose own highly popular and radically empiricist philosophy did the most to undermine Peirce's – Peirce was also able to present occasional lectures at Harvard and other institutions.

The get-rich scheme that may well have sealed Charles and Juliette's financial fate was the Peirces' use of the inheritance bequeathed by Benjamin Peirce to purchase land in rural Pennsylvania, where they built a grand country home that they named *Arisbe*. They hoped to transform *Arisbe* into a kind of intellectual resort, in today's terms a boutique hotel for academics. The plan never came to fruition, and instead squandered Peirce's inheritance even as it generated growing debt and unpaid property taxes. By the time that Charles died in 1914 at his beloved *Arisbe*, the Peirces had come to depend entirely on the charity of Charles' friends (especially William James), his family, neighbours, and even a local baker who handed over bread he couldn't sell.

In this context, it is a tribute to Peirce's scholarly *askesis* that he could accomplish as much as he did, in so many fields. Indeed, his most productive writing, certainly from the point of view of semiotic theory, was completed from the 1890s onwards, arguably the most traumatic period of his life. One can only wonder what would have happened if Peirce had kept his position at Johns Hopkins.

As I began to note earlier, today Peirce has been restored to the pantheon of the great philosophers thanks to the work of a series of scholars who have found and edited his voluminous work. Even this task almost came to naught in the five decades after Charles died. Juliette, who survived Charles by 20 years, sold Charles' papers and books to the Harvard philosophy department as soon as Peirce died. There, Josiah Royce, one of Peirce's most avid fans, set about editing the papers with the help of an assistant, W. Fergus Kernan, who by all accounts did a remarkable job of organising a chaotic collection. Unfortunately, when Royce himself died, the department assigned the editing task to a succession of individuals who in some cases undermined the work begun by Royce and Kernan. It is rumoured that a significant number of documents were either lost or stolen.[12] Whatever the validity of such claims, no doubt the very late recognition of Peirce's brilliance owes much to the Harvard philosophy department's disinterested, to not say chaotic management of the collection after Royce's death.

Nathan Houser, one of the leading members of the collective that began work to re-edit Peirce's texts in the 1980s, noted in 1989 that, of the 1200 or possibly many more of Peirce's books that Harvard received, fewer than 25 were included

---

[12] For an account of the history of the editing process, see Nathan Houser, 'The Fortunes and Misfortunes of the Peirce Papers' in Michel Balat and Janice Deledalle-Rhodes (Eds.) *Signs of Humanity*, Vol. 3 (Berlin: Mouton de Gruyter, 1992), 1259-1268. Online version available at < http://www.iupui.edu/~arisbe/menu/library/aboutcsp/houser/fortunes.htm>[Accessed 28 February 2018].

in the dedicated Peirce collection at the Houghton Library. Most of the rest remained, at least by that date, in the open stacks of Harvard's library system, where in Houser's words, 'they may be borrowed and carried away by patrons that who may be entirely oblivious that the marginal comments are by one of the world's great thinkers'. The books included Peirce's sixteen volume collection of the writings of Galileo, his copy of Bayle's Dictionary, the *Opus Majus* of Roger Bacon (in two volumes), and Schelling's *Transcendental Idealism*.[13] Thrilling as the prospect of coming across such books might be to some Peirce fans, it quite literally speaks volumes about the fate of Peirce's life – and about philosophy as late as the 1980s at Harvard University – that, more than a *century* after his social and economic fortunes began to decline, one of the world's foremost universities still had not bothered to devote the resources and the energy required to preserve the work of one of its greatest alumni.

---

[13] Nathan Houser, 'The Fortunes and Misfortunes of the Peirce Papers'.

PART 3

Assemblages of Wildlife Observation
Sociological-Philosophical Perspectives

After an especially dry January and February, the waters of the Gatún Lake are lower than usual, exposing the rocks and the soil that give Barro Colorado Island its name (in Spanish, *barro colorado* means red clay). As we walk along one of paths that cross the Smithsonian Tropical Research Institute's (STRI) field station, trying not to step on leaf-cutter ants (*Atta spp.*), our guide bids us to stop, and points to a spot along the shoreline, some 20 or so metres below the path. Lying quietly along the water's edge is a specimen of the American crocodile (*Crocodylus acutus*), a species that, as I noted in Part 2, can grow to over four meters in length and weigh close to a tonne. This specimen is over three meters long.

The visitors participating in my research and I are lucky twice over: lucky to see the specimen in fairly close proximity, but also lucky that, in so far as the Gatún Lake's water levels are especially low, only a couple of centimetres of water cover the rocks on which the crocodile is resting. This means that we can observe the creature's entire body, as opposed to the more commonly seen head protruding above the waterline (as per the picture opposite, which was taken during another tour). Alas, as is often the case with encounters of this kind, its joy is short-lived. Almost as soon as we stop to get a better look and to take pictures, the crocodile turns and disappears into the green depths of the Gatún.

Amid the excited commentary that follows, one US visitor's response to the encounter gives a good sense of the aspects of wildlife observation that I would like to consider in this part of the volume. After the crocodile vanishes, the visitor turns and muses, half to himself, but also to those standing nearest to him: 'It's amazing to see the wildebeest jumping off the cliffs… the crocs just wait for 'em, they're so huge…'. In the course of the ensuing discussion, it becomes apparent that the tourist is referring to the Nile crocodiles (*Crocodylus niloticus*) in eastern Africa, which wait in ambush for the common wildebeest (*Connochaetes taurinus*) and other ungulates that cross the Mara River during their seasonal migration across the Serengeti plain. It also becomes clear that the visitor himself has no firsthand experience of that region or of its crocs; instead, he's seen the crocs and the wildebeest on television. Long before 2007 (which is when the above encounter took place), scenes of large Nile crocodiles attacking wildebeest had become one of the signature scenes of wildlife TV, and clearly, the scenes made a lasting impression on this visitor.

In Part 2, I explained how and why signs play a key role in wildlife observation. The encounter I have just described might well be used to make much the same point: if the visitor interpreted the crocodile resting along Barro Colorado's shoreline with reference to *C. niloticus*, it was because he had seen signs of the latter crocodilian species on TV, and no doubt interpreted them as more or less purely indexical signs which he now transposed to Barro Colorado. The tourist was, in this sense, engaging in a textbook example of what I described in the General Introduction as *media-based dynamics of transmediation* – the process whereby someone interprets something by transposing a media representation, or an aspect of such a representation from the proximal context of everyday media use to a context such as the one described above.

In this selfsame semeiotics, as in the encounter more generally, we nevertheless find at least two sets of reasons for moving beyond semeiotic analysis, *sensu stricto*. First, if the tourist in question *did* interpret the crocodile on Barro Colorado with reference to the crocodiles in the TV series, it was because *organisations* such as the BBC or the National Geographic Society produced and circulated the signs of *C. niloticus* attacking wildebeest. Moreover, if the tourist was able to visit Barro Colorado, and to interpret a specimen of *C. acutus* in a biological reserve, it was thanks to the intervention of several other kinds of organisations: tour operators, airlines, hotels, and park authorities (in this case STRI, the scientific institution which acted as the custodian of the Barro Colorado biological reserve). In their different ways, all these would have played a key role in enabling certain kinds of observational experiences and practices.

Then again, and second, if the visitor did watch TV, and did travel for leisure purposes, it was because he partook in practices linked not just to organisations, but to broader social institutions and cultural formations. If, for example, the visitor travelled from the United States to Panama, it was to engage in tourism. In a similar vein, if the visitor was indeed influenced by media representations, it was because he engaged actively in/with the process of mass communication, itself a cultural formation that goes far beyond the individual organisations that generate media contents. The same kind of point might be made about wildlife observation itself: however much particular organisations were involved, the decision to engage in the practice reflects an itself cultural valorisation of wild animals, and of the practice of observing such animals *in situ*.

The more general point that I am making is this: however much semeiotic processes are involved in – and indeed make possible – the observation of wild animals, any theoretical approach to wildlife observation must also consider wildlife observation as a practice that is mediated by particular institutions, and cultural formations. The idea is not to oppose this kind of analysis to the semeiotic, but on the contrary to link the two, and to this end, in this chapter I will introduce a form of inquiry that does just that: cultural sociology. I should clarify straight away that, as I will use the term, cultural sociology refers not just to the research conducted by those who would describe themselves as such (or indeed as sociologists of culture), but also to the theories that many such researchers have often drawn on. The theories in question include the kind produced by social and symbolic anthropologists, social semioticians and critical discourse analysts, as well as by philosophers who have made it their business to explain the interrelation between signs (or what might also be described as symbolic forms), codes, social institutions and cultural formations as materialised in particular forms of everyday practice.[1]

Important as such a perspective is in elucidating wildlife observation regarded as a practice, it raises some issues which the rest of the chapters that make up this part of the study will problematise. The issues include what I will characterise as a one-way conception of the relation between cultural practice and embodiment; a tendency to adopt a culturalist, and 'identitary' logic when explaining all manner of practices; a tendency to adopt an organicist perspective when explaining the interrelation of elements in more or less complex ensembles; and last but certainly not least, a tendency to exclude from the analysis all manner of beyond-human formations, and more specifically, beyond-human animals.

So while Chapter 8 will introduce a cultural sociological perspective on wildlife observation, Chapter 9 will offer a critique of that perspective's approach to embodiment – a critique that starts from the feminist proposals of Toril Moi and Simone de Beauvoir, and then considers in some detail the phenomenology of Maurice Merleau-Ponty. Chapter 10 will then turn to the philosophy of Gilles Deleuze, and of Deleuze and Félix Guattari in order to consider the complex interrelation between individuals, affects, and assemblages. The aim will be to develop an account of what might be termed a bio-cultural (short for bio-socio-cultural) conception of wildlife observation.

---

[1] I myself am not primarily a cultural sociologist, and so I hope that will allay any concerns that I am engaging in a kind of academic imperialism; terms such as social semiotics, symbolic anthropology, or communication and culture studies might also have been employed, but I find that cultural sociology comes closest to representing this section's point of departure.

8
Cultural Practice

Sociology as a scientific field of inquiry was founded by Auguste Comte (1798-1857), who conceived it as a *physique sociale*, i.e. a mode of inquiry driven by the kind of positivist philosophy and politics that Comte advocated. To this day, many sociologists still engage in what is, in effect, a version of the 'policy science' that Comte advocated. The sociologists in question employ positivist research methods to determine the causes of this or that *behaviour*, in order to propose policies designed to eliminate, or at least curb, one or another *social disfunction*. A case in point, the kind of research conducted by many criminologists.

However, as I began to explain in Chapters 1 and 2, in the second half of the twentieth century (in some cases long before), many scholars on both sides of the Atlantic began to reject research based on what I described with Leszek Kolakowski as the rules of phenomenalism, nominalism, objectivism, and consilience. US and French colonialism in the postwar period constituted an important motivation for the rise of an anti-positivist sentiment amongst many scholars and activists around the world. But the events that occurred in France itself during May 1968 also played an important role in the rise of the more critical Continental philosophies. In that famous month, students and workers in Paris and elsewhere in France arose to reject the war in Vietnam, and more generally, the capitalist order which made that war possible. The uprising was so successful that it nearly brought down the government of President Charles de Gaulle; de Gaulle is even said to have briefly fled France at the height of the protests. The fact that the 'revolution' could occur at all – but also that it ultimately failed to secure at least a conventional regime change – set many scholars in France and beyond thinking about the nature of social and cultural relations in advanced capitalist societies.

As they searched for critical alternatives, many sociologists (as well as anthropologists, linguisticians and semioticians, and to a lesser extent psychologists and psychoanalysts) turned to the work of thinkers who had been either forgotten, or were dismissed by the headlong embrace of positivist and behaviourist forms of research. An account of the alternatives that were found is far beyond the scope of this chapter. In the following pages I would nevertheless like to give some sense of the profound intellectual reorientation that led to the emergence of what I described earlier as cultural sociology, and with it, the critical social sciences and humanities more generally.

Chapter 8 Cultural Practice

To begin this account, it is convenient to mention one particularly influential figure who many of the critical thinkers turned to: Max Weber (1864-1920),[1] another of the great founders of modern sociology, but a scholar who proposed a very different approach from the one developed by Comte. According to Weber,

> Sociology (in the sense in which this highly ambiguous word is used here) is a science concerning itself with the interpretive understanding of social action and thereby with a causal explanation of its course and consequences. We shall speak of "action" insofar as the acting individual attaches a subjective meaning to his behavior – be it overt or covert, omission or acquiescence. Action is "social" insofar as its subjective meaning takes account of the behavior of others and is thereby oriented in its course.[2]

By this account, action not only involves meaning, but an orientation to others, i.e. it is inter-subjective. When Weber refers to meaning, he does not mean 'correct' meaning, or meaning that is true in some metaphysical sense; he refers instead to the 'the actual existing meaning in the given concrete case of a particular actor, or to the average or approximate meaning attributable to a given plurality of actors'.[3]

As far as Weber is concerned, sociologists – and many after him will argue critical social scientists more generally – must interpret such meaning by way of what Weber describes as 'ideal types'. Here 'ideal' refers not to the connotation of perfection, but to the sense of *idea*: the sociologist must proceed by classifying certain elements of social action in order to produce types that explain the action. This kind of process is unavoidable in social inquiry (and no doubt in *all* kinds of inquiry), and explains why sociology is ineluctably interpretive, no matter what philosophy or epistemology is invoked: 'An ideal type is formed by the one-sided accentuation of one or more points of view and by the synthesis of a great many diffuse, discrete, more or less present and occasionally absent concrete individual phenomena, which are arranged according to those one-sidedly emphasized viewpoints into a unified analytical construct…'[4] These proposals established an early precedent for the *hermeneutic* turn in the critical social sciences (here as

---

[1] See for example, Max Weber, *Economy and Society: An Outline of Interpretive Sociology*, edited by Guenther Roth and Claus Wittich, translated by Ephraim Fischoff, Hans Gerth, A. M. Henderson, Ferdinand Kolegar, C. Wright Mills, Talcott Parsons, Max Rheinstein, Guenther Roth, Edward Shils, and Claus Wittich (Berkeley: University of California Press, 1968).

[2] Max Weber, *Economy and Society*, p. 4.

[3] Max Weber, *Economy and Society*, p. 4.

[4] Max Weber, *On the Methodology of the Social Sciences*, translated and edited by Edward Shils and Henry A. Finch (New York: Free Press, 1949), p. 90.

elsewhere in this book, the critical social sciences are those opposed to positivist philosophy).

The hermeneutic is the first of three at once theoretical and methodological turns which I will consider in this chapter. The etymological roots of 'hermeneutic' are to be found in the Greek verb *hermemeuein*, usually translated as 'to interpret'. The term may be linked back to the Ancient Greek myth of Hermes, the messenger of the gods who had the thankless task of translating into everyday terms the meanings of the gods – meanings that, in and of themselves, were incomprehensible to mere mortals. Today scholars use the term hermeneutics to refer to a body of philosophical investigations concerned with developing theories of interpretation, or what the philosopher Paul Ricoeur (1913-2005) defines as 'the theory of the operations of understanding in their relation to the interpretation of texts'.[5]

As I will use the term, the notion of a hermeneutic turn describes the shift from positivist philosophy towards forms of research that are reflexive with respect to the role of symbolic forms – and this not just in the analysis of the production of such forms in everyday life, but in the consideration of the methodology of the research itself. From this perspective, the hermeneutic turn may be regarded as a result of the influence of hermeneutic philosophy, *sensu stricto*, but also of fields such as structuralist linguistics, semiology, and symbolic anthropology; analytic forms of philosophy of language, including Speech Act Theory (one of whose leading scholars, John Searle, I referred to in Part 1); forms of sociology influenced by Max Weber and a long line of post-Weberian scholars; and then again, the many interdisciplinary, and indeed almost *anti*-disciplinary forms of research that emerged between the 1960s and early 1980s, with a specially prominent role played by the first wave of what became known as Cultural Studies. What all of these have in common is an acknowledgement that symbolic forms play a constitutive role in social practice in general, and in the activities of researchers in particular.

This point requires further elucidation. To begin with, one may understand the 'symbolic' in 'symbolic forms' as a reference to the importance not just of *symbolism*, but of social conventions that are themselves meaningful. In cultural sociological inquiry, the notion of symbolic forms thereby has a much wider reference than that of linguistic forms, or any explicitly inscribed modes of communication. It includes all objects and forms of action that entail meaning-making. The sociol-

---

[5] Paul Ricoeur, 'The Task of Hermeneutics', in *Paul Ricoeur: Hermeneutics & the Human Sciences*, edited and translated by John B. Thompson (Cambridge: Cambridge University Press, 1981), 43-62, p. 43. For more general introductions, see for example Richard E. Palmer, *Hermeneutics: Interpretation Theory in Schleiermacher, Dilthey, Heidegger, and Gadamer* (Evanston, Ill: Northwestern University Press, 1969); and John B. Thompson, *Critical Hermeneutics: A Study in the Thought of Paul Ricoeur and Jurgen Habermas* (Cambridge: Cambridge University Press, 1981).

ogist John B. Thompson defines symbolic forms as 'a broad range of actions and utterances, images and texts, which are produced by subjects and recognized by them and others as meaningful constructs. Linguistic utterances and expressions, whether spoken or inscribed, are crucial in this regard, but symbolic forms can also be non-linguistic or quasi-linguistic in nature'.[6]

In this study, I will prefer to use the concept of the sign as defined in Part 2 with Charles Peirce. Here it should nonetheless be acknowledged that, via the concept of symbolic forms, cultural sociologists foreground the interpretive challenges that emerge when it is understood, as Weber proposed, that action is fundamentally intertwined with meaning. As Thompson puts it, 'the object domain of social-historical inquiry is not only a concatenation of objects and events which are there to be observed and explained: it is also a subject domain which is made up, in part, of subjects who, in the routine course of their everyday lives, are constantly involved in understanding themselves and others, in producing meaningful actions and expressions produced by others'.[7]

In so far as this perspective is accepted, it becomes clear that it is naive to adopt the kind of phenomenalism critiqued with Kolakowski in Chapter 1. It is also naive to assume that the researcher can secure a critical distance from whatever is being investigated by adhering to the positivist rule of objectivism. From a hermeneutic perspective, the problem is not simply to achieve a *bonne distance* from what is researched; it is to arrive at a more critical understanding of the own social and cultural investment in whatever is being researched, and the methods used to research it. Whatever else this requires, after the hermeneutic turn cultural sociologists will argue that it entails the explanation and interpretation of the manifold roles that signs (or symbolic forms) play in everyday cultural practice. Signs work to determine not only how we *interpret* the world – perhaps an uncontroversial, if often overlooked verity – but how the social world itself is organised.[8]

Thanks in part to the hermeneutic turn, cultural sociology adopts what is, broadly speaking, a *semiotic* conception of culture, and cultural practice. To more fully understand such a perspective, it is necessary to consider a second turn: what may be described as the *cultural* turn in the critical social sciences. Whereas Weber spoke of *social action*, by the last quarter of the twentieth century, cultural sociologists and the scholars whose work they drew on more typically referred to *cultural practice*. One reason for this shift lay in the research of twentieth century anthropologists and ethnographers, who showed sociologists why the cultural

---

[6] John B. Thompson, *Ideology and Modern Culture* (Cambridge: Polity, 1990), p. 59.

[7] John B. Thompson, *Ideology and Modern Culture*, p. 21.

[8] For an account of this kind of perspective, see for example Ernst Cassirer, *An Essay on Man: An Introduction to a Philosophy of Human Culture* (New Haven, CT: Yale University Press, 1944), or Nelson Goodman, *Ways of Worldmaking* (Indianapolis, Indiana: Hackett, 1978).

ought to be regarded as a key dimension of any form of social action, and by implication of its analysis. To explain how and why they did so, a brief explanatory excursus is required that provides some indication of the changing meanings of the concept of culture.

Raymond Williams, himself a historian and a sociologist of culture, provides a useful genealogical account of the concept, which as Williams notes, has a long history of changing meanings.[9] Amongst the earliest meanings of culture we find the tending of something in an agrarian context (e.g. crops and animals). This connotation was later extended by European aristocrats to refer to 'the culture of minds', and with it to the explicitly class-bound notions that emerged in the eighteenth century, i.e. culture as that which is 'cultivated', as in 'High Culture'. Yet Williams notes that in that same century, the philosopher Johan Gottfried von Herder railed at the arrogance of the defenders of a notion of a superior European culture (as represented especially via the concept of *civilisation*), suggesting that it was necessary to speak of cultures in the *plural*: 'the specific and variable cultures of different nations and periods, but also the specific and variable political and economic groups within a nation'.[10] It was a version of this understanding that would inform the rise of the more critical forms of anthropology, and which continues to determine some of the predominant everyday meanings of the notion of culture to this day.

The importance of Herder's approach notwithstanding, in the second half of the twentieth century, cultural sociologists began to adopt not only the kind of pluralistic understanding of culture defended by Herder, but an understanding that widened the frame to speak of something like a 'whole way of life',[11] in T.S. Eliot's evocatively idealist expression, or the equally evocative, but distinctly materialist 'whole social order' suggested by Williams himself.[12] As Williams explains, for cultural sociologists (or as he would have described them, sociologists of culture), the analysis of culture must be an analysis of cultural *practice*, i.e. a form of analysis that starts from an understanding that involves an amalgam of the two mentioned connotations (idealist and materialist), but engages in an important theoretical turn when it recognises that cultural activities are not simply *derived* from an established order, as implied by structuralist approaches, or indeed by economicist versions of Marxist theory (e.g. the Marx of *The Capital*). Everyday culture is made and remade via practices that are themselves constitutive of the order. To speak of cultural practice is thereby to refer to an activity that

---

[9] Raymond Williams, 'Culture', in *Keywords: A Vocabulary of Culture and Society*, revised edition (London: Fontana Press, 1983), pp. 87-93.

[10] Raymond Williams, 'Culture' in *Keywords*, p. 89.

[11] Thomas Stearns Eliot, *Notes Towards the Definition of Culture* (London: Faber, 1962 [1948]), p. 31.

[12] Raymond Williams, *Culture* (London: Fontana Press, 1981), p. 12.

is not only the *bearer* of the meanings and mores of an established order, but one that also *produces* or on occasion *contests* such an order.

In Chapter 1, I employed a famous analogy developed by the anthropologist Clifford Geertz to explain what's wrong with the rule of phenomenalism. Here it might be noted that Geertz is one of the earlier advocates of an approach to culture that foregrounds the importance of symbolic forms. Geertz famously invokes Max Weber to propose what he describes as a semiotic theory of culture: 'Believing, with Max Weber, that man [sic] is an animal suspended in webs of significance he himself has spun, I take culture to be those webs, and the analysis of it to be therefore not an experimental science in search of law but an interpretive one in search of meaning'.[13] Geertz's definition establishes a particularly clear and close interrelation between what I am describing as the hermeneutic and the cultural turns; the interpretation of culture requires a hermeneutic turn, but equally, any hermeneutic (in the sense of a mode of interpretation) must itself be cultural in so far as social groups interpret their worlds from the perspective of culturally specific 'webs of significance'. Raymond Williams materialises this kind of approach when he defines culture as the '*signifying system* through which necessarily (though among other means) a social order is communicated, reproduced, experienced, and explored'.[14]

Building on Williams' work, the literary theorist and cultural critic Terry Eagleton suggests a way of specifying the concept of culture in a manner that is particularly pertinent to this study. Culture, Eagleton argues, must (also) be understood as a *collective unconscious:* 'that vast repository of instincts, prejudices, pieties, sentiments, half-formed opinions and spontaneous assumptions which underpins our everyday activity, and which we rarely call into question. In fact, some of these assumptions run so deep that we probably could not query them without some momentous change in our way of life, one which would make them fully perceptible for the first time'.[15]

No doubt Eagleton is right to suggest that some of the assumptions do run 'so deep' that they are unconscious in the strongest sense of this expression. In this study, I will nevertheless assume that it is equally appropriate to speak of culture as a collective *unselfconscious*, where the unselfconscious refers to all those aspects which individuals or social groups are likely to be unaware of thanks to their more or less implicit character, and thanks to the fact that they have become so much a matter of *habit* that they tend to elude reflexivity and self-reflexivity,

---

[13] 'Thick Description: Toward an Interpretive Theory of Culture', in *The Interpretation of Cultures* (New York: Basic Books, 1973), 3-30, p. 5.

[14] Raymond Williams, *Culture*, p. 13. Italics in the original.

[15] Terry Eagleton, *Culture* (London: Yale University Press, 2016), p. 49.

which is to say, a questioning or self-analysing disposition.[16] This definition is not meant to suggest that the unselfconscious always remains that, or can never be questioned; on the contrary, unlike the *unconscious*, the unselfconscious can be, and often *is* questioned even without the help of psychoanalysis – for example, when circumstances change, or when people interrupt their own practice in order to think more critically about what they do. From this perspective, the concept of the unselfconscious is meant to make clear that much of what one does – and in this study, much of what one observes, and equally if not more importantly the *way* that one observes something – involves a logic which cannot be adequately explained by way of a wholly voluntaristic account of practice.

In later chapters we will see that Williams' qualification that signifying systems are *one* among other means of communication is an important one, and indeed I will explain that, over time, empirical, and one might even say beyond-semiotic aspects of cultural practice have come to preoccupy more and more scholars (cf. Chapters 9 and 10, below, and also Chapters 13 and 14 in Part 4). From this perspective, an account like Eagleton's puts too much emphasis on the 'ideational', or on 'subjectivity', however much it recognises that both are ineluctably linked to relations of production, and to cultural materialities more generally.

Implicit to the account thus far is a third turn, a third displacement which I will describe as the *cultural-materialist* turn. The turn in question actually involves multiple theoretical and methodological shifts; however, for the propaedeutic purposes of this chapter, it will suffice to unify them under one heading. Simplifying somewhat, cultural materialism suggests that it is necessary to link what Geertz describes as 'webs of significance' to the *material* conditions which shape how subjects or actors engage in all forms of practice. However, at least in the context of the earlier cultural sociology, 'material' refers not so much to the physical properties of things, but to the notion that all forms of practice reflect, however directly or indirectly, particular relations of production, or more broadly, what Williams describes as *social orders*.

Cultural sociology owes much in this sense to the writings of the 'young' Marx. For many researchers in this field, as for the so-called New Left more gen-

---

16 Readers familiar with the Freudian notion of the *preconscious* will see a similarity with the concept of the unselfconscious (See for example, Sigmund Freud, 'The Unconscious', translated by Cecil. M. Baines, in Philip Rieff, *General Psychological Theory: Papers in Metapsychology, Sigmund Freud* [New York, Colliers, 1963(1915)], 116-150.) I distinguish between the two as follows: the preconscious is entirely open to attention; access to the unselfconscious requires more sustained reflection. Someone may unthinkingly put a key in the door, their mind far away, focussed on other things. When asked, 'Did you put the key in the door?', the answer will automatically be yes of course. When, by contrast, someone says that they are going out to enjoy nature, they may be asked if they mean that nature is only really only found outdoors, or in a 'wild place'. The answer is bound to require rather more thought. The unselfconscious is thus something that lies between the preconscious and the unconscious.

erally, the most valuable legacy of Marx, philosophically speaking, is not to be found in *The Capital* (1867) – a work that reflects Marx's famous turn towards economic determinism – but in earlier works such as *Critique of Hegel's "Philosophy of the Right"*(1843), *Economic and Philosophic Manuscripts of 1844* (1844), and *The German Ideology* (1845). [17] It is via this earlier Marxist writing, and via the work, published a century or so later by the Frankfurt School (including philosophers such as Theodor Adorno and Walter Benjamin), and at a later stage by cultural theorists such as Pierre Bourdieu, Judith Butler, Michel Foucault, Stuart Hall, Donna Haraway, and Raymond Williams (to mention just some of authors referenced in this work), that researchers in the broad field of cultural sociology find ways of materialising practice along critical cultural lines. The key insight is that all cultural forms reflect, however complexly and indirectly, the interests, the social positions, the contradictions, the asymmetries in power, but also the ambiguities and the heteroglossic nature of cultural formations. Any analysis of cultural forms, including the analysis of symbolic forms or signs, must thus reveal the workings of the aforementioned aspects.

What, then, can a cultural sociological perspective such as I have begun to describe contribute to a study of wildlife observation amongst tourists? Wildlife observation may be regarded as a cultural practice in so far as the social actions that it entails both produce, and reproduce the 'signifying systems' that Williams refers to, and more concretely, the 'instincts, prejudices, pieties, sentiments, half-formed opinions and spontaneous assumptions' that Eagleton invokes. In the case of wildlife observation amongst tourists, this is perhaps most clearly the case in so far as it is assumed that some animals are wild while others are not, that some places are wildernesses while others are not, and that both (the wild animals and the wildernesses) have such an inherent value as to justify a leisure practice organised around viewing the animals and their habitats. Much the same point can be made about tourism more generally: it may seem that travelling and engaging in sightseeing purely for the purposes of leisure are self-evidently 'good things'. But of course, they, like wildlife observation, are contingent not just on certain cultural valorisations, but on both specific and generic social institutions, and with them, certain social orders – not least, the late capitalist opposition of work and leisure time, work and leisure spaces.

Having offered a very general introduction to cultural sociology, I would now like to offer a more precise account of the contribution that such a perspective can make to this study by describing several key concepts which are particularly germane to the present investigation. The concepts in question are those of organisation, consumption, field, habitus, code, discourse, genre, and cultural formation. In their different ways and levels, these concepts enable the researcher to

---

[17] For a useful if hermeneutically-inclined account of the earlier Marxist theory, see for example Paul Ricoeur, *Lectures on Ideology and Utopia*, edited by George H. Taylor (New York: Columbia University Press, 1986).

better explain how it is that observation comes to be instituted. Indeed, taken together, the following conceptualisations can serve as the basis for an account of what might be termed the cultural institution of wildlife observation. By this I mean the process whereby wildlife observation is generated as a specific kind of practice, and given forms that reflect, however complexly, the interests, values and 'world views' associated with specific and generic institutions.

Organisations

Let's begin with a deceptively simple aspect of wildlife observation amongst tourists: as I began to explain at the start of this chapter, wildlife observation is enabled, but also regulated, by institutions. To start with, institutions may be regarded as actual *organisations*, viz., groups of people working to complete particular tasks according to some form of division of labour, with a more or less hierarchical set of relations, and with certain rules and resources. The cultural institution of wildlife observation goes far beyond organisations, but certainly organisations in the strict sense of the word tend to play an important role in the context of wildlife observation engaged by people on holiday, i.e. by tourists.

This volume has offered several examples of encounters between wild animals and tour groups visiting Barro Colorado Island, and those tour groups offer a first way of illustrating the significance of organisational activity to wildlife observation. It may be noted that when one observes wildlife as part of a tour group, it is several sets of eyes, ears, and noses – several bodies – that work to look out for wildlife. In so far as the group is led by an experienced guide who requires the group to adhere to certain rules, then the observational practice may also come to be regulated, as well as hierarchically structured. Of course, the organisation of tour groups can vary significantly, but it is fair to say that, whatever the specifics of each group's practices, there will be a *collective* observational process for reasons that are both quantitative and qualitative: there will be multiple observers, but there will also be a form of interaction that is itself social. A guide will offer information in accordance with a certain knowledge, and certain codes of communication, and both aspects may then both prompt and structure interventions from the rest of the visitors. This being the case, the tour group, regarded as a micro-organisation, will constitute a space for discursive mediation, and for a more or less heteroglossic exchange. Simplifying somewhat, heteroglossia refers to the presence of multiple voices, individual or discursive, in a certain text or context.

In Volume 2, there will be the opportunity to analyse tour groups and their interactions in some detail. For now I would like to consider the roles of larger institutions, and especially, of institutions that affect wildlife observation from afar. For every tour group that traverses the trails of tropical forests, there are likely to be numerous additional organisations that will have played a role in enabling and/or regulating the tour. The following list includes some of the more salient kinds of organisations:

1) *Representational:* organisations that work to represent tropical forests (or indeed any other biome), and which in so doing alert audiences of the existence those forests, and to generate expectations as to the kinds of observational experiences they may afford. Examples: the nature media, and those that produce travel shows or magazines (e.g. the BBC, the National Geographic, Conde Nast, etc.).

2) *Promotional*: organisations that work to promote particular forests, or parts of forests *as tourist destinations.* Examples: tour operators and any dedicated websites, but also any park authorities that promote their own forests as visitor destinations.

3) *Travel Logistics and Support:* organisations that provide the physical means with which to reach the destination, and then accommodation once the tourists reach the destination (if any is required). Examples: airlines and hotels.

4) *Regulatory*: organisations that work to generate, and/or to enforce any regulations that apply to the visiting practice. Examples: Ministries of tourism, park authorities, visitor programmes.

This list is neither exhaustive, nor made up of mutually exclusive categories. On the contrary, one of the characteristics of tour operators has long been to sell 'package tours' in which/for which one organisation engages in the first three types of activity (representation, promotion, and travel logistics), and more often than not lobbies authorities devoted to the regulation of tourism, or even generates its own regulations. In the case of Barro Colorado, by 2007 one such modality of visiting the island was the one offered by Lindblad Expeditions, which took groups of 'upmarket' tourists on live-aboard vessels to Barro Colorado and to other destinations in Panama and Costa Rica. Lindblad described itself as an organisation devoted to 'adventure travel' which had teemed up with the National Geographic Society to sell what both institutions described as Lindblad-National Geographic *expeditions* – 'the world's *ultimate, authentic expedition experience*'.[18] The actors of this modality of tourism deliberately blurred the boundaries between tourism and one nature media's discourse in order to offer what was a more or less explicit conflation of arm-chair and actual travel to sites deemed to be of outstanding ecological significance. There was thus an integration of representation, promotion, and logistics, as well as the generation of a kind of self-contained reg-

---

[18] Lindblad Expeditions/National Geographic, 'Why Lindblad-National Geographic?', at <http://www.expeditions.com/why-us/> [Accessed 5 March 2016]. Italics in the original text.

ulatory context, embodied in the do's and don'ts of living aboard one of the Lind-blad Expeditions' ships.

These examples to one side, in general it may be suggested that wildlife obser-vation amongst tourists is likely to reflect the interventions of multiple organisa-tions, no one of which will control the entire visiting process (and this even after acknowledging that of course each individual visitor will have a degree of auton-omy as individual). In such a multi-organisational context, some organisations are likely to exercise a much greater influence over one or more aspects of the overall visiting dynamic. In keeping with this, some are likely to have greater conse-quences for the observational process *per se*. This is particularly true for media organisations, and/or any organisations that produce representations of forests for the benefit of audiences, at least some of which may go on to become first-time visitors in those forests (or kinds of forests). As I began to explain in the introduc-tion to this volume, my ethnographic research suggests that the nature media play a particularly important role when it comes to wildlife observation amongst tourists. For this reason, I would like to analyse this role in some detail in this volume; an analysis of other kinds of organisational roles will be offered in Vol-ume 2.

To begin with, it may be noted that the media in general may play a significant role in virtually all kinds of tourism. This is hardly surprising in so far as the phe-nomenon of mediazation now plays such an important role in so many spheres of modern cultural practices. With the sociologist John B. Thompson, we may define mediazation as 'the general process by which the transmission of symbolic forms becomes increasingly mediated by the technical and institutional apparatuses of the media industries'.[19] If mediazation plays a particularly important role in poli-tics and in generating a consumer society, it also may shape tourist practices. The sociologists John Urry and Jonas Larsen suggest that what they describe as the *tourist gaze* is 'largely preformed by and within existing mediascapes';[20] in this context, 'much tourism becomes, in effect, a search for the photogenic'.[21] There is, as the authors put it, a hermeneutic circle; what is often sought in a holiday is 'a set of photographic images which have already been seen in brochures, TV pro-grammes, and social networking sites'.[22] This account is in keeping with what I explained in the introduction: mediazation may also influence the forms of obser-vation that many tourists expect to engage in whilst visiting tropical forests.

Urry and Larsen point to one way in which the media may exercise a certain influence over wildlife observation: by producing and circulating a certain set of

---

[19] John B. Thompson, *The Media and Modernity: A Social Theory of the Media* (Cambridge: Polity Press, 1995), p. 46. See also John B. Thompson, *Ideology and Modern Culture*, p. 3-4.

[20] John Urry and Jonas Larsen, *The Tourist Gaze 3.0* (London: Sage, 2011), p. 179.

[21] John Urry and Jonas Larsen, *The Tourist Gaze 3.0*, p. 178.

[22] John Urry and Jonas Larsen, *The Tourist Gaze 3.0*, p. 179.

images, the organisations establish, in effect, not just a 'mediascape', but with it, a set of criteria with which to imagine a destination, and imagine oneself visiting that destination. As suggested in the Prologue, often the images that are produced by the nature media are so intense, and so idealise tropical forests that the criteria in question generate what are often unattainable expectations amongst first-time visitors. To a visitor 'on the ground', no tropical forest looks like the one depicted by *Planet Earth II*.

Now it may be assumed that this kind of dynamic is all down to an economic imperative: that media organisations must try to expand audiences by generating a kind of eye candy. While such a logic does play a role and will be discussed below, here it might be noted that mediazation entails rather more complexity than is acknowledged by any economically determinist approach.

First, even now at least some media organisations continue to be public entities (in the economic sense of 'public'), and so have a remit that is not as constrained by the profit motive as is typically the case with private organisations (though admittedly, in the times of neoliberalism, even public corporations such as the BBC have become increasingly driven by economicist discourses).

Second, media organisations tend to be large institutions with complex bureaucracies whose logic cannot itself be reduced to a matter of economics. In the case, for example, of the BBC's Natural History Unit (cf. Volume 2), leading producers often have science backgrounds, and regard their work as a form of popular science. Even producers that don't have such backgrounds tend to adhere to production criteria that are at least as much a matter of craft discourses and logics of field (more on discourse and field, below) as they are on any requirement to generate higher ratings.

Important as these aspects are, perhaps the most significant way in which the nature media, and the process of mediated communication go beyond narrowly economic determinations involves what John B. Thompson describes as *space-time distanciation*.[23] By way of a preamble, it should be acknowledged that large organisations typically enjoy a big difference in access to resources when compared to tour groups. This may work to establish a significant asymmetry between those who produce representations of a certain destination, and their audiences. One possible consequence is that media organisations (or organisations that produce similar kinds of representations) may have the economic power, if not to impose, then certainly to promote their own structural interests in ways that most end users cannot. Until the early twenty-first century, this was particularly true with respect to large organisations' capacity to deploy complex technologies of communication, and the bureaucracies that were typically required to use them.

Such asymmetries and partialities notwithstanding, in the case of the nature media, as in that of other organisations that represent destinations or interact with users from afar, there is likely to be a *break* between the production of messages,

---

[23] John B. Thompson, *Ideology and Modern Culture,* pp. 168.

and the reception of those messages by what are potentially widespread, and heterogeneous groups. This break has particularly complex implications for communicational practices. According to Thompson, the media's transmission of signs (or as Thompson calls them, symbolic forms) 'necessarily involves the detachment of this form, to varying degrees, from the context of its production: it is distanced from this context, both spatially and temporally, and inserted into new contexts which may be located at different times and places.'[24] This has at least two implications: first, the producers must take measures to try to reduce the structural uncertainties generated by the break. Historically, this has often led media producers to engage in generic forms of communication, and to embrace certain forms of populism, with significant implications for the selection, and the representation of subject-matter. It is no coincidence that, for example, wildlife TV producers have long favoured representing apex predators, and scenes involving predation. Producers have traditionally also sought to pre-dispose audiences to engage in certain kinds of reading practices via a variety of extra-textual markers (e.g. trailers, advertisements, interviews with sympathetic users, etc.).

However, any such strategies cannot in themselves eliminate the aforementioned break, the aforementioned distanciation. And indeed, the second implication is that the products generated by the organisations will always undergo socially and culturally situated forms of reception. Strictly speaking, there is never a homogeneous, undifferentiated 'mass' (as in mass communication or mass tourism). In practice, this means that individuals and social groups appropriate products in ways that render them more or less continuous with the own discourses, the own cultural predispositions. This too, contradicts any narrowly economicist, or indeed conspiratorial analysis of the interrelation between organisations and the users of those organisations' products. There is always at least some space for a transformative reception of media messages.

In the context of mediated representations of wildlife, a good example of this kind of process involves the Oscar-winning film *The March of the Penguins* (Wild Bunch/National Geographic Films 2005). As I have noted elsewhere,[25] after this film premiered, a controversy arose when it was reported that Christian fundamentalists in the U.S. were treating this film, a feature-length documentary produced by the filmmaker Luc Jacquet and the National Geographic, as evidence of 'monogamy' in nature, and as evidence of the thesis of 'Intelligent Design' in the universe.[26] The documentary itself made no such claims, though it is significant that, at least in the English-language version, the film did not offer any interpretation that was explicitly based on the theory of evolution.

---

[24] John B. Thompson, *Ideology and Modern Culture*, pp. 168-169.

[25] Nils Lindahl Elliot, *Mediating Nature* (London: Routledge, 2006), pp. 215-216.

[26] See Jonathan Miller, 'March of the Conservatives: Penguin Film as Political Fodder' in *New York Times* online 13 September 2005 <http://www.nytimes.com/2005/09/13/science/13peng.html>[Accessed 9 March 2014].

Now Thompson's analysis centres on the kinds of communications initiated by media organisations, but as I have started to suggest, similar dynamics occur when tourist, or tourist-related organisations engage with their own end users from afar. While some of the interaction may involve face-to-face encounters, many will entail space-time distanciation, and so will also entail what is for all practical purposes a form of mediated communication. For example, tour operators have long used brochures, travel programmes and more recently travel websites not only alert tourists to the existence of the destinations, but try to promote particular ways of consuming those destinations.

In the process, such organisations, like dedicated media organisations, have exploited the relative invisibility of the 'back of the house' – an invisibility that arises thanks in part to the gap between the production and the reception of products – to conceal their motivations, and/or controversial aspects of the production or promotional process. Here again, media organisations offer powerful examples. Historically, BBC producers have captured (and in some cases killed) specimens in order to film them. As I have noted elsewhere, in 2003 a court in Western Australia found that Michael Linley, an award-winning wildlife TV producer, was guilty of trying to smuggle 27 species of frogs, lizards and snakes (217 animals in total) from Australia to the UK in suitcases. Linley was fined AUS $10,000, and the judge that heard the case was reported to have commented that it was 'hard to understand' why a man with Linley's professional standing 'would do such an incredibly stupid thing'.[27] Actually, the case probably had little or nothing to do with stupidity; Linley, who was reportedly caught thanks to a tip-off, was almost certainly engaging in the nature media's 'business as usual'. By his own account, the plan was to film the creatures under 'controlled conditions' in a studio back in Britain.[28] If the autobiographies published by several leading producers are anything to go by,[29] removing animals from habitats to film them in studios is by no means as exceptional a practice as some would like to believe. The same is true for another practice: using animals in captivity, usually in zoos, as surrogates for wild animals. For example, at the time that I was conducting research for this study, the British press discovered that what were made to look

---

[27] BBC News online, 'Fine for Reptile-Smuggling Briton' [sic] 12 December 2003 <http://news.bbc.co.uk/1/hi/uk/3313385.stm> [Last accessed 27 July 2014].

[28] BBC News online, 'Fine for Reptile-Smuggling Briton'.

[29] See for example, the account provided by Colin Willock, *The World of Survival: The Inside Story of the Famous TV Series* (London: Andre Deutsch, 1978); by Christopher Parsons, *True to Nature* (Cambridge: Patrick Stephens, 1982), or more recently by David Attenborough, *Life on Air* (London: BBC Books, 2002).

like images of a polar bear giving birth to young in the Arctic were actually scenes shot in a Dutch zoo.[30]

But here again, the practices in question are not the exclusive purview of media organisations. Tour operators and hotels may promote the ostensibly green credentials of the own accommodations whilst engaging in practices that are anything but that. Guides, if not entire visitor programmes, may also take steps to ensure that certain wildlife are more easily viewed via baits and other forms of conditioning, and in the process may end up altering the delicate ecology of one or another species, one or another ecosystem. This has long been the scandal of what some might describe as 'inauthentic', or today 'fake', ecotourism.

Issues of this kind point to a fundamental ambiguity when it comes to the role of organisations vis-a-vis the practice of wildlife observation. On the one hand, it can be argued that the organisations are *enablers* of wildlife observation amongst tourists, and in some respects surely they are. Without the representational or promotional activities of certain media, let alone the activities of airlines, hotels, or park authorities, it is difficult to imagine many visitors travelling long distances to stay near, and observe wildlife in tropical forests. On the other hand, the obverse side of this institutional coin is that organisations may also work, deliberately or not, to promote forms of observational practice that are in some sense misleading, or that are ultimately detrimental to an environment or a part of that environment. In some cases, this kind of role may be obvious; in many, it is so thoroughly dissimulated (or normalised), as to escape critical attention. One way or the other, this kind of issue brings to the forefront the need to reconsider classic conceptions of *consumption*, and it is to this concept and process that I would now like to turn.

Semeiotic Goods and Consumption

Earlier, I suggested that it is problematic to explain the intervention of organisations in wildlife observation, and in tourism more generally in economically determinist ('economicist') terms. But of course, the obverse danger is to so minimise the significance of the economic, which is really to say the capitalist impera-

---

30 In the BBC/Open University *Frozen Planet* (2011) series. See Anita Singh, 'Frozen Planet: Sir David Attenborough denies misleading viewers over 'faked' polar bear birth', in Telegraph online, 12 December 2011, <http://www.telegraph.co.uk/culture/tvandradio/bbc/8950895/Frozen-Planet-Sir-David-Attenborough-denies-misleading-viewers-over-faked-polar-bear-birth.html>[Last accessed 15 November 2016]. See also Robert Mendick and Edward Malnick, 'BBC accused of routine 'fakery' in wildlife documentaries', in Telegraph online, 18 December 2011, <http://www.telegraph.co.uk/culture/tvandradio/bbc/8963053/BBC-accused-of-routine-fakery-in-wildlife-documentaries.html>[Last accessed 15 November 2016].

tive as to fall for an idealising interpretation of organisational roles. From a cultural sociological perspective, one way of avoiding both tendencies is to analyse the fate of *symbolic goods*, or what I will describe as *semeiotic* goods. Simplifying somewhat, and echoing once again the kind of analysis offered by John B. Thompson,[31] semeiotic goods are signs which undergo a process of commodification, i.e. they are given an economic value in order to be bought and sold in a market. To speak of semeiotic goods is to recognise that, thanks to a capitalist modus operandi, objects in the broadest semeiotic sense of the term (cf. Part 2) may be, and indeed routinely are given an economic value, viz. a value which is by no means inherent to them. Objects are *transformed* into goods. As a result, beyond-human animals, and indeed any other beyond-human object one may care to think of, may become the objects of an economically cybernetic rationality, i.e. monetary quantification is used to try to establish control for economic purposes. From an anthropological perspective, it is possible to suggest that commodification also works to incorporate all manner of objects into a capitalist cosmos, a capitalist cultural imaginary, where things matter not so much for what they are, as for what they can fetch.

The above analysis notwithstanding, in so far as the objects entail some form of semeiosis, then commodification cannot itself be a purely economic affair. If this is a point that Marxist scholarship has always made with reference to political economy, cultural sociology raises an analogous point with reference to the importance of the cultural semiotic turn referred to earlier. Meaning is not itself a matter of economy in the capitalist sense of economy (which is not to say that it cannot be co-opted for such an end). In so far as *signs* are involved, then both less than, and more than an economic relation is involved. Less than in so far as no object is, as noted earlier, economic in itself; but also more than in so far as signs, and semeiotic forms more generally must be overcoded by way of a logic of commodification.

Here it is pertinent to return to the example with which Part 3 began. The crocodile that the visitor observed along the shoreline of Barro Colorado was not in itself a commodity, or a semeiotic good. Moreover, by 2007, it was clear that the Smithsonian Tropical Research Institute was not itself an organisation devoted to the commodification of wildlife on Barro Colorado Island. And yet, by virtue of charging visitors a fee to visit Barro Colorado, and by virtue of the fact that the visitors were tourists who had paid significant sums to travel to Panama, and to visit the island (along with other destinations), the crocodile was itself transformed, however implicitly, into a kind of semeiotic good. The use of the passive voice is deliberate, for there was no one individual which effected this transformation; rather, it involved a conglomeration of organisations, which together worked, however inadvertently or implicitly – and however

---

[31] John B. Thompson, *Ideology and Modern Culture*, p. 154-156.

independently – to economically valorise if not the particular crocodile itself, then certainly the opportunity to observe the crocodile and other wildlife.

Thus far the analysis has focussed on the production side of commodification. But of course, for an animal or environment to be commodified, there must be producers *and consumers*, and I would now like to turn to the question of the consumption of semeiotic goods. At least in mainstream economic theory, the concept of consumption is often employed to refer to the actual purchase of goods and services. If one goes by this approach, then consumption only occurs at the point of manifest acquisition, if indeed there *is* an actual purchase. While this may be a convenient way of tabulating consumption, and may offer an accurate numerical way of characterising the acquisition of goods or services for purely utilitarian purposes, it constitutes an overly simplistic way of characterising those forms of consumption for goods and services that are acquired indirectly, and/or are acquired for reasons to do with desire and pleasure. In the latter scenarios, consumption involves the *social reception* of goods and services as part of a process which may begin long before any payment is made, and indeed continue long *after* any payment has been made.[32] Whatever else they may be, such processes are themselves semeiotic ones.

In tourism we find a case in point. Tourism typically requires the acquisition of transportation services in so far as it involves long distance travel. But of course, the displacement is not simply a matter of an empirical action involving the transfer of one or more bodies from point 'A' to a point 'B'. Before any actual travel occurs, a form of virtual travel may begin which involves the representation of destinations in brochures, TV programmes, or any other media; in such cases, the travellers may imagine themselves in the destinations long before they leave 'home'. It is, in this sense, possible to speak of an *imaginary* displacement: if people travel 'somewhere else' during their holidays, it is more often than not because they imagine, before they board the first plane, train, bus, or car, that doing so will enable them not only to leave everyday life behind, but to experience a different geography, or what is *imagined* to be a different geography. This kind of displacement is an absolutely integral aspect of tourism regarded as a form of semeiotic consumption. As John Urry explains, '…there is an anticipation, especially through daydreaming and fantasy, of intense pleasures,

---

[32] For a critique of the classic economic understanding of consumption, see for example Mary Douglas and Baron Isherwood, *The World of Goods: Towards an Anthropology of Consumption* (Harmondsworth: Penguin Education 1979). As the authors put it, '…the very idea of consumption itself has to be set back into the social process… Consumption has to be recognized as an integral part of the same social system that accounts for the drive to work, itself part of the social need to relate to other people, and to have mediating materials for relating to them… Goods, work, and consumption have been artificially abstracted out of the whole social scheme.' p. 4.

either on a different scale or involving different senses from those customarily encountered'.[33]

As this account begins to make clear, in marked contrast to the kind of consumption theorised by classical economics, cultural sociology starts from a theoretical displacement that is analogous to the one performed via the concept of the semeiotic good: it is necessary to speak of *semeiotic* consumption, i.e. the consumption of *meanings*. To be clear, this is a dimension of *any* form of consumption; if nothing else, one must be able to identify a product in order to buy it, and identification is contingent on the recognition of form, which is a simple way of conceiving the semeiotic process (there is thus a risk of pleonasm when referring to semeiotic consumption). The last point notwithstanding, some forms of consumption hinge on the acquisition of goods and services which are sought *primarily* for their meaningful character, such that any use-value becomes more or less incidental. When, for example, someone buys a DVD with a TV series about crocodiles or rain forests, the plastic box, and the DVD itself are relatively unimportant (at least until they malfunction...); what matters are the programmes which the DVD makes available. In this context, the physical materiality is a means to an end to a different kind of materiality: the kind of materiality that Raymond Williams and other cultural sociologists invoke when referring to social signifying systems.

Now it might seem that semeiotic consumption is an ecologically benevolent activity, at least when compared with the kind of consumption that hinges more directly on use-value. And indeed, in their different ways, many nature media producers and ecotourism advocates such as the International Ecotourism Society (TIES), have long argued that the kind of observational practices they promote are beneficial for the environment. To cite one prominent example of such claims amongst nature media producers, as late as 2016, the website of Wildscreen, a nonprofit entity which organises trade fairs and represents the interests of the wildlife filmmaking industry, quoted Sir David Attenborough (one of its patrons) as saying that 'Films and photographs are major elements in the battle to protect our imperilled natural world.'[34] At about the same time, TIES, arguably the most authoritative and certainly the oldest nonprofit ecotourism advocacy organisation, was itself claiming that at least authentic ecotourism is neither consumptive nor extractive.[35] In fact, there are good reasons for rejecting any such idealisation. On the one hand, semeiosis is itself *always material* in so far as it

---

[33] John Urry, *The Tourist Gaze*, 2nd edition (London: Sage Publications, 2002), p. 3.

[34] Sir David Attenborough, in the homepage of Wildscreen. See <http://www.wildscreen.org.uk/about/[Accessed 29 September 2016].

[35] In 2015 TIES produced a press release which redefined ecotourism, but continued to insist that at least authentic ecotourism was non-consumptive and non-extractive. See TIES, Press Release 'TIES Announces Ecotourism Principles Revision', 7 January 2015 <http://www.ecotourism.org/news/ties-announces-ecotourism-principles-revision>[Accessed 15 February 2015].

involves the production of forms – a process that requires at the very least the energy to think or to express something. To be sure, when the meaning-making occurs via media that come in plastic boxes, or that require the mobilisation of the large teams of producers used to make BBC wildlife TV series – or indeed that use energy-hungry servers of the kind deployed by Google, let alone the mobile phones made with metal, plastic, and ceramic elements – there can be no pretence that semeiotic consumption is not directly linked to extractive activities, and so that it somehow escapes the ecological issues raised by consumer culture more generally.

On the other hand, when certain meanings promote mass tourism – and of course, ecotourism has become just that (cf. Volume 2) – then the very act of travelling constitutes not just a consumptive, but also an extractive activity if only because the means of transportation and accommodation themselves require what are typically fossil-fuel based forms of energy, as well as building materials taken from the ground. Even if we disregard the often hidden materialities of semiotic consumption, much research now suggests that even the most strictly managed forms of ecotourism may also disrupt the natural patterns of behaviour of certain species.[36]

It is nonetheless possible to get a sense of the complex politics at stake when it is noted that, by the late 2010s, neoliberal politicians' drastic cutbacks of state funding for all kinds of ecologically-oriented programmes meant that many environmental ministries, park administrations and NGO's had come to rely on wildlife tourism and ecotourism to fund conservation projects. In many cases, it is thus no longer a choice between tourism/no tourism, but rather a matter of finding the best ways to mitigate the impacts of tourists' presence.[37]

---

[36] This much had begun to be evident by the early 2000s, if not long before. See for example, Anil Ananthaswamy, 'Beware the Ecotourist', *New Scientist*, 6 March 2004, pp. 6-7. For a more recent discussion of the impacts, see Daniel T. Blumstein, Benjamin Geffroy, Diogo S. M. Samia, and Eduardo Bessa (Eds.) *Ecotourism's Promise and Peril: A Biological Evaluation* (Cham, Switzerland: Springer 2016). One of the impacts was a shift in many species from diurnality to nocturnality – a shift that in some areas was almost certainly made more likely by tourism/ecotourism. See Kaitlyn M. Gaynor, Cheryl E. Hojnowski, Neil H. Carter, Justin S. Brashares, 'The Influence of Human Disturbance on Wildlife Nocturnality', *Science*, 360:6394 (15 June 2018), 1232-1235.

[37] For a discussion of this issue, see Ralf Buckley, 'Endangered Animals Caught in the Tourist Trap', in *New Scientist*, 10 October 2012 < https://www.newscientist.com/article/mg21628860-200-endangered-animals-caught-in-the-tourist-trap/>[Accessed 20 July 2018].

Fields

Thus far, I have explained how both economic, and more-than-economic aspects must be taken into account in the analysis of the relation between wildlife observation and organisations. I would now like to move beyond organisations, to consider categories that go beyond the organisation, *sensu stricto*. In advanced capitalist societies, the agents of organisations typically work according to the rules of their organisations, but also those of the broader *fields* that structure the practices of all similar, or similarly positioned, organisations. Earlier I referred to Wildscreen, an organisation that promotes the wildlife filmmaking industry, and that industry may be conceptualised as a field – what I describe as a part of the nature media field. A similar point can be made about TIES and the *ecotourism* industry; the latter industry is also a field that entails multiple organisations and their agents.

As these examples begin to suggest, it is necessary to conceptualise field, and in this study I will draw on the proposals of the sociologist Pierre Bourdieu (1930-2002). According to Bourdieu, a field may be defined 'as a network, or a configuration, of objective relations between positions'; the positions in question 'are objectively defined, in their existence and in the determinations they impose upon their occupants, agents or institutions, by their present and potential situation (*situs*) in the structure of the distribution of the species of power (or capital) whose possession commands access to the specific profits that are at stake in the field, as well as by their objective relation to other positions (domination, subordination, homology, etc.)'.[38]

Several aspects of this conceptualisation require elucidation. The first is that Bourdieu rejects the classic opposition between the so-called objective and subjective aspects of social action (or what this study describes as cultural practice). From his perspective, the problem is to show how *both* aspects are not only related, but *constitutive of each other*. A field is not a field in the abstract, but a dynamic process constantly made and remade by individuals as they partake in practice. This individuality, or what a positivist might describe as the 'subjective' element of action, is part of what makes the field, and vice-versa: the field at one and the same time shapes the individuality. As the sociologist Loïc Wacquant puts it, a field is like a *magnetic* field in the sense that it imposes an invisible pattern of practice on anyone, or anything, that enters the context in question. 'A field is a

---

[38] Pierre Bourdieu, 'The Purpose of Reflexive Sociology (The Chicago Workshop)', in Pierre Bourdieu and Loïc J. D. Wacquant, *An Invitation to Reflexive Sociology* (Cambridge: Polity & Blackwell, 1992), 61-215, p. 97.

patterned system of objective forces ... a *relational configuration endowed with a specific gravity* which it imposes on all of the objects and agents that enter in it'.[39]

In the context of wildlife observation, a question may be raised as to the extent to which fields really do impose their gravity on *all* the objects and agents that enter it. Clearly, the wildlife encountered in situ by tourists are mostly *not* a part of the tourist field – though it might be noted that this may change dramatically if visitors are allowed to feed the animals, and/or the animals become dependent on tourism.

Further complexity occurs in so far as individuals may be, and tend to be a part of more than one field. Amongst other things, this may generate contradictions, or at least weaken the pull of any one field vis-a-vis those individuals. Whatever the case, we may agree with Bourdieu that, especially in those cases where individuals identify strongly with one or another field (the example of a professional field is a particularly good one), the 'magnetism' that Loïc Wacquant refers to may certainly be a powerful one.

Two examples may help to illustrate this notion. The first involves the different Smithsonian guides who led the groups with whom I engaged in participant observation on Barro Colorado's trails. The guides in question were employed by STRI's Visitors Programme, which provided instruction as to how the guides should lead groups, including what sorts of topics to discuss, and how to respond to certain kinds of wildlife. As might be expected, this meant that each tour tended to adhere to what might be characterised as an informal 'curriculum': the guides followed one of two trail options (Donato or Fausto, cf. Volume 2), and referenced certain ongoing research, institutional histories, wildlife, etc.

In keeping with this approach, it might be assumed that the guides adhered to strictly local guiding criteria. However, much of the advice provided by STRI's Visitors Programme regarding techniques of communication was itself influenced by the kind of principles, codes, and practices developed by the heritage interpretation field in the US, which includes organisations such as the National Association for Interpretation (NAI). Then again, over the course of my ethnographic research, it became clear that the guides, most of whom had undergraduate, or in some cases postgraduate science degrees, tended to engage particularly closely with those aspects of Barro Colorado's ecology that they were most familiar with thanks to their training in particular academic fields/subfields. For example, those with a knowledge of entomology tended to spot and could say a lot more about insects, if not invertebrates more generally. This was not, however, merely a matter of the possession of specialised knowledge; it was also to do with the internalisation of what was itself a field-based *disposition*: entomologists were

---

[39] Loïc J. D. Wacquant, 'The Structure and Logic of Bourdieu's Sociology', in Pierre Bourdieu and Loïc J. D. Wacquant, *An Invitation to Reflexive Sociology*, 1-59, p. 17. Italics in the original.

more 'comfortable' talking about invertebrates, botanists about plants, and so forth.

A second example may help to clarify the last point. Whilst visiting a biological field station in Costa Rica's *La Selva* reserve, I had the opportunity to discuss my research with a botanist, who told me about a significant change that had occurred in the way in which the researcher observed tropical forests after taking a course in phytopathology – simplifying somewhat, the study of plant diseases. Where once the botanist walked through the forest and looked out for particular species of plants, after taking the course, the own observational practice was transformed: now the botanist couldn't help *but* focus on signs of disease in plants.

While this example doesn't involve the observation of wildlife, a first implication for wildlife observation lies precisely in the fact that someone was *not* intent on observing wildlife – at least not in the everyday meaning of the expression, which refers to wild *animals*. It might be argued that wildlife observation takes primacy in tropical forests because 'we humans' are innately *bound* to pay more attention to animals, especially those likely to pose a threat, or to be a source of food (cf. the sociobiological arguments put forward by E. O. Wilson, and critiqued in Chapter 2). But here was a researcher that was instead entirely preoccupied with plants – a preoccupation that emerged thanks in part to an immersion in a *field* of research (botany).

A second implication requires a clarification about the botanist's experience. One of the things the researcher mentioned was that, try as they might, after taking the course, it was not possible to shake off an inclination to 'look out' for signs of phytopathological conflict. To put the matter somewhat instrumentally, something akin to an observational switch had been thrown that would not go back to its original setting. This illustrates the extent to which what is involved in field relations is not just knowledge, in some rationalist sense of the word, but a *disposition* that is only partly conscious, only partly voluntary. Indeed, in many cases, the dispositions associated with field may be entirely unselfconscious.

As the last point begins to make clear, fields are not simply bodies of knowledge which may be taken or left like courses. Nor are they like pastures, with fixed boundaries and locations – which is not to deny that some places may be temporarily or durably associated with, and indeed particularly strategic to, certain fields. In so far as a field is created by interrelating trajectories, then the field 'goes' wherever the trajectories take it. This accounts at once for a field's relative invisibility – there is of course no 'fencing' – and for the capacity of one field and its agents to shift, or even to subsume, other fields.

The absence of physical boundaries does not mean that there is no regulation, or no barriers to entry. On the contrary, fields may be, and in practice are just as strongly policed as some physical places are; typically, the agents of fields establish filters in the form of qualifications and other requirements which exclude those who don't meet the criteria. Even when certain criteria are not actually set, a field may be exclusive by virtue of involving activities which require a degree of economic, cultural or social capital that many prospective entrants would not be able to match.

I will return to these forms of capital in a moment. First, another clarification is required: the concept of field does not imply a simple homogeneity of interests on the part of all those who constitute a field. On the contrary, Bourdieu explains field with reference to the analogy of a game, albeit one that is not primarily a matter of deliberate or explicit rules. Like a game, a field involves stakes – what Bourdieu describes as *enjeux* – which are the product of competition amongst the 'players' in the field. While the players do have a certain investment in the game, and as Bourdieu puts, 'are taken in by the game', they may oppose one another, 'sometimes with ferocity'; but this only to the extent that 'they concur in their belief (*doxa*) in the game and its stakes; they grant these a recognition that escapes questioning'.[40]

This competition has as its aim, if one can speak in such teleological terms, the accumulation of capital. Bourdieu argues that an agent's role in a field is always fundamentally an interested one, if only because the agent will typically partake in the interactions in a field in such a manner as to maintain or increase their capital, or indeed to change the rules of the game such that a certain kind of capital will acquire more value than it had hitherto.[41] While this conception clearly relies on an economic metaphor, Bourdieu does not merely refer to *economic* capital, i.e. material wealth such as money, property, etc. He also refers to *cultural* capital (simplifying, an agent's knowledge and skills) and somewhat confusingly in terms of the nomenclature chosen, to *social* capital (the prestige accorded by more or less specialised forms of recognition, such as awards, rank, etc.).[42] From this perspective, the agents within a field work to establish differential positions, contingent on the amount of one or another, or more likely a certain *combination* of different kinds of capital.

Perhaps the most obvious way in which this logic applies to wildlife observation involves the fields constituted by the most specialised, and we might say most 'professionally' avid wildlife observers – for example, the more devoted birdwatchers known in Anglo-American circles as 'birders' or 'twitchers'. Amongst birders, the own standing in the field does not mainly depend on material wealth (economic capital). The key elements are ornithological knowledge (a form of cultural capital) and the prestige associated with, say, having observed very rare birds in 'exotic' locations (a form of social capital).

---

[40] Pierre Bourdieu, 'The Purpose of Reflexive Sociology (The Chicago Workshop)', p. 98.

[41] Pierre Bourdieu, 'The Purpose of Reflexive Sociology (The Chicago Workshop)', pp. 98-99.

[42] Pierre Bourdieu, 'Forms of Capital', in John G. Richardson (Ed.) *Handbook of Theory and Research for the Sociology of Education*, translated by Richard Nice (London: Greenwood Press, 1986) 241-258.

Chapter 8 Cultural Practice

It might be objected that such considerations do not really apply to wildlife observation amongst more generalist tourists whilst visiting a tropical forest. In fact, a strong case can be made that the ways in which guides and visitors interact on the trail still have much to do with a logic of field, and differing amounts and forms of capital. To begin with, if and when tourists arrive in a biological reserve with a visitors programme, they enter a field constituted by the trajectories of the guides, and of all those who contribute to the design of the kind of informal 'curriculum' I mentioned earlier. From this perspective, the tourists enter and become a part, however temporarily, not just of the biological reserve, but of the field-based social relations that regulate observational as well as other kinds of practices on the reserve. Then again, the manner in which the visitors do so, and their relations to guides or any other figures endowed with a certain authority, are likely to be linked to capital. So it is, for example, that the guide will be expected to show that they possess more cultural capital than the visitors, at least when it comes to finding wildlife, and explaining its ecology. Moreover, the 'respect' or deference shown to the guide may well hinge on social capital, including markers of prestige such as science degrees, past experiences on the trail, etc. The same logic may apply to the visitors themselves, who are likely to partake in a visitors programme in a manner that reflects *their* capital. This includes, but should not be reduced to their economic capital (and so what kind of holiday and tour they can afford). Also important are their cultural capital (and so, for example, what kinds of questions they will be able to ask guides); and their social capital (for example, during the ethnographic research that I conducted, it was apparent that many visitors spent quite a lot of time 'comparing experiences', and this arguably generated a space for comparing social capital).

Now the example with which Part 3 began offers perhaps the best illustration of the way in which a field can influence actors who are not, strictly speaking, a part of it. The visitor in question interpreted a crocodile on Barro Colorado with reference to the kinds of representations produced by the nature media field; despite not being a media producer, the visitor's ways of looking were clearly influenced by the aforementioned field. The last point notwithstanding, the opposite point can also be made: having highlighted the field-bound nature of the tours organised by the Smithsonian's Visitors Programme, on occasion said programme found itself having to engage with visitors who made it clear from the start that they had no interest in the kinds of capital (social or cultural) valued by the Smithsonian. In a few cases this meant that the visitors actively refused to abide by the rules established by the Visitors Programme for visiting the island.

The last point shows the folly of trying to establish *a priori* what fields will be the most important for all tourists visiting a tropical forest. The lived space of tourism (cf. Part 4) is often a particularly heterogeneous one, and so likely to include visitors who constitute/inhabit a variety of fields and indeed cultural groups. This verity notwithstanding, the instituted space associated with tourism in national parks and nature reserves does entail a certain regularity that is generated by the interaction of several concrete fields. In Volume 2 I will describe the roles of four fields which, together with that of the nature media, seem particularly important: the scientific; the conservation(ist); what I described earlier as 'heritage interpretation'; and of course, that of tourism itself – a field made strange by both its transient, multi-spatial, and indeed heterotopic qualities (cf. Part 4).

Habitus and Hexis

Bourdieu's account of field raises a question: how is it that individuals can move across fields, but remain 'true' to their life histories (however unselfconsciously), while adapting with varying degrees of success to new contexts, and with them, new logics of field? In so far as Bourdieu's theory can answer this question, it does so via his concept of *habitus*, which may be defined provisionally as a habitual disposition, and with it, a form of practice, or better yet, an *orientation* to practice that develops over a lifetime.

Bourdieu himself defines *habitus* by way of a statement that seems almost like a magical incantation: *habitus* as 'durable, transposable dispositions, structured structures predisposed to function as structuring structures, that is, as principles which generate and organize practices that can be objectively adapted to their outcomes without presupposing a conscious aiming at ends or an express mastery of the operations necessary to attain them'.[43]

In this account, it should be clarified that *habitus* is not about the 'application of ideas', or an entirely voluntary form of practice arising from the conscious mastery of skills; we return to the importance of the unconscious or unselfconscious. But nor is *habitus* the 'mindless' or 'mechanical' practice associated with ostensibly *blind* habit. As Bourdieu puts it, 'I said habitus so as *not* to say habit – that is, the generative (if not creative) capacity inscribed in the system of dispositions as an *art*, in the strongest sense of practical mastery, and in particular as an *ars inveniendi*'[44] – the 'art of invention', or a capacity for innovation that *starts* from that which has been experienced, but develops it in ways that enable application to new contexts.

As theorised by Bourdieu, *habitus* typically entails a gradual process of inculcation, and a durable appropriation – though Bourdieu, like many other sociologists, acknowledges the crucial importance of the first two to three years of any person's life. This process has consequences not just for the own beliefs, but for the own *bodily* disposition, or what Bourdieu describes as *hexis*: 'Bodily hexis is political mythology realized, *em-bodied*, turned into a permanent disposition, a

---

[43] Pierre Bourdieu, *The Logic of Practice*, translated by Richard Nice (Cambridge: Polity Press, 1990 [1980]), p. 53.

[44] In Pierre Bourdieu and Loïc J. D. Wacquant, 'The Purpose of Reflexive Sociology (The Chicago Workshop)', p. 122.

durable way of standing, speaking, walking, and thereby of feeling and thinking.'[45] *Hexis* names something that sociologists and anthropologists alike have long noted with respect to social class: one 'shows' one's class (always in the sociological sense of the expression) via the ways in which one holds one's body, moves across a place, etc. An analogous analysis might be offered with respect to a variety of professional dispositions. Consider, for instance, the *hexis* of 'old school' military officers, or of all those who devote themselves to certain sports, to certain kinds of dance, etc.. In these and many other examples, there is a bodily inculcation and expression of *habitus*.

In this study, two forms of *hexis* of a kind not usually considered by sociologists seem particularly relevant: the first is the one that results from a life spent tracking animals, be it as a guide, a scientist, park ranger, etc.. The second is that of a person who, by contrast, has always lived in a metropolitan area, and has learned to insulate themselves from the noise, the smells, and more generally the 'stimuli' (for want of a better word) associated especially with very densely inhabited cities, or cities where there are few or no regulations concerning emissions, noise pollution, etc.

In the case of the former, the individual effectively cultivates what will seem at least to the city-dweller like a remarkable mixture of local knowledge and perceptual acuity. It is a combination because the *hexis* in question is not simply a matter of 'knowing more' or, say, 'listening more carefully' (however important both dispositions might be). Instead, it involves developing something akin to what is popularly described as a 'sixth sense', but is probably more accurately described as an at once deliberate, but also at least partly unselfconscious ability to focus the own attention on cues which disturb, at times very subtly, a certain pattern, one might even say a certain order of environmental feedback. Obvious examples are the forest that suddenly goes silent, or conversely, the distress calls of some creature.

It might seem that, by contrast, the individuals who live in the city have no such skills. In fact, they are likely to have developed, *ceteris paribus*, a disposition that is relevant to *their* everyday environment. Consider, for example, the individual who in the last moment stops before crossing a road, just as a hitherto unnoticed cyclist whizzes by... Or the parents that, despite the hubbub of everyday household noise, detect the otherwise faint cry of a baby that has awoken in another room.

After acknowledging examples of this kind, in so far as cities are filled with what may be characterised as unsubtle cues – loud noises, strong smells, landscapes full of strong visual imagery, etc. – then a significant part of the own *hexis* is likely to entail closing down, or at least drastically filtering, environmental indexes. Perhaps the most literal example of this tendency is the habit, which

---

[45] Pierre Bourdieu, *The Logic of Practice*, pp. 69-70. Italics in the original.

emerged thanks in part to the advent of mobile media, to wear earphones whilst travelling in public transport, or walking along streets.

The consequences for wildlife observation of these two different dispositions can be illustrated by way of a personal anecdote. When I first began to accompany tourist groups along Barro Colorado's trails, I was as utterly unaware of subtle signs of wildlife as any first-time visitor to a tropical forest. However, just three months later, my capacity to perceive and interpret such signs had not only improved dramatically, but travelled back with me to the UK. One evening, whilst out walking with family just as night started to fall on the edge of England's Exmoor National Park, I detected a fallow deer (*Dama dama*) that was standing stock-still just above us, on the verge between the country road and some woods. This particular specimen's melanistic colouration made it nearly invisible in the growing gloom, and it was using, as deer often do, immobility as a defence measure against what it would have regarded as potential predators. Had I not just spent several months trying to detect fauna in Barro Colorado's forest, most likely I would have failed to notice the deer, and would have strolled right past. I had, in effect, acquired at least something of the *hexis* required of guides who partake in wildlife observation with tourists.

As I have already noted, for Bourdieu *habitus* and *hexis* entail much longer term processes, starting from infancy. However, both processes (for that is what they are) need to be regarded not as purely transcendental phenomena that are set or fixed for once and for all, but as dynamics that are at least somewhat permeable to changes in context. If habitus and hexis are learned, the learning does not suddenly stop after age two or three; older age may slow down changes in the own disposition, but in principle we may of course all be affected by a significant change in circumstances.

Codes (or Coding Orientations)

Earlier, I noted that Bourdieu suggests that each field has certain rules. One way of conceiving such rules is by way of a theory of codes. This is not something that Bourdieu himself does, perhaps because he believes that doing so would detract from the specificity of the rules that apply to each context. As he makes clear in one of his works, there is also always the danger that a theory of codes works inadvertently to promote the official language, the official versions of the world.[46] However, if that is the logic which one applies to decide what concept to theorise, then one would have to question *all* concepts, including those of field or *habitus*, and researchers would very quickly find themselves stuck in the kind of empiri-

---

[46] See Pierre Bourdieu, *Language and Symbolic Power,* edited by John B. Thompson and translated by Gino Raymond and Matthew Adamson (Cambridge: Polity Press, 1991), pp. 43-49.

cism that denies *a priori* the validity of *any* kind of abstraction (i.e. abject nominalism).

Before presenting an alternative, it should be clarified that a theory of codes is not primarily about explicit rules of the kind associated with, say, public prohibitions found in biological reserves, e.g. stay on the trails, do not touch the animals, etc. Nor is it about 'dress codes' of the kinds that actors may acquire more or less consciously. Instead, a socio-cultural theory of codes attempts to explain the workings of mostly, if not entirely *tacit* rules, which is to say rules that are likely to remain unselfconscious to those that follow them.

It should also be clarified that, while the notion of a rule suggests an inflexible regulation or principle, the rules that govern practice in particular fields must themselves entail at least a degree of flexibility in order to accommodate variations in context. Put differently, just as the signifying systems that Raymond Williams refers to are made and remade in practice, the same point can be made of the rules that guide those practices; in some sense, the rules are 'made up' as individuals produce trajectories (we return by this route to *ars inveniendi*). Once again, this is by no means an entirely, or even mostly conscious or voluntary process, which is not to deny that on occasion it may become just that.

So how are we to theorise such rules-that-are-not-rules? There are probably as many theories of code as there are scholars that have researched the subject. In this study I adapt a theory first proposed by the sociologist Basil Bernstein for the analysis of pedagogic practice.[47] This choice reflects my sense that Bernstein's approach has a much broader application than the narrowly or 'officially' pedagogical (e.g. formal education). Simplifying somewhat, *any* cultural practice entails a pedagogic process in so far as it must be taught and/or learned. This being the case, any cultural practice entails what is bound to be, at the very least, a tacit pedagogic process – what I will describe in the conclusions of Part 3 as a nonformal pedagogic mode. Aspects of Bernstein's theory of codes may thus be incorporated as part of a sociological account of observational practices. However, his theory needs to be modified in order to acknowledge what Peirce describes as dynamical objects, and the importance of indexicality. Accordingly, codes may be conceived as sets of mostly tacit rules that regulate, or attempt to regulate

1) what objects an individual or group attends to (where object is understood, as it is throughout this study, in the broadest semeiotic terms introduced in Part 2);
2) how those objects are categorised; and
3) how the resulting classifications are combined with other elements of practice, *in* practice, and so in particular contexts.

---

[47] See for example, Basil Bernstein, *The Structuring of Pedagogic Discourse: Class, Codes and Social Control Vol. IV* (London: Routledge, 1990).

These aspects may be called *indexing*, *classification*, and *framing* rules, respectively. What I understand as a coding *orientation* involves all three elements. The following is a more detailed account of each of the elements, followed by a clarification regarding the character of their inter-relation, and an ontological issue that requires elucidation in its own right.

Indexing rules direct the subject to attend to some objects and not others. Indexing rules act as 'pointers'; they say, or rather *say without saying*, 'Look at (or otherwise attend to) x!' I will, however, explain in a moment that this aspect of code must take into account the fact that attention is not only directed by socially constructed rules; on occasion, it may also be a result of the actions of the objects themselves (this is the point I made in Part 2, via Umberto Eco, about the importance of recognising *dynamical* objects).

Classification rules regulate the categorisation, i.e. they provide principles at once of identity and difference for whatever object is indexed, i.e. 'x is y', but at the same time, 'x is not z'. Put differently, classification entails all boundary-making activities. So classification rules are what lead a guide to say, 'That is a Coati', yet in so doing, to understand that the indexed creature is, for example, *not* a primate, yet *is* related to Procyonids ('racoons'), etc. This example entails a more or less explicit linguistic/scientific classification, but it should be noted that this aspect of codes, like the previous, also applies to non-linguistic forms of practice and their boundaries – for example, those that separate off trails that are available for tourist use from those that are 'off limits', or ways of relating to wildlife that are 'allowed' and 'not allowed', e.g. touching, feeding, etc. To reiterate an earlier point, it also refers to classifications that are likely to be partly if not wholly implicit; for example, what wildlife are deemed worthy of observation, and which are not, which animals are deemed to be *wild* and which are not.

As described by Bernstein, classification rules can be stronger or weaker, depending on the extent to which actors deem two or more objects to be more or less separate, more or less distinct. A seemingly trivial example involves the difference between different species, or 'mere' subspecies. I say seemingly trivial because of course, scientific careers may well be at stake; for those who make the case for a subspecies (and so a weaker classification between two sets of animals), as for those who argue in favour of a different species altogether (the stronger classification), the matter is anything *but* trivial. Indeed, it is possible to understand how this kind of issue may even lead to war when one considers politically charged geographical examples such as the one of Northern (or northern) Ireland, with Catholics arguing that northern Ireland is part of Ireland (weaker classification), and Protestants insisting that Northern Ireland is, politically speaking, *entirely* separate from Ireland (stronger classification). As this example begins to make clear, classification rules (also) establish categorial principles for what constitutes a legitimate or an illegitimate interpretation for a given context.

Framing rules regulate the enactment, the actualisation or realisation of a certain modality of indexing and classification. They provide the orientation for the actual engagement of practice in space and time – be it a matter of representational practice, physical displacements, the organisation of objects in a certain place, or any other form of practice. The reference to a certain place leads to a fundamental qualification: if all practices are, by their very nature, contextual, the framing rules are the aspect of codes that is most sensitive to context; and indeed, a key role of the framing rules is to 'adapt' indexing and classification to a particular context.

At this point an example may help not only to clarify what is entailed by framing, but also how it relates to indexing and classification. Say, for instance, that a guide on a trail with a group of visitors points to a crocodile, as the guide did in Part 3's opening example (indexing). That the guide does so reflects the social fact that this species is one of a relatively small number of creatures that guides, tourists and all those who partake in a certain kind of wildlife observation have come to *classify* as being worthy of indexing (it has been deemed, shall we say, 'point-out-able'). This being the case, the guide will then be looking out for these and other similarly classified creatures, and when she finds them, she will not only point to them, but will do so in a particular way, e.g. she will physically point to a specimen, and may then proceed to interpret it in a particular way. It should be reiterated that framing is not just about linguistic framing. The word crocodile is only one element of a practice that will include aspects such as how loudly or quietly one says 'Crocodile!', how close one gets to the crocodile, whether one tries to photograph or hunt the crocodile, etc.[48]

In keeping with the ideal-realist, and pragmaticist philosophy of Charles Sanders Peirce – but in some respects, also with the radically empiricist ecological psychology of James Gibson (cf. Part 1) – I reject the assumption, evident in Bernstein's work, that the whole world is, in effect, always already 'socially constructed', which is another way of saying pre-interpreted, by/for humans. On the one hand, and as explained in Part 2, there are circumstances in which the wildlife may do at least a part of the indexing thanks to a sudden movement or appearance, the production of loud calls or strong smells, etc. On the other hand, there are circumstances in which an individual may encounter a creature and be at a complete loss as to how to classify and frame it. For example, the creature may defy known principles of classification, or may make such a sudden, or otherwise surprising appearance that, for a split second, the observer is not able to classify, let alone frame the object in question. In the conclusions to Part 2 I offered an example drawn from a crocodile attack, and also from encounters with the large, turkey-like crested guans (*Penelope purpurascens*), which on Barro Colorado and elsewhere have a habit of erupting into flight in ways that tend to surprise visitors and guides alike.

---

[48] In this sense, Bernstein's choice of 'framing' to describe these aspects is somewhat unfortunate; while the term is handily evocative, a more useful terminology might be indexing/classifying/*enacting*.

Part 3 Sociological-Philosophical Perspectives

So, contrary to what may be suggested by the kind of analysis developed thus far, in the case of wildlife observation, as in observational practices more generally, the objects of observation can and do play a role in their own 'construction', their own *presentation*. Indeed, on occasion they may either trigger, or interrupt, a habitual coding orientation. The importance of this possibility is difficult to overemphasise; however much coding orientations *do* establish frames for wildlife observation, there is no guarantee that the wildlife, or their habitat, will 'respect' such frames. This being the case, in some cases an encounter will not involve what is best described as mediate modes of the kind that I described in Part 2, and which are often taken for granted by cultural sociologists.

Discourses

The concept of framing is a very general one. This being the case, it requires specification by way of additional concepts. I will begin with the concept of discourse, which in cultural sociology has a meaning that is quite different from the one employed in everyday life, where discourse more often than not refers to spoken or written communication. The meaning is also quite different from the one preferred in some linguistic fields, where the concept refers to language in use.[49] In this study the concept is approached from a perspective that builds on the work of Michel Foucault, and also on post-Foucaultian scholars such as Basil Bernstein[50] and the social semioticians/critical discourse analysts Gunther Kress and Robert Hodge.[51] In keeping with the kind of approaches developed by the mentioned scholars, we may define discourse provisionally as a set of statements that produce, reproduce, or at times contest the meanings and values of social institutions.

A full account of the complexities of discourse as understood by Foucault, let alone by his many followers, is far beyond the scope of this chapter. A number of scholars have noted that there is a significant difference between the concept as Foucault develops it in, say, *The Archeology of Knowledge*,[52] and as he does in his

---

[49] See for example, Guy Cook, *Discourse* (Oxford: Oxford University Press, 1989).

[50] See for example, Basil Bernstein, *The Structuring of Pedagogic Discourse*.

[51] See for example, Robert Hodge and Gunther Kress, *Social Semiotics* (Cambridge: Polity Press, 1988), or Robert Hodge and Gunther Kress, *Language as Ideology*, 2nd edition (London: Routledge, 1993). See also, the compendium of papers on the field published by Michael Toolan (Ed.) in *Critical Discourse Analysis: Critical Concepts in Linguistics, Volumes I-IV* (London: Routledge, 2002).

[52] Michel Foucault, *The Archeology of Knowledge and the Discourse on Language*, translated by A. M. Sheridan Smith (New York: Barnes & Noble, 1993 [1969 & 1971]).

later genealogical investigations (e.g. in *Discipline and Punish*,[53] or in the *History of Sexuality*[54]). Here it will suffice to note that in the earlier accounts, the emphasis is on the structures that are internal to discourse. By contrast, in 'The Orders of Discourse'[55] – Foucault's inaugural lecture at the prestigious Collège de France in Paris – the emphasis shifts to linking the analysis of the internal features of discourse to social forces that are in some sense external to the discourse itself.

My own definition of discourse combines both of these aspects: discourses are particular modalities of framing, as per the account I offered above of coding orientations. As such, discourses 'carry' particular forms of indexing and classification. They encourage subjects to index some objects and not others, and to classify those objects in some ways and not others. In so doing, they produce, reproduce, or at times contest the coding orientations of specific organisations, or more likely of certain *fields*. The last point notwithstanding, in time discourses may become so widespread – and may incorporate so many other discourses – that it becomes almost impossible to attribute them to a particular organisation or field. A case in point, the neoliberal discourse that today dominates so many of the officially recognised fields of interaction.

It may be helpful to begin with some examples of discourse, thus defined, to then consider nuances and complexities. A lawyer's, a doctor's, an engineer's, or a scientist's characteristic professional way of representing an object (as always, in the widest semeiotic sense of the term) constructed within each of the mentioned fields may all be regarded as examples of particular discourses: legal discourse, medical discourse, scientific discourse, etc. In each case, the subjects who produce/reproduce the discourse will not only employ a more or less specialised vocabulary, but will focus on certain kinds of phenomena which they will classify in particular ways. The classification will be expressed (and in the process produced/reproduced) by way particular forms of communication, particular forms of interrelation, and so particular modalities of framing (as per my earlier account of coding orientations).

As suggested above, discourse is thus not so much a matter of the use of specific terms, or even a 'jargon', as it is of a certain *pattern* of representation, a certain logic, albeit with an important caveat: here logic is not logic in the conventional and formal sense of the expression, e.g. syllogisms which are or seem to be entirely explicit and so open to itself logical scrutiny. Rather, the logic in question is a relatively hidden one, one that is only uncovered via a certain kind of analysis. This being true, critical discourse analysis is not like the more conventional linguistic analysis, which can point to, and consider for instance the structuring of

---

[53] Michel Foucault, *Discipline and Punish: The Birth of the Prison*, translated by Alan Sheridan (New York: Vintage Books, 1977 [1975]).

[54] Michel Foucault, *History of Sexuality, Volume I: An Introduction*, translated by Robert Hurley (New York: Pantheon Books, 1978[1976]).

[55] Michel Foucault, 'The Orders of Discourse', *Social Science Information*, 10 (2)(1971), 7-30.

statements into groups, which can in turn be subdivided into words, and the words into morphemes (as per the kind of functional grammar developed by, say, Michael Halliday[56]). This is an issue that Foucault grappled with in his earlier writings about discourse. In *The Archeology of Knowledge*, he explains in some detail how he arrives at a certain theory of discourse as part of an attempt to develop what is, in effect, a new, and more critical form of historiography – one that might explain a certain 'unity' in the linguistic statements found in historical archives, but not a unity that is based simply on the object that is referred to by the statements, the overt connections between the statements, or indeed a simple continuity in their conceptualisation or themes.[57] Instead, Foucault suggests that what is actually at stake is a *system of dispersion* and with it a certain *regularity* – an order that involves correlations, positions, functionings, as well as transformations of statements. So by statements, Foucault does not primarily refer to sentences, the locutionary aspects of speech acts, or any kind of formal unit of grammatical construction (much as these would have to be employed to produce discourse); rather, he refers to signs that have a function that is internal to a discourse, and which can only be recognised by way of the discursive analysis.[58]

Although Raymond Williams might not have called it that, the following is an example of a discourse analysis as applied to the everyday use of the concept of Nature in the English language.[59] Williams shows that while the term 'nature' has a nominal continuity over centuries, it is actually one of the more complex words in the language, and one that has a long history of *changing* meanings. In *Keywords*, he describes 'three areas of meaning':

> (i) the essential quality and character *of* something; (ii) the inherent force which directs either the world, or human beings, or both; (iii) the material world itself, taken as including or not including human beings. Yet it is evident that within (ii) and (iii), though the area of reference is broadly clear, precise meanings are variable and at times even opposed. The historical development of the word through these three senses is important, but it is also significant that all three senses, and the main variations and alternatives within the two most difficult of them, are still active and widespread in contemporary usage.[60]

---

[56] Michael A. K. Halliday, *An Introduction to Functional Grammar* (2nd edition) (London: Edward Arnold, 1994).

[57] Michel Foucault, *The Archeology of Knowledge*, pp. 31-35.

[58] Michel Foucault, *The Archeology of Knowledge*, pp. 79-87.

[59] See Raymond Williams, 'Nature', in *Keywords: A Vocabulary of Culture and Society*, 219-224. See also Raymond Williams, 'Ideas of Nature', *Problems in Materialism and Culture* (London: Verso, 1980), 67-85.

[60] Raymond Williams, 'Nature', p. 219. Italics in the original.

What is perhaps the key discursive feature – and problem – with the later meaning of the term is its character and use as an *abstract singular*. Whereas the first meaning ('i') is a specific singular, the *nature of something*, in the case of the second and third meanings, it is now 'the **nature of** all things having become singular **nature** or **Nature**'.[61] As Williams explains, the problem with this is that 'a singular name for a real multiplicity of things and living processes may be held, with an effort, to be neutral, but I am sure it is very often the case that it offers, from the beginning, a dominant kind of interpretation.'[62] The task of the discourse analyst is precisely to uncover such 'dominant kinds of interpretation'.

Williams' analysis might be extended to suggest that, at least in the case of the modern, and romantic version of the discourse of nature, nature is represented by way of a series of dualisms, all of which work as a *discursive formation* to try to define a certain orientation to the world of objects that are indexed via the concept (see Table 3, below). In Foucault, a discursive formation refers to a regularity, a system of dispersion across multiple statements. As Foucault puts it, 'Whenever one can describe, between a number of statements, such a system of dispersion, whenever, between objects, types of statement, concepts, or thematic choices, one can define a regularity (an order, correlations, positions and functionings, transformations), we will say, for the sake of convenience, that we are dealing with a *discursive formation*.'[63]

On the basis of a matrix including the dualisms shown in Table 3,

Culture	Nature
Human	Nonhuman
Mind	Body
Male	Female
Subject	Object
I/us	It/them
Domestic	Wild
Proximal	Distal/Remote
Civilised	Uncivilised[64]

Table 3: Nature-Culture Dualism Homologies

---

[61] Raymond Williams, 'Nature', p. 220. Bold lettering in the original.

[62] Raymond Williams, 'Ideas of Nature', p. 69.

[63] Michel Foucault, *The Archeology of Knowledge,* p. 38.

[64] For a genealogical account of these dualisms, see Nils Lindahl Elliot, *Mediating Nature.*

it is possible to understand how, for example, historically an equivalence was established between discovering the 'secrets' of both women and nature (as per the some of the original metaphors of modern science); how it might seem that, to really experience nature, it is necessary to go to 'remote' places (as is typically the case of wildlife tourism amongst city dwellers); or how racist European colonists have long represented the ways of aboriginal cultures as being 'closer to nature'.

The analysis I've just offered provides a first, very general sense of a critical understanding of discourse, and of critical discourse analysis. A number of points need to be made to clarify the workings of discourse, thus conceived. To begin with, it should be reiterated that discursive processes (for that is what they are, processes) tend to be entirely implicit or opaque to their subjects, at least until they are subjected to critical scrutiny. Discourses do not work by way of the overt logic that, say, Charles Peirce might invoke to try to explain a philosophical issue, e.g. argument in the strictly logical sense of the term (e.g. syllogisms). Instead, what might be characterised as discursive power (or what Bourdieu would describe as *symbolic* power) will typically entail a combination of largely tacit, unselfconscious, or even unconscious forms of indexing, classifying, and framing a phenomenon, as backed by the practices of specific or generic institutions and fields, and with them a certain configuration of economic, cultural, and social capital.

It is for this reason that, in his later analyses[65] Foucault shifts from the analysis of the 'internal' structures of discourse to the analysis of the social forces that motivate its forms. Returning to the example of modern discourses of Nature, the analyst might, for instance, seek to show how nature treated as an abstract singular might be deployed by patriarchal institutions to justify the imposition of masculinist orders on women. In Chapter 2 I noted how the sociobiologist Edward O. Wilson argues that one of the 'basic human patterns' is that 'adult males are more aggressive and are dominant over females'.[66] In this kind of discourse, there is a double abstract singularity: if nature is the nature of Darwinian adaptation, humans are the beings conceived as a single species driven by such adaptation, but with a dichotomous divide that performs the same singularising feat again, this time with respect to male and female humans. At each step of the way, an abstract singular is generated to produce a generalisations that, if accepted, not only keep both men and women in 'their' place, but exclude those who cannot be classified as being either men or women.

As this example begins to explain, critical discourse analysis requires, amongst other aspects, procedures with which to deconstruct whatever indexing, classification and framing may take place in a particular context or set of contexts. How-

---

[65] See for example, Michel Foucault, 'The Orders of Discourse', *Social Science Information*, 10 (2)(1971), 7-30.

[66] Edward O. Wilson, 'Human Decency is Animal', in *New York Times Magazine*, 12 October 1975.

ever, the analyst must then also link such dynamics to the interests, however un-selfconscious, of particular institutions and their predominant actors. This may seem like a form of conspiracy theorising. In fact, another key theoretical shift entailed by critical discursive analysis is known as the *decentering of the subject*. It is often assumed that individuals have ideas, and that their ideas are, if not entirely their own, then certainly the outcome of more or less voluntary, and conscious forms of reasoning. This is one of the keys to conspiracy theories: one or more individuals work covertly to further their interests. By contrast, a critical approach to discourse analysis is premised on a critique of such voluntaristic notions of subjectivity. Critical discourse analysts argue that, in a manner of speaking, it is *ideas that have individuals,* or what they would more likely describe as *subjects*. According to this perspective, ideas, regarded as elements or aspects of discursive activity, at once offer, and attempt to impose ways of conceiving both the represented objects, and the *self* as it relates to those objects. Put differently, discourse involves not just subjects, but *subjectivity*, and indeed, *subjection*.

An ecologist may, for example, think that they have had a genial insight with respect to some forest dynamic; but however much it may be true that they *have* combined explanatory elements in a more or less innovative way, their thinking will always reproduce at least some aspects of the coding orientations, and with them the ideas of multiple other minds working in specific fields and institutions – a notion that is encapsulated in the common sense suggestion that there is nothing new under the sun. Of course, the scientist may be, and *will* be aware of *some* of the ideas that have influenced the own research, but the researcher (or indeed any other kind of analyst or writer) will never be aware of *all* of the explicit and implicit ways in which their indexing, classification and framing have been shaped by the writing of others, the *labour* of others. Foucault famously alludes to this problematic when in 'The Orders of Discourse', he invokes Molloy in Samuel Beckett's *Molloy*: '... I must say words as long as there are words, I must say them until they find me, until they say me – heavy burden, heavy sin; I must go on; maybe it's been done already; maybe they've already said me; maybe they've already borne me to the threshold of my story...'[67]

Understood in this manner, discourse entails a fundamental ambiguity. To begin with, it may be *enabling* in the way that legal, or engineering, or scientific discourses are always meant to be, and often *are*, at least for some subjects – a 'power to'. That said, any discursive logic will at one and the same time constitute a principle of partiality, and so a rationale of and for exclusion, if not for 'irrationality'. An example of this power to exclude is that of a doctor who quizzes a patient who is clearly ill, but is unable to provide the doctor with what is, from the perspective of *medical discourse*, a recognisable and diagnosable set of symptoms. The doctor tries, in effect, to get the patient to communicate in ways that conform to

---

[67] Michel Foucault, 'The Discourse on Language', in *The Archeology of Knowledge & The Discourse on Language*, p. 215.

the patterns of indexing, classification and framing of the medical discourse. When this fails, the doctor may conclude that the patient is not *really* ill, and has instead some kind of 'psychological condition'. This is not a matter of a conspiratorial attitude on the part of any individual doctor (although on occasion it may be). Instead, it is more likely to be the outcome of a process of field-based acculturation, with individuals learning to interpret the world through a discursive lens that then prevents them from considering alternatives.

An example that is closer to this investigation is the way in which park authorities invoke the sacrality of this or that species, this or that ecosystem to exclude local inhabitants from what were once their own lands. From the perspective of the conservationists who adhere to such a discourse, it is their mission, indeed their *obligation* to try to prevent anyone from destroying a habitat or killing specimens of an endangered species. In this context, anyone who explicitly opposes, or secretly transgresses this sacred mission is likely to be described via labels that reflect the logic of the discourse. The category of poacher is, in this sense, one charged with particular violent histories of dis-possession, and dis-placement. I might note that Barro Colorado is not exempt from this kind of history; as I have noted elsewhere, US colonial authorities in what was then the Canal Zone evicted tens of thousands of people from their own homes, and on the island itself some small-scale farmers were bought out by the scientists who came to manage the new biological reserve (cf. Volume 2).[68] At the time that this happened, such actions would have seemed entirely justifiable within the aegis of the subjectivities promoted by a mixture of colonial, neo-colonial, and scientific discourses.

The reader will have noticed that in this account of discourse, there is a contiguity vis-a-vis the conceptualisation of coding orientations. This is deliberate; building on the work of Basil Bernstein, I have suggested that discourse enables subjects to enact a particular modality of framing – and with it, a particular way of indexing and classifying objects. What is distinctive about this mode of framing is that it employs linguistic means of expression. Put simply, discourse occurs when a certain coding orientation is at once produced, and reproduced (or on occasion contested) via a linguistic medium, or a linguistic form of communication acting in tandem with one that is not (what we may describe as the *paralinguistic*). This constitutes a change vis-a-vis some post-Foucaultian forms of discourse analysis, which suggest that discourse is produced via any and all forms of semiosis, and indeed via any and all forms of more-than-semiotic practice – leading some to conclude that, in effect, *everything is discourse*, or at least, everything is *discursive*.

One reason for adopting such a *discursivist* approach is a pragmatic one, in the sense defined in Part 2 of this study, i.e. linguistic expression has consequences that typically go far beyond the linguistic medium, making it necessary to reject

---

[68] See Nils Lindahl Elliot, 'A Memory of Nature: Ecotourism on Panama's Barro Colorado Island', *Journal of Latin American Cultural Studies*, 19:3(2010), 237-259.

any neat distinction between the linguistic and the non-linguistic, the semiotic and the more-than- (or indeed less-than-) semiotic. Foucault himself seems to take for granted a much broader application of the concept when he refers, for example, to the design of jails (as in the case of the panopticon, in *Discipline and Punish*).

The last point notwithstanding, one risk of such an approach is that the analyst may be tempted to treat as 'statements' events which do not have the consistency, even within a system of dispersion, of anything like linguistic phenomena. A universalising conception of discourse may also encourage the conflation of a discursive regime and the everyday appropriation of such a regime. We return by this route to the kind of objection raised by Bourdieu vis-a-vis codes: Bourdieu would object that the kind of approach to discourse developed by many post-Foucaultian scholars, if not by Foucault himself, decouples meaning-making from the micro or macro determinations of fields. In so doing, it idealises social processes and may, ironically, end up doing some of the work for the *agents* of official or dominant discourses.

Accurate as this kind of critique may be for some forms of discourse analysis, if taken too far it may overlook the fact that one of the characteristics of modern cultures is precisely that certain coding orientations, as produced/reproduced by discourse, *do* cut across fields, and indeed allow their agents to *colonise* fields, and do so at least in part via linguistically inscribed communications. Linguistic messages inscribed in durable media have a certain coherence, a certain structure and a 'communicability' that can, after acknowledging objections of the kind raised by Bourdieu, convey certain meanings, and so promote certain forms of interrelation *across* far-flung contexts. Without some recognition of this phenomenon, we cannot explain the rise of such diverse phenomena as global religions, or indeed successive waves of economic globalisation.

One way to address issues of the kind I've just raised is to distinguish not only between linguistic and paralinguistic forms of communication, but also between non-discursive, pre-discursive, para-discursive and discursive phenomena *sensu stricto*. To clarify what these distinctions entails it may be useful to go back to the modes of observation I described in the conclusions of Part 2 (the immediate, dynamical, and mediate modes of observation). What I characterised as the immediate mode is clearly non-discursive in so far as it is all about firstness. In turn, the dynamical mode may be either non-discursive or pre-discursive in so far as language begins to shape the observer's response in what is nevertheless a form of observation that is primarily driven by secondness. Finally, the mediate mode will be, with some possible exceptions, both discursive and para-discursive in so far as it will involve discourse working in relation to – in 'parallel' to – non-discursive aspects. The exceptions involve forms of observation that are more or less entirely based on reasoning. Peirce, for example, argues that mathematical reasoning also involves a form of observation; '[i]t is observational, in so far as it makes constructions in the imagination according to abstract precepts, and then observes these imaginary objects, finding in them relations of parts not specified in the precept of construction. This is truly observation, yet certainly in a very peculiar sense; and no other kind of observation would at all answer the purpose of mathematics'.[69] In most if not all other cases, the fact that observation involves perceptual systems, and of course objects of perception that may not themselves be discursive, means that the discursive co-exists with the non-discursive, however much the discursive may constitute a kind of filter, an important way of framing the observational process.

The last points serve to clarify how I will relate discourse to observational practice. At least from a post-Foucaultian perspective, observation is always mediate, and discourses always mediate the observational practices. I reject this stance for the reasons explained in Part 2. However, it is still possible to acknowledge that discourses may and in practice *do* tend to shape *mediate* modes of wildlife observation, and may and in practice *do* work to try to 'colonise' any instances of dynamical and immediate observation. When, for example, someone sees something they've never seen before, they are bound to start generating informal hypotheses as to what they've just seen, and chances are that they will do so from the perspective of one or another field, one or another discourse.

I am not suggesting that merely to utter words is already to engage in discursive practice. A person might, for example, exclaim 'What was that?' without necessarily reproducing a particular *discursive* mode of indexing, classification, and framing (beyond that of the language itself, which is of course never neutral). Hence my suggestion of the category of the *pre-discursive*. By contrast, when a guide identifies the phenomenon, and provides information about its natural history or ecology, then they *will* re/produce not only a certain coding orientation, but with it, a particular linguistic modality of framing the object, i.e. a discourse.

So discourse relates to observation as a kind of linguistic filter for the observational practice. I am, however, recognising that in practice it will be difficult, if not impossible to establish any neat boundaries as to where discourse begins and ends vis-a-vis non-linguistic modes of semeiosis. The absence of such boundaries is not, however, a justification for assuming that everything is discursive. I say a *linguistic* filter, but as I explained earlier, indexing, classification, and framing, at least as developed via discursive statements, will always reflect the partiality and interests of one or more institutions and fields, as well as those of the empirical subjects who produce the discourse. This *instituted* and *instituting* partiality is nonetheless likely to remain hidden from at least some subjects thanks to the largely implicit, and unselfconscious character of the coding orientation re/produced by the discourse. For example, *many of the tourists who partook in my ethnographic research were unaware of a paradox: while they wished to experience a tropical forest 'firsthand', many arrived knowing, or thinking that they*

---

[69] Charles Sanders Peirce, *CP* 1.240. (For Peircian referencing conventions, see footnote 16, Chapter 5.)

knew, what they would find, what they would *observe*, and *how* they would observe it, thanks in no small part to the discourses re/produced by the representations of tropical forests circulated by the nature media. Accordingly, Barro Colorado could, would, *should* in effect be a *jungle* of the kind portrayed by the nature media's linguistic and para-linguistic signs, and one teeming visibly with what I have described as the *charismatic* fauna. This being the case, while the visitors wanted to see wildlife 'firsthand', and 'with their own eyes', a case can be made that they actually arrived on the island expecting to see what they had already been shown by the nature media (we return to the point raised by Urry and Larsen about the tourist gaze). This kind of dynamic may be taken as a metaphor for the way in which discourses may come to permeate someone's observational practices, and consciousness more generally.

Genres and Techniques of Observation

Discourse constitutes one way of specifying the framing aspect of coding orientations. Another way of doing so is via the concept of *genre*. In everyday life, genre is typically treated as a word for *type* – for example, we may refer not only to a literary genre or a genre film, but also to zoos, national parks, wildlife tourism, birdwatching, etc. The last examples notwithstanding, in literary, communication and cultural studies, genre is theorised as involving far more than just a type, if by that one means a kind of label. Genres are thought to act as devices (in the French sense of *dispositif*, or the Spanish sense of *dispositivos*) for the adjustment, or better yet, the attempted *synchronisation* of social interaction. Such synchronisation is particularly important in contexts that involve mass communication, where, as I began to note earlier, space-time distanciation generates a gap between those who produce and those who produce messages. However, a similar phenomenon occurs even in the context of activities not immediately associated with mass communication, or 'communication from afar'; in museums as in zoos, in botanic gardens as in nature parks, genres play a key role by providing subjects with more or less shared coding orientations with which to engage in acts of mutually understood semeiosis, or what might also be described as mutually assured construction. When someone says they're going to a zoo, or to a nature reserve, the identification of the experience will serve to begin to explain what kind of activity is likely to be engaged. From this perspective, genres are best understood at once as horizons of expectation, and as 'strategies of practicability'[70] – ways of ensuring, or at any rate *attempting* to ensure that all those who interact can do so

---

[70] I modify in this way the term discussed by Jesús Martín-Barbero: strategies of 'communicability' in Jesús Martín-Barbero, *Communication, Culture and Hegemony: From the Media to the Mediations* (London: Sage, 1993), p. 223.

according to forms that are at least partly shared, understood, and/or appropriately corresponded.

I say attempting to ensure because genres do not guarantee, in and of themselves, that mutually assured construction *will* occur. There is always the possibility that some individuals or social groups may not understand regardless, or indeed that some will do their best to deliberately undermine generic forms. It is also the case that some groups may recognise, but not be able to realise (i.e. partake) in some forms of generic practice. It nonetheless seems clear that, if nothing else, genres may reduce the likelihood of misinterpretation based on the misrecognition of form. For example, to say to someone that a wildlife TV programme is a documentary is to attempt to ensure, amongst other things, that the person will interpret the programme as being 'factual'; by contrast, to say that a TV show is a drama suggests the need for a different interpretive strategy, e.g. one that accepts that a story is being told which is fictional. Then again, some genres such as 'reality shows' attempt to blur the boundaries between existing genre classifications, deliberately confusing the boundary that ostensibly separates fact from fiction. One way or another, genres do not simply provide formal prescriptions for framing; their specificity lies not in 'content' or even in 'form' (approached as a purely taxonomical matter) but in the *pragmatics* of communicative practice.

How, then, does this conceptualisation relate to wildlife observation by tourists visiting tropical forests? A first, and relatively obvious application concerns the genres of tourism. Terms such as 'wildlife tourism' or 'ecotourism' all evoke not just *types* of tourism, but potential destinations, as well as objects to gaze upon or otherwise experience in those destinations. A related, but somewhat more subtle application of the concept of genre involves types of observation, or what are more accurately described as *techniques of observation*. As I explained in Part 1, in everyday life and in some positivist fields, perception tends to be treated as a matter of purely physiological, and indeed innate activity – the assumption being that individuals observe objects in much the same way regardless of context. But as I've suggested in this chapter, a cultural sociology of observation insists on the importance of cultural context, as specified by institutions, fields, codes, and discourses. The category of genre further specifies the conceptualisation of framing by noting that even within a field, or as part of a coding orientation, there may be different ways of engaging in the observational process – there may be not only different genres, but with them, different techniques of observation.

Consider, for example, the difference between the kind of observation engaged, within wildlife tourism, by 'generalist' travellers, and by birdwatchers – amongst the latter, especially the 'birders' or 'twitchers' that I referred to earlier. The generalist wildlife observers are typically content to observe a variety of taxa, though within the confines of so-called charismatic fauna (cf. Part 4). This being the case, a tour of a tropical forest will take the form of a guided excursion for which each encounter, each observational event will entail a relatively informal, to not say casual examination of whatever specimens are encountered and/or chosen for closer observation. Indexing, classification, and framing rules will be relatively weak, in the sense that they will entail lightly policed categories and boundaries at least with respect to what wildlife is sought, and how it is observed. More often than not, the members of the tour will be free to point out any aspect of the forest that interests them, and to raise a more or less broad spectrum of questions about them.

This may be contrasted with the kind of strongly policed coding orientations of birders or twitchers, for whom the repertoire of charisma will be reduced not only to avian species, but to those which are considered to be either rare, or interesting for reasons to do with detailed characteristics of one or another genus, species, or sub-species. As part of this shift, the discursive interaction will be far more confined to a discussion not just of species of interest, but to more or less specialised commentary ('observations') regarding the particularities of one or another type of bird, one or another *specimen.*

The point I am leading up to is that each kind, each *genre* of tour will entail quite different techniques of observation. In the introductory chapter, it was suggested that the notion of a technique of wildlife observation could be understood as a structured, and *structuring* way of engaging in observational practice. On the basis of the concepts introduced in this part of the study, the notion can be specified as follows: a technique of observation is a particular way of indexing, classifying and framing an object, or a set of objects (always in the broadest semeiotic sense of the expression), contingent on an ensemble of aspects, including whatever object is being observed; any organisation, field, discourse, genre or technology with which, in which or from which the object is observed; and of course the capacities that individuals or groups bring to bear, and co-develop as part of the realisation or 'use' of the technique. Simplifying somewhat, particular techniques of observation tend to 'go' with particular genres.

Both genres and techniques of observation entail the kind of ambiguity that I noted earlier with respect to organisations, fields or discourses: a genre-based way of observing something may be enabling to some observers, in some respects; but that same enablement may be, and *must* be the beginning of exclusion of some kinds of indexing, classification, and framing, and by implication, of other genres and techniques of observation. Even if generic ways of framing may help to develop and sustain more or less synchronised horizons of expectation between participants in a certain practice, if those ways become so sedimented that they come to seem natural, then at least some may come to regard them as being sacred, in the anthropological sense of the term. When this happens, genres may aid and abet ideological relations if only because they make it difficult for newcomers to contest what has become, in effect, a genre orthodoxy.

In the Prologue, I described the *Planet Earth II* TV series, which, despite being broadcast a decade later than its original namesake, *Planet Earth* (2016 vs. 2006, respectively) employed very similar techniques of observation (and with these, narrative techniques). As part of such techniques, a discourse was employed which mentioned only very briefly the challenges posed by anthropogenic climate change – a tendency that has long been criticised by environmental activists, and which in *Planet Earth II's* case generated a sharp newspaper article penned by another BBC producer[71] (I will return to that article in the Postscript). It is, however, not too difficult to imagine why the notoriously conservative BBC Natural History Unit would be reluctant to change genre rules that have long excluded detailed, and critical consideration of the kinds of institutional forces that are responsible for climate change, and indeed for virtually all other anthropogenic threats to the environment. There is always the risk of offending conservative audiences, whose tendency to idealise nature has often stood in the way of recognising the reality of environmental change. Then again, such conservatism might find an ironic common ground with more liberal audiences who seek the solace of programmes devoted to representing an ostensibly timeless, and 'non-political' Nature in the midst of social upheaval. Rather more prosaically, the TV producers have long worried that references to particular environmental histories might affect the shelf-life of their products.[72] Or in so far as many of the producers in the field of natural history TV have science degrees, many may believe that 'politics should be kept out of nature'.

Whatever the precise motivations, any choice to include or exclude the overt politics of nature in wildlife documentaries would have to be based on a certain calculus of its impact on a particular representation's relation to the genre – a relation based on more or less durable coding orientations, the drastic change of which might lose audiences, and with them economic capital for the producers. I've focussed on the example of natural history documentaries, but of course the latter calculations apply to any genre, and for any more or less established technique of observation.

Cultural Formations

Thus far, I've approached each of the concepts introduced by this chapter separately. To conclude the chapter, I'd like to propose a concept that links the various categories: that of *cultural formation*. As I will employ this term, the concept refers to a cluster, or better yet what I will describe in Chapter 10 as an assemblage of institutions, fields, coding orientations, discourses, genres, techniques and tech-

---

[71] See Martin Hughes-Games, 'The BBC's Planet Earth II Did Not Help The Natural World', *Guardian online*, <https://www.theguardian.com/commentisfree/2017/jan/01/bbc-planet-earth-not-help-natural-world> 1 January 2017 [Accessed 2 January 2017].

[72] See the analysis of market conditions offered by Simon Cottle, 'Producing Nature(s): On the Changing Production Ecology of Natural History TV', in *Media, Culture, and Society*, 26:1(2004), 81-102.

nologies. Examples of such formations are tourism, mass communication, and science. A cultural formation is cultural in so far as it involves, in Raymond Williams' terms, signifying systems through which a social order is communicated, reproduced, experienced, and explored.[73] However, the communication, production, experience and exploration occur in, and *across* multiple institutions, fields, coding orientations, discourses, genres, techniques and technologies.

The concept of cultural formation can be illustrated by returning to the example with which this part of the study began. On the face of it, a person was merely looking at a crocodile along the edge of Barro Colorado's shoreline. In fact, that person was only able to reach that place, and then interpret the crocodile in the way that he did thanks to an 'ensemble of ensembles', viz., a number of different organisations (tour operators, airlines, hotels, park authorities, etc.); fields (tourist, but also scientific, environmental, etc.); discourses (again, tourist, scientific, environmental, etc.); and genres (e.g. that of the holiday, of 'wildlife observation', but also of 'wildlife TV'). What may have seemed like a 'simple act of perception' was thus actually the outcome of a remarkably complex concatenation of elements – a concatenation which cannot be explained adequately merely by focussing on one or another organisation, one or another field, discourse, or genre. Nor can it be adequately explained by taking for granted that the ensemble works as an organic whole; as I will explain in Chapter 10, it is necessary to approach such wholes from the perspective of assemblage theory.

[73] Raymond Williams, *Culture*, p. 13.

9

Body, Biopower and Situation

I've now discussed in some detail the various ways and levels in which cultural practice might shape wildlife observation. But the reader will note that, with the exception of *hexis*, none of the concepts that I mentioned allowed for the fact that practice is *embodied*. To be sure, even the concept of *hexis* takes for granted something like a one-way system, with culture flowing from social orders to shape the body. From this perspective, the body, understood provisionally as the body of 'flesh and bone', or better yet what might be termed the *organismic* body, is left bereft of any capacity to return favours, so to speak: it is the recipient of culture, but has no capacity of its own to shape culture.

This kind of stance is the obverse of the one generally adopted by sociobiological and neuroscientific approaches (cf. Chapters 2 and 3). From a neuroscientific perspective such as Francis Crick's, it's neurons all the way down, never mind Mind or Culture. For Edward O. Wilson, culture is a kind of developmental afterthought (or perhaps I should say 'after-body'), a set of rules mostly pre-determined by bodily adaptation. While I argued in Part 1 that at least some forms of neuroscience would appear to be premised on a deeply paradoxical decorporealisation of perception, in general both sociobiology and neuroscience err on the side of 'de-culturalisation'.

Is it possible, or indeed desirable, to find some kind of halfway house? One alternative might be the one proposed by ecological psychologists, and more specifically by James Gibson (cf. Chapter 4), who, despite opposing both neurological and adaptationist reductionism, reminds us time and again that perception is a matter of the whole body – not just this or that 'sense', but a *perceptual system* that includes receptors, organs, and a body capable of the all-important but so often overlooked locomotion. The head that holds the brain and the eyes is not static, let alone free-floating; when we look at something, we not only turn our eyes and our head, but we use our whole bodies to position ourselves in a place that better enables the observation. And of course, even as we stop to look, listen, or smell something, the lungs continue to expand and contract even as a fluid flows through the blood vessels, cells grow and die.

Notwithstanding these and numerous other insights, ecological psychology is itself unacceptably reductionist in so far as it completely excludes representation and institutionalisation from what Gibson quite literally regards as direct perception. What, then, might be a different alternative?

Beginning the late 1980s and continuing throughout the 1990s and even the early 2000s, theories of embodiment became all the rage amongst many poststructuralist scholars. However, during this period many in sociology shared in Bourdieu's inclination to conceive embodiment in terms of that 'political mythology realized, *em-bodied*'.[1] And those with a Foucaultian inclination were, if anything, even more radically constructivist. Until the late 1990s if not beyond, many assumed that everything, including the body, is subjected, if not by discourse, then by the kind of cultural descent described by Foucault as part of his turn to genealogical inquiry. While this study rejects that stance, it may be helpful to consider its core assumptions in order to prepare the ground for a different approach. To this end, I will begin by examining a Foucaultian take on embodiment, as developed initially by Foucault himself, and then by Judith Butler, one of the feminist scholars who engages most closely with Foucault's account of the relation between body and cultural process.

It is pertinent to begin by acknowledging the importance of Foucault's genealogical turn. I will describe this theoretical and methodological displacement in some detail in Volume 2, so here it will suffice to describe it in very general terms. For Foucault, genealogy entails the *problematisation* of a contemporary practice (contemporary, that is, to the time when the research begins) by way of historical inquiry. Genealogy may thereby be characterised as a 'history of the present', but one for which the aim is *not* to show how a practice is the outcome of the kind of logic expressed by a diagrammatic tree of past relations, or by a neat chain of causes and their effects. It is also not conceived as the outcome of voluntary or entirely conscious actions by reflexive, and self-reflexive individuals. Instead, Foucaultian genealogy seeks to explain how it is that particular forms of practice emerge from the activities of subjects operating across multiple spaces and times, and in relation to manifold institutions, discourses, genres, techniques and technologies. Such elements shape practice by way of at least partly unselfconscious motivations, by way of nocturnal connections between things or processes thought *not* to be connected, but also, and conversely, by epistemic *breaks* in what might seem to be continuous, and linear sequences of events.[2]

In *History of Sexuality*,[3] Foucault applies this approach to explain modern understandings of sexuality. In a context in which it is widely taken for granted that sexuality is determined by biology, or by the nexus of biology and complexes of the kind described by psychoanalytic theory (cf. Chapter 10), Foucault proposes an account that theorises what he describes as 'bio-power' (in keeping with cur-

---

[1] Pierre Bourdieu, *The Logic of Practice*, pp. 69-70. Italics in the original.

[2] For an introduction to the difference between Foucault's 'archeological' and 'genealogical' phases, see for example Lois McNay, *Foucault: A Critical Introduction* (New York: Continuum, 1994). See also Robert Castel, '"Problematization" as a Way of Reading History', in Jan Goldstein (Ed), *Foucault and the Writing of History* (Oxford: Blackwell, 1994), 237-252.

[3] See for example, Michel Foucault, *History of Sexuality, Volume I: An Introduction*, translated by Robert Hurley (New York: Pantheon Books, 1978[1976]). See also, Michel Foucault, 'Nietzsche, Genealogy, History', translated by Donald F. Brouchard and Sherry Simon, in James D. Faubion (Ed.), *Aesthetics, Method, and Epistemology: Essential Works of Foucault 1954-1984, Volume 2* (Harmondsworth: Penguin, 1998 [1971]), 369-392.

rent academic practice, henceforth I will spell the term without the hyphen). Biopower is the process whereby modern institutions come to acquire both knowledge of, and a certain control over the relation between the human body and the environment. In so doing, the institutions not only manage to ameliorate what have hitherto been the grave and omnipresent threats of starvation or plague, but also develop ever more extensive and intensive ways of trying to control the body for these, as well as for more manifestly political purposes. As Foucault puts it, 'If one can apply the term *bio-history* to the pressures through which the movements of life and the processes of history interfere with one another, one would have to speak of *bio-power* to designate what brought life and its mechanisms into the realm of explicit calculations and made knowledge-power an agent of transformation of human life'.[4] It is not, Foucault clarifies, that life 'has been totally integrated into techniques that govern and administer it'; on the contrary, 'it constantly escapes them'. And indeed, outside the 'Western world, famine exists, on a greater scale than ever' and 'the biological risks confronting the species are perhaps greater, and certainly more serious, than before the birth of microbiology. But what might be called a society's "threshold of modernity" has been reached when the life of the species is wagered on its own political strategies.'[5]

In this context, the problem for Foucault – and indeed for many of the scholars that will follow his lead – is not to adumbrate the ways in which biology does indeed escape total incorporation into the social apparatuses that might govern and administer it – that would perhaps be too 'obvious', and in any case, 'romantic'. Nor is it to acknowledge the ways in which bodies may mark cultural processes – from a Foucaultian perspective that would be to revert to biologism, and with it essentialism. Instead, the challenge is to reveal how 'descent' (being at once a part of, and an outcome of a more or less ancient system of affiliation and power) 'inscribes itself in the nervous system, in temperament, in the digestive apparatus' and 'appears in faulty respiration, in improper diets, in the debilitated and prostrate bodies of those whose ancestors committed errors'.[6] To this list, we may add the ways in which descent inscribes the sexualities of individuals and groups – part of the subject of Foucault's *History of Sexuality*.

This kind of approach, which Foucault developed in research spanning from the early 1970s to the early 1980s, was decades ahead of medical research that has belatedly acknowledged that one's health is also determined by so-called 'social factors'. If this is true for any culture, it is likely to be particularly true for cultures that are based on vast inequalities in access to economic, cultural and social capital. It is nonetheless possible to take issue with the way in which Foucault, and especially some of his followers have made it difficult, to not say impossible to

---

[4] Michel Foucault, *History of Sexuality*, p. 143. Italics in the original.

[5] Michel Foucault, *History of Sexuality*, p. 143.

[6] Michel Foucault, 'Nietzsche, Genealogy, History', p. 375.

conceive of relations between bodies that are not themselves entirely 'bio-pow-ered', i.e. relations not entirely institutionalised in the cultural sociological sense of the term, which is to say 'culturalised', pre-interpreted (to use a hermeneutic expression) and disciplined. If this kind of problem is already evident in Fou-cault's famous rejection of 'nature',[7] a similar stance appears with respect to the body. Even if Foucault does acknowledge that human bodies are not entirely cul-turalised (or to use his vocabulary, subjected), it is clear that his emphasis is very much on revealing the manifold ways in which a certain subjection *does* prevail.

After Foucault died, some post-Foucaultian scholars pushed this kind of ap-proach to what might be regarded as its logical extreme, i.e. showing how even those aspects of the body that Foucault treated as being somehow biological are themselves not only subjected, but *conceived as being biological* thanks to an itself discursive process. In the early 1990s, perhaps the most important exponent of this perspective was the feminist philosopher Judith Butler, who in her book *Gen-der Trouble* suggests that *both* sex and gender are cultural constructs, and that there is, in effect, no such thing as a 'biological body' as opposed to a 'cultural body', a sexual body as opposed to a gendered body.

To explain this perspective, a brief contextualisation is required. Amongst Eng-lish-speaking feminists, the sex/gender distinction has long provided a key dis-cursive basis with which to challenge patriarchal attempts to fix social roles on the basis of sex. The distinction is often attributed to the psychologist and sexologist John Money, who, together with other scholars in the 1950s, proposed to distin-guish between 'sex roles' and 'gender roles', the latter involving what today might be described as social construction.[8] As taken up by the 1970s by feminist scholars and activists, the distinction was used to make the case that, while sex might be understood as the biologically given aspects of the body in a context of sexual dimorphism, gender should be conceived as a matter of culturally con-structed roles assigned to either sex via particular social institutions. At this his-torical juncture – and perhaps today again – the battle was to show that there are no automatically, and biologically-determined links between sex and gender, and to reveal the many systematically arbitrary ways in which patriarchal institutions use the conflation of sex and gender to dominate women (at that point, many feminists took for granted the prevalence of sexual dimorphism, to the exclusion of gay, lesbian and transexual orientations). From the perspective of the so-called

---

[7] By one account, during a holiday in the southern Alps, Foucault's friend Jacqueline Verdeaux invited Foucault to survey magnificent landscapes along their travel route, and Foucault 'made a great show of walking off toward the road, saying, "My back is turned to it"'. See Didier Éribon, *Michel Foucault*, translated by Betsy Wing (London: Faber & Faber, 1993), p. 46.

[8] See for example John Money, Joan G. Hampson, and John L. Hampson, 'Imprinting and the Establishment of Gender Role' *A.M.A. Archives of Neurology and Psychiatry*, 77 (1957), 333-336.

second-wave of feminist theory and activism, there might be some differences between the bodies of men and women, but that should not, and indeed does not naturally or automatically translate into similarly fixed conceptions of gender.

In this context, a Foucaultian perspective, as developed and eventually superseded by Judith Butler, provided a welcome critique of any simplistic opposition of sex and gender, albeit in the direction of a more radically constructivist interpretation. In *Gender Trouble*, Butler initially adopts a Foucaultian perspective when she notes that 'For the most part, feminist theory has assumed that there is some existing identity, understood through the category of women, who not only initiates feminist interests and goals within discourse, but constitutes the subject for whom political representation is pursued.'[9] As far as Butler is concerned, this is a good example of the kind of process that Foucault critiques when he points out that juridical systems of power (including those that claim political or other forms of representation on behalf of one or another group) *produce* the subjects they subsequently come to represent.[10] In the case of feminism, the assumption that feminists represent *women*, and that *women* denotes a common identity, a stable signifier (the Saussurean concept of sign continues to predominate in this context), may be questioned on various grounds. Even if one can be said to be a woman, that is not *all* one is. To be sure, gender is not a transhistorical category (i.e. one that is independent of historical context or contingency), let alone one that can be neatly separated from 'race', class, ethnicity, sexual orientation or other identities. In the context of this kind of critique – and an analogous one of similarly univocal discourses on patriarchy – Butler asks if 'the construction of the category of women as a coherent and stable subject' is not 'an unwitting regulation and reification of gender relations', and if such a reification is not contrary to feminist aims.[11]

As part of a genealogical inquiry into the matter, Butler suggests that a key part of the discursive basis for 'women' relies on the very distinction between sex and gender. As I noted above, amongst feminists the distinction was originally intended to *dispute* biologically determinist accounts of gender roles. However, Butler is concerned that the dyad treats sex, and by implication the body, at once as a matter of a biological given, and as a *tabula rasa* awaiting inscription by *both* sexist and feminist discourses. 'Taken to its logical limit, the sex/gender distinction suggests a radical discontinuity between sexed bodies and culturally constructed genders. Assuming for the moment the stability of binary sex, it does not follow that the construction of "men" will accrue exclusively to the bodies of males or that "women" will interpret only female bodies'; 'even if the sexes appear to be unproblematically binary in their morphology and constitution' – a

---

[9] Judith Butler, *Gender Trouble: Feminism and the Subversion of Identity* (London: Routledge, 1990), p.1.

[10] Judith Butler, *Gender Trouble*, p. 2.

[11] Judith Butler, *Gender Trouble*, p. 5.

stance that Butler takes a leading role in questioning – 'there is no reason to assume that genders ought also to remain as two'.[12] 'The presumption of a binary gender system', Butler argues, 'implicitly retains the belief in a mimetic relation of gender to sex whereby gender mirrors sex or is otherwise restricted by it.'[13] This relation, Butler suggests, is secured by 'casting the duality of sex in a prediscursive domain'; instead, sex ought to be understood as a discursive production in its own right, i.e. not sex as the prediscursive but sex as 'the effect of the apparatus of cultural construction designated by *gender*'.[14]

As I began to explain earlier, Butler links this kind of displacement to the conception of the body. If there is, as it were, a political shape to 'women' – one that precedes and prefigures the political elaboration of their interests and epistemic point of view – it may well be one that treats the sexed body as the ground, surface or site of cultural inscription.[15] If this is the case, Butler asks '[w]hat circumscribes that site as "the female body"? Is "the body" or "the sexed body" the firm foundation on which gender and systems of compulsory sexuality operate'?[16]

Butler's answer is a clear 'no', especially when the body is regarded as 'a passive medium that is signified by an inscription from a cultural source figured as "external" to that body'.[17] As part of her critique, Butler points out the perils of adopting a Cartesian stance (a stance that I will describe in Chapter 10) that treats the body as 'mute facticity', even as consciousness is regarded as being transcendent, and as 'radically immaterial'. Butler believes that Foucault himself is guilty of this kind of idealism when he appears to suggest that there is a body prior to inscription, a medium that is treated as a blank page. 'By maintaining a body prior to its cultural inscription, Foucault appears to assume a materiality prior to signification and form.'[18]

Turning to the social anthropologist Mary Douglas's classic text *Purity and Danger*,[19] Butler proposes as an alternative that 'the very contours of "the body" are established through markings that seek to establish specific codes of cultural coherence'; 'what constitutes the limit of the body is never merely material, but the surface, the skin, is systematically signified by taboos and anticipated transgressions; indeed, the boundaries of the body become, within her analysis [Mary Douglas', albeit as interpreted by Butler], the limits of the social *per se*.' A post-

---

[12] Judith Butler, *Gender Trouble*, p. 6.

[13] Judith Butler, *Gender Trouble*, p. 6.

[14] Judith Butler, *Gender Trouble*, p. 7. Italics in the original.

[15] Judith Butler, *Gender Trouble*, pp. 128-129.

[16] Judith Butler, *Gender Trouble*, p. 129.

[17] Judith Butler, *Gender Trouble*, p. 129.

[18] Judith Butler, *Gender Trouble*, p. 130.

[19] Mary Douglas, *Purity and Danger: An Analysis of Concepts of Pollution and Taboo* (London: Routledge, 1996[1966]).

structuralist appropriation of Douglas' view 'might well understand the boundaries of the body as the limits of the socially *hegemonic*'.[20]

The last point brings us back to the issue with which this chapter began: in the kind of approach developed by Butler, if not by Foucault himself, the physical materiality of the body becomes either a vanishing quality, or one shaped entirely by one or another form of biopower. Butler quotes Foucault when he suggests that the body is 'a volume in perpetual disintegration', and I can think of no better way of expressing what happens to the physical materiality of body in the writing of Butler and those who follow her, if not Foucault's, lead. This is deeply ironic, for having started by criticising – rightly in my view – univocal conceptions of the body that treat it as a kind of *tabula rasa*, the scholars arguably end up performing precisely this kind of rationalising operation: the body becomes, in effect, little more than an expression of regulatory power. What is lost, in *Gender Trouble* as in many other poststructuralist texts with otherwise admirably sharp critiques, is the possibility that the body – or rather, the *bodies*, be they female, male, queer, transexual, human or beyond-human or none of the above – may have a materiality that is not only discontinuous vis-a-vis the apparatuses of hegemonic social control, but capable of resisting inscription in ways that are not themselves 'cultural' or 'social'.

How to avoid, then, an understanding of bodies that is not riven by the abject materialism of the neuroscientists and sociobiologists on the one hand, and what has to be regarded as the *false materialism* of at least some of the poststructuralist scholarship?

According to the feminist philosopher Toril Moi, one alternative may be found in the proposals of the existentialist and phenomenological philosophers of the 1940s and early 1950s. In order to explain this turn, Moi notes that 'The widespread tendency to criticize anyone who thinks that biological facts exist for their "essentialism" or "biologism" is best understood as a *recoil* from the thought that biological facts can ground social values. Instead of denying that biological facts ground any such thing, ... poststructuralists prefer to deny that there *are* biological facts independent of our social and political norms'.[21] Moi is left with the impression 'that poststructuralists believe that if there *were* biological facts, they would indeed give rise to social norms. In this way, they paradoxically share the fundamental belief of biological determinists ... In their flight from such unpalatable company they go to the other extreme, placing biological facts under a kind of mental erasure'.[22] Moi sums up the logic of this erasure as follows: if it is claimed that biological facts lead to social norms, then this kind of claim can be undermined simply by striking out biological facts. Echoing the realism that I attributed

---

[20] Judith Butler, *Gender Trouble*, p. 131. Italics in the original.

[21] Toril Moi, 'What is a Woman? Sex, Gender and the Body in Feminist Theory' in *What is a Woman and Other Essays* (Oxford: Oxford University Press, 2000), 3-120., p. 41.

[22] Toril Moi, 'What is a Woman?', p. 42.

to Charles Peirce in Part 2, Moi notes that many poststructuralists 'believe that in order to avoid biological determinism one has to be a philosophical nominalist of some kind. In their texts, philosophical realism becomes a *politically* negative term'. This is absurd; 'to avoid biological determinism all we need to do is to deny that biological facts justify social values, and even the most recalcitrant realist can do that.'[23]

To move away from the nominalism in question, Moi proposes to re-engage with the writing of Simone de Beauvoir (1908-1986),[24] a feminist philosopher, novelist and activist long dismissed by many poststructuralist scholars for the supposed essentialism of her claims regarding the relation between body and gender. In response to this kind of claim, Moi points out that de Beauvoir actually opposes an essentialist conception of gender, while still acknowledging a fundamental role for the biologically given, or at any rate, the biologically *received*. And indeed, in *The Second Sex* de Beauvoir states very clearly that while biological facts do not establish a fixed or inevitable destiny for women, they nonetheless *are* an essential *element* of a woman's situation:

> ... biological data are of extreme importance: they play an all important role and are an essential element of woman's situation ... Because the body is the instrument of our hold on the world, the world appears different to us depending on how it is grasped, which explains why we have studied these [biological] data so deeply; they are one of the keys that enable us to understand woman. But we refuse the idea that they form a fixed destiny for her. They do not suffice to constitute the basis for a sexual hierarchy; they do not explain why woman is the Other; they do not condemn her forever to this subjugated role.[25]

As de Beauvoir puts it, 'The body is not a *thing*, it is a situation: it is our grasp on the world and a sketch of our projects'.[26] Moi underscores that the body as situation is not to be confused with the more widely accepted notion that bodies are always *in* some situation. For de Beauvoir, both claims are valid, but non-reducible to each other.

So how, Moi asks, can de Beauvoir maintain that biology is in some sense fundamental to a woman's situation, but at the same time, that biology is not her destiny? (a similar point might of course be raised with respect to any other gender

---

[23] Toril Moi, 'What is a Woman?', p. 43.

[24] Simone de Beauvoir, *The Second Sex*, translated by Constance Borde and Sheila Malovany-Chevallier (New York: Vintage Books, 2009 [1949]).

[25] Simone de Beauvoir, *The Second Sex*, p. 68.

[26] Simone de Beauvoir, *The Second Sex*, p. 68. Italics in the original.

classification). The answer, she suggests, lies in the existentialist and phenomenologist orientation of de Beauvoir's conception. According to de Beauvoir,

> As Merleau-Ponty rightly said, man is not a natural species: he is a historical idea. Woman is not a fixed reality but a becoming; she has to be compared with man in her becoming; that is, her *possibilities* have to be defined: what skews the issues so much is that she is being reduced to what she was, to what she is today, while the question concerns her capacities; the fact is that her capacities manifest themselves clearly only when they have been realized: but the fact is also that when one considers a being who is transcendence and surpassing, it is never possible to close the books.[27]

As Moi interprets de Beauvoir, the latter is acknowledging that human transcendence is always incarnate(d): 'My body is a situation, but it is a fundamental kind of situation, in that it founds my experience of myself and the world. This is a situation that always enters my lived experience'; and this is why 'the body can never be just brute matter to me'.[28] Moi refers to two oft-quoted passages, one from the existentialist philosopher Jean-Paul Sartre (1905-1980), and the other from the phenomenologist philosopher Maurice Merleau-Ponty (1908-1961), that sum up what it means to say that the body *is* situation, even as it is *in* situation: 'existence precedes essence' (Sartre[29]), and 'the body is our general medium for having a world' (Merleau-Ponty[30]).

The famous, and oft-quoted statement by Sartre is meant to challenge essentialism; contrary to the notion that nature determines how each individual exists – effectively, the stance of sociobiology and in a somewhat different way, neuroscience (cf. Chapters 2 and 3, respectively) – Sartre argues that the individual shapes its own life through acts of consciousness which are fundamental to any human essence. I will return to this kind of stance in Chapter 10. In this chapter I would like to consider in some detail the statement by Merleau-Ponty, which emerges in the research that Merleau-Ponty presented in two of his earliest works, viz. *The Structure of Behavior*,[31] and *Phenomenology of Perception*, which was quoted earlier. Especially in the second of these texts, we may find a way of approaching the relation between embodiment and observation that refuses a *positivist* real-

---

[27] Simone de Beauvoir, *The Second Sex*, p. 68. Italics in the original.

[28] Toril Moi, 'What is a Woman?', p. 63.

[29] Jean-Paul Sartre, *Existentialism is Humanism* (New Haven, CT: Yale University Press, 2007[1946]).

[30] Maurice Merleau-Ponty, *Phenomenology of Perception,* translated by Colin Smith (London: Routledge, 2002 [1945]), p. 169.

[31] Maurice Merleau-Ponty, *The Structure of Behavior*, translated by Alden L. Fisher (Boston, MA: Beacon Press, 1967 [1942]).

ism (or what Merleau-Ponty describes as a scientific realism), but also at least aspects of the drift towards a 'new idealism' that is found in the work of some post-structuralists – what Merleau-Ponty in his own time describes as a transcendental idealism, and as an *intellectualism* in a reference to the philosophy of Immanuel Kant and his many followers.

By way of a contextualisation, Maurice Merleau-Ponty is regarded as one of the great phenomenological philosophers. However, the kind of phenomenology that Merleau-Ponty practices is in some respects quite different from that of Peirce (cf. Part 2). Merleau-Ponty takes his lead, at least to begin with, from the founder of Continental phenomenology, Edmund Husserl (1859-1938). Husserl describes his phenomenology (what he calls transcendental phenomenology) as a kind of neo-Cartesianism – a return to the philosophical principles of René Descartes (cf. Chapter 10), despite the fact that Husserl rejects virtually all of the doctrinal content of Cartesian philosophy.[32] What nevertheless links Husserl to Descartes is a shared conviction that critical thought should proceed on the basis of the kind of radical doubt proposed by Descartes. Husserl suggests that Descartes arrives at the wrong conclusions when applying this doubt, but Descartes is right to suggest that the 'method of doubt' should be employed by the philosopher in order to refuse anything as existing unless 'it can be secured against any conceivable possibility of becoming doubtful'.[33] In this context, the phenomenologist tries to describe things precisely as they are experienced, but this *not* on the basis of the Cartesian (and later positivist) notion that we should distinguish between the inner or subjective *appearance* of things on the one hand, and the outer or objective *realities* of the world of the other. If phenomenologists are rigorous enough, and apply the method of doubt appropriately, then they will be able to produce scientific accounts of phenomena (phenomena understood in this context as anything that presents itself to someone). Not scientific in the sense of the physical or empirical sciences, but scientific in the sense of the German '*Wissenschaft*', viz., a systematic form of investigation that may be validated by other similarly rigorous and objective scholars employing the same methods.

Like Husserl, Merleau-Ponty is concerned to establish a method of inquiry that is systematic, and which rejects not just Cartesian dualisms or the positivism that stems from these, but also the idealism of Immanuel Kant. Key to this is an approach to phenomenology that puts far more emphasis on embodiment and perception.

As Merleau-Ponty puts in the Preface of *Phenomenology of Perception*, published in 1945 and long the work he was best known for, 'I am not the outcome or the meeting-point of numerous causal agencies which determine my bodily or psy-

---

[32] Edmund Husserl, *Cartesian Meditations*, translated by Dorion Cairns (The Hague: Martinus Nijhoff, 1977 [1931]), p. 1.

[33] Edmund Husserl, *Cartesian Meditations*, p. 3.

chological make-up. I cannot conceive myself as nothing but a bit of the world, a mere object of biological, psychological or sociological investigation. I cannot shut myself up within the realm of science'; on the contrary, Merleau-Ponty argues that all his knowledge of the world, even his scientific knowledge, 'is gained from my own particular point of view, or from some experience of the world without which the symbols of science would be meaningless'.[34] Indeed, Merleau-Ponty argues that scientific points of view, 'according to which my existence is a moment of the world's, are always both naïve and dishonest, because they take for granted, without explicitly mentioning it, the other point of view' – 'that of consciousness, through which from the outset a world forms itself around me and begins to exist for me'.[35]

Merleau-Ponty is nonetheless clear that the intellectualism of the Kantian stance does not offer a better alternative. Kant's famous 'Copernican revolution', presented in *The Critique of Pure Reason*,[36] suggests that the problem is not to explain how objects determine concepts – a stance widely adopted amongst intellectuals after Newton – but instead how objects, or the experience in which alone they can be known (as given objects), conforms to the forms projected by the concepts. Like Nicolaus Copernicus, who initially tried to explain the celestial motions by assuming that the celestial host revolves around the observer (only to find that it is the other way around), Kant suggests that the more critical way of explaining reason is to adopt what today we would describe as a constructivist stance: as Kant puts it, 'experience is itself a species of knowledge which involves understanding; and understanding has rules which I must presuppose as being in me prior to objects being given to me, and therefore being *a priori*.'[37] Simplifying greatly, it is not the objects that determine how we reason, but reason that determines how we conceive the objects.

As far as Merleau-Ponty is concerned, however much it recognises the importance of consciousness, a constructivist stance such as Kant's is problematic if it too, assumes an opposition between consciousness and the world (however much it inverts the agency). 'The real', Merleau-Ponty argues, 'has to be *described*, not constructed or formed', and this means that 'I cannot put perception in the same category as the syntheses represented by judgements, acts or predications [as Kant does]. My field of perception is constantly filled with a play of colours, noises and fleeting tactile sensations which I cannot relate precisely to the context of my clearly perceived world, yet which I nevertheless immediately "place" in the

---

[34] Maurice Merleau-Ponty, *Phenomenology of Perception*, p. ix.

[35] Maurice Merleau-Ponty, *Phenomenology of Perception*, p. ix.

[36] Immanuel Kant, *The Critique of Pure Reason*, translated by Norman Kemp Smith (London: Macmillan, 1990[1929]).

[37] Immanuel Kant, *The Critique of Pure Reason*, p. B xvii.

world, without ever confusing them with my daydreams'[38] (as Descartes claims that we might, cf. Chapter 10).

In his earliest works, what Merleau-Ponty advocates is thus a phenomenology understood as 'pure description' – a stance on one level very much like Peirce's – but which starts from a conception of the body that treats it as an integral aspect of any imaginative faculties. It is this conception of the body that interests me for the purposes of this study. The body itself, Merleau-Ponty suggests, must not be conceived as 'an assemblage of organs juxtaposed in space'[39] but as something for which the subject is in 'undivided possession': 'I know where each of my limbs is through a *body image* in which all are included'.[40] This body image is not the result of associations established through experience, but 'a total awareness of my posture in the intersensory world, a "form" in the sense used by Gestalt psychology',[41] i.e. a matter of grasping 'wholes' whose form is in some sense external to their parts conceived in an atomistic manner.

Merleau-Ponty further suggests that the body image is a dynamic one – and this for reasons that are not entirely unlike those put forward by James Gibson as part of his critique of static conceptions of perception[42] (cf. Chapter 4). As Merleau-Ponty puts it, 'it is clearly in action that the spatiality of our body is brought into being, and an analysis of one's own movement should enable us to arrive at a better understanding of it. By considering the body in movement, we can see better how it inhabits space (and, moreover time)'; 'movement is not limited to submitting passively to space and time, it actively assumes them, it takes them up in their basic significance'.[43] Or as Merleu-Ponty explains, '... my body appears to me as an attitude directed towards a certain existing or possible task. And indeed its spatiality is not, like that of external objects or like that of "spatial sensations", a *spatiality of position*, but a *spatiality of situation*.'[44] 'The word "here" applied to my body does not refer to a determinate position in relation to other positions or to external co-ordinates, but the laying down of the first co-ordinates, the anchoring of the active body in an object, the situation of the body in face of its tasks'.[45]

––––––––––––––––––

[38] Maurice Merleau-Ponty, *Phenomenology of Perception,* p. xi. Italics added to the original.

[39] Maurice Merleau-Ponty, *Phenomenology of Perception,* p. 112.

[40] Maurice Merleau-Ponty, *Phenomenology of Perception,* pp. 112-113. Italics in the original.

[41] Maurice Merleau-Ponty, *Phenomenology of Perception,* p. 114.

[42] Interestingly, Gibson does not make Merleau-Ponty a part of his major works; assuming that someone at some point would have alerted Gibson to the proximity between the two scholars' approaches, it is tempting to speculate that this may have something to do with Merleau-Ponty's categorical rejection of a positivist perspective, and with it the kind of empiricism that Gibson very much adheres to.

[43] Maurice Merleau-Ponty, *Phenomenology of Perception,* pp. 117.

[44] Maurice Merleau-Ponty, *Phenomenology of Perception,* pp. 114-115. Italics in the original.

[45] Maurice Merleau-Ponty, *Phenomenology of Perception,* p. 115.

Merleau-Ponty's references to situation clarify the point made earlier by Toril Moi via Simone de Beauvoir, regarding the body not just *in* a situation, but *as* situation. As Merleau-Ponty explains, '... if my body can be a "form" and if there can be, in front of it, important figures against indifferent backgrounds [a reference to Gestalt theory], this occurs in virtue of its being polarized by its tasks, of its *existence towards* them, of its collecting together of itself in its pursuit of its aims;' 'the body image is finally a way of stating that my body is in-the-world'; moreover, 'one's own body is the third term, always tacitly understood, in the figure-background structure, and every figure stands out against the double horizon of external and bodily space'.[46]

Returning to the question of movement, or what Merleau-Ponty describes as 'motility', Merleau-Ponty echoes Husserl when he suggests that we must understand motility 'as basic intentionality. Consciousness is in the first place not a matter of "I think that" [as in Kant] but of "I can"'.[47] When we move a hand towards an object, the movement is not simply a matter of a displacement, let alone displacement in an 'empty space'; in the action of the hand 'is contained a reference to the object, not as an object represented, but as that highly specific thing towards which we project ourselves, near which we are, in anticipation, and which we haunt. Consciousness is being-towards-the-thing through the intermediary of the body'.[48]

In this early Merleau-Ponty one thereby finds an approach to embodiment that underscores not only the role of consciousness, but the body as a fundamental *medium* (to return to the earlier quote) for consciousness. In effect, the perceiving subject is a body whose sensuous character cannot be separated from an act of perception, and by implication, the practice of *observation*. From this perspective, to speak of the 'mind's eye' is wholly misleading, if by this is meant a conception of visual perception that is mainly if not entirely a matter of an ideal (as in 'ideas-based') subjectivity. The kind of subjectivity that interests Merleau-Ponty is the one that not only involves a living, breathing body, but one for which a body *as* situation determines what the individual perceives, if not how it does so.

This perspective can be used to interpret the example with which Part 3 began. From the kind of sociological perspective advocated, however differentially, by Bourdieu, Foucault, or Butler, the visitor observing the crocodile must be regarded as an actor that inhabits first and foremost the world constructed via a field, discourse, or any other kind of disciplinary or administrative apparatus. In the terms I introduced in Chapter 8, the visitor's way of observing the crocodile, and with it his bodily hexis are more or less completely determined by the coding orientation, and with it the discourse that is circulated by the field constituted by the

---

46 Maurice Merleau-Ponty, *Phenomenology of Perception*, p. 115. Italics in the original.

47 Maurice Merleau-Ponty, *Phenomenology of Perception*, p. 158-159.

48 Maurice Merleau-Ponty, *Phenomenology of Perception*, p. 159-160.

producers of natural history documentaries (and any other related field[s]). As the visitor's commentary appears to confirm, to look at, to observe the crocodile is to do so entirely via the kind of viewpoint constructed by the nature media.

With Merleau-Ponty, it may be argued by contrast that, to adopt such a stance is to take for granted a process that remains in some sense fundamental to the visitor's interpretation. That process has to do with the bodily situation of the tourist, who not only is there, on Barro Colorado, one member amongst several of a group – what we might describe as body *in* situation – but whose consciousness is itself embodied – the body *as* situation. I may refer to a discursive or field-based *elsewhere* when interpreting the crocodile, but to interpret that crocodile in that place I have to be *there* – 'in', as, and thanks to a body which begins to determine an attitude, a disposition merely by being a, which is really to say, *my,* body.

If, then, it is true that what one perceives is always contingent on the kind of social and cultural constraints described by the cultural sociologists (or the kind of discourse they draw on), it is also fundamentally a matter of being able to *do* certain things with, or as, one's body. This approach acts as a powerful correction of the inclination of at least the earlier cultural sociology to focus on the social and cultural situation that is thought to entirely determine bodies (or at any rate, *embodiment*) in the manner described by Bourdieu via the concept of *hexis*, or in a somewhat less totalising manner by Foucault via *biopower*.

The power of these insights notwithstanding, Merleau-Ponty's early proposals raise some problems of their own. On the one hand, Bourdieu and Foucault, who would have been very familiar with Merleau-Ponty's thought, would doubtless object to what they themselves would see as a certain idealisation of perception via individual consciousness. Where the early Merleau-Ponty puts all the emphasis on body as situation, in their different ways Bourdieu and Foucault would put more emphasis on *situation*, full stop.

On the other hand, Merleau-Ponty himself admits in his later work that the focus on consciousness actually prevents him from engaging in a critical manner with the relation between perception and the nature (or perhaps I should say Nature) which it starts from. To explain how this is the case, in the remaining pages of this chapter I would like to turn to Merleau-Ponty's later work, which is found in texts that he produced between the early 1950s and his sudden death in 1961.[49] During this period, Merleau-Ponty introduces a significant change in the orientation of his research. In a series of lectures presented between 1956 and 1960 at the Collège de France,[50] as well as in the essay 'Eye and Mind'[51] and in the unfin-

---

[49] Merleau-Ponty was struck down by a stroke, when the philosopher was 53.

[50] Maurice Merleau-Ponty, *Nature: Course Notes from the Collège de France*, translated by Robert Vallier (Evanston, ILL: Northwestern University Press, 2000 [1968]).

[51] Maurice Merleau-Ponty, 'Eye and Mind', in Ted Toadvine and Leonard Lawlor (Eds.) *The Merleau-Ponty Reader* (Evanston, IL: Northwestern University Press, 2007 [1961]), 351-378.

ished *The Visible and the Invisible*,[52] Merleau-Ponty effectively abandons the focus on consciousness (however embodied) in order to interrogate the relationship between the perceiver and the perceived as part of an ontology of nature. If in his earlier studies the problem is to find ways of making the body a central aspect of conscious perception, in the later research this perspective is in some sense turned on itself to ask how perception can take place in a world full of bodies – both human and beyond-human.

It is possible to represent this shift via one oft-quoted passage found in the working notes at the end of *The Visible and the Invisible*, in which Merleau-Ponty suggests that 'The problems posed in *Ph.P.* [*Phenomenology of Perception*] are insoluble because I start there from the "consciousness"-"object" distinction------ […] Starting from this distinction, one will never understand that a given fact of the "objective" order could entail a given disturbance of the relation with the world – a massive disturbance, which seems to prove that the whole "consciousness" is a function of the objective body'[53] (Merleau-Ponty is referring, amongst other possibilities, to a cerebral lesion). So long as one starts from consciousness, and focusses on the explanation of consciousness, then however much this account takes into account bodily situation, it is difficult not to avoid a certain psychologism.

To start from consciousness poses an additional issue that is particularly pertinent for the present study: so long as the problem is to explain the attitude of *human* consciousness, then the investigation is likely to remain anthropocentric through and through. A case can be made that anthropocentrism, like anthropomorphism, are inevitable (cf. Part 4). But even so, it is clear that some forms of research, some approaches are likely to compound this tendency by utterly disregarding the question of beyond-human bodies that are themselves subjects and agents.

Taken out of context, this kind of shift might well be interpreted as a return to the kind of scientific realism that Merleau-Ponty criticises in *Phenomenology of Perception*, and in his first work, *The Structure of Behavior*. In fact, already in *The Structure of Behaviour* we find evidence that what is really at stake is a turn to a very different kind of conceptualisation of biology. And indeed, Merleau-Ponty's writing is influenced, amongst many other thinkers, by the work of Jakob von Uexküll,[54] a biologist and proto-biosemiotician who, despite being a practicing scientist, develops a non-positivist account of the nature of nature. This account

---

[52] Maurice Merleau-Ponty, *The Visible and the Invisible*, edited by Claude Lefort and translated by Alphonso Lingis (Evanston, IL:Northwestern University Press, 1968 [1964]).

[53] Maurice Merleau-Ponty, *The Visible and the Invisible*, p. 200.

[54] See for example, Jakob von Uexküll, *Theoretical Biology*, translated by D. L. MacKinnon (New York: Harcourt Brace, 1926); and Jakob von Uexküll, *A Foray Into the Worlds of Animals and Humans* with *A Theory of Meaning*, translated by Josep D. O'Neil (Minneapolis: University of Minnesota Press, 2010 [1934 & 1940, respectively]).

has at its centre the way in which each species', and indeed each creature's world gravitates around a kind of 'plan', and with it an *umwelt*, a surrounding-world or 'self-world' that is in some sense monadic, and unique. I will discuss Uexküll's work in Part 4; here it suffices to note that, thanks in part to the work of Jakob von Uexküll and other similarly critical biologists, Merleau-Ponty realises that human consciousness is not the end-all that phenomenological inquiry makes it out to be. Instead, there may even be a continuity between human and beyond-human bodies that deserves theorisation in its own right. For Merleau-Ponty, that theorisation requires the development of an entirely new ontology of nature.

A detailed account of that ontology, which Merleau-Ponty was unable to complete before he died, is beyond the scope of this chapter.[55] Here it will suffice to make two points. First, Merleau-Ponty is clearly impressed by Uexküll's insights (more specifically with his concept of *umwelt*) and shares his opposition to the kind of mechanistic cosmology provided by physics and chemistry.[56] As Merleau-Ponty puts it already in *The Structure of Behavior's* critique of behaviourism, science is not about 'dealing with organisms as the completed modes of a unique world *(Welt)*, as the abstract parts of a whole in which the parts would be most perfectly contained. It has to do with a series of "environments" and "milieu" *(Umwelt, Merkwelt, Gegenwelt)* in which the stimuli intervene according to what they signify and what they are worth for the typical activity of the species considered'. Equally, 'the reactions of an organism are not edifices constructed from elementary movements, but gestures gifted with an internal unity. Like that of stimulus, the notion of response separates into "geographical behavior" – the sum of the movements actually executed by the animal in their objective relation with the physical world; and behavior properly so called – these same movements considered in their internal articulation and as a kinetic *melody* gifted with a meaning'.[57] We see here how Merleau-Ponty links what I will describe in Part 4 as a melodic metaphor[58] to the kind of analysis of movement that I referred to earlier. Movement is not to be understood purely along the lines of physics, any

---

[55] This ontology has received detailed attention from a number of different authors. See for example, Renaud Barbaras, 'Merleau-Ponty and Nature', *Research in Phenomenology*, 31 (2001), 22-38. See also, Toady Levine, *Merleau-Ponty's Philosophy of Nature* (Evanston, IL: Northwestern University Press, 2009).

[56] For an analysis of this influence, see Brett Buchanan, *Onto-Ethologies*: *The Animal Environments of Uexküll, Heidegger, Merleau-Ponty and Deleuze* (Albany, NY: Suny Press, 2008).

[57] Maurice Merleau-Ponty, *The Structure of Behavior*, pp. 129-130. Italics added to the original.

[58] A metaphor which he borrows either from Henri Bergson (cf. Chapter 14), or from Uexküll himself (as per the interpretation of Brett Buchanan in *Onto-Ethologies*), but which has an older history in nineteenth century thought. Then again, as I noted in Chapter 5, already in 1898 William James was speaking of 'belief being the demicadence which closes a musical phrase in the symphony of our intellectual life'.

more than a song can be understood simply by measuring the frequency of each note; it entails a purposiveness, and with it a certain cohesion that must be explained *as* cohesion – the song can only be properly understood as melody, and not just as a succession of notes. However, as part of the same theoretical displacement, the problem is not only to explain this cohesion, this 'unity in its own right', but how it relates to a multiplicity of other unities – yet this without reverting to any of the following perspectives, each of which Merleau-Ponty rejects: first, that of a 'dead' cosmos populated by mindless machines driven by a Newtonian physics, or by the dictates of natural selection (the predominant metaphors in the physical sciences); second, that of a vitalist cosmology which assumes that all organisms are 'some huge animal whose organs our bodies would be, as for each of our bodies, our hands, our eyes are the organs';[59] third, that of an empty space, with each corpuscular body moving in something like a vacuum (cf. the critique of James Gibson in Chapter 4, but also the critique of Henri Lefebvre in Chapter 12); or fourth, one that establishes a neat distinction between the perceiving subject and the perceived object.

What, then, is the alternative? The subject/object dichotomy presupposes not just a neat boundary, but a veritable chasm between the perceiver and the perceived, the observer and the observed. It is most obvious in, and indeed utterly integral to positivist philosophy; what I described in Chapter 1 as the rule of objectivism implies that a prerequisite for scientific knowledge is a distance, indeed a 'good' distance between the researcher and the researched. Ironically, even as positivist scientific methods in some sense take for granted this chasm, they must work quite hard to produce, and to preserve it: the objectivity of scientists is thought to hinge on their ability to maintain intact *the objectivity of the object* by way of a series of principles designed to prevent the contamination of the object by the subject, and vice-versa. If the boundary between subject and object is blurred or lost, then scientific inquiry must fail. From this kind of perspective, it might well be assumed that a similar chasm applies to *any* observational process, with the observer surmounting the chasm that separates it from the observed.

Merleau-Ponty's account subtly reworks the observer-observed, subject-object dichotomy by asserting that there is an essential *continuity* between the referents of the two terms. As I began to explain in Chapter 4, for ecological psychologists such as James Gibson and Eleanor Gibson, this matter is resolved by recourse to what is, in effect, a kind of pedagogy of perception: from infancy each individual learns about the affordances of different objects, different kinds of objects. Merleau-Ponty wants to explain a phenomenon that is prior, and which goes beyond conscious human learning: how can we even *begin* to engage in anything like 'perceptual learning'? If *Phenomenology of Perception* starts to tackle this question by way of phenomenological *description*, and by positing an integral relation be-

---

[59] Maurice Merleau-Ponty, *The Structure of Behavior*, p. 142.

tween subjectivity and embodiment, in *The Visible and the Invisible* Merleau-Ponty goes one step further to argue for the *intertwining* of continuity and discontinuity, proximity and distance, subjectivity and objectivity, but all this according to a logic that is not corpuscular, or even purely 'physical'. We can only detect the presence of some thing thanks to its sensible qualities (the visible); but at the same time, we can only really relate to those qualities thanks to their cohesion or unity – a unity whose essence can only be *intuited* (and so is the *in*visible). As Merleau-Ponty puts it in the notes he makes in preparation for *The Visible and the Invisible*, the invisible is firstly 'what is not actually visible, but could be (hidden or inactual aspects of the thing – hidden things, situated "else where" – "Here" and "elsewhere")'; secondly, 'what, relative to the visible, could nevertheless not be seen as a thing (the existentials of the visible, its dimensions, its non-figurative inner framework)'; thirdly, 'what exists only as tactile or kinesthetically, etc.'; and finally 'the λίκτα, the Cogito'.[60]

> The sensible is precisely that medium in which there can be *being* without it having to be posited; the sensible appearance of the sensible, the silent persuasion of the sensible is Being's unique way of manifesting itself without becoming positivity, without ceasing to be ambiguous and transcendent. The sensible world itself in which we gravitate, and which forms our bond with the other, which makes the other be for us, is not, precisely qua sensible, "given" except by allusion-----The sensible is that: this possibility to be evident in silence, to be understood implicitly, and the alleged positivity of the sensible world (when one scrutinizes it unto its roots, when one goes beyond the empirical-sensible, the secondary sensible of our "representation," when one discloses the Being of Nature) precisely proves to be an *ungraspable*, the only thing finally that is seen in the full sense is the totality wherein the sensibles are cut out. Thought is only a little further still from the *visibilia*.[61]

The openness of the world to perception – or to what I describe as *observation* – lies paradoxically in its invisibility, in the fact that anything we see or otherwise sense has additional aspects which we don't see, we can't sense, and so we must explore to perceive. The same is true even of our own perceptual system, which in some way hides behind itself: we cannot see ourselves seeing, we cannot fully feel ourselves touching. This is true even when someone touches a part of the own body:

> To touch and to touch oneself (to touch oneself = touched – touching) They do not coincide in the body: the touching is never exactly the touched. This

---

[60] Maurice Merleau-Ponty, *The Visible and the Invisible,* p. 257.

[61] Maurice Merleau-Ponty, *The Visible and the Invisible,* p. 214. Italics in the original.

does not mean that they coincide "in the mind" or at the level of "consciousness." Something else than the body is needed for the junction to be made: it takes place in the *untouchable.* That of the other which I will never touch. But what I will never touch, he does not touch either, no privilege of oneself over the other here, it is therefore not the *consciousness* that is the untouchable.[62]

The concept that Merleau-Ponty proposes to account for the very possibility of this play of visibility/invisibility is that of *flesh.* The flesh is not matter understood as corpuscles of being which would add up like so many atoms or molecules to form a body. But nor is it some 'psychic' or even 'social' material that comes into being by things that act on the body, let alone a representation in or for the mind. It is instead something like an 'element', but element as the term was used by the classic philosophers to refer to water, air, earth, and fire: 'that is, in the sense of a *general thing,* midway between the spatio-temporal individual and the idea, a sort of incarnate principle that brings a style of being wherever there is a fragment of being. The flesh is in this sense an "element" of Being'.[63]

This notion of the flesh must also engage with the question of otherness in perception, and with it multiplicity: how is it that there may be a certain cohesion, and indeed a certain purposefulness, *across* bodies, across things? If in his earlier writing Merleau-Ponty borrows Uexküll's (or is it Bergson's?) melodic metaphor, in the later Merleau-Ponty the primary image shifts to that of one hand touching the other hand in a 'chiasmatic' relation: if flesh is a way of naming unity in multiplicity, the concept of *chiasm* explains interrelation *across* unities, unities in interrelation both despite, and thanks to, difference.

The term chiasm's etymology goes back to the Greek χίασμα, 'crossing', itself from the Greek χιάζω 'to mark with an X'. In physiology, the optic chiasm refers to that part of the brain where the optic nerves partly cross; this structure allows the visual cortex to receive the same visual field from both eyes. In rhetorical studies, *chiasmus* refers to a reversal of *grammatical* structures, A:B/D:C; for example 'Never kiss a fool, and never let a kiss fool you'. For Merleau-Ponty, both kinds of 'crisscrossing' offer metaphors for the potential reversal of roles that he sees as being fundamental to the relation of the perceiver to the perceived, the sentient and the sensible. For example, if one's hands can move, touch and feel textures, it must be because the hands have not only some 'shape', but because they can initiate an exploration, and so must be capable of opening up to a tactile world. But that can only happen if one's hand, even as it is felt from within, can also touch and be touched 'from without', i.e. it is itself tangible to one's other hand, or to any other body. 'Through this *crisscrossing* within it [the hand] of the

---

[62] Maurice Merleau-Ponty, *The Visible and the Invisible,* p. 254. Italics in the original.

[63] Maurice Merleau-Ponty, *The Visible and the Invisible,* p. 139. Italics in the original.

touching and the tangible, its own movements incorporate themselves into the universe they interrogate, are recorded on the same map as it; the two systems are applied upon one another, as the two halves of an orange'.[64] A hand touching another hand partakes of, and is part of the same body, and we might say by extension, the same flesh; but one hand is not the other hand, and when one hand touches the other, the touching is not the touched – the two, as Merleau-Ponty points out, never exactly coincide in the body. The body is in this sense of two 'sides', two 'leaves': the body as sensible, the body as sentient – the body a thing amongst things that can be seen and touched, but at the same time, the body that can itself see, and palpate those things.

Perception is thus not simply the result of a consciousness that seeks to perceive. It is the result of some thing that arises from, even as it stays among, other things which may themselves do the same in turn. In more technical terms, the visible, taken in the broadest sense of something that appears to any of the senses, is *transcendent*. To return to the earlier quote, 'the sensible appearance of the sensible, the silent persuasion of the sensible is Being's unique way of manifesting itself without becoming positivity, without ceasing to be ambiguous and *transcendent*'.[65] Here '[w]ithout becoming positivity' may be taken as a critique of the modern scientific reduction of objects into so many 'physical' givens, as per the rule of phenomenalism (Chapter 1).

What, then, are the implications of these two major shifts vis-a-vis the kind of approach that I outlined in Chapter 8?

As developed in *The Structure of Behavior* and in *The Phenomenology of Perception*, Merleau-Ponty's phenomenology fills in a big blank in the earlier cultural sociology – that which concerns the role of the body 'in' perception, the body 'in' observation. If, as cultural sociologists would insist, all perception, all *observational practice* is situated in so far as it occurs in the aegis of fields, institutions, codes, discourses, genres and cultural formations more generally, the same is true again, but perhaps on a more fundamental level, with respect to the situation of observation in the body itself. To return to the point made by Moi via de Beauvoir, the body is not just situat*ed*, but is *situation in its own right*. Foucault and his followers may argue, rightly in my view, that even *that* situation is affected by biopower. But in so far as the analysis of biopower gets in the way of recognising the kind of dynamics described by Merleau-Ponty, then ironically that analysis too, ends up reinventing the wheel of the mind-body dualism: even if mind is no longer considered along the lines of an enlightened discourse of rational sovereignty, any attempt to sever observational practices from their most material, and embodied moorings ironically reintroduces precisely the kind of idealism that a *cultural* materialism is meant to contest – hence my earlier reference to a false materialism on the part of Butler and those who adopt a similarly poststructuralist stance.

---

[64] Maurice Merleau-Ponty, *The Visible and the Invisible*, p. 133. Italics added to the original.

[65] Maurice Merleau-Ponty, *The Visible and the Invisible*, p. 214. Italics added to the original.

The later Merleau-Ponty – the Merleau-Ponty of the *Nature* lectures and of the unfinished *The Visible and the Invisible* – pushes this kind of approach much further by situating the body in a nature that is not already culturalised, and which is held together by a *flesh* – not the kind of flesh made conceivable by physics or chemistry, but a chiasmatic entity where a kind of mutualism prevails: I can observe, but I can also *be* observed; I can perceive *because* I myself can be perceived.

This last point may seem overly abstract until one thinks of what happens in any encounter with a creature that has an *umwelt* capable of detecting the presence of, and reacting to, the human observer: it is not just that humans – in this study's case, tourists – observe wildlife; the wildlife may observe them too. From Merleau-Ponty's perspective, this is no coincidence; on the 'level' of flesh, if we can observe wild animals it is because they can observe us. But of course, this is partly what makes it difficult to observe them: in so far as a wild animal fears us, it may seek to remain *entirely* invisible. It is only after this dynamic is explained, that it may nevertheless be acknowledged that the observer may seek to in some sense 'supplement' the observed with the kinds of images circulated by the media. That this can happen raises questions about the nature of embodiment that Merleau-Ponty did not himself investigate, and which arguably require a different approach – one which does not treat bodies as purely, or indeed necessarily 'organismic' entities.

10
Mode, Affect and Assemblage

Implicit to Merleau-Ponty's analysis is the notion that the bodies involved in observational practices are all, in one way or another, of 'flesh and bone' – or at any rate of *flesh* in Merleau-Ponty's sense of the term. They are thus bodies with organs, limbs, perceptual systems, intuitions, purposiveness, etc. If, however, we return to the encounter with *C. acutus*, it is evident that it involved rather more than organismic bodies. As I explained at the start of Part 3, the visitor's observational process entailed media-based dynamics of transmediation (cf. Introduction to Volume 1). This meant that the visitor's observations were not only mediated from afar, but by what might be described as non-organismic bodies (e.g. the represented Nile crocodiles, but also the film cameras and editing suites, the discourses and genres of the nature media, television screens, etc.). To be sure, even if we were to stay with the immediate observational circumstances, the guide provided what might well be described as a *body of knowledge* with which to interpret the specimen of *C. acutus*.

The importance of the non-organismic bodies to wildlife observation may suggest that cultural sociology – or at any rate, a poststructuralist theory of the kind proposed by Butler – is actually the more critical one when it comes to wildlife observation amongst tourists. Alas, anyone wishing to test such an approach might wish to have a dip in the Gatún, and do so close a live, and hungry specimen of *C. acutus*; doing so might well reveal that, however much a variety of kinds of bodies are indeed involved in wildlife observation, some have a capacity to bite in a way that others don't. More generally, however much the media might play a role in wildlife observation, it seems clear that if *in situ* forms of wildlife observation are still valued and sought, it is because there is something about the *haecceity* of actual animals, and of actual encounters, that matters even in a world as dominated by the mediazation of culture as the modern.

In this, the last chapter of Part 3, I would like to turn to the work of two philosophers who may help us to sort out some of the issues raised by these and other considerations involving the complexities of bodily interrelations. The philosophers are Gilles Deleuze, writing on his own,[1] and Deleuze writing with Félix Guattari.[2]

By way of a brief contextualisation, Gilles Deleuze (1925-1995) is widely regarded as one of the most influential twentieth-century philosophers. While his doctoral dissertation *Difference and Repetition* is often described as his *magnum opus*, at least amongst anglophone social and cultural researchers, in the early 2000s it was probably his joint work with Guattari – especially the jointly authored *One Thousand Plateaus* – that initially received the most attention.

Where Deleuze was very much an academic – however critical his works, and active his engagement with current affairs via his writing and university lectures – Félix Guattari (1930 - 1992) is often described as being first and foremost a political activist, with a gift for bringing diverse people together to advance progressive causes. The causes in question included opposing France's colonial role in the Algerian War; seeking the transformation of the treatment of the people in the mental health institutions in which he worked (Guattari was also a practicing psychoanalyst, and at one stage was a student of Jacques Lacan); or towards the end of his life, engaging with environmental(ist) politics. It says a lot about Guattari that even as he could engage with the world as practically as he did, he co-wrote with Deleuze what is arguably one of the more complex philosophical works of the twentieth century – the two volumes of *Capitalism and Schizophrenia*. To be sure, Guattari also produced his own philosophical works.[3]

Foucault once famously suggested that the twentieth century would come to be known as the Deleuzian century (philosophically speaking),[4] and at least in the

---

[1] See for example Gilles Deleuze, *Difference and Repetition*, translated by Paul Patton (New York: Columbia University Press, 1994 [1968]); Gilles Deleuze, *Expressionism in Philosophy: Spinoza*, translated by Martin Joughin (New York: Zone Books, 1990 [1968]); and Gilles Deleuze, *Spinoza: Practical Philosophy*, translated by Robert Hurley (San Francisco: City Light Books, 1988 [1970]).

[2] See for example, Gilles Deleuze and Félix Guattari, *Anti-Oedipus: Capitalism and Schizophrenia*, translated by Robert Hurley, Mark Seem, and Helen R. Lane (Minneapolis, MN: University of Minnesota Press, 1983[1972]); and Gilles Deleuze and Félix Guattari, *A Thousand Plateaus: Capitalism and Schizophrenia*, translated by Brian Massumi (London: Continuum Press, 1988[1980]).

[3] See for example, Félix Guattari, *Molecular Revolution: Psychiatry and Politics*, translated by Rosemary Sheed (Harmondsworth: Penguin, 1984 [1972 & 1977]); Félix Guattari, *Schyzoanalytic Cartographies*, translated by Andrew Goffey (London: Bloomsbury Press, 2013 [1989]); Félix Guattari, *The Three Ecologies*, translated by Ian Pindar and Paul Sutton (London: Continuum, 2000 [1989]; and Félix Guattari, *Chaosmosis: An Ethico-Aesthetic Paradigm*, translated by Paul Bains and Julian Pefanis (Bloomington, IN: Indiana University Press, 1995 [1992]).

[4] See Michel Foucault, *Language, Counter-Memory, Practice: Selected Essays and Interviews* (Ithaca, NY: Cornell University Press, 1980), p. 165.

case of the critical social sciences and humanities, he was not far wrong. In the late 1990s and early 2000s, so many scholars began to embrace especially Deleuze's reworking of the concept of affect that a number of researchers began to speak of an 'affective turn'.[5] Since then numerous introductions, critiques, and applications of Deleuzian, and Deleuzian/Guattarian philosophy have been published not only in academic volumes across several fields, but even in the popular press.

Regrettably, it is often taken for granted that philosophically speaking, it is only really Deleuze's thought that matters. In fact, by Deleuze's own account Guattari was an intellectual co-equal, and so privileging Deleuze's contribution is only justified in those cases where Deleuze develops certain concepts whilst writing on his own – a case in point, his development of what I will describe as a philosophy of difference. In recognition of this point, I will sometimes refer only to Deleuze's work; in other cases, I will use the somewhat tedious, but accurate 'Deleuze, and Deleuze and Guattari'.

Back, then, to Deleuze and Guattari's philosophy. As I've begun to suggest, that philosophy is extraordinarily complex, not least because much of the work is written in a way that is designed to refuse any sense that it is *about* something (I will come back to this seemingly absurd claim at a later stage). Even so, Deleuze and Guattari's work offers tools for the analysis of almost any phenomenon one might care to study. I use the metaphor of tools advisedly; as I noted in the introduction to this volume, Deleuze famously suggested that the work of scholars can be approached in much the way that one does a toolbox,[6] and I will be doing just that with several aspects of Deleuzian/Deleuzian and Guattarian philosophy: in this part of the study, I will consider the concepts of body, affect, assemblage, and desire; in Part 4 I will consider the notion of 'becoming', and more specifically, of 'becoming-animal'.

A properly contextualised account of the place of these concepts within Deleuzian/Guattarian philosophy could easily span the length of this work, and then some. In this chapter I will simply begin by considering the proposals of two early modern philosophers whose work provides especially important referents (positive or negative) for those aspects of Deleuzian/Guattarian philosophy that I will engage with, and without which it is unlikely that Deleuze and Guattari's contributions may be understood. The first is René Descartes (1596-1650), and the

---

[5] See for example Patricia Ticineto Clough (Ed.) *The Affective Turn: Theorizing the Social* (Durham, NC: Duke University Press, 2007). See also Melissa Gregg and Gregory J. Seigworth (Eds.) *The Affect Theory Reader* (Durham, NC: Duke University Press, 2010).

[6] Gilles Deleuze, 'Intellectuals and Power', in *Desert Island and Other Texts (1953-1974)*, (Cambridge: MA: MIT Press, 2004), 206-213, p. 208. In this text, which records an interview with Michel Foucault, Deleuze says: 'Yes, that's what a theory is, exactly like a tool box. It has nothing to do with the signifier ... A theory has to be used, it has to work. And not just for itself.'

second is Baruch Spinoza (1632-1677). (In Chapter 14 I will also refer briefly to another of the great influences of Deleuze's writing, the philosopher Henri Bergson.) In the Preface, I noted that even as this volume would offer what might seem like a somewhat 'slow' propaedeutic account of the theories to be employed, it would also ask readers to come up quite quickly to speed with some of the more complex contemporary philosophies, and this is likely to be particularly true in the case of the introduction to the Spinozan inspiration of Deleuzian philosophy. I would thereby like to invite readers without a background in philosophy to persevere across the following pages; I expect that all will become clear towards the end of this chapter, if not before.

Let's begin, then, with Descartes. It should be explained from the outset that Deleuze, and Deleuze and Guattari's philosophy flatly rejects much of what Descartes has to propose; however, to more fully understand *Spinoza*, the philosopher that Deleuze describes as the 'prince' of philosophers, it is necessary to begin with Descartes. Doing so has the added advantage that one may more fully comprehend how and why the modern mind-body dualism, and with it a mechanistic conception of the body, emerged. Indeed, if there is any need to re-engage with our humanly corporeality, howsoever it may be conceived, it is at least in part because Descartes so successfully proposed a way of separating human thought from embodiment.

The very idea that anyone might wish to do that may seem ridiculous until one understands the context in which Descartes came up with his famous, and for the time revolutionary distinction between things to do with the mind, and things to do with the body, or more precisely things to do with thought and things to do with extension (*res cogitans*, and *res extensa*). Having grown dissatisfied with the philosophy taught in most European schools and universities during his time, Descartes embarked on a quest for an alternative account of the ultimate truths of the cosmos, and of human beings in particular – an alternative metaphysics. As Descartes puts it at the beginning of his *Meditations* (1641), 'Some years ago I was struck by how many false things I had accepted as true in my childhood, and how doubtful were the things that I subsequently built by them'.[7] The edifice that Descartes refers to is that of a scholastic interpretation of Aristotelian philosophy, which posited that all knowledge arises via the senses. Descartes worried about this kind of approach because he was keenly aware that the senses can deceive the fully awake person, but also, that somebody fast asleep may have dreams so vivid that they can have the appearance of a manifestly sensual reality. In both scenarios, one finds good reason to be deeply mistrustful with respect to the ostensibly self-evident route to truth preferred by abject empiricists. This being the case, Descartes decides to try to find an alternative metaphysics – and this even after allowing that some truths afforded by sensory experience are difficult, if not

---

[7] René Descartes, *Meditations on First Philosophy*, translated by Desmond M. Clarke (Harmondsworth: Penguin, 2010 [1641]), p. 10.

impossible to doubt: 'for example, the fact that I am here, sitting by the fire, wearing a dressing-gown, holding this page in my hand and other things like that'.[8]

As part of his quest for a new foundation for critical reason, Descartes embarks on what is, in effect, an exercise in discursive purification. He deliberately tries to assume that all of his opinions, all of his knowledge, even that knowledge most obviously and incontrovertibly 'proved' via sensory perception, is 'completely false and imaginary' – a stance that he should maintain 'until at length, as if I were balanced by an equal weight of prejudices on both sides, no bad habit would any longer turn my judgement from the correct perception of things'.[9] Of course, this would be a difficult, to not say impossible task for anyone; to complicate matters further, Descartes wrote during a period when an incipient modernity had still not really questioned the reality of evil spirits and sorcery. So it was entirely plausible that there might be 'malicious demons' that might try to sabotage Descartes' best attempts to distinguish between what is unquestionably true, and what is merely a figment of dreams, or of perceptual mistakes.

Descartes' famous solution is represented by the expression *cogito ergo sum*, i.e. to seize on thought itself as a kind of foundational proof of existence. Even a malicious demon cannot undermine the fact that, if there is some unidentified deceiver, all-powerful and supremely cunning who is deliberately and constantly fooling the philosopher, 'it is indubitable that I also exist, if he deceives me'; 'having weighed up everything adequately, it must finally be stated that this proposition "I am, I exist" is necessarily true whenever it is stated by me or conceived in my mind'.[10]

But what is 'I', in the *I am, I exist*? Again, as part of the exercise of purification, it is not the body, or the process by means of which the senses appear to furnish truths about the world. In a famous passage, Descartes notes that wax that may one moment seem to have a clearly perceptible form will come undone in another (e.g. if placed next to a fire). A similar transformation might well occur in the case of virtually any other object which the senses appear to perceive clearly and distinctly.

In another example, Descartes notes that, were he to look out of the window and see men crossing the square, he would normally say that he saw the men themselves, just as he would say that he saw the wax. 'But what do I see apart from hats and coats, under which it may be the case that there are automata hidden? Nonetheless, I judge that they are people. In this case, however, what I thought I saw with my eyes I understand only by the faculty of judging, which is in my mind.'[11]

---

[8] René Descartes, *Meditations*, p. 12.

[9] René Descartes, *Meditations*, p. 15.

[10] René Descartes, *Meditations*, pp. 17-18.

[11] René Descartes, *Meditations*, p. 25.

Descartes suggests that it is this faculty that not only defines the 'I', but distinguishes humans from nonhuman creatures, and furnishes the proper epistemological basis for that which he prizes above all as a mathematician, and as a philosopher of nature: *clear and distinct ideas*. The following passage is worth quoting in some length in as much as it provides the clearest of insights to the kind of logic which centuries later would still pervade aspects of positivist philosophy, and even some forms of ideologically critical inquiry:

> ... when I distinguish the wax from its external forms and consider it as if it were bare and without its clothes on, then, although there may still be a mistake in my judgement, I cannot perceive the wax correctly without a human mind. [...] What shall I say, however, about this mind itself, or about myself, for I do not yet admit that there is anything in me apart from a mind? What, I ask, am 'I' who seems to perceive this wax so distinctly? Do I not know myself much more truly and certainly and also more clearly and distinctly? For if I judge that the wax exists from the fact that I see it, it would certainly follow much more clearly, from the fact that I see it, that I myself exist. For it may be the case that what I see is not really wax; it might even be true that I have no eyes, by which to see anything; but obviously it cannot be the case, while I see or while I seem to see ... that I myself am nothing as long as I am thinking.[12]

Three aspects of this passage may be highlighted. A first is the intellectualism of the whole exercise (Descartes is, after all, known as a rationalist philosopher). A second, related one is the invitation not just to mistrust, but effectively to banish the most material *signs* as an epistemological basis for knowing the world. Elsewhere Descartes is clear that knowledge produced via the senses certainly has its place in craft-like activities, and even furnishes an important part of the basis for scientific knowledge; but from a metaphysical perspective, he is convinced that one must mistrust signs, and instead use one's intellectual powers, one's powers of judgement, to try to arrive at the *essence* of things. A third aspect is the emphasis on all that is clear and distinct; while Descartes may not be the first to propose a quest for clarity and distinctness as an end in its own right, henceforth many philosophers of nature, and the scientists that will follow them will learn to dismiss accounts that are not driven by this ethic, this *aesthetic*; it will take centuries for scholars to push back at this notion, and to suggest that another role of science, and for critical thinking more generally, may be to search for, and/or to *articulate*, complexity – but this without simply trying to reduce it to clear and simple 'components'.

Now it would be an exercise in individualism and rationalism in its own right to suggest that it was Descartes' philosophical discourse, and his philosophical

---

[12] René Descartes, *Meditations*, pp. 25-26.

discourse alone, that generated what has come to be known as the modern mind-body dualism. Clearly, the mind-body dualism required an echo in the writing of numerous other philosophers. Equally if not more importantly, cultural historians have pointed out that it is no coincidence that it was the mind, associated with the upper body, that would carry the day amongst at least a part of Europe's emergent bourgeoisie. By cultivating that association, the elite in question could (further) distance themselves from those popular culture groups amongst whom precisely the opposite ethic and aesthetic prevailed. I refer here to what Mikhail Bakhtin,[13] and later Jesús Martín-Barbero[14] have characterised as a *grotesque realism*, which presupposes a coding orientation for which the essence of humanity is not the mind, as for Descartes, but the grotesque, the 'body-world and the world of the body'.[15] The grotesque, Martín-Barbero suggests, 'is a world view which gives value to what are commonly considered the lowest elements – the earth, the belly – posed in direct contrast to the higher things – the heavens and the human countenance. The grotesque values the lower regions because "the lowest is always a beginning"'.[16] Martín-Barbero adds that, in contrast to a rationalistic realism that emphasizes the completeness and isolation of objects – we return by this route to Descartes' predilection for the 'clear and distinct' – grotesque realism presupposes a world for which the essence of the body lies precisely in those parts that open up *communication* to and with the world: the nose, the mouth, the genitals, the breasts, the phallus, and the anus. 'An obscenity is so valuable precisely because through it we can express the grotesque: the realism of the body'.[17]

In Descartes' time, the upper body would also have offered a key metaphor for the early modern patriarchal relations that were pervading not just philosophy, but the society more generally. As part of the shift from agrarian communism to early forms of capitalism, a single 'man of the house' became the *head* of the family, and it was his job to use the own *mind* to arrive at rational decisions concerning the domestic economy and its relations to the world beyond the home, just as it was the woman's job to use the lower body to procreate, and to engage in domestic labour. In the broader society, this abjectly sexist division of labour was echoed by the notion of the king as head of the *body politic*, a position famously illustrated in the 1651 cover of Thomas Hobbes' *Leviathan*, which literally showed the king's head on a body made up of countless individuals.

As feminist and environmental historians have long argued, this imagery and the social relations that it represents should not be interpreted in a biologically

---

[13] Mikhail Bakhtin, *Rabelais and His World*, translated by H. Iswolsky (Bloomington, IN: Indiana University Press, 1984[1965]).

[14] Jesús Martín-Barbero. *Communication, Culture and Hegemony.*

[15] Jesús Martín-Barbero. *Communication, Culture and Hegemony, p. 65.*

[16] Jesús Martín-Barbero. *Communication, Culture and Hegemony, p. 65-66.*

[17] Jesús Martín-Barbero. *Communication, Culture and Hegemony, p. 66.*

reductionist manner, e.g. as a raw expression of the kind of adaptationist processes conceived by the evolutionary psychologists (cf. Chapter 2). Instead, they are best understood as an early modern expression of the rise of the patriarchal relations that were imposed on women and on the often matriarchal forms of governance found in ancient kinship groups. One interpretation of the Holy Inquisition – still very active during Descartes' time in northern Europe – is that it was principally about a kind of pedagogy for patriarchal relations. It is no coincidence, from this perspective, that the vast majority of individuals burnt at the stake or otherwise pursued were women.

Nor was this process entirely separate from the scientific revolution. On the contrary, according to the feminist and environmental historian Carolyn Merchant, 'The mechanists transformed the body of the world and its female soul, source of activity in the organic cosmos, into a mechanism of inert matter in motion, translated the world spirit into a corpuscular ether, purged individual spirits from nature, and transformed sympathies and antipathies into efficient causes. The resultant corpse was a mechanical system of dead corpuscles, set in motion by the Creator, so that each obeyed the law of inertia and moved only by external contact with another moving body.'[18] Where earlier organic cosmogonies gendered nature with images of what was often a benevolently maternal figure, the discourse of mechanism re-gendered nature with images of men subjecting women to sexual violence.

I have considered these developments in historical context elsewhere,[19] and so will not consider them in any detail here. If I raise them again in this book it to provide a context not only for Descartes' *cogito*, but also for the mechanistic conception of organisms, and of the world more generally – a mechanism that I will return to in Chapter 14. For now, I would like to turn to the second of the early modern philosophers that I mentioned earlier, Baruch Spinoza (also known as Benedict de Spinoza). Like Descartes, Spinoza was himself very much a rationalist philosopher, and indeed in some respects he was an avid follower of Cartesian philosophy. The last point notwithstanding, he developed a metaphysics that flatly rejects the notion that mind and matter may be conceived as two wholly separate *substances* – the philosophical term for the fundamental entities that constitute reality – and by implication that mind and body are essentially different entities. (It seems that, like Husserl and Merleau-Ponty centuries later, despite strongly disagreeing with Descartes, Spinoza continued to be amazed by his thought.)

From an early age, Spinoza rejected the kind of transcendental god favoured by institutional religion (not least, the Roman Catholic Church). When he made his views known amongst the Jewish community of Amsterdam (where he was born), he got into trouble with its authorities. At the age of 23, Spinoza was ex-

---

18 Carolyn Merchant, *The Death of Nature: Women, Ecology and the Scientific Revolution* (San Francisco, CA: Harper-Collins, 1980), p. 195.

19 See Nils Lindahl Elliot, *Mediating Nature*, esp. chapters 3 and 4.

communicated from the mentioned community – a deeply ironic event when it is considered that many of that community's members were migrants who had themselves been banished from the Iberian peninsula by its Roman Catholic monarchy.

After his excommunication, things took a sharp turn for the worse in 1670, when Spinoza published his *Theological-Political Treatise,*[20] which offered a critique of Judaism and of organised religion more generally. Although Spinoza published the work under a pseudonym, he was quickly identified as the author, and as Deleuze notes, 'few books occasioned as many refutations, anathemas, insults, and maledictions: Jews, Catholics, Calvinists, and Lutherans – all the right-thinking circles, including the Cartesians themselves – competed with one another in denouncing it. It was then that the words "Spinozism" and "Spinozist" became insults and threats'.[21] The vitriol was such that one man tried to stab Spinoza to death.

Important as the *Treatise* was in establishing Spinoza's credentials as a 'heretic', the more revolutionary manuscript, and the one in which Spinoza most clearly rejects the Cartesian metaphysics is *The Ethics,*[22] which Spinoza did not dare to publish within his own lifetime (it was published after his death in 1677). Spinoza is nevertheless thought to have completed a version of the work by the time that he published the *Treatise,* and to have discussed it with some philosophers. This would explain why news of the importance of *The Ethics* had already spread amongst philosophers even before Spinoza died. When the work was finally published, scholars were treated to a way of thinking that was in some respects centuries ahead of its time.

Where Descartes opposes mind and body, Spinoza adopts a 'substance monism' that argues that there is, in fact, just *one* substance: God, albeit an 'intellectual' god, a god that might well be regarded, at least from an official Christian or Judaic perspective, as no god at all. As Spinoza famously puts it at the start of *Ethics,* the god in question is 'a being absolutely infinite, that is, a substance consisting of an infinity of attributes, of which each one expresses an eternal and infinite essence'.[23] The *absolute* infinite, Spinoza explains, must not be confused with what is infinite in its own kind. So we can, for example, imagine the infinity of extension (e.g. the universe considered spatially), but that infinity is not to be confused with the *absolute* infinity of God him/itself. This absolute infinite is the one

---

[20] Benedict de Spinoza, *Theological-Political Treatise,* edited by Jonathan Israel and translated by Michael Silverthorne and Jonathan Israel (Cambridge: Cambridge University Press, 2007[1670]).

[21] Gilles Deleuze, *Spinoza: Practical Philosophy,* p. 10.

[22] Baruch Spinoza, *The Ethics,* in Edwin Curley (Ed. and translator) *The Spinoza Reader: The Ethics and Other Works,* (Princeton, NJ: Princeton University Press, 1994[1677]), 85-265.

[23] Baruch Spinoza, *The Ethics,* p. 85.

real substance, in the sense of something that cannot be subdivided into something else, as per the Cartesian criterion for what constitutes a true substance.

Later in the *Ethics,* having engaged in a detailed proof of the verity of this conception, Spinoza argues that '... except God, no substance can *be* or, consequently, be *conceived*. For if it could be conceived, it would have to be conceived as existing. But this (by the first part of this demonstration) is absurd. Therefore, except for God no substance can be or be conceived' and '[f]rom this it follows most clearly, first, that God is unique, that is ... that in Nature there is only one substance, and that it is absolutely infinite'; and second, 'that an extended thing and a thinking thing are either attributes of God, or [by way of the axiom that whatever is, is either in itself or in another[24]] affections of God's attributes'.[25]

Again, this is Spinoza contradicting the primacy of the Cartesian dualism – we return to *res extensa* and *res cogitans* – and replacing it with just one substance, an infinite god. In keeping with the times, Spinoza still refers to this substance in theological terms, but it is clear that, if he ran into trouble with all of the different religious authorities in his social milieu, it was no doubt because many sensed that this kind of conception in some sense denied anything like the authoritarian gods that were conceived then as now. Indeed, a case can be made that Spinoza's monadic substance was a clear precursor to the most modern conceptions of nature – conceptions which we have seen themselves continue to sneak in divisions and hierarchies in a manner that is absent in Spinoza.

Now as part of this approach, Spinoza proposed an alternative to the mind-body dualism which will be particularly interesting to Deleuze, and long before him, to Charles Peirce (we return by this route to the doctrine of synechism, cf. Chapter 5). As Spinoza puts it, '... the mind and the body are one and the same thing, which is conceived now under the attribute of thought, now under the attribute of extension. The result is that the order, *or* connection, of things is one, whether Nature is conceived under this attribute or that; hence the order of actions and passions of our body is, by nature, at one with the order of actions and passions of the mind'.[26]

This conceptualisation has important consequences for how one conceives the passions, or all that which Spinoza will conceptualise via the category of affect. As Spinoza points out at the start of Part III of *Ethics* (which is devoted to the subject of affect), most of those who have written about the affects have done so in the assumption that the affects follow *not* the '*common laws of Nature, but of things which are outside of Nature. Indeed they seem to conceive man in Nature as a dominion within a dominion*'.[27] This may be interpreted as a criticism of what today we

---

[24] Baruch Spinoza, *The Ethics*, p. 86.

[25] Baruch Spinoza, *The Ethics*, pp. 93-94.

[26] Baruch Spinoza, *The Ethics*, p. 155. Italics in the original.

[27] Baruch Spinoza, *The Ethics*, p. 152. Italics in the original.

would describe as the speciesism of Descartes, who suggests that, among all living creatures, only humans have mind (a claim with some nuances that I will consider in Chapter 14). By contrast, Spinoza argues that in so far as 'nothing happens in Nature which can be attributed to any defect in it, for Nature is always the same, and its virtue and power of acting are everywhere one and the same', so it must be with the affects: '[t]he affects ... of hate, anger, envy, and the like, considered in themselves, follow with the same necessity and force of Nature as the other singular things. And therefore they acknowledge certain causes, through which they are understood, and have certain properties as worthy of our knowledge as the properties of any other thing, by the mere contemplation of which are pleased'.[28]

But what precisely *is* affect, then? Spinoza offers a deceptively simple definition: 'By affect I understand affections of the body by which the body's power of acting is increased or diminished, aided or restrained, and at the same time, the ideas of these affections. [...] *Therefore, if we can be the adequate cause of these affections, I understand by the affect an action; otherwise, a passion.*'[29] Or as Spinoza puts in the 'Postulates' of the third book of the *Ethics*, 'The human body can be affected in many ways in which its power of acting is increased or diminished, and also in others which render its power of acting neither greater nor less'; equally, 'The human body can undergo many changes, and nevertheless retain impressions, *or* traces, of the objects ... and consequently, the same images of things'.[30] Spinoza adds that '*The body cannot determine the mind to thinking, and the mind cannot determine the body to motion, to rest, or to anything else (if there is anything else)*'.[31] This would appear to contradict Spinoza's own critique of the mind-body dualism, but presumably the contradiction is avoided by recognising different attributes and modes, a point I will return to below.

Now today Spinoza's outlook may seem almost commonsensical in the light of everything that has been said about nature, and about the mind-body dualism, not least by numerous ecologists influenced by Spinozan philosophy over the last several decades.[32] It is, however, important not to lose sight of the fact that Spinoza was responding not just to Descartes, whose dualistic philosophy would dominate much philosophical discourse for centuries to come, but to what was already in the 1600s a poor common sense: the notion, as Spinoza paraphrases it, that 'the body now moves, now is at rest, solely from the mind's command, and that it does a great many things which depend only on the mind's will and its art

---

[28] Baruch Spinoza, *The Ethics*, p. 153. Italics in the original.

[29] Baruch Spinoza, *The Ethics*, p. 154. Italics in the original.

[30] Baruch Spinoza, *The Ethics*, p. 154. Italics in the original.

[31] Baruch Spinoza, *The Ethics*, p. 155. Italics in the original.

[32] Amongst them, the philosopher-ecologist Arne Naess. See for example, Arne Naess, *Ecology of Wisdom* (Harmondsworth: Penguin Classics, 2016).

of thinking'.[33] In response to this common sense, Spinoza notes in what has become a famous passage, that

> ... no one has yet determined what the body can do, that is experience has not yet taught anyone what the body can do from the laws of Nature alone, insofar as Nature is only considered to be corporeal, and what the body can do only if it is determined by the mind. For no one has yet come to know the structure of the body so accurately that he could explain all its functions – not to mention that many things are observed in the lower animals which far surpass human ingenuity, and that sleepwalkers do a great many things in their sleep which they would not dare to awake. This shows well enough that the body itself, simply from the laws of its own nature, can do many things which its mind wonders at.[34]

It goes almost without saying that this verity remains true to this day, despite the best funded efforts of tens of thousands of physiologists and neuroscientists – what indeed *can* the human body, the human *bodies*, do?

How, then, do these insights relate to the philosophy of Deleuze and Guattari? For centuries, Spinoza was ignored by many mainstream philosophers, but in the late twentieth century this state of affairs changed dramatically thanks largely to Deleuze, who effectively led a revival of 'Spinozism' by publishing influential works about the early modern philosopher.[35] Deleuze, and later Deleuze and Guattari's reworking of Spinozan philosophy, and in particular of Spinoza's concept of affect became the basis for what I described earlier as the 'affective turn' in the Anglo-American critical social sciences and humanities. To be clear, I will not be concerned with the fate of the concept of affect in cultural sociology (*sensu stricto*), or in cultural geography, cultural studies and other fields whose researchers have used affect to try to explain pre-personal, or even pre-cultural dimensions of social practice.[36] Instead, I will stay closer to the Deleuzian *partitura*, and will consider its implications for the conception of both human and beyond-human bodies (I say Deleuzian, and not Deleuzian and Guattarian, because at least in the case of the concept of affect, it is Deleuze's philosophy that plays the more important role).

So how does Deleuze reinterpret Spinoza? To explain how Deleuze turns affect into a tool in his own toolkit, a detailed theoretical excursus is required which

---

[33] Baruch Spinoza, *The Ethics*, p. 155.

[34] Baruch Spinoza, *The Ethics*, pp. 155-156.

[35] See Gilles Deleuze, *Expressionism in Philosophy: Spinoza*; and Gilles Deleuze, *Spinoza: Practical Philosophy*.

[36] See for example Brian Massumi, *Parables for the Virtual: Movement, Affect, Sensation* (Durham, NC: Duke University Press, 2002). See also Patricia Ticineto Clough (Ed.) *The Affective Turn*, and Melissa Gregg and Gregory J. Seigworth (Eds.) *The Affect Theory Reader*.

some readers may find daunting, but which is unavoidable if the philosophical and theoretical implications are to be adequately grasped. A glimpse of what Deleuze will propose may be seen in the shift that occurs in Merleau-Ponty's later work (cf. Chapter 9), and which I described as a turn away from consciousness, and towards a chiasmatic relation between organisms and objects. Perception is the result of some thing that arises from, even as it stays among, other things which may themselves do the same in turn. From this perspective, the visible, taken in the broadest sense of something that appears to any of the senses, is a *transcendent* thing. But that thing remains resolutely embodied, one might say 'enfleshed' (after Merleau-Ponty). This begins to give a sense of what Deleuze himself refers to as his 'transcendental empiricism'. What follows is an account of how he arrives at this stance via the writings of Spinoza.

As Deleuze explains to Martin Joughin, the translator of *Expressionism in Philosophy: Spinoza*, 'What interested me most in Spinoza wasn't his substance, but the composition of finite modes'.[37] Put simply, Deleuze isn't interested in the Spinozan God (as noted earlier Deleuze himself was atheist), but in all that for Spinoza is the 'expression' of God: the attributes, and the modes. According to Spinoza, the attributes are 'what the intellect perceives of a substance, as constituting its essence'.[38] For their part, modes are 'the affections of a substance, *or* that which is in another through which it is also conceived'.[39] It is tempting to simplify and to suggest that the attributes are the characteristics of substance, and modes the attributes' concrete instantiation. Accordingly, examples of attributes are thought and extension, and examples of modes are any concrete, 'real life' expressions of either attribute: this or that thought, this or that geography, etc. That, however, would be to miss a far more subtle point that Deleuze makes via Spinoza, and which concerns the nature of the *relation* between substance and attribute, attribute and mode. To explain what that relation entails, it is necessary to say something about Deleuze's novel approach to ontology and metaphysics.

As I have suggested elsewhere in this volume, by the second half of the twentieth century, metaphysics had come under such sustained attack by positivists – and also by critical thinkers like Michel Foucault – that the very term had become synonymous with ideological forms of rationalisation. For positivists, metaphysics stands for the kind of speculative, 'unscientific' reasoning that is responsible for everything from superstition to religion to ideology; for Foucault, metaphysics is a purely discursive exercise, i.e. a socially interested, to not say abjectly partial attempt to define the nature of being for political purposes (however conscious or unconscious, and however much with a capital or small 'p') – a quest for

---

[37] Martin Joughin, 'Translator's Preface', in Gilles Deleuze, *Expressionism in Philosophy: Spinoza*, p. 11.

[38] Baruch Spinoza, *The Ethics*, p. 85.

[39] Baruch Spinoza, *The Ethics*, p. 85. Italics in the original.

essences that conceals, however inadvertently, the historical contingency of all manner of *representations* of the nature of being.

By contrast, Deleuze rehabilitates metaphysics via an ontology grounded in difference. Where a more traditional ontology sets out as an effort to *discover* 'what there is', for Deleuze ontology is a matter of the generation of a web of concepts with which to *interpret* the expressive potential of what is. If difference is really to be recognised, the philosopher cannot begin by dividing up the cosmos into parts which are given a more or less fixed identity, and are rendered subordinate to other parts. This is arguably what Descartes does when he privileges mind over body, and as I will explain in Part 4, humans over beyond-human animals. As Deleuze explains it, Spinoza's substance monism paves the way for a shift away from this kind of transcendentalism in so far as it acknowledges only one substance, the 'intellectual' God, and an infinity of modes, each empirically itself (what Deleuze will describe as a transcendental empiricism).

Now it might seem from parts of Spinoza's *Ethics* that the only attributes that Spinoza recognises are thought and extension. But Deleuze argues that for Spinoza there are in fact an infinity of attributes because 'God has an absolutely infinite power of existing, which cannot be exhausted either by thought or by extension'.[40] The same must thus be true for modes. And indeed, as Spinoza puts it in the first part of *Ethics*, '*From the necessity of the divine nature there must follow infinitely many things in infinitely many modes, (i.e., everything which can fall under an infinite intellect)*'.[41] Spinoza further suggests that 'the intellect infers from the given definition of any thing a number of properties that really do follow necessarily from it (that is, from the very essence of the thing)'; and that 'it infers more properties the more the definition of the thing expresses reality, that is, the more the reality of the essence of the defined thing involves ... each of which also expresses an essence infinite in its own kind, from its necessity there must follow infinitely many things in infinite modes...'[42]

Put very simply, this conception allows at once for just one substance, and so just one nature – or as Deleuze would put it, 'the plane of immanence' – even as it recognises that nature is utterly heterogeneous in so far as it is made up of an infinite set of finite modes, which is to say an infinite number of individuals, or one might also say of *bodies*, each with its own capacities, but always in a relation to other bodies with their own immanent capacities.

Important as this shift is, it is accompanied by another that is equally, if not more important: Spinoza, as read by Deleuze, treats the one substance, and also its attributes and modes, as being wholly *positive* in the sense that 'God is cause of

---

[40] Gilles Deleuze, *Spinoza: Practical Philosophy*, p. 52.

[41] Baruch Spinoza, *The Ethics*, p. 97. Italics in the original.

[42] Baruch Spinoza, *The Ethics*, p. 97.

himself in no other sense than that in which he is cause of all things. *Rather he is cause of all things in the same sense as cause of himself.*'[43] Or as Deleuze also puts it,

> The cause of itself is approached first of all in itself; this is the condition for "in itself" and "through itself" to take on a perfectly positive sense. Self-causality is, as a consequence, no longer asserted in *another sense* than efficient causality; rather is [sic] efficient causality asserted in the same sense as self-causality. Thus God produces as he exists: on the one hand, he produces necessarily; on the other, he necessarily produces with in the same attributes that constitute his essence.[44]

What is at stake for Deleuze is not the kind of 'conceptual' difference that is associated with what has come to be known, say, as a politics of identity, and which proceeds by defining what the subject is *not*, e.g. '*a* is *x*'. To suggest, for example, that a crocodile is a reptile is to identify it by means of a class which, strictly speaking, no one crocodile 'is'. Any thing, or more technically any *mode* is *itself* before it is anything else. This effectively inverts the kind of ontology that would lead someone to suggest, for example, that nature is culture before it is nature.[45]

This shift makes way for another closely related one: whereas the traditional Christian concept of God would have us believe that either God *creates* something different from itself (thereby re-establishing a difference in substances), or that something different *emanates* from God – but this such that the emanation goes from something superior to something inferior – at least the Spinoza represented by Deleuze conceives instead of the relation as entailing *expression* – yet expression treated as being *immanent* to the one substance. Expression is not representation or *mimesis*; it is not 'about' something else (that would be to return to an ontology of identity). Instead, expression means that the creative process 'stays as one', or 'continues to belong to' the one, *even as it generates something different*. This is what explains the title of one of Deleuze's books on Spinoza (*Expressionism in Philosophy: Spinoza*), and what leads Deleuze to suggest in that same work that 'The significance of Spinozism seems to me this: it asserts *immanence as a principle* and frees expression from any subordination to emanative or exemplary causality. Expression itself no longer emanates, no longer resembles anything. And such a result can be obtained only within a perspective of univocity. God is cause of all things in the same sense that he is cause of himself; he produces as he formally

---

43 Gilles Deleuze, *Expressionism in Philosophy*, p. 164.

44 Gilles Deleuze, *Expressionism in Philosophy*, p. 164-165.

45 In a study of the representation of tropical natures, the historian Nancy Stepan makes precisely that suggestion: 'nature is always culture before it is nature'. See Nancy Stepan, *Picturing Tropical Nature*, (London: Reaktion Books, 2001), p. 15.

exists...'.[46] Throughout the rest of the volume I will refer to this principle as the 'principle of immanence'; doing so will allow me to refer more easily to the notion that things are firstly what they are in themselves; *they express themselves*, as opposed to being an expression of something else. It might be noted that this is why Deleuze, and later Deleuze and Guattari, are so keen to deny that their work is *about* something; what they mean by this is that they do not set out to *represent* the world in the manner that, say, positivist scientists, or indeed constructivist social scientists, might. Precisely for this reason, Deleuze and Guattari can be said to be 'new materialists' in so far as the ontological focus shifts from relations of identity to the things in themselves.

Here is how Deleuze interprets affect in relation to this conceptualisation:

> It has been remarked that as a general rule the affection (*affectio*) is said directly of the body, while the affect (*affectus*) refers to the mind. But the real difference does not reside there. It is between the body's affection and idea, which involves the nature of the external body, and the affect, which involves an increase or decrease of the power of acting, for the body and the mind alike. The *affectio* refers to a state of the affected body and implies the presence of the affecting body, whereas the *affectus* refers to the passage from one state to another, taking into account the correlative variation of the affecting bodies.[47]

Simplifying somewhat, the concept of affect acts as a kind of hinge between one body's and another's affections. Affect is all that by which a body's power to act is increased or diminished, aided or restrained, by another body. The crescendos or diminuendos are all a matter of each individual's immanent capacities, *but as enhanced or curtailed by those of other individuals*, with the same happening again and again, in a chain re-action, or indeed, a chain 're-passion'. The key point to be made here is that Deleuze, and later Deleuze and Guattari, are in a sense having their ontological cake, and eating it too: almost in the same breath, they are suggesting that individuals are what they are (and not what they are not!: principle of immanence), yet at the same time, that in so far as what individuals are what they can *do* (one might say: principle of pragmatism) – and what they can do is always linked to what other things *enable* them or *prevent* them from doing – then they have to be thought in relation to those other individuals as well (one might say: principle of affectivity).

However, and crucially, this affectivity is not understood as a set of frozen 'states', of fixed entities affecting each other as per a mechanistic world view. There is, instead, a permanent flux; everything is always *becoming*, however minutely, something else. As Deleuze puts it,

---

46 Gilles Deleuze, *Expressionism in Philosophy*, p. 180. Italics added to the original.

47 Gilles Deleuze, *Spinoza: Practical Philosophy*, p. 49.

Affects are becomings: sometimes they weaken us in so far as they diminish our power to act and decompose our relationships (sadness), sometimes they make us stronger in so far as they increase our power and make us enter into a more vast or superior individual (joy). Spinoza never ceases to be amazed by the body. He is not amazed at having a body, but by what the body can do. Bodies are not defined by their genus or species, by their organs and functions, but by what they can do, by the affects of which they are capable – in passion as well as in action. You have not defined an animal until you have listed its affects. In this sense there is a greater difference between a race horse and work horse than between a work horse and an ox.[48]

In *Spinoza: Practical Philosophy*, Deleuze spells out what all of this means from the point of view of the nature/culture divide, and so is also worth quoting at length:

It should be clear that the plane of immanence, the plane of Nature that distributes affects, does not make any distinction at all between things that might be called natural and things that might be called artificial. Artifice is fully a part of Nature, since each thing, on the immanent plane of Nature, is defined by the arrangements of motions and affects into which it enters, whether these arrangements are artificial or natural. Long after Spinoza, biologists and naturalists will try to describe animal worlds defined by affects and capacities for affecting and being affected.[49]

Deleuze then uses the biologist's (and proto-biosemiotician) Jakob von Uexküll's conceptualisation of the functional cycle of ticks as an example. The tick is an animal

... that sucks the blood of mammals. He [Uexküll] will define this animal by three affects: the first has to do with light (climb to the top of a branch); the second is olfactive (let yourself fall onto the mammal that passes beneath the branch); and the third is thermal (seek the area without fur, the warmest spot). A world with only three affects, in the midst of all that goes on in the immense forest. An optimal threshold and a pessimal threshold in the capacity for being affected: the gorged tick that will die, and the tick capable of fasting for a very long time. Such studies as this, which define bodies,

---

48 Gilles Deleuze, 'On the Superiority of Anglo-American Literature', in Gilles Deleuze and Clare Parnet, *Dialogues II*, translated by Hugh Tomlinson and Barbara Habberjam (London: Continuum, 2006 [1977]), pp. 27-56, p. 45.

49 Gilles Deleuze, *Spinoza: Practical Philosophy*, p. 124.

animals, or humans by the affects they are capable of, founded what is today called *ethology*. The approach is no less valid for us, for human beings, than for animals, because no one knows ahead of time the affects one is capable of; it is a long affair of experimentation, requiring a lasting prudence, a Spinozan wisdom that implies the construction of a plane of immanence or consistency. Spinoza's ethics has nothing to do with a morality; he conceives it as an ethology, that is, as a composition of fast and slow speeds, of capacities for affecting and being affected on this plane of immanence. That is why Spinoza calls out to us in the way he does: you do not know beforehand what good or bad you are capable of; you do not know beforehand what a body or a mind can do, in a given encounter, a given arrangement, a given combination.[50]

I will have more to say about Uexküll in Part 4. Here it may be noted that, in Deleuze's interpretation of Uexküll, the functional cycles become affects, i.e. they are considered as a matter of interrelation between two individuals, two bodies. Accordingly, the task of the analyst is no longer simply that of discerning what species, conceived auto-ecologically and as essential types, are or are not capable of doing with their bodies. That would be to return to an ontology of identity. Instead, the task is to explain how any of the affections of individuals, on any level, are at once related to, and transformed by, the affections of other individuals thanks to a morphogenesis, one might even say a morphogenetic *trajectory* (cf. Part 4) which is always unique.

The danger of course with this kind of approach is that it may appear to deny that individuals entail any degree of generality, any shared attributes of the kind discovered via biological research. Taken too far, this stance results in the most abject nominalism: anything we say about something is in some sense false because there is no continuity between the representation, and the object or body.

The philosopher Manuel DeLanda,[51] a scholar that is very close to Deleuzian thought, explains why and how this problem can actually be approached from a realist perspective. In the case of species, DeLanda refers to the argument put forward by the biologist and philosopher Michael Ghiselin: that species are best conceived not as *kinds*, but as *individuals* in their own right whose commonality does not constitute a higher ontological category than the individual organisms that compose it.[52] Rather, species come to be species thanks to a historical process of agglomeration and differentiation (we return by this route to morphogenesis).

---

[50] Gilles Deleuze, *Spinoza: Practical Philosophy*, p. 124-125.

[51] See Manuel DeLanda, *Intensive Science and Virtual Philosophy* (London: Continuum, 2002).

[52] Manuel DeLanda, *Intensive Science*, p. 46, referring to the work of Michael T. Ghiselin, *Metaphysics and the Origin of Species* (Albany, NY: State University of New York Press, 1997).

According to DeLanda, this suggests that species are, in fact, 'just another individual entity, [albeit] one which operates at larger spatio-temporal scales than organisms, but an individual entity nevertheless'.[53] In a word, species are themselves 'bodies' (a point I will return to in a moment). Where an essentialist and typological way of thinking about species tries to isolate a limited number of fixed, and relatively unchanging traits underlying any variability, a Deleuzian perspective on Spinoza emphasises a 'flat ontology', i.e. an ontology that recognises unique, singular individuals, differing in spatio-temporal scale but not in ontological status.[54] In this context, what matters is to show how the morphogenesis has occurred, as distinct from simply (or indeed complexly) establishing a hierarchy of types and pointing out the exceptional status of those individuals that don't 'fit', viz., the 'exceptions'. In some sense we come back by this route to Sartre's existence precedes essence.

Before moving on to another Deleuzian/Guattarian concept, I should clarify an additional point. When DeLanda suggests that a species can be regarded as an individual, he is echoing another suggestion by Deleuze: that in Deleuze's own words, '[a] body can be anything; it can be an animal, a body of sounds, a mind or an idea; it can be a linguistic corpus, a social body, a collectivity.'[55] Deleuze is not, in this sense, only interested in what might be described as 'organismic' bodies; on the contrary, his philosophy problematises Merleau-Ponty's assumption that a body is an organism, whoever chiasmatically it relates to other organisms. Just as the dualism between mind and body must be problematised, the same is true for related dichotomies such as those of the human and the nonhuman, the natural and the unnatural, the natural and the artificial. If we really are to oppose the mind-body dualism, then we have to accept that ideas are embodied, but equally, that bodies may themselves be ideas. This perspective is not a million miles away from the kind of Peircian pan-semeiotic cosmology that I referred to in Part 2.

I would now like to turn to a concept that follows on from affect: that of *assemblage*. To introduce this concept, it may be useful to return to the encounter with the crocodile. Each of the individuals in that encounter was itself, and itself only. Yet, by dint of the circumstances generated by the tourists' trajectories, as well as the creation of a biological reserve on Barro Colorado, the arrival of the Smithsonian, the devolution of the Canal Zone to the Republic of Panama, the start of tourism on the island and so forth, an encounter occurred in which, and during which, the individuals in question came to affect each other. This is most obviously illustrated by the fact that the tourists were excited at the discovery of the crocodile, and by the fact that the crocodile felt compelled to make itself scarce. Even as this happened, at least one of the visitors interpreted the scene (if one can call it

---

[53] Manuel DeLanda, *Intensive Science*, p. 47.

[54] Manuel DeLanda, *Intensive Science*, p. 47.

[55] Gilles Deleuze, *Spinoza: Practical Philosophy*, p. 127.

that), by drawing on a body of representations which he clearly regarded as being every bit as real as the crocodile in front of him. Clearly, all of the different elements were related if nothing else by the visitor's act of perception/interpretation – by his observational practice. It would, however, be a mistake to assume that this meant that all of the different bodies were part of some grand scheme, some great organismic 'whole' with wholly interdependent parts. A first way of understanding assemblage is precisely in terms of an ensemble of elements which are at least partly *independent* of each other – but which can and nevertheless do *affect* each other (in the Spinozan/Deleuzian sense of affect).

It might be thought on the basis of this first account that assemblage is the 'system', and the individuals, its 'parts'. However, assemblage entails something quite different from a 'system' – at least a system conceived organically or mechanically. Put simply, what one individual can do depends at least in part on additional individuals, but this according to a logic that is *not* organic. Each individual has a certain autonomy, a certain independence, which allows it to do certain things, regardless of what other individuals do. Assemblage theory, as DeLanda calls it,[56] attempts to think through the enormous implications of this insight.

This is, in some respects, an old problem for sociologists, and one that has been revisited by numerous scholars over the decades, if not centuries. It may thus help to engage in a relatively detailed explanatory excursus to clarify both the continuities and discontinuities between the kind of perspective developed by some of the earlier cultural sociologists, and the Deleuzian/DeLandian proposals. For example, in *Ideology and Modern Culture*, the sociologist John B. Thompson raises this kind of question as part of his broader theory of ideology (a concept that both Deleuze and his friend Michel Foucault reject out of hand, perhaps because they associate it with the older Marxist notion of an inversion of reality). Thompson refers critically to what he describes as a 'consensual theory of social reproduction' – one whose advocates assume that 'the ongoing reproduction of social relations depends in part on the existence of values and beliefs which are collectively shared and accepted by individuals, and which thereby bind individuals to the social order'.[57] While there are some important differences across the writings of those who defend this kind of conception – Thompson distinguishes between a 'core' consensual theory, and a 'differentiated' consensual theory – Thompson notes that they fail to address at least two inter-related issues: first, the extent to which individuals are not merely 'products' of a 'cultural system' or the result of a neatly homogenising process of 'socialisation' (we return by this route to the kind of issue raised by Williams with reference to the relation between culture and social orders, cf. Chapter 8); and second, the extent to which 'in the course of their everyday lives individuals typically move through a multiplicity

---

[56] Manuel DeLanda, *A New Philosophy of Society: Assemblage Theory and Social Complexity* (London: Bloomsbury, 2006).

[57] John B. Thompson, *Ideology and Modern Culture*, p. 87.

of social contexts and are subjected to conflicting social pressures and processes. Rejecting one set of values and norms may coincide with accepting another, or may facilitate their participation in social activities which serve, *ipso facto*, to reproduce the status quo.'[58] In so far as consensus, or as Thompson also puts it, 'symbolic glue' or 'social cement' is over-emphasised, then theories of social order (or indeed of ideology) fail to acknowledge that, in Thompson's words, '[t]he reproduction of the social order does not require some deep underlying consensus concerning values and beliefs, so long as there is sufficient *dissensus* to prevent the formation of an effective oppositional movement'.[59]

As developed by Deleuze, by Deleuze and Guattari, and later by DeLanda, the concept of assemblage offers what is both a more radical, and a more general way of interpreting this problem. As Deleuze defines it, an assemblage is 'a multiplicity which is made up of many heterogeneous terms and which establishes liaisons, relations between them, across ages, sexes and reigns – different natures. Thus, the assemblage's only unity is that of co-functioning: it is a symbiosis, a "sympathy"'.[60] Assemblage is thereby *not* a case of a 'sum' of parts generating a whole, or even the kind of part-whole relation conceived by Gestalt psychologists (or the early Merleau-Ponty). In such theories, as indeed in those that adhere to what Thompson describes as the 'core' or 'differentiated' consensual theories of social reproduction, there is a mistaken adoption of what Manuel DeLanda describes as the *organismic metaphor*.[61] DeLanda does not merely refer to the now ancient tendency to compare all manner of social entities to the human body (such that actors may speak, for example, of the 'organic unity' of a work of art, or that businesspeople may refer to the 'organic growth' of a company). Instead, DeLanda, and before him Deleuze and Guattari refer to conceptions of entities that, in one way or another, take for granted that wholes are necessarily made up of parts that are *essentially related*. As DeLanda puts it, 'The basic concept ... is what we may call *relations of interiority*: the component parts are constituted by the very relations they have to other parts of the whole. A part detached from such a whole ceases to be what it is, since being this particular part is one of its constitutive properties'.[62]

For example, David Marr's computational model of vision (cf. Chapter 3) conceives visual perception on the basis of an organismic model in so far as it focusses on the workings of the retina in relation to the central nervous system (however re-conceived it may be as a *computational* system), but also in so far as it

---

[58] John B. Thompson, *Ideology and Modern Culture*, p. 89.

[59] John B. Thompson, *Ideology and Modern Culture*, p. 90. Italics added to the original.

[60] Gilles Deleuze in Gilles Deleuze and Claire Parnet, 'On the Superiority of Anglo-American Literature', p. 52.

[61] Manuel DeLanda, *A New Philosophy of Society*, p. 8.

[62] Manuel DeLanda, *A New Philosophy of Society*, p. 9.

assumes a system with completely interdependent elements. Marr does recognise an element of *analytical* 'part-independence' in his model in so far as he separates 'computational theory' from 'representation and algorithm', and this in turn from 'hardware implementation'. But vision as a whole is conceived organically in so far as Marr posits a *function* for human vision (in the way that a function might be proposed for the vision of flies), and crucially in as much he argues that such a function (in the human case, discerning shapes) is fulfilled by way of a representational sequence generated by what are themselves essentially interdependent organs and processes.

Were we to conceive wildlife observation in a similar manner, then we would have reify it by treating the observational process as a whole that is itself made up of a complex set of *interdependent* parts, all of which work in essential consort to produce a wholly unified, and indeed univocal way of perceiving, if not also of conceiving and interpreting wildlife in tropical forests (or in any other biome). In Chapter 2 I explained that this is the kind of ontology that underlies Edward O. Wilson's biophilia hypothesis, and I will explain in Part 4 that, ironically, it is also one of the hallmarks of a certain version of mechanism. Simplifying somewhat, when it comes to wildlife observation, so long as one sticks to a logic of interiority, then any human or beyond-human response to a body must ultimately be explained, if not in a Barlowesque (cf. Chapter 3), then in a biophiliac manner (cf. Chapter 2). By contrast, the Deleuzian/Guattarian/DeLandian approach to assemblage reveals the folly of an approach that takes for granted relations of interiority. Not all visitors respond to wildlife in the same way; and indeed, not all humans are compelled to observe that which English-speaking Moderns characterise as 'wildlife'. And of course, not all the members of beyond-human species react to *humans* in the same manner, or even in a manner that is determined in advance by their own 'species-being'.

In recognition of such differences, Deleuze and Guattari replace the organic model with one premised on what DeLanda describes as *relations of exteriority*: as DeLanda puts it, an assemblage involves relations between component parts that 'may be detached from it and plugged into a different assemblage in which its interactions are different. In other words, the exteriority of relations implies a certain autonomy for the terms they relate ... Relations of exteriority also imply that the properties of the component parts can never explain the relations which constitute a whole'.[63]

This account begins to explain how affect and assemblage go hand in hand, or perhaps one should say *body to body*. If affect entails immanence plus relations across affections, assemblage entails the affects of multiple bodies plus 'part independence', or what might also be described as *inter-independence*. The key aspect of the reconceptualisation is that each of the parts is not like a gear in a machine that can do nothing on its own; real gears not only link different parts of a ma-

---

[63] Manuel DeLanda, *A New Philosophy of Society*, pp. 10-11.

chine, but can be removed and used to perform a similar, or indeed a wholly different function in a different machine. The gear does what it does, or what it can do, because it is a gear; but what it does, and what it can do, depends at the same time on what machine it forms a part of. Change the machine, and you will change the function of the gear; change the gear, and you will (also) change the machine.

I think I have now said enough about Deleuze, and Deleuze and Guattari's philosophy to be able to consider some of its implications for the present study. A first set of implications may be explained by returning to Edward Wilson's account of biophilia and snakes (cf. Chapter 2). In keeping with his essentialist approach, Wilson assumes from the outset that encounters with snakes will involve what he describes as an innate tendency to react, if not with 'fear and fascination', then certainly in a manner that reflects the over-determination generated by an 'epigenetic rule' – a universal code for humans that sets out, in advance, what each individual is most likely to, one might also say 'condemned' to *learn* after they are five years old. By contrast, while an approach of the kind described by Deleuze and DeLanda will recognise that what humans and snakes can do with their bodies is *of course* in some sense 'species-specific' thanks to a more or less shared morphogenesis, what *this* or *that* human – this or that *snake* – can do with *its* body is 'individual-specific', i.e. the result of a morphogenesis that does not, *did not* stop hundreds of thousands or millions of years ago. On the contrary, the morphogenesis continues as a flow of change right up to the moment of the encounter between the human and beyond-human individual(s) – and of course continues beyond, assuming that all members of the encounter survive.

So it is, for example, that someone devoted to studying snakes, who has spent their life handling highly venomous species, may well be expected to respond to a snake's presence quite differently from a very young child, or indeed, from an adult who is only generally familiar with the potentially extreme danger of snakes. So the child might want to play with a very colourful coral snake (e.g. *Micrurus* spp.), and conversely, someone who doesn't know that such snakes have very small, and fixed fangs might treat it as if it could strike in the way that, say, a snake with relatively large and hinged fangs (i.e. a viper) can. Exactly the same principle will apply to any other human, or as I will explain in Part 4, to any beyond-human animal; depending of course on its '*umwelt*', a creature that has never encountered humans, or indeed that does not know humans as a threat, is likely to respond quite differently to encounters with humans from those that have been attacked by humans recently, let alone for generations.

A related implication can be teased out with respect to what I described in the introductory chapter as hybrid geographies. A tropical forest, or indeed any species in that forest, described purely as a *type* whose features materialised millennia ago, may well tell us something about many of the forest's or its species' 'key aspects'. To deny this verity would be to accept, in effect, that a scientific discourse on tropical forests is as valid as, say, a creationist discourse. However, conceived in isolation from changes produced by humans locally or across the planet, such forests and their species will almost certainly be represented in a manner that overlooks important aspects of their contemporary nature. A good example of this issue may be found in the difference between accounts of biodiversity in forests that are based on island biogeography, and those that, by contrast, are based on theories and processes of habitat fragmentation.[64]

In each example, what comes to the fore is a conception of individual bodies that at once recognises the immanence of any affections, but also the link, precisely via those self-same affections, to other bodies, and to the necessarily processual interrelation that results: from this perspective, it is not so much being as *becoming*, a point that I will expand on in Chapter 14 as part of Deleuze and Guattari's analysis of animals.

The last point returns us to the question with which I began this chapter. If it is true that what any one body can do is at least partly a matter of its relation with additional bodies, and those bodies are not all 'organismic', then the analysis has to, and *can* take into account the interrelation between bodies 'with organs' and non-organic bodies such as those that go by the name of techniques and technologies. In the case of tourists, the obvious example is those bodies used to detect, photograph or video animals in tropical forests. This is, however, not just a matter of 'use'; the reconceptualisation afforded by affect and assemblage also has implications for where the human body begins and ends. If we go by the common sense notion of the body as a 'bag of bones', i.e. the organism that, spatially-speaking, begins and ends with the largest organ – the skin – then there can be, and should be no discussion as to what constitutes at least this kind of body. From there it is but an essentialist skip and hop to the suggestion that, if we know a body's type, we will know what it can do. If, by contrast, we adopt the kind of radically pragmatic perspective favoured by Deleuze, Guattari and DeLanda, then we may suggest that what a body can do is determined by itself, and by its relation to other bodies, including bodies such as film cameras, editing suites, TV screens, and so forth. Such cyborg-like 'additions' may mean that a body with them can do rather different things than one without them, or one that has not interiorised them in the way that modern audiences tend to have done. It is this that is proven by the example of the tourist who, in a manner of speaking, is capable of observing *C. niloticus* and wildebeest in the absence of either species, thereby creating what is in effect (in *a*ffect?), something like Serengeti-Upon-Gatún.

One issue that may be raised with this kind of approach is whether or not Deleuze and Guattari go too far down the road of flux, of becoming, and the kind

---

[64] For an account of the differences, see William F. Laurance, 'Beyond Island Biogeography Theory: Understanding Habitat Fragmentation in the Real World', in Jonathan B. Losos & Robert E. Ricklefts (Eds.) *The Theory of Island Biogeography Revisited* (Princeton: Princeton University Press, 2010), 214-236.

of pragmatism mentioned above. Are we really to believe that in the human as in the beyond-human world, nothing is set, nothing is fixed? A first answer is affirmative: *strictly speaking*, nothing is entirely fixed. Going back to Sartre, essence does not precede existence. Anyone in doubt as to the wisdom of this stance could do worse than consider the evident historical fallibility of numerous biological theories, and of ongoing discoveries regarding what even well-researched species can do (cf. Chapter 14).

That said, Deleuze, and Deleuze and Guattari are by no means conjuring a world in which there are no continuities, let alone no constraints on bodies (of whatever kind). On the contrary, the concepts of affect and assemblage underscore the extent to which bodies are both enabled *and limited* by other bodies, and by other *ensembles* of bodies. And indeed, in *A Thousand Plateaus*, Deleuze and Guattari clarify that assemblages constitute ways of staking territories in, and across *strata*. The latter is Deleuze and Guattari's concept for a certain congealment, a certain 'coagulation' of relations. Strata are 'phenomena of thickening on the Body of the earth, simultaneously molecular and molar: accumulations, coagulations, foldings'.[65] In keeping with this perspective, Deleuze and Guattari would be the first to recognise that bio-social orders slow rates of change, and impose, or at any rate attempt to impose, limits on what bodies can do. As part of this kind of analysis, Deleuze and Guattari also recognise the key role that desire plays in what humans can do, and I would like to devote the remainder of this chapter to a consideration of this aspect of the philosophers' work.

A word, firstly, about the importance of desire to the present investigation. While I have used the term from time to time, I have not theorised it. Were this to remain the case, it would constitute a particularly glaring omission for a work that is, after all, about a leisure practice (wildlife observation regarded as tourism) that in some sense hinges on desire. It is not possible to understand tourism without explaining a desire to travel, and to experience geographies of alterity, and the same thing is true for wildlife observation of the kind engaged by tourists in tropical forests and all other biomes.

There are, of course, numerous theories of desire, many of them positivist, and many biologically reductionist along neuroscientific, adaptationist, or analytical-philosophical lines.[66] If I have chosen Deleuze and Guattari's, it is for two reasons: first, and as might be expected, there is a goodness of fit between their theory and the rest of the Deleuzian/Guattarian philosophy. But second, it is also the case that Deleuze and Guattari's theory works well in a context in which I myself have rejected a sociobiological account of wildlife observation; according to such an

---

[65] Gilles Deleuze and Félix Guattari, *A Thousand Plateaus*, p. 502.

[66] For an encyclopaedic entry on the kinds of issues raised by, and distinctions made via positivist philosophical perspectives, see for example Timothy Schroeder, 'Desire', Stanford Encyclopedia of Philosophy, <https//:plato.stanford.edu/entries/desire>[Accessed 2 July 2018].

account, wildlife observation would be no more (and no less) than some kind of atavism (cf. Chapter 2). From a sociobiological perspective one would have to argue that the desire to observe wildlife arises in response to an evolved 'need' to be 'close to wild animals', or at any rate, 'close to nature'. By contrast, Deleuze and Guattari argue in favour of precisely the opposite perspective: needs arise from desire, but desire is in some sense always 'up to date'; it always reflects, however directly or indirectly, however actively or passionately (we return to Spinoza on affects), the changing 'states of assemblage'.

So how do Deleuze and Guattari conceptualise desire, and before that, why are they interested in conceptualising it? It is pertinent to begin by noting that, in the wake of the events of May 1968, Deleuze and Guattari were keen to try to explain what it was that led people to oppose (or not) repressive orders of the kind embodied by Charles De Gaulle or Richard Nixon, the Algerian or the Vietnam War. They argued that what was at stake – and what needed to be explained and opposed – was not only *state* authoritarianism, but what one might describe as the 'micro-fascism', the micro-authoritarianism that every person potentially has *within them*. As Michel Foucault puts it in the preface to the English-language translation of *Anti-Oedipus*, the first volume of *Capitalism and Schizophrenia* (published in French in 1972), 'the strategic adversary is fascism ... not only historical fascism, the fascism of Hitler and Mussolini – which was able to mobilize and use the desire of the masses so effectively – but also the fascism in us all, in our heads and in our everyday behavior, the fascism that causes us to love power, to desire the very thing that dominates and exploits us.'[67]

For Deleuze and Guattari the question of desire – individual, but also collective – is one of the keys to understanding what happened in 1968, and how best to oppose fascism. And a theory of desire should start from a Freudian psychoanalytic conception of the unconscious. Simplifying somewhat, according to Freud the unconscious includes instincts or drives, affects and ideas, a part of which the individual mind censors (and so remains unaware of) because the elements in question are too threatening to be contemplated consciously. However, the fact that these aspects are censored does not stop them from silently shaping the thoughts, actions and desires of the individual. On the contrary, the drives as well as the repressed thoughts and emotions live on in the form of sublimated, and entangled desires.[68]

If this kind of dynamic occurs in the individual, Freud suggests that a similar dynamic takes place in societies. In *Civilization and Its Discontents*,[69] Freud notes

---

[67] Michel Foucault, 'Preface', in *Anti-Oedipus*, xi-xiv, p. xiii.

[68] See for example, Sigmund Freud, 'The Unconscious', translated by Cecil. M. Baines, in Philip Rieff, *General Psychological Theory: Papers in Metapsychology, Sigmund Freud* (New York, Colliers, 1963[1915]), 116-150.

[69] Sigmund Freud, *Civilization and Its Discontents*, translated by David McLintock (London: Penguin, 2012 [1930]).

that desire has a complex relation to social order: on the one hand, civilisation is generated by people partly to protect themselves from the dangers of external threats, or indeed from the anarchy that is thought to result from the absence of social order. But it also allows people to achieve aims which would otherwise be unrealisable. As evidence Freud points to all of the technical advances associated with modern cultures by the late 1920s (which is when he wrote the book).

On the other hand, such aims can only be achieved by controlling the instincts and drives of a society's members, and according to Freud this sets up an unavoidable contradiction – a contradiction between drives and regulation – which leads to a kind of permanent collective neuroticism, and with it collective sublimations and entangled expressions of desire. As Deleuze and Guattari put it, 'the way a bureaucrat fondles his [sic] records, a judge administers justice, a businessman causes money to circulate; the way the bourgeoisie fucks the proletariat; and so on. And there is no need to resort to metaphors, any more than for the libido to go by way of metamorphoses. Hitler got the fascists sexually aroused. Flags, nations, armies, banks get a lot of people aroused'.[70]

Thus far, I've emphasised the continuity between a psychoanalytic and a Deleuzian/Guattarian conception of desire. However, Deleuze and Guattari take critical distance from Freud's and other psychoanalytic scholars' Oedipal turn: '[T]he traditional logic of desire is all wrong from the very outset: from the very first step that the Platonic logic of desire forces us to take, making us choose between *production* and *acquisition*. From the moment that we place desire on the side of acquisition, we make desire an idealistic (dialectical, nihilistic) conception, which causes us to look upon it as primarily a lack: a lack of an object, a lack of the real object.'[71]

In Freudian psychoanalysis, this kind of approach is expressed via the notion that what starts out as a desire for the mother (as part of the Oedipal complex) becomes transmogrified into a general desire for that which is forbidden – one 'acquires' a desire for what one cannot have. Even the famously complex analysis of Jacques Lacan, which introduces numerous changes to Freudian theory, posits a kind of lack: the child entering the mirror stage – simplifying greatly, Lacan's term for the formation of the Ego, via an identification with one's own specular image, be it via an actual mirror, or the approving/disapproving commentary of significant others – becomes the object of the libidinal economies of those others. Parents, siblings, extended family, friends, etc. encourage the child to be more like them, and so the child learns to recognise itself not in the terms of its own desires, but in those of others' desires: *be more like me/us* – a psycho-social version of the identitary logic I referred to earlier, *a*(you) is *x*(us). If this process is sufficiently

---

[70] Gilles Deleuze and Félix Guattari, *Anti-Oedipus*, p. 293.

[71] Gilles Deleuze and Félix Guattari, *Anti-Oedipus*, p. 25. Italics in the original.

severe – today we might say *austere* – the own desire is utterly repressed in favour of meeting the desires of the others.[72]

As I began to note above, part of what worries Deleuze and Guattari about this kind of approach is the danger of a conceptual idealism. They quote the philosopher Clément Rosset when he suggests that, every time that the emphasis is put on desire as a lack of an object, 'the world acquires as its double some other sort of world, in accordance with the following line of argument: there is an object that desire feels the lack of; hence the world does not contain each and every object that exists; there is at least one object missing, the one that desire feels the lack of; hence there exists some other place that contains the key to desire (missing in this world)'.[73]

Deleuze and Guattari are also concerned about the negativity of the resulting conception. So long as desire is theorised via a lack, it cannot really be productive in any way. By way of an alternative, they propose a conception of desire that is very much in keeping with the philosophy of difference, and the principle of immanence, but now inflected by the materialism of the earlier Marx.

Why Marx? One critique of Freudian psychology is that is so focussed on familial dynamics that it leaves out more general social determinations; Deleuze and Guattari believe that Freud's discovery of the unconscious is both valid and useful, but it must be materialised. As Deleuze and Guattari note, 'There is no particular form of existence that can be labeled "psychic reality." As Marx notes, what exists in fact is not lack, but passion, as a "natural and sensuous object"'.[74] Here Deleuze and Guattari are referring to Marx's famous critique of Hegel's idealism, as expressed in the *Economic and Philosophical Manuscripts*. As Marx himself puts it,

> As a natural being, and as a living natural being, he [the human] is on the one hand equipped with natural powers, with vital powers, he is an active natural being; these powers exist in him as dispositions and capacities, as drives. On the other hand, as a natural, corporeal, sensuous, objective being, he is a suffering, conditioned, and limited being, like animals and plants. That is to say, the objects of his drives exist outside him as objects independent of him; but these objects are objects of his need, essential objects, indispensable to the exercise and confirmation of his essential powers. To say that man is a corporeal, living, real, sensuous, objective being with natural powers means that he has real,

---

[72] See Jacques Lacan, 'The Subversion of the Subject and the Dialectic of Desire in the Freudian Unconscious', *Écrits: A Selection*, translated by Alan Sheridan (London: Routledge, 1977[1960]), 293-325.

[73] Clement Rosset, *Logique du Pire* (Paris: Presses Universitaries de France, 1970), p. 37, in Gilles Deleuze and Félix Guattari, *Anti-Oedipus*, p. 26.

[74] Gilles Deleuze and Félix Guattari, *Anti-Oedipus*, p. 27.

sensuous objects as the object of his being and of his vital expression, or that he can only express his life in real, sensuous objects.[75]

The key insight is this: for Deleuze and Guattari, '[d]esire is not bolstered by needs, but rather the contrary; needs are derived from desire: they are counter-products within the real that desire produces. Lack is a countereffect of desire; it is deposited, distributed, vacuolized within a real that is natural and social. Desire always remains in close touch with the conditions of objective existence; it embraces them and follows them, shifts when they shift, and does not outlive them.'[76] In effect, desire is like a series of habits (or rather, the far more complex Deleuzian concept of 'passive syntheses'[77]) that generate 'partial objects, flows, and bodies', and that work as 'units of production'.[78] 'Desire does not lack anything; it does not lack its object. It is, rather, the subject that is missing in desire, or desire that lacks a fixed subject; there is no fixed subject unless there is repression. Desire and its object are one and the same thing: the machine, as a machine of a machine. Desire is a machine, and the object of desire is another machine connected to it.'[79]

Three points require further elaboration if this remarkable statement is to be adequately contextualised. The first is that, as suggested earlier, Deleuze and Guattari are accepting the Freudian concept of the *unconscious*, but not the Oedipal turn, which they think is a mistaken one, precisely in so far as it substitutes an idealisation for the productiveness of the unconscious: 'The great discovery of psychoanalysis', Deleuze and Guattari argue, is 'that of the production of desire, of the productions of the unconscious'; however, once Oedipus enters the picture, this discovery comes to be buried under a new kind of idealism: 'a classical theater [is] substituted for the unconscious as a factory; representation [is] substituted for the units of production of the unconscious; and an unconscious that [is] capable of nothing but expressing itself – in myth, tragedy, dreams – [is] substituted for the productive unconscious'.[80]

If this idealisation is left out, then Freud's unconscious may be conceived as a *'desiring-production'*, a desire to couple, to connect. According to Deleuze and Guattari, this force is best illustrated not by way of the neurotic on the couch – the archetypal repressed figure for Freudian and Lacanian psychoanalysis – but by way of the ostensively inappropriate actions and feelings made manifest in schiz-

---

75 Karl Marx, *Economic and Philosophical Manuscripts,* in *Marx/Engels Collected Works,* Vol. 3 (London: Lawrence & Wishart, 1997), p. 336.

76 Gilles Deleuze and Félix Guattari, *Anti-Oedipus,* p. 27.

77 Gilles Deleuze, *Difference and Repetition.*

78 Gilles Deleuze and Félix Guattari, *Anti-Oedipus,* p. 26.

79 Gilles Deleuze and Félix Guattari, *Anti-Oedipus,* p. 26.

80 Gilles Deleuze and Félix Guattari, *Anti-Oedipus,* p. 24.

ophrenic episodes. Distinguishing what they describe as the schizophrenic *process* from the *illness* of schizophrenia, Deleuze and Guattari find in the former the purest expression of the unconscious, and with it a potential for productive disruption and *dis*-order – the kind illustrated by artists with schizophrenia. From this perspective, schizophrenia is the unconscious fully expressing itself, and revealing its powers of production. By contrast, Deleuze and Guattari are only too well aware – presumably thanks not least to Guattari's own practice in the field of psychoanalysis – of the horrors of the *illness* of schizophrenia, a condition which they believe is produced by the standard psychiatric treatments of their time, which result in the pathologies that, ironically, have come to signify schizophrenia 'itself'.

Here is a second elaboration: as I began to suggest earlier, the reference to production in desiring-production signals the recontextualisation of desire in an (early) Marxist framework. For Marx, non-alienated labour is one in/for which the producers can enjoy production as a confirmation of their personal and humanly powers, even as they meet the needs of others. It is this kind of production which capitalism represses by way of wage labour, and with it the *division* of labour that results not just in class differences, but in the splitting of activities into themselves alienating tasks. Marx puts it thus in 'Comments on James Mill, *Éléments D'économie Politique*':

> Exchange, both of human activity within production itself and of human product against one another, is equivalent to species-activity and species-spirit, the real, conscious and true mode of existence of which is social activity and social enjoyment. Since human nature is the true community of men, by manifesting their nature men create, produce, the human community, the social entity, which is no abstract universal power opposed to the single individual, but is the essential nature of each individual, his own activity, his own life, his own spirit, his own wealth.

Conversely,

> The more diverse production becomes, and therefore the more diverse the needs become, on the one hand, and the more one-sided the activities of the producer become, on the other hand, the more does his labour fall into the category of labour to earn a living, until finally it has only this significance and it becomes quite accidental and inessential whether the relation of the producer to his product is that of immediate enjoyment and personal need, and also whether his activity, the act of labour itself, is for him the enjoyment

of his personality and the realisation of his natural abilities and spiritual aims.[81]

A final aspect that requires elaboration concerns Deleuze and Guattari's use of the concept of machine. Except for a reference in the Postscript, I myself will not be using this concept in the way that Deleuze and Guattari do, but I should at least say something about what the philosophers mean when they suggest that 'Desire and its object are one and the same thing: the machine, as a machine of a machine'. It may be helpful to preface this concept by noting that it emphatically does *not* reproduce philosophical mechanism, or indeed a mechanistic conception of causality, let alone a machine regarded as a technical device (though, confusingly, the latter may also act as a machine in the sense conceived by Deleuze and Guattari). Instead, '[a] machine may be defined as a *system of interruptions* or breaks (*coupures*). These breaks should in no way be considered as a separation from reality; rather, they operate along lines that vary according to whatever aspect of them we are considering. Every machine, in the first place, is related to a continual material flow (*hyle*) that it cuts into.'[82] The examples that Deleuze and Guattari offer are doubtless meant to be at once shocking and humorous, at least to the stereotypical bourgeois sensibility: '[A machine] functions like a ham-slicing machine, removing portions from the associative flow: the anus and the flow of shit it cuts off, for instance; the mouth that cuts off not only the flow of milk but also the flow of air and sound; the penis that interrupts not only the flow of urine but also the flow of sperm'.[83] Deleuze and Guattari add that 'Each associative flow must be seen as an ideal thing, an endless flux, flowing from something not unlike the immense thigh of a pig. The term *hyle* in fact designates the pure continuity that any one sort of matter ideally possesses.'[84]

As Deleuze and Guattari note, this kind of process raises questions of the sort 'What flow to break? Where to interrupt it? How and by what means? What place should be left for other producers or antiproducers ... ?'[85] They address such questions via the suggestion that each machine has a code 'built into it, stored up inside it'; '[t]his code is inseparable not only from the way in which it is recorded and transmitted to each of the different regions of the body, but also from the way in which the relations of each of the regions with all the others are recorded'.[86] For

81 Karl Marx, 'Comments on James Mill, Éléments D'économie Politique', written in 1844 but first published in German in 1932. The quote is from the version published at Marxists.org/archive/marx/works/1844/james-mill [Accessed 1 June 2018].

82 Gilles Deleuze and Félix Guattari, *Anti-Oedipus*, p. 36.

83 Gilles Deleuze and Félix Guattari, *Anti-Oedipus*, p. 36.

84 Gilles Deleuze and Félix Guattari, *Anti-Oedipus*, p. 36.

85 Gilles Deleuze and Félix Guattari, *Anti-Oedipus*, p. 38.

86 Gilles Deleuze and Félix Guattari, *Anti-Oedipus*, p. 38.

example, '[a]n organ may have connections that associate it with several different flows; it may waver between several functions, and even take on the regime of another organ – the anorectic mouth, for instance'.[87]

So what Deleuze and Guattari are referring to when they speak of desire as a machine is a 'device' that at once makes an entity paradoxically *present* by interrupting the flow of the cosmos, even as it, or the code it 'contains' regulates the manner in which it does so. In keeping with Spinoza's undermining of the nature-culture dualism, such machines may equally be meat slicers made by humans or the muscle of the cow being sliced, or indeed the desire of the individual who operates the meat-slicing machine, or who eats the cow. As will begin to be clear, the philosophers are taking the Cartesian, and more generally the modern notion of a machine and in some sense turning it against itself; far from being the lifeless, 'inert' mechanism that pervades the metaphysics of modern biological science, what there is instead is a conceptual mongrel that couples (in the Deleuzian/Guattarian spirit of affect one should perhaps say *copulates*) what sociologists used to call a code (or what I myself have described as a coding orientation), with any body (human, beyond-human, non-organismic, etc.) that serves to produce desire (or desiring-production). Its 'outputs' are more 'couplings', more affects.

What, then, are the implications of this kind of approach for this study? As suggested at the start of this discussion, contrary to a sociobiological discourse that would have us believe that there is, say, a transcendental human longing for experiences with wildlife (cf. Chapter 2), Deleuze and Guattari allow us to theorise wildlife observation – amongst tourists, but also amongst any other groups, including biologists – in a manner that links the practice to affects and to assemblages, and to the desires generated with/by them. If organismic bodies are desiring bodies, but if those bodies change, however minutely, whenever an assemblage changes, then according to Deleuze and Guattari, desires too, will change. Put simply, 'new' bodies go with 'new' desires. So it is that someone who has never particularly paid attention to birds may join a group of birders 'for social purposes' but may within a short time find themselves lusting after a sighting of this or that species. Or that someone who had never thought of visiting a tropical forest, sees the representations of one across several TV programmes, and ends up organising a trip to Central America just to experience 'such a forest' firsthand, first-eye.

Alas, when that person reaches the actual forest, they may find that the media have, quite inadvertently, positioned them to experience a lack: the lack of a telegenic forest, the lack of a forest that is teeming with charismatic animals that are totally visible in Merleau-Ponty's sense of the play of visibility and invisibility. This is not a contradiction of Deleuze and Guattari's theory of desire; as I noted earlier, lack is a counter effect of desire, '[d]esire always remains in close touch with the conditions of objective existence; it embraces them and follows them,

---

[87] Gilles Deleuze and Félix Guattari, *Anti-Oedipus*, p. 38.

shifts when they shift'. By this account, the nature media may themselves be re-garded as a machine – one that captures the visible, or perhaps one should say slices off the invisible 'parts' of tropical forests, in either case according to a code premised on manufacturing absolute visibility. For the audiences brought up with this machine and its associated techniques of observation, observing wildlife in the absence of an invisibility slicer may generate, if not a sense of disappointment, then at least an identitary logic of the kind suggested by the tourist with which Part 3 began: a (the crocodile in front of them) is x (the crocodiles depicted by the nature media). In this context, it is not difficult to understand why, if some aspect of an excursion contradicts the kind of identity established by the ubiquitous me-dia of mass communication, then at least some tourists would experience a sense of dis-illusion.

Conclusions to Part 3

Let's go back to the encounter with the crocodile. It would, of course, be possible to use Peircian semeiotic theory to analyse several aspects of that encounter. However, such an analysis would say little or nothing about all the different kinds of bodies that might affect such an encounter. For this reason, I've suggested that it's necessary to move beyond semeiotic phenomenological analysis, *sensu stricto*, to consider what might be described as the bio-cultural dimensions of wildlife observation amongst tourists visiting tropical forests.

In Chapter 8, I focussed on the kinds of bodies that have long been of interest to cultural sociologists (and social anthropologists): those that are constitutive of cultural formations. From this kind of perspective, whether or not tourists attend to crocodiles, and *how* they do so – how they *observe* them – depends on organisations, fields, coding orientations, discourses, genres, and the techniques and technologies of observation associated with them. If the individuals that made up the group of tourists that I joined on Barro Colorado were interested in travelling there, and observing certain kinds of animals, it was thanks to the fact that they were engaging in a particular kind of tourism, and as part of that process, were influenced by the representations produced by institutions both proximate and distal: by STRI's Visitors Programme, as embodied by guides, but also by media institutions such as the BBC, via its famous wildlife documentaries.

One way of summing up this kind of approach is with reference to what may be described as an instituted *dialectic of attention and observation*: observation presupposes *attending* to something, but what one attends to depends in turn on socio-cultural forces that are quite separate from the observed object, and which go beyond the individual observer. What I described at the end of Part 2 as mediate modes of wildlife observation entail such a dialectic in far as they are at least partly shaped by the practices of particular organisations such as STRI or the BBC, and by the coding orientations, the techniques and technologies of observation, and of course the power relations associated with certain fields, discourses, and genres.

This raises a question: how precisely do individuals and groups internalise particular techniques of observation, and with them, characteristic dialectics of attention and observation? This is a question that can only really be answered by way of concrete, empirical ethnographic research (cf. Volume 2). I would, however, like to at least mention a key part of any such process: there must always be a pedagogy of observation (or in this case, a pedagogy of *wildlife* observation). Wildlife observation, and more specifically techniques of wildlife observation, must be taught and/or learned; however, the pedagogic process is not necessarily a formal, or even an informal one – on the contrary, it is most like to be non-formal.[1]

This point requires a brief explanatory excursus. In the case of a formal pedagogy, there is a clearly defined pedagogic space, with what are themselves typically well-defined and hierarchical roles (educator/educand), and more or less explicit contents to be taught and learned over time (i.e. an explicit curriculum). For its part, an informal pedagogy is one where a de facto educator (an individual or body which is not usually identified as an *educator*) smuggles a pedagogic dynamic to a space that is not itself conceived by the users as being primarily educational. So, for example, a guide 'offers information' to tourists who don't necessarily expect to be taught anything, and who certainly don't adopt the identity of 'educands'. But even this kind of mostly tacit pedagogic relation may be contrasted with what I describe as a non-formal pedagogic mode, which may be characterised by way of a paradoxical apophthegm: in the non-formal mode, *nothing is taught and nothing is learned, but something is taught and something is learned.* Nothing is consciously, or *deliberately* taught, and nothing is deliberately or *consciously* learned, but something may nevertheless be taught and/or learned. This is what some might describe as learning 'by osmosis', or 'by example'. No one deliberately or consciously taught the tourist in the example at the start of this part of the volume to gaze upon crocodiles in the way that he did, and I very much doubt that he himself learned to do so in a deliberate or conscious manner. What I am describing as a non-formal pedagogy of observation thereby lacks all of the real or assumed clarities not just of the formal, but even of the informal mode of pedagogic practice. There is not only no *manifest* pedagogic place, but also, there is not even a more or less covert attempt to teach something to someone, as often happens in the case of the informal mode.

The more general point is this: where a sociobiological approach to perception would argue that perception is mostly if not entirely a matter of a transcendental nature, in the form of an innate mechanism involving one or another perceptual system, in this study I argue that wildlife observation entails multiple pedagogic dynamics. That said, many, if not most of these dynamics are likely to be non-formal, and will typically obey a multiplicity of influences, contingent on groups' cultural practices and everyday involvement with institutional spheres and forces.

In Volume 2 I will offer some more concrete examples of how such a pedagogy might work in the context of tourism in a tropical forest such as Barro Colorado's. For now I would like to move on to the kinds of issues that I presented via Beauvoir, Merleau-Ponty, Deleuze, and Guattari. In so doing, it is relevant to note that, if I have taken the time to describe a cultural sociological approach, and to use many of its categories, it is because I still believe that this kind of perspective has

---

[1] I first offered an account of a non-formal pedagogy in Nils Lindahl Elliot, *Mediating Nature*.

something to contribute to the analysis in so far as it continues to be necessary to recognise the importance of the human social and cultural aspects of wildlife observation – aspects which clearly *are* shaped by fields, discourses, genres, and institutional techniques of observation.

The last point notwithstanding, a study of wildlife observation cannot simply focus on the mentioned aspects. It must also take into account the confluence of body, biology, sociality, and culturality, and this requires several theoretical and methodological displacements.

The first displacement involves giving bodies a centrality that they lack in the earlier cultural sociology, but also an agency, and indeed a *kind* of agency that is itself absent from both mechanistic conceptions, and from much cultural theory. With Simone de Beauvoir and Maurice Merleau-Ponty it is possible to highlight the importance of the body *as situation*, and to build especially on Merleau-Ponty's suggestion that the body is not simply the outcome or the meeting-point of causal agencies which determine one's bodily or psychological make-up; nor is it an ensemble of organs juxtaposed in space. Instead, it is something for which the subject is in 'undivided possession': 'I know where each of my limbs is through a *body image* in which all are included'.[2] This body image is a dynamic one; as Merleau-Ponty puts it, 'it is clearly in action that the spatiality of our body is brought into being'; '[b]y considering the body in movement, we can see better how it inhabits space (and, moreover time)'. '[M]ovement is not limited to submitting passively to space and time, it actively assumes them, it takes them up in their basic significance'.[3] '... [M]y body appears to me as an attitude directed towards a certain existing or possible task. And indeed its spatiality is not, like that of external objects or like that of "spatial sensations", a *spatiality of position*, but a *spatiality of situation*.'[4]

This body *as* situation is clearly fundamental to any act of wildlife observation; one observes wildlife at once in, and from that situation – a situation that is dynamic both for the reasons that the early Merleau-Ponty explains (and which are in some respects similar to the ones pointed out by James Gibson) and for others which I will consider in Part 4. Here it suffices to note that, just as Peirce refers to dynamical *objects*, it is possible to refer to dynamical *bodies*. Such bodies not only have their own existence, but a capacity to produce their own signs, and to engage with other bodies in ways that affect the way in which those other bodies produce signs in turn. One's body constitutes the fundamental situ-ation, but is of course at the same time a part of an interrelation, a 'mutualism' of the kind described by the later Merleau-Ponty via the concept of chiasm: I can

---

[2] Maurice Merleau-Ponty, *Phenomenology of Perception*, pp. 112-113. Italics in the original.

[3] Maurice Merleau-Ponty, *Phenomenology of Perception*, pp. 117.

[4] Maurice Merleau-Ponty, *Phenomenology of Perception*, pp. 114-115. Italics in the original.

observe other bodies, but my own body can also *be* observed, or to put the matter differently, my body can observe *because* it too, can be observed.

Implicit to what I have just suggested are two more displacements, effected this time by way of the philosophy of Gilles Deleuze, and of Deleuze and Félix Guattari. Fundamental to at least the earlier cultural sociology is both the assumption, and explanation of what with Cornelius Castoriadis we might describe as the logic of *legein* – as Castoriadis explains, an identitary and ensemblist logic. *Legein* is 'distinguish-choose-posit-assemble-count-speak', and as such is 'at once the condition and the creation of society, the condition created by what it itself conditions'.[5] It is a fundamental cultural operation. For the earlier cultural sociology, *legein* is, in one way or another, what makes the world go round. Via a Spinozan-inspired principle of immanence, Gilles Deleuze questions the primacy of *legein*, and suggests in effect that things are what they are before they are something else; in a manner of speaking, they 'cause themselves'. So long as one does not recognise this absolutely fundamental point, then one must fall back on the kind of transcendentalism that is at the root of identitary philosophies. From the Deleuzian perspective, when there is an encounter between a tourist and a crocodile, the tourist is not first and foremost a tourist, and the crocodile is not first and foremost a crocodile; each individual is itself, and the same is true for the encounter itself. The names, the representations, come afterwards.

The last point notwithstanding, the principle of immanence must be rescued from a hermetic monism by noting that all bodies have certain capacities, certain *affections* which always stand in relation to the affections of other bodies. Building still on Spinoza, the conjunction and mutual effect of these capacities is what Deleuze describes as affect. Affect may be regarded as a kind of homological equivalent, within Deleuze's system, of Peirce's phenomenological secondness; to understand the tick, we need to understand not just its own affections, but those capabilities as a reaction to another body (in this case, its host). Affect is thus a kind of hinge between bodies, between the haecceities of bodies.

The other element contributed by affect is the one highlighted by what has come to be known as the 'affective turn'. This turn is about all those aspects of relations that are pre-personal, pre-conscious, and some would argue pre-semiotic – or at least, pre-*semiological*. The key to understanding affect from this perspective is the unconscious, and desire. In the earlier cultural sociology, any reference to the unconscious and to desire is typically interpreted along psychoanalytical, or post-psychoanalytical lines. (To this day, much of the work on desire relies on the theory of Jacques Lacan.) By contrast, Deleuze and Guattari sever the unconscious and desire from a 'theory of lack'. As suggested in Chapter

---

[5] Cornelius Castoriadis, *The Imaginary Institution of Society*, translated by Kathleen Blamey (Cambridge: Polity Press, 1987[1975]), p. 223. I also refer to this part of Castoriadis' work in Nils Lindahl Elliot *Mediating Nature*, p. 69.

10, for Deleuze and Guattari, '[d]esire is not bolstered by needs, but rather the contrary; needs are derived from desire'.[6] This perspective stands as another important corrective vis-a-vis the kind of transcendental conception proposed by sociobiologists like Edward O. Wilson, for whom any desire to see wildlife would have to be interpreted along adaptationist lines. With Deleuze and Guattari, we can say that the desire to observe wildlife is one that goes with the very formations, the very assemblages that promote wildlife observation. Tourists do not travel because they have nomadism in their genes, or because they are genetically predisposed to seek contact with other organisms; they travel because they desire to partake in a cultural practice that came to be promoted in Europe in the late eighteenth and nineteenth centuries, and eventually became established as a genre of leisure which is today encouraged by capitalist institutions the world over. The aforementioned desires, and the promotion of tourism and its characteristic pleasures have worked together to generate a 'need' for travel, and not vice-versa. Exactly the same directionality holds true for wildlife tourism, or an ecotourism focussed on observing wildlife. Any 'need' to observe wildlife is not itself a transcendental one of the kind theorised by some sociologists; not all people will desire to travel even when exposed to the messages of the tourist industry, and even those that do will not do so in a mechanical, or automatic way. The logic of assemblages is such that the individuals maintain what I have described as an 'inter-independence' with/from other individuals.

The last point brings us to a final shift vis-a-vis the earlier forms of cultural sociology. When, for example, a researcher uses Bourdieu's powerful concept of field, four theoretical operations are likely to occur. The first is that the concept will encourage the researcher to think of the logic of that field, and that field alone. The second is that the researcher is likely to regard the field's actors as being in some sense locked into that field's logic (we return to the metaphor of 'magnetism'). The third is that it will be very tempting to treat the actors in the field as relating according to an organic logic, or what DeLanda describes as relations of interiority. The fourth operation is not specified anywhere, but is likely to occur in so far as the researcher does not question the anthropocentrism of the early cultural sociology: fields, and their operations, only really apply to human actors. Everything else is either ignored, or made subservient to human dynamics.

By contrast, the concept of assemblage assumes from the start that some of the bodies involved in fields, or field-like spaces, may not be human. Indeed, according to Deleuze a body can be anything. It can, of course, be a human or beyond-human animal, but as Deleuze notes, it can also be a body of sounds, a mind, an idea, a linguistic corpus, a social body, or any collectivity.[7] Clearly, it is not just organismic bodies that are involved in encounters with wildlife; the

---

[6] Gilles Deleuze and Félix Guattari, *Anti-Oedipus*, p. 27.

[7] Gilles Deleuze, *Spinoza: Practical Philosophy*, p. 127.

'bodies with organs' involved in wildlife encounters may also be affected by the representations produced by the nature media (or other media), by any mechanical devices used *in situ* by tourists, and then again, by sounds, or indeed the emanations that the wildlife may generate (for example, in this volume's conclusions, I will refer to the smells produced by peccary). To make the point once again, the tourist in the opening example of this part of the volume interpreted the signs produced by the crocodile *in situ* by way of signs produced elsewhere by the nature media.

Equally if not more importantly, assemblage theory is clear that the bodies that partake in its interrelations continue to operate according to a principle of immanence, and so are not simply 'interdependent'. Even as they relate to each other, a logic of exteriority continues to apply, with each body having, as the old expression puts it, a life of its own.

Finally, as I will explain in Part 4, even as assemblage entails a logic of territorialisation, Deleuze and Guattari are clear that there is a constant trans-territorialisation (as I describe it), and that a body can be a part of more than one assemblage at once. From this perspective, nothing simply 'stays in place', and in any case, one can be in more than one 'place' at a time.

I suggest that the three concepts in their interrelation – body, affect, and assemblage – provide a means of describing the extraordinary circumstances that people may find themselves in when, having grown up and lived their lives thousands of miles away, they travel to immerse themselves, however temporarily, in a setting quite literally replete with relations which they are not a part of (at least not on anything like a permanent basis), which may and indeed necessarily will affect them, but which are neither discursive nor field-based, even if they *are* semeiotic. I refer, of course, to a tropical forest. One may argue, as Bourdieu no doubt would have, that even tourists visiting a tropical forest must act in accordance with a logic of field – if nothing that else, they may be shepherded by guides who invite them to accept the rules, regulations and modalities of framing of a particular visitors programme. But such a field-based logic tells us little or nothing about actors of the kind found in the forest itself, or indeed of the ways in which all the different participants interrelate. To understand such interrelations, it is necessary to understand relations of exteriority. And it is also necessary, of course, to understand the forest and its animals, and in Part 4 I will turn to forms of analysis that foreground ecology, geography, and beyond-human animality.

Appendix 2: Deleuze and Guattari, Hjelmslev and Peirce

There is a discontinuity between Parts 2 and 3 – or at any rate between the intellectual 'heroes' of each section – that ought to be acknowledged. Given that Part 3 already has covered so much philosophical ground, I have decided to engage with the discontinuity in this appendix.

The discontinuity is this: in Part 2, I extolled the virtues of a Peircian semeiotics. Alas, at least in the two volumes of *Capitalism and Schizophrenia*, Deleuze and Guattari declare themselves in favour of the semiotic theory of Louis Hjelmslev[1] (1898-1965). Hjelmslev is in many respects much closer to Saussure than he is to Peirce, and so it is necessary to consider the implications of this discontinuity for the present study.

To start with it may be noted that Deleuze and Guattari's choice of Hjelmslev seems like a contradictory one, given the fact that in numerous passages of their books, the authors take critical distance from semiology and from structuralism. Hjelmslev can be classified as a post-Saussurean theorist in so far as his own work starts from the Saussurean dyadic or bilateral conception of the sign. Indeed, someone not well-versed in Hjelmslevian theory might argue that all that Hjelmslev's theory does is refine Saussure's signified/signified conjunction by renaming its aspects and adding additional tiers or strata: the signified comes to be renamed as the *content*, and the signifier as the *expression*, even as each of these elements acquires an additional duality, that of substance and form. So, for example, the chain in the English language 'I do not know' constitutes not only a formed content (a certain meaning as formed by a sequence of words), but also a formed expression (each word as shaped by its spelling, the choice of font, etc.). These give form to a substance of content (the thought to be conveyed) and a substance of expression, the actual materiality of the forms used to express the meanings (e.g. ink, or in the case of words on a computer screen, the pixels, the play of light and dark, etc.). In the above example, which Hjelmslev himself employs, 'I do not know' may be contrasted with the French *je ne sais pas* or the Finnish *en tiedä* or indeed the Eskimo (sic) *nalvara*.[2] In each case, different signs – different conjunctions, or as Hjelmslev puts it, different *functions* and their respective *functives* of content-form and expression-form – are used to express thoughts which nonetheless may have a common content and expression *purport* – the latter being Hjelmslev's term for the 'pre-semiotic', that which the semiotic gives form to. It should be noted that, in keeping with the Saussurean perspective, Hjelmslev is

---

[1] See for example, Louis Hjelmslev, *Prolegomena to a Theory of Language*, translated by Francis J. Whitfield, revised English edition (Madison: University of Wisconsin Press, 1961 [1943]).

[2] Louis Hjelmslev, *Prolegomena to a Theory of Language*, p. 50.

very clear that the signs which give form to the purport are arbitrary – indeed, he goes so far as to say that 'all terminology is arbitrary'.[3]

Hjelmslev is also clear that the distinction between content and expression is in some sense an analytical one: 'There will never be a sign function without the simultaneous presence of both these functives' [...] 'The sign function is in itself a solidarity. Expression and content are solidary – they necessarily presuppose each other. An expression is expression only by virtue of being an expression of a content, and a content is content only by virtue of being a content of expression. Therefore – *except by artificial isolation* – there can be no content without an expression, or expressionless content'.[4]

As I interpret it, this approach thereby remains a structuralist one, albeit with one important difference. Saussure was not interested, epistemologically or ontologically speaking, in anything beyond signified and signifier, or systems of such conjunctions. Hjelmslev, by contrast, does acknowledge the importance of the purport. But this is not the reason, at least not the main reason, why in *Anti-Oedipus*, Deleuze and Guattari suggest that Hjelmslev's linguistics 'stands in profound opposition to the Saussurean and post-Saussurean undertaking'; the reason is that Hjelmslev's theory 'abandons all privileged reference' and this '[b]ecause it describes a pure field of algebraic immanence that no longer allows any surveillance on the part of a transcendent instance, even one that has withdrawn. Because within this field it sets in motion its flows of form and substance, content and expression. Because it substitutes the relationship of reciprocal precondition between expression and content for the relationship of subordination between signifier and signified'.[5]

It is true that in *Prolegomena*, Hjelmslev does criticise the tendency of scholars to treat language not as an investigative end in itself, but as a means: 'means to a knowledge whose main object lies outside language itself, although it is perhaps fully attainable only through language, and which can be gained only on other assumptions than those implied by language'.[6] He quite rightly defends the study of language for its own sake, *but in so doing defends the structuralist mantra*: 'Linguistics must attempt to grasp language, not as a conglomerate of non-linguistic (*e.g.*, physical, physiological, psychological, sociological) phenomena, but as a self-sufficient totality, a structure *sui generis*. Only in this way can language in itself be subjected to scientific treatment without again disappointing its investigators and escaping their view'.[7] As far as I can tell, this means that Hjelmslev very

[3] Louis Hjelmslev, *Prolegomena to a Theory of Language*, p. 58.

[4] Louis Hjelmslev, *Prolegomena to a Theory of Language*, p. 48-49. Italics added to the original.

[5] Gilles Deleuze and Félix Guattari, *Anti-Oedipus*, p. 242.

[6] Louis Hjelmslev, *Prolegomena to a Theory of Language*, p. 4.

[7] Louis Hjelmslev, *Prolegomena to a Theory of Language*, pp. 5-6. Italics in the original.

much agrees, at least in this sense, with the Saussurean ideal of setting any connexions between signs and beyond-signs to one side. So it is perplexing, to say the least, that Deleuze and Guattari seem to be suggesting that Hjelmslev is performing that operation *contra Saussure*.

Whatever the ins and outs of this issue, Deleuze and Guattari would like to use Hjelmslev's model as the basis for an account of the articulation of *all* strata. According to Deleuze and Guattari, and despite what Hjelmslev suggests, content and expression, substance and form, are not only linguistic in scope or origin.[8] On the contrary, Hjelmslev's *purport*, or what Deleuze and Guattari prefer to describe as *matter*, refers to 'the unformed, unorganized, nonstratified, or destratified body in all its flows: subatomic and submolecular particles, pure intensities, prevital and prephysical free singularities' – what Deleuze and Guattari describe as the 'Body without Organs' – with content referring to 'formed matters', and expression for 'functional structures', each needing to be considered from the point of view of its substance and content.[9] Every stratum, and indeed everything that is stratified on earth, entails this double articulation.[10]

How, if at all, does this enthusiasm for Hjelmslev tally with a study that adopts a Peircian perspective? There is no mention of Peirce in *Anti-Oedipus*, and only a brief mention in *A Thousand Plateaus*. In the latter work, Deleuze and Guattari use the Peircian concepts of symbol, icon, and index in passing, but without acknowledging Peirce in the main text. The acknowledgement of Peirce occurs in the footnote, and it is a rather unpromising footnote in so far as the authors arguably make a mistake that has long dogged accounts of Peirce written by those more knowledgeable of Saussurean or post-Saussurean semiotics: Peirce's index, icon, and symbol, Deleuze and Guattari say, 'are based on signified-signifier relations'.[11] This amounts to a triple whammy. First, it establishes a false equivalence between Peirce's conception of the sign, and Saussure's. Second, it appears to reduce Peirce's semeiotic to a semiotics of *re*presentation, which is an unfortunate reduction, however much it may apply to that particular trichotomy. Third, it fails to acknowledge the link between Peirce's semeiotic and his phenomenology, which as suggested in Part 2, is arguably *all about* the kind of ontological relations that Deleuze, and later Deleuze and Guattari, pursue. One has to wonder, in this sense, if the authors were simply unfamiliar with the details of Peirce's more general philosophy. After all, and as I noted in the Appendix 1, as late as the 1980s Peirce was still a relatively unknown figure thanks to the long-term consequences of the Harvard philosophy department's early neglect of the Peirce archive.

---

8 Gilles Deleuze and Félix Guattari, *A Thousand Plateaus*, p. 43.

9 Gilles Deleuze and Félix Guattari, *A Thousand Plateaus*, p. 43.

10 Gilles Deleuze and Félix Guattari, *A Thousand Plateaus*, p. 44.

11 Gilles Deleuze and Félix Guattari, *A Thousand Plateaus*, p. 531, n. 41.

Appendix 2 Deleuze and Guattari, Hjelmslev and Peirce

Evidence both for and against this interpretation may be found in Gilles Deleuze's *Cinema 1: The Movement Image*[12] and in *Cinema 2: The Time-Image*.[13] In the first work, published approximately a decade after *A Thousand Plateaus*, Deleuze appears to have changed his mind about Peirce, and makes use of his classification and typology of signs as part of a philosophical analysis of cinema. He not only suggests that 'Peirce is the philosopher who went the furthest into the systematic classification of images', but also makes reference to Peirce's phenomenological categories of firstness and secondness; indeed, one of Deleuze's key cinematographic categories, the 'affection-image', is characterised as being entirely a matter of firstness – or rather, Deleuze says that it 'only *refers* to firstness'.[14] So Deleuze appears to have moved beyond the interpretation of *A Thousand Plateaus*, yet there continue to be signs (dare I say), that Deleuze is not fully aware of the significance of Peirce's phenomenology, and of its relation to Peirce' semeiotic. For example, Deleuze suggests that firstness and secondness are categories that Peirce uses to distinguish between two different sorts of *images*.[15]

At first glance, this seems like a very interested reduction (Deleuze is analysing cinema). However, in *Cinema 2*, Deleuze clarifies what he means by saying that 'Peirce begins with the image, from the phenomenon or from what appears'.[16] I am not certain that it is wise to establish an equivalence between images, and what appears, in general. When, for example, a peccary ejects volatile compounds (cf. Conclusions to Volume 1), it produces (amongst other types of signs) rhematic indexical sinsigns (cf. Chapter 7). Should we regard such sinsigns, such emanations, as an image? The worry has to be that in his *Cinema* books, Deleuze is 'visualising' Peirce, for reasons that are comprehensible given the subject matter, but problematic to anyone who regards Peirce's signs as being rather more than a matter of visuality.

[12] Gilles Deleuze, *Cinema 1: The Movement-Image*, translated by Hugh Tomlinson and Barbara Habberjam (London: Athlone, 1986 [1983]).

[13] Gilles Deleuze, *Cinema 2: The Time-Image*, translated by Hugh Tomlinson and Barbara Habberjam (London: Athlone, 1989 [1985]).

[14] Gilles Deleuze, *Cinema 1*, p. 99. Italics added to the original.

[15] Gilles Deleuze, *Cinema 1*, p. 98.

[16] Gilles Deleuze, *Cinema 2: The Time-Image*, translated by Hugh Tomlinson and Barbara Habberjam (London: Athlone, 1989 [1985]).

PART 4

Geographies of Wildlife Observation
Ecological-Geographic Perspectives

Halfway through our walk, there is a glimmer, or perhaps I should say a squeak of hope. Having trudged up and down steeply sloping trails without seeing much more than trees, we suddenly hear some thrashing, and bird-like noises high up in the forest's canopy. As we crane our necks to try to see what is going on, our guide informs us that it's Geoffroy's spider monkeys (*Ateles geoffroyi*). But as we squint up the trees' spindly boles, we can only hear the monkeys; the foliage of the canopy trees makes it impossible to see the creatures.

The tour group has just about given up on spotting the monkeys when a member says in a *sotto voce* that might as well have been a shout: 'There!' And indeed, thanks to a gap between two trees, we see several monkeys cross a patch of light as they use implausibly long limbs to brachiate, and then to jump from the branches of one tree to the next. Alas, as so often happens during encounters with wild animals in tropical forests, the monkeys soon become invisible again; we've come, we've heard, and finally we've also *seen*, but all that is left is the memory of a glimpse.

This encounter, which took place on Barro Colorado Island, in some ways exemplifies the kind of event that many tourists hope to experience in tropical forests across the world. For many travellers, the ideal is to see wild animals in their own habitats, going about their business as they have since times immemorial. In this kind of encounter, the geography of observation acquires a special salience: for wildlife tourists as for ecotourists, an important part of the justification for travel lies precisely in the opportunity to experience a certain place, and with it an ecology that is conceived as being, if not pristine, then relatively undisturbed by human intervention.

As I explained in this volume's introductory chapter, in general most of the tourists who partook in my ethnographic research would have liked to have had longer, and closer looks at these or any of the other 'charismatic' animals. But at least this group of travellers had a chance to experience the presence of the spider monkeys first-hand, and in a real 'jungle', a real 'rainforest'. In the Republic of Panama, Barro Colorado is arguably one of the better places to fulfil that ideal, if only because the island has what is probably the best preserved tropical moist forest, and the biggest trees in that country.[1] Moreover, unlike many other national parks in Panama (if not Central America more generally), the island's fauna has been protected by a large group of forest rangers for several decades. This being the case, spider monkeys and other species found on the island are often less likely to flee from humans.

---

[1] Richard Condit, Roland Pérez, and Nefertaris Daguerre, *Trees of Panama and Costa Rica* (Princeton, NJ: Princeton University Press, 2011), p. 16.

And yet, both Barro Colorado Island and its spider monkeys have a surprising environmental history – one which forces both tourist and analyst alike to reconsider labels such as 'pristine', 'intact' or 'little disturbed'. As I also noted in the introduction to this volume, Barro Colorado was once a hill along the lower Chagres River Valley. Between 1910 and 1913, US engineers building the Panama Canal dammed the river close to its mouth on the Caribbean, and over the next three years, they flooded the valley by stages, creating the giant Gatún reservoir. In the process, Barro Colorado and a few other hilltops became islands, and so forest fragments.[2]

The history of Barro Colorado's spider monkeys is also surprising. As noted by the ecologists Katherine Milton and Mariah Hopkins, *A. Geoffroyi* are only found on Barro Colorado thanks to the fact that, beginning the late 1950s, Martin Moynihan, the third director of the island's biological field station, acquired at least 18 orphaned spider monkeys from animal traders in Panama's markets, released them on the island, and instructed a caretaker to feed them until they were able to look after themselves.[3] *Ateles Geoffroyi* is found elsewhere in Panama, but it seems that the species had been hunted to extinction in the vicinity of Barro Colorado by the time that the island was designated a 'nature park' in 1923. Five of the reintroduced monkeys survived (one male and four females), and went on to reproduce, forming a group that by 2018 numbered 23 monkeys.[4]

Both of these sets of events – Barro Colorado's environmental history, and the reintroduction of the spider monkeys to the island – point to an ecology and geography that confound at least two equally problematic stances vis-a-vis Barro Colorado, and perhaps any other forest that has undergone a process of fragmentation. On the one hand, and to borrow the terms proposed by the philosopher Kate Soper, it is not possible to accept at least a naively *nature-endorsing* stance vis-a-vis the island, i.e. one that interprets the island as if it were 'little disturbed' by anthropogenic developments. Then again, I would argue that it would be equally, if not more problematic to adopt a naively *nature-sceptical* stance, i.e. one that en-

---

[2] For an account of the making of the Panama Canal, see David McCullough, *The Path Between the Seas: The Creation of the Panama Canal 1870-1914* (London: Simon & Schuster, 1977).

[3] Katharine Milton and Mariah E. Hopkins, 'Growth of a Reintroduced Spider Monkey (*Ateles geoffroyi*) Population on Barro Colorado Island, Panama', in Alejandro Estrada, Paul A. Garber, Mary Pavelka & Leandra Luecke (Eds.) *New Perspectives in the Study of Mesoamerican Primates: Distribution, Ecology, Behavior, and Conservation* (New York: Springer, 2005) 417-435.

[4] Katharine Milton, (2018) 'Spider Monkeys', Rainforest Connection website, Montclair State University/PRISM, 18 November 2018. <https://www.montclair.edu/prism/2018/11/27/spider-monkeys/>[Accessed 17 April 2019].

tails a culturalist stance with respect to Barro Colorado's ecology and geography.[5] If one adopts the first stance, then one cannot explain the consequences of fragmentation, or isolation from the 'mainland'. Then again, if one adopts the second stance, one cannot explain how it is that for the spider monkeys I mentioned above – as for all of the rest of the mammals on the island – life is still dominated by the need to find food and a social place amongst the local group, even as the creatures try to avoid predation by the region's larger felids or birds of prey. In the case of the spider monkeys, being able to rest, play, or groom each other on the roof of a biological field station arguably does very little to change that.

How, then, should one approach the ecology and geography of Barro Colorado – or indeed of all of those ecosystems across the world which, in one way or another, echo the peculiar mixture of nature and modern human culture found on this island?

In this, the last part of Volume 1, I will consider the strengths and limitations of several different types of approaches, each of which is heterogeneous in its own right. In Chapter 11, I will consider Barro Colorado from the perspective of ecology and physical geography. In Chapter 12, I will turn to a field that, beginning the late 1980s, became known as the 'new' cultural geography. Chapter 13 will then consider Actor-Network-Theory, and with it the turn to what, beginning the early 2000s, became known as hybrid geographies. Finally, in Chapter 14, I will focus on the question of the natures of animality, and to do so will consider perspectives formulated by several scholars located in different fields. The scholars in question are the biologist Jakob von Uexküll, the philosophers Gilles Deleuze and Félix Guattari, and the philosopher and feminist STS scholar Donna Haraway.

---

[5] Kate Soper, *What is Nature? Culture, Politics and the Non-Human* (Oxford: Blackwell, 1995), p. 4.

11
Tropical Moist Forests

In situ wildlife observation of the kind practiced by tourists in tropical forests, or indeed in any other biome requires fauna to observe, and an intact, or at any rate more or less intact, ecosystem. Both of these aspects are not only important to the practice, but absolutely indispensable; without fauna, and without the ecosystem, there can be no *in situ* wildlife observation. This being the case, a study of wildlife observation such as the present one must not only recognise these features, but do so in a way that gives them the theoretical place that they deserve. One way of doing so is to turn to the research that offers the most detailed account of both the fauna and the ecosystems in question, viz., research of the kind produced by physical scientists.

This raises a practical problem: there are many aspects of the ecology and geography of tropical forests that could, and indeed perhaps ought to be considered, and a vast amount of literature about those aspects. The researcher is thus forced to decide what knowledge to draw on, and how to do so. In this study, a first answer is that whatever research is chosen must be able to specify the nature of the setting being considered by the research. While terms such as 'tropical forest' or 'rain forest' may seem self-explanatory, we will see that they actually are very broad terms which require elucidation.

A second answer is that the information must be directly relevant to the practice being considered: that of tourists engaging in wildlife observation. So while researchers in the ecological and physical geographic fields may consider virtually anything from the most ancient of telluric structures (geomorphology) to the relation between forest physiognomy and phenomena such as evapotranspiration (hydrology and climatology), in this chapter I will mainly be concerned with those aspects of the ecosystem that most directly impact on the observational *affections* and *affects* (cf. Chapter 10) of tourists attempting to engage in the mentioned practice. This being so, the problem is not to characterise tropical forests in the manner of, say, an interdisciplinary ecological study that provides a comprehensive account of an ecosystem's features.[1] Instead, it is to select those aspects that have the most important bearing on wildlife observation of the kind (or kinds) being considered, and then to explain how they may affect visitors' observational practices.

Let's begin, then, with a conventional account of the kind of forest this study is concerned with. Tropical forests occur in the equatorial region, i.e. the area be-

---

[1] See for example, Egbert G. Leigh Jr, A. Stanley Rand, & Donald M. Windsor (Eds.) *The Ecology of a Tropical Forest.* See also Lucinda A. McDade, Kamaljit S. Bawa, Henry A. Hespenheide, and Gary S. Hartshorn (Eds.), *La Selva: Ecology and Natural History of a Neotropical Rain Forest* (Chicago: University of Chicago Press, 1994).

tween the Tropic of Cancer (23° 26′ N) and the Tropic of Capricorn (23° 26′ S). This is a band determined by the northernmost and southernmost latitudes at which the sun can appear directly overhead during the summer solstices (June 21 for the Tropic of Cancer, and December 22 for the Tropic of Capricorn). In this study I will be particularly concerned with the tropical forests that occur in the Americas, viz., those generally known amongst scientists as the *neotropical* forests.

Of course, within the mentioned latitudes it is possible to find several different kinds of forests, thanks not least to the very pronounced differences in elevation generated by mountain ranges such as the South American Andes. I am, however, most concerned with wildlife observation in what are often described as the *lowland* forests, i.e. those that generally occur at altitudes no higher than 1000 meters or above sea level (< 3,300 feet). I am also most concerned with what are popularly described as tropical *rain* forests, though I will explain below that this classification also requires specification.

Now such forests have long been characterised with reference to statistics that specify climatological features such as mean rainfall, temperature, relative humidity, etc. One commonly cited set of figures suggest a minimum of 2000 mm of rainfall per annum, average daytime temperatures in the upper 20s C, and air humidity in the region of 75 to 90%. If one goes by such figures, one immediate implication is that even in the deep shade of mature forests, climatological conditions may be quite arduous at least for those visitors not used to walking in very hot and humid landscapes. The last generalisation notwithstanding, there may be significant climatological variations across different regions, and at times even within one region, especially where precipitation is concerned. Equally if not more importantly, over the last decades weather patterns in many regions have been severely disrupted by the increasingly catastrophic climate change.

We are on somewhat firmer ground when it comes to characterising the floristic structure of tropical rain forests. In such forests, relatively tall, evergreen and broadleaved trees constitute the dominant life form. Unless the forest is interrupted by one or another topographic feature, the trees' structures, and their characteristic arrangement in any one area will constitute the predominant feature of the landscape. In a mature, and relatively undisturbed lowland rain forest, the observer will see what may well seem like an endless succession of quite straight and slender boles, interspersed with some much larger trees, some with flaring or buttressed roots, and others with so-called 'prop' roots. Palm trees are also likely to be plentiful, as are ferns and vines; many of the latter may grow as lianas, but others may be bole climbers, or stranglers. The branches of at least the larger trees will typically be studded with bromeliads, orchids and other epiphytical plants. One immediate implication of this kind of floristic structure for tourists is that it drastically reduces the distance at which even comparatively large animals can be seen.

Visibility is further reduced by the fact that tropical rain forests have closed canopies. The trees are typically so tightly packed together that they form one seemingly never-ending cover of dense green foliage, and as noted earlier this also works to drastically limit the amount of solar radiation that reaches the ground level. While treefall or other factors may produce breaks in the cover, generally speaking the ground level remains deeply shaded even when the sun reaches its zenith on a clear day.

Ecologists often describe the vertical structure of tropical rain forests as entailing four strata, each with its own more or less characteristic fauna and flora (or parts of flora). According to this kind of perspective, moving from the lowest to the highest levels, we have firstly the *forest floor*, which as noted above receives very little light, and is a repository for quickly decaying plant and animal matter; then the *understory*, which lies between the forest floor and the canopy, and contains seedlings and saplings of canopy trees, as well as specialist, and typically shade-tolerant shrubs, herbs and small trees; then the *canopy*, which is formed by the crowns of the more mature trees; and finally, the *emergent* layer, which is made up of exceptionally tall trees that grow above the general canopy. The tallest emergent trees in tropical rain forests may exceed 60 meters in height, but canopy trees more typically range between 25 and 45 meters.

While this account of forest structure may make sense from a purely ecological perspective, from the perspective of the observational practices of tourists keen to observe wildlife, it makes more sense to distinguish between just three strata: the ground, the understory, and what I will describe as an *'overstory'*. The distinction between the understory and the overstory is based on the ecological psychologist James Gibson's *principle of reversible occlusion*, which refers to the fact that, as an observer moves, surfaces that are temporarily out of sight may come into view, and vice-versa. 'If locomotion is reversible, as it is, whatever goes out of sight as the observer travels comes into sight as the observer returns and conversely'.[2] While this applies *in and as a principle* to the observer's perception of all surfaces in a tropical moist forest, to the (relatively) unaided observer it will only apply to the understory – or to be more precise, to *much* of the understory, in so far as there will still be, for example, parts of trees, or folds of plants below the canopy level that will not be visible unless one can climb up to them. By contrast, what I describe as the *overstory* is that part of the forest where, thanks to the closed canopy, no amount of unaided locomotion, no amount of unaided *repositioning* (short of climbing the trees themselves, which most tourists would be unable to do without suitable equipment or assistance) will afford a view of any surface occluded by that barrier. From this perspective the canopy provides such an opaque and consistent barrier to at least *visual* perception that it prevents the observer from seeing anything above it, and in many cases, anything *in* it. From a tourist's perspective, this is unfortunate in so far as many of the larger and/or more popular fauna that may be encountered in tropical forests dwell precisely in that part of the for-

---

[2] James J. Gibson, *The Ecological Approach to Visual Perception* (London: Psychology Press, 2015 [1979]), p. 69.

est. In such circumstances, the visitor will have to rely on the other perceptual systems – especially that of hearing – to try to detect the presence of animals in the overstory.

This kind of observational context explains why it was so difficult to observe the spider monkeys that I referred to at the start of this part of the volume. Spider monkeys are generally canopy-dwelling species, and indeed, throughout my stay on Barro Colorado it was typically only possible to see them, or see them clearly, if and when they crossed a gap between trees, or when they moved close to STRI's research station, and used the station's structures and its nearby trees for resting or foraging.

As I began to suggest earlier, despite tropical forests' location within the equatorial region, closed canopies mean that at least mature, and undisturbed forests' interiors are relatively dark – even at midday, the forest ground level may be characterised as a *penumbral* space. From this perspective, the popular image that identifies tropical rain forests with bright green and 'lush' 'jungles' which explorers must hack their way through is potentially a very misleading one. First, to an observer moving at ground level, tropical forests are more likely to seem, if not monochromatic, then certainly not brilliantly polychromatic. In the Prologue, I referred to representations of the BBC *Life on Earth II* series, and noted that both of the sequences chosen by the BBC for the presentation of its then-new 4K Hybrid Log-Gamma system involved scenes produced in neotropical forests. This might seem like a natural choice given the extraordinary colours that can be found in both the flora and fauna of such forests. In fact, from the point of view of a visitor walking through the forest, the predominant impression is likely to be that of a variety of somewhat greyish greens, as well as grey or brown or otherwise relatively dull tree trunks interspersed, relatively rarely, with brighter splashes of green (e.g. where the sun is able to penetrate the canopy), and even more rarely, with the reds, yellows and other colours of flowers or fauna that feature so prominently in the media representations (cf. Prologue).

Then again, and second, the popular image of tropical forests as 'impenetrable' *jungles* is also a misleading one. In so far as a very high proportion of solar radiation is filtered before it reaches the ground, the germination and growth of shade-intolerant species is suppressed. Unless a tree falls or some other factor allows abundant solar radiation to reach the ground, then mature lowland tropical forests are typically places where it is relatively easy for an observer to walk more or less unimpeded. In ecological and botanical studies, the concept of jungle is usually reserved for areas of forests that *have* been disturbed (be it by natural causes, or human intervention). In such areas, a newly opened gap in the canopy will allow much more sunlight to reach the ground, and to promote fast-growing, and shade-intolerant plant species. Where this happens, it is true that the environment will soon become a tangle of young trees, shrubs and vines that may well be inaccessible to all but the most determined, and suitably equipped of travellers. However, at least in the depths of mature and relatively undisturbed forests, this kind of physiognomy is the exception, as opposed to the rule implied by many popular representations.

As the reader will have begun to surmise, the kind of landscape, or we might also speak of *forestscape* that tourists are likely to find when they visit a lowland tropical forest is very different from the one typically represented by the nature media. While there may be some places that do involve memorable vistas, more often than not visitors will encounter a monotony of seemingly endless boles, and a dense canopy. To compound the sense of a lack of differentiation, the trees themselves will be difficult to identify, a point that requires elaboration in its own right as it constitutes such an important feature of tropical forests.

In temperate forests, it is often the case that specimens of a single species grow next to each other, and so the appearance of the forest landscape is largely determined by one or at most a handful of species in the case of coniferous forests, and no more than a few tens of species in the case of broadleaved forests. By contrast, in tropical forests the observer will find that any given tree will typically be surrounded by trees of different species. The leading theory for this phenomenon is known as the Janzen-Connell Hypothesis.[3] Simplifying somewhat, this hypothesis posits that host-specific pathogens, herbivores and other organisms that attack trees are more likely to be avoided by a species if it can disperse seedlings to areas removed from the parent tree. The result is that any one forest patch is likely to have several different kinds of trees, as opposed to groves of one species.

The observer will also find that there are likely to be many more species in such patches thanks to the famous biodiversity of tropical forests. The ecologist Egbert G. Leigh provides a sense of that diversity when he notes that a hectare (a 100-by-100 meter square) of mature broadleaf forest in Maryland was found to include 16 species amongst its 351 trees over ten centimetres in trunk diameter (a standard for censuses in forest ecology); by contrast, an area of the same size on Barro Colorado had 91 species amongst 425 such trees.[4] Perhaps an even more striking contrast may be obtained by comparing much larger areas. Leigh notes, for example, that North, Central and Eastern Europe, with an area of some 6,000,000 km², had 124 species of trees, while 78,500 km² in Panama had 2,870 species of trees.[5]

---

[3] The hypothesis is named after two researchers who arrived at the same conclusion independently. See Daniel H. Janzen, 'Herbivores and the Number of Tree Species in Tropical Forests', *The American Naturalist*, 104 (1970) 501-528; and J. H. Connell, 'On the Role of Natural Enemies in Preventing Competitive Exclusion in Some Marine Animals and in Rain Forest Trees.', in P.J. Den Boer and G.R. Gradwell (Eds.) *Dynamics of Population* (Wageningen: Pudoc, 1971), 298-312.

[4] Egbert G. Leigh, Jr. *A Magic Web: The Tropical Forest of Barro Colorado Island* , 2nd edition, (Panama City, Panama: Smithsonian Tropical Research Institute, 2012), p. 37.

[5] Egbert G. Leigh, Jr. *A Magic Web*, p. 38.

Despite this astonishing difference in floral biodiversity, another kind of difference across temperate and tropical forests may well mean that at least the inexpert observer is unlikely to be fully aware of its extent. In the forests of the temperate zones, even an amateur can quite easily acquire a working knowledge of the visual characteristics of local trees and plants. By contrast, differences between many species of trees and plants in tropical forests may be so subtle that experts may struggle to go beyond an identification on the level of genus. By one account, even experienced botanists may be unable to identify as much as 25% of the plants encountered during explorations of some of the more remote regions of neotropical forests.[6]

It may be noted finally that if the flora in tropical forests is diverse, the same is true for the fauna. In a famous example, the myrmecologist Edward O. Wilson once noted that one sample of invertebrates produced by spraying poison on a single leguminous tree (presumably with a commercial pesticide; Wilson refers to a 'bug bomb') at the Tambopata Reserve in Perú found 43 species of ants in 26 genera – roughly the equal of the entire ant fauna across the British isles. Wilson further noted that a study of beetles in one hectare of Panamanian rain forest estimated it had 18,000 species, compared to 24,000 known to exist across the entire United States and Canada.[7]

The question of biodiversity becomes rather more complex when one takes into account additional factors such as the difference between alpha, beta and gamma diversity,[8] or indeed what is described as the *functional* biodiversity of species.[9] Even so, it is clear that there remains a massive difference in the sheer diversity of species across the tropical and temperate forests, and this raises the question as to why this is the case. The *raison d'etre* is bound to involve many different factors, ranging from the age of the tropical forest biome as a biome (tropical forests are thought to be the oldest biome and the one least disrupted by ancient climate change), to the sheer heterogeneity of the tropical forest environ-

---

[6] Richard Condit, Roland Pérez, and Nefertaris Daguerre, *Trees of Panama and Costa Rica*, p. 11.

[7] Edward O. Wilson, *The Diversity of Life* (Cambridge, MA: The Belknap Press, 1992), p. 198.

[8] See for example, Richard Condit, Nigel Pitman, Egbert G. Leigh Jr., Jérôme Chave, John Terborgh, Robin B. Foster, Percy Núñez, Salomón Aguilar, Renato Valencia, Gorky Villa, Helene C. Muller-Landau, Elizabeth Losos, and Stephen P. Hubbell, 'Beta Diversity in Tropical Forests', *Science*, 295:5555 (25 Jan 2002), 666-669.

[9] See for example, Vesna Gagic, Ignasi Bartomeus, Tomas Jonsson, Astrid Taylor, Camilla Winqvist, Christina Fischer, Eleanor M. Slade, Ingolf Steffan-Dewenter, Mark Emmerson, Simon G. Potts, Teja Tscharntke, Wolfgang Weisser, and Riccardo Bommarco, 'Functional identity and diversity of animals predict ecosystem functioning better than species-based indices', *Proceedings of the Royal Society B*, 282:1801 (22 February 2015) <http://rspb.royal-societypublishing.org/content/282/1801/20142620>[Accessed 1 June 2017].

ment.[10] Even so, Leigh suggests that the diversity of tropical forests must be related to climate: species diversity in tropical forests is greater than that of temperate-zone forests because tropical forests' year-round warm and wet climate make possible a continuous, and large production of vegetable matter. Leigh contrasts Harvard Forest, in Massachusetts, which produces the equivalent of 28 tonnes of sugar per hectare per year, and which functions for about half the year, with the Amazonian forest's production of 75 tonnes per hectare, the result of a forest that is productive year-round.[11] As Leigh explains, '[t]he vastly more stable climate and food supply of tropical habitats are what most enhance tropical-forest diversity', and indeed, the continual availability of at least some fruit, flowers, and leaves provides livings for creatures that are unknown in a temperate forest such as the Harvard: fruit-eating monkeys and kinkajous in the trees; coatis, agoutis, pacas, and spiny rats on the ground; a remarkable diversity of fruit-eating bats and birds; leaf-eating creatures such as sloths, tapirs, untold numbers of herbivorous insects, and then again the creatures that prey on any of the above.[12] In this highly competitive world, specialisation allows the members of species to exploit and defend their food and habitat more effectively, and so this dynamic may well be the engine that drives tropical diversity.[13]

This kind of account suggests an exuberance of wildlife in rain forests. Anyone who has watched wildlife documentaries on TV or read articles about tropical forests in wildlife magazines may well come away with the impression that the faunal species diversity as it applies to vertebrates, and more specifically to larger mammals, reptiles and other popular taxa will be evident to the casual visitor/observer. Put simply, tropical rain forests are not only *bound* to be teeming with life, but must be teeming *visibly* especially with what I will theorise in Chapter 14 as the *charismatic* fauna. In fact, given the floristic structure I characterised earlier, visitors are unlikely to be able to observe even a tiny fraction of the faunal diversity that may be found in tropical forests, and this even before the filter of charisma is added.

Later in this chapter I will explain that another reason for this is found in crypsis, and other aspects of the observable creatures' auto-ecology. Before that I would like to specify further what has thus far been a very general account of tropical rain forests. In some respects this generality is justified by the significant continuities that exist across tropical rain forests right across the world. Egbert

---

[10] The authors of one study classify over 30 hypotheses for what is known as the 'latitudinal gradient' pattern of species diversity, viz. that generally locations in lower latitudes have more species than those in higher latitudes. See M.R. Willig, D.M. Kaufman, and R.D. Stevens, 'Latitudinal Gradients of Biodiversity: Pattern, Process, Scale, and Synthesis', *Annual Review of Ecology, Evolution, and Systematics*, 34 (2003), 273–309.

[11] Egbert G. Leigh, Jr. *A Magic Web*, p. 41.

[12] Egbert G. Leigh, Jr. *A Magic Web*, p. 45.

[13] Egbert G. Leigh, Jr. *A Magic Web*, p. 45.

Leigh, referring to the botanist Andreas Schimper – who came up with the notion of a tropical *rain* forest[14] – notes that anyone travelling west along the latitudinal line that crosses New York City would encounter very different North American climates and vegetations: for example, the wet summers near New York support mesic deciduous forest, but west of the Mississippi, a traveller would find vast grasslands where sporadic fires historically precluded trees. By contrast, the lowlands along the equatorial line have no such variety of climate and vegetation. With the exception of east Africa, the traveller will encounter some form of what is generally described as a rain forest (or 'rainforest'), i.e. what I characterised earlier as a broad-leaved, evergreen forest with a closed canopy. At least the *lowland* rain forests in Latin America, Africa, Australia resemble each other more than do, say, the coniferous rain forests of western Washington and the cove forest of the Great Smokies.[15]

One key factor in this similarity is evapotranspiration, the loss of water from the forest that results from a combination of evaporation and plant transpiration. As I noted earlier, this study will not be centrally concerned with the intricacies of evapotranspiration and related phenomena. Here it suffices to note that evapotranspiration depends on factors such as the amount of solar radiation, wind speed, humidity and air temperature. It also depends on plant or crop type and its leaf area, the seasonal or phenological stage of the plants, and the soil moisture that is available to plant roots. Leigh notes that the most decisive similarity amongst lowland tropical climates is an annual evapotranspiration of 1400mm per year, and this feature is closely associated with similarities in the physiognomy of lowland equatorial forests right around the world.[16]

After recognising such similarities, it is necessary to acknowledge that there are nevertheless a number of factors that may lead to a certain heterogeneity across the lowland tropical forests. A particularly important one is the seasonality of rainfall; in some areas, there are significant variations in rainfall across the year. This variation is generally the result of oscillations in a low pressure area that girdles the planet in the equatorial region, and which is known as the Intertropical Convergence Zone, or ITCZ (the name refers to the fact that the northern and southern hemispheres' trade winds converge in the zone). The phenomenon takes the form of a band of very cloudy, and rainy weather that oscillates seasonally by as much as 40° latitude. The oscillation is caused by the fact that the sun warms up most the latitudinal area that is closest to its midday zenith; however, this area fluctuates thanks to the planet's 'wobble'. While there is typically a lag of some weeks, the ITCZ is usually found close to the area with the maximum theoretical

[14] A. F. W. Schimper, *Plant-Geography Upon a Physiological Basis* (Oxford: Clarendon Press, 1903).

[15] Egbert Giles Leigh, Jr, *Tropical Forest Ecology*, p. 46.

[16] Egbert G. Leigh, *Tropical Forest Ecology*, p. 59.

warming.[17] The effect on climate is to produce a pronounced variation in the amount of cloud cover and precipitation; when the lower pressure band is overhead, the affected region tends to experience very cloudy conditions and the raining season; when it is not, the skies tend to be much clearer and a dry season prevails.

In some areas, this leads to a climate that goes from very wet seasons to very dry seasons. Forests in areas with such a variation are known as monsoon or seasonal forests. Contrary to the popular image of endless evergreen, broadleaved forests where the appearance of the forest does not seem to change, some lowland tropical forests are semi-deciduous or deciduous, and so have tree species that lose their leaves during the dry season.

Here again, the forest on Barro Colorado offers a useful example. During the period from 1925 to 2005 (a convenient time frame, as my own research began in 2007), the average monthly rainfall in January, February, March and April was only 57.2 mm.[18] By contrast, the monthly average for the remaining months (the raining season) was 301 mm. The first figure is somewhat skewed because monthly rainfall in April was, at 94 mm, three times the average for the driest month in the dry season (February, at 31 mm). As this difference shows, on Barro Colorado the dry season for the mentioned period was indeed very dry, and the wet season, quite wet. I say *quite* wet because considerably wetter forests can be found just a few tens of kilometres to the north of the island, along Panama's Caribbean coastline, and even more so just a couple of hundred kilometres to the south. Indeed, one way of putting Barro Colorado's rainfall in perspective is by contrasting the accumulation of precipitation over a year (on average, approximately 2500mm, for the mentioned data set) with that of what some believe to be the wettest place on earth, in the vicinity of the village of Puerto López, a part of the López de Micay municipality in the Pacific lowlands of Colombia's Cauca province, approximately 850 kilometers (530 miles) to the south-southeast of Barro Colorado as the crow flies. Here annual precipitation from 1960 to 2012 was on average almost 13,500 mm. Within the mentioned data set, even the driest year (1980) was

---

[17] Donald M. Windsor, A. Stanley Rand, William M. Rand, 'Características de la Precipitación en la Isla de Barro Colorado', in Egbert G. Leigh Jr., A. Stanley Rand, and Donald M. Windsor (Eds.) *Ecología de un Bosque Tropical; Ciclos Estacionales y Cambios a Largo Plazo* (Washington, D.C.: Smithsonian Institution, 1990), 53-74, p. 56.

[18] Statistic derived from information provided by Steven Paton, *2005 Meteorological and Hydrological Summary for Barro Colorado Island* (Panama City, Panama: Smithsonian Tropical Research Institute 2005) p. 10.

still well over 6000 mm,[19] more than twice as much as Barro Colorado's precipitation (for comparison, between 1981 and 2010, the wettest parts of the UK, e.g. Snowdonia and the Lake District, received on average about 4000 mm of rainfall per annum.[20]). Differences of this kind suggest the need for a more delicate classification of tropical forests, and the botanist and climatologist Leslie Holdridge[21] arrives at such a classification by way of a scheme that articulates latitude, altitude, precipitation, evapotranspiration and temperature to distinguish, even within the lowland tropical forests, between very dry, dry, moist, wet, and *rain* forests *sensu stricto* (see Figure 3, next page).

In this study, I am concerned with wildlife observation in what may be classified, according to Holdridge's schema, as tropical moist, wet and rain forests. In keeping with ecological convention, henceforth I will use the expression 'moist forests' (sometimes also known as 'humid' forests) as a general way of referring to all three types.

Returning to Barro Colorado, perhaps the most obvious sign of the seasonality of its forest is that, during the island's dry season, many canopy trees lose their leaves. As Leigh notes, beneath those trees, the forest floor can look very much like the ground on a dry day in the late autumn in a temperate broadleaved forest, with a crackling leaf litter that in some places may be knee deep. From the perspective of a tourist visiting such a forest during the dry season, this change has two general consequences. The first is that a moist forest, *sensu* Holdridge, is likely to contradict the ideal of a 'lush' 'rainforest'. And indeed, in the surveys I conducted with visitors on Barro Colorado, many tourists expressed surprise at finding such a 'dry' forestscape.

A second consequence concerns the sighting of especially the larger wildlife. As I began to suggest above, one of the key factors that limits the observation of some of the fauna is the dense barrier constituted by the canopy foliage. However, in so far as at least some trees in seasonal forests may lose their leaves in the dry season, then visitors may have a partial view of the higher reaches of the forest, especially in those places where a *group* of trees have lost their leaves. This gain is likely to be offset by the feature I described earlier, namely that the trees in

---

[19] Statistics obtained from Colombia's IDEAM (Instituto de Hidrología, Meteorología y Estudios Ambientales) by Christopher C. Burt, 'New Wettest Place on Earth Discovered?', *Weather Underground*, 18 March 2013 <www.wunderground.com/blog/weatherhistorian/comment.html?entrynum=135>, archived at <https://web.archive.org/web/20141031053635/http://www.wunderground.com/blog/weatherhistorian/comment.html?entrynum=135>[Accessed 3 May 2018]. Previously the record was thought to be north in the Chocó province, in the town of Lloro, which in Spanish means 'I cry'.

[20] Met Office, <https://www.metoffice.gov.uk/learning/precipitation/rain/how-much-does-it-rain-in-the-uk>[Accessed 1 June 2018].

[21] See Leslie Holdridge, 'Determination of World Plant Formations From Simple Climatic Data'. See also Leslie Holdridge, *Life Zone Ecology*.

one patch of tropical forest will, as a rule, *not* all belong to one same species. This being the case, one tree may lose its foliage, but may be surrounded by others that don't, and this will act to limit any gains in visibility that result from one tree's deciduousness.

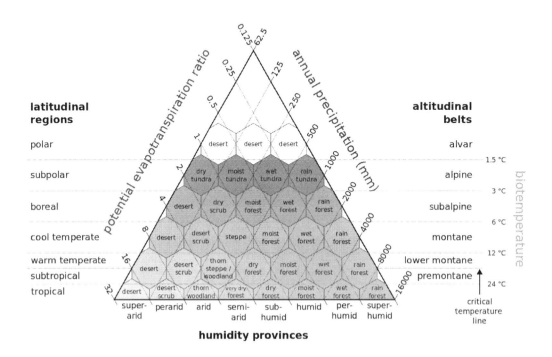

Figure 3: Holdridge's Life Zone System, as represented by Peter Halasz.[22]

Then again, during the dry season fruit and new leaves are much less abundant, and so many animals may be forced to forage further to find food;[23] in some cases this too, may aid visibility. But this aspect may itself be mitigated by another: during the dry season, relatively few birds and mammals breed,[24] and so there may be less avifaunal and mammalian activity overall. As a result, by 10 am on a hot, dry day – the very time of day when some of the more sedate tours of many tropical forests have only recently reached trails – the forest may seem devoid of any of the charismatic fauna.

---

[22] Holdridge's Life Zone schema as represented by Peter Halasz. Image reproduced in black and white (from colour original) as per Creative Commons licence, Attribution-ShareAlike 2.5 Generic (CC BY-SA 2.5). Image appears in Wikipedia, Holdridge Life Zones <https://en.wikipedia.org/wiki/Holdridge_life_zones>[Accessed 1 July 2018].

[23] Egbert G. Leigh Jr. *Tropical Forest Ecology*, p. 149.

[24] Egbert G. Leigh Jr., 'Introduction', in Egbert G. Leigh Jr., A. Stanley Rand, and Donald M. Windsor (Eds.) *The Ecology of a Tropical Forest*, pp. 12-13.

Thus far I have focussed on variations in the *physiognomy* of lowland tropical moist forests as they relate especially to rainfall. But of course, other aspects may also be employed to differentiate tropical forests, and these too, may have significant consequences for wildlife observation. One of these is whether the forest is one that floods, or is partially submerged for extended periods of time, i.e. a wetland forest.

Perhaps the most widely recognised type of wetland forest is the mangrove forest, which occurs along muddy coasts, tidal estuaries, and salt marshes in tropical and subtropical zones. The key botanical feature that gives this kind of forest its name is a remarkable family of plants (*Rhizophoraceae*) that thrive in the brackish water of tidal zones. Mangrove forests have such unique features that they may be classed as a separate biome, and indeed a separate study would be necessary to consider at least some of the peculiarities of this kind of forest, which is typically only observed by tourists from boats.

There are, however, other kinds of wetland forests which do occur as part of the biome of tropical forests – for example, so-called *riparian* forests, i.e. wetland forests that occur along rivers and lakes. In the case of South America, Ghillean Prance distinguishes between several kinds of such forests, including the seasonal *várzea* and *igapó*.[25] The *várzea* forests are inundated during the raining season by nutrient-rich whitewater rivers such as the main stem of the Amazon, the Juruá, and the Madeira. In such rivers, the water has a near-neutral pH and relatively high concentrations of dissolved solids (e.g. alkali-earth metals and carbonates) which give them their milky-brown colour. The name *várzea* refers to the large fringing floodplains where these rivers deposit their sediments – sediments which make the *várzea* forests very fertile.[26] By contrast, the *igapó* forests are flooded by blackwater rivers such as the Río Negro, which have clear waters with an acidic pH range (4-5), lower electrical conductivity, and low amounts of suspended matter but high amounts of humic acids (acids formed by the decomposition of organic matter). The blackwater rivers are named after *their* floodplains (*igapó*), and are generally much less fertile;[27] it is no coincidence that much of the human development of the Amazonia has taken place in *várzea* floodplains.

In such seasonally flooding forests, an area of forest floor that is completely dry and looks, at least to the lay observer, like any other lowland tropical moist

---

[25] Ghillean T. Prance, 'Notes on the Vegetation of Amazonia III. The Terminology of Amazonian Forest Types Subject to Inundation', *Brittonia*, 31:1 (January 1979), 26–38.

[26] Wolfgang J. Junk and Maria T. F. Piedade, 'An Introduction to South American Wetland Forests: Distribution, Definitions and General Characterization', in Wolfgang J. Junk, Maria T. F. Piedade, Florian Wittmann, Jochen Schöngart, & Pia Parolin (Eds.) *Amazonian Floodplain Forests: Ecophysiology, Biodiversity and Sustainable Management* (Dordrecht: Springer, 2010) 3-26, p. 15.

[27] Wolfgang J. Junk and Maria T. F. Piedade, 'An Introduction to South American Wetland Forests', p. 15.

forest during the dry season may come to seem more like a 10-meter deep aquarium during the raining season. This change generates its own possibilities and difficulties vis-a-vis wildlife observation.

First and most obviously, flooding may block access to some areas of the forest. Then again, it may also *facilitate* access to those with suitable craft. And while the flooding may force some of the larger animals to retreat to drier areas, any decrease in dispersion may facilitate wildlife observation, contingent on having timely access to local knowledge regarding the whereabouts of any wildlife concentration (and again, suitable transportation). Indeed, some of the charismatic species may congregate in flooded areas where they may then become somewhat easier to find – examples are Amazonian Manatees (*Trichechus inunguis*), River Dolphins (*Inia geoffrensis*) and the Black Caiman (*Melanosuchus niger*). It might be noted in passing that the flooding may also have the effect of bringing the observer closer to the canopy.

As these points make clear, while floods constitute in principle a fundamental barrier to human access, in some tropical forests flooding may actually *facilitate* wildlife observation, contingent on having timely knowledge and suitable craft. The aforementioned conditions, as allied to the possibility of observing large tracts of forest in a relatively effortless manner (while floating in a canoe or riding in a boat with an outboard motor), may well *increase* some tourists' capacity to attend to the forest and its organisms. A similar point might be made with respect to forests where there is no extensive flooding, but where the existence of navigable rivers or lakes may have the same implications for wildlife observation. Especially in the case of more sedentary forms of wildlife tourism and ecotourism, the possibility of navigating along bodies of water – of course with a substantial break in the canopy – may allow tourists not just access, but a certain *ease* of access, and with it views not typically associated with walks along trails within the forests themselves.

Whether an area of forest floods or not will depend not just on the hydrology of a region, but also on its elevation relative to river banks, and this aspect brings us to the geomorphology of tropical forests, another important factor for tourism. Lowland forests are often found in the relatively flat flood plains of heavily silted, and slow-moving rivers. Where climatological and hydrological conditions vary relatively little over large areas, this may ensure a certain regularity across wide swathes of territory. However, even forests in lowlands may have hills, valleys, ravines, gorges and other topographical features with consequences not just for the landscape, but for wildlife observation.

On the one hand, the topographic features may work to protect plants from, or indeed expose them to more sunlight, wind, or varying amounts of moisture. This may result in forest patches with quite different conditions, and thereby fauna and especially flora that are themselves quite different from those found just a short distance away. In some cases, topographical differences may be such that they act to isolate species, generating the conditions for the emergence of endemic species. Perhaps the best example of both phenomena are the famous hills and *tepui* or table-top mountains found along the Guiana Shield (it may be noted that, strictly speaking, the *tepui* should not be classified as being 'lowland' by virtue of having an altitude that ranges from 1000 to 3000 meters).

Then again, even relatively small differences in elevation may also have a decisive impact on visibility. Headlands or promontories, waterfalls and any other features that lead to breaks in the forest cover may facilitate the observation of canopy-dwelling creatures, especially when conditions allow the observer to gaze down on an area, and/or the observer has access to powerful binoculars, telephoto lenses, etc. On the downside, even differences of a few tens of meters in elevation may produce slopes that force an observer to expend significantly more energy whilst walking along a trail, and this may end up excluding some visitors from some trails.

Returning to the case of Barro Colorado, prior to the flooding of the Chagres River Valley, the island was a hill that, like many other hills in central Panama, formed thanks to differing rates of erosion across dense volcanic and sedimentary rocks.[28] This process resulted in a broad and flat top surrounded by steep slopes (20° to 30°) with a series of narrow ridges and deep ravines that radiate from the hilltop to give the island its crenelated shoreline.[29] Despite having a maximum elevation of only 120 meters or so above sea level (and of these, only about 95 meters above the Gatun's waterline), one consequence of this topography is that tourists are forced to go up and down relatively steep hillsides when using the trails designated for visitor use. As a result, it takes a lot more effort to walk along trails, and this may work to reduce the amount and the kind of attention that especially the less fit visitors may be able to pay to the forest in their search for wildlife.

Large-scale geomorphological features may also have climatological implications. The climate of a region is affected by its location relative to landmasses and large bodies of water. It might be thought, for example, that a relatively narrow stretch of land, such as the Panamanian isthmus at the point where the Panama Canal crosses it (approximately 80 kilometres, or 50 miles, across), would generally have much the same weather patterns, and identical forests at least in the lowland areas. In fact, as one moves south along the Panama Canal (i.e. from the Caribbean towards the Pacific), forests tend to get drier; indeed, the deciduous forests along the Pacific coast on average only get 1800mm of annual rainfall, whereas those on the Caribbean get 3000mm.[30] The main reason for this difference is that

---

[28] William E. Dietrich, Donald M. Windsor, & Thomas Dunne, 'Geology, Climate, and Hydrology of Barro Colorado Island', in Egbert G. Leigh Jr et al. (eds.) *The Ecology of a Tropical Forest*, 21-46, p. 22.

[29] William E. Dietrich, Donald M. Windsor, & Thomas Dunne, 'Geology, Climate, and Hydrology of Barro Colorado Island', pp. 22.

[30] Egbert Giles Leigh, Jr., *Tropical Forest Ecology*, p. 3.

moist air from the Caribbean is blown onto Panama's Caribbean coast where it rises to produce orographic rainfall – simplifying somewhat, rain that falls from clouds produced when the topography forces moist air to rise, and which may fall at any time of day. Further inland the rainfall is produced by convective thunderstorms, i.e. the rainfall is produced with the forest's evapotranspiration, and so tends to fall between 14.00 and 18.00.[31] On the Pacific side, the rain shield created by the continental divide, also known as Panama's *Cordillera Central*, produces an even drier climate, and with it tropical *dry* forests.

Of course, the formation of isthmuses and other large-scale geomorphological processes also have *biogeographical* implications. Geomorphological factors may play an important role not just in the distribution, but in the evolution of taxa. For example, islands tend to have more endemic species thanks to their isolation from the larger landmasses (a famous example being the Galápagos finches [*Geospiza* spp.] that so fascinated Darwin), but deserts, mountain ranges and even rivers may have similar effects, enabling the evolution of certain species to diverge from those of populations only a relatively short distance away. Conversely, the appearance of new bridges across landmasses may lead to the confluence of species from the different landmasses.

Here the Panamanian isthmus offers a particularly good example. The isthmus is a land bridge that is thought to have formed approximately three million years ago, when continental drift led the North and South American landmasses to collide.[32] After the land bridge was formed, the isthmus became an extraordinary biotic corridor which enabled the exchange of species hitherto isolated by the seaway that once separated North and South America. As part of this exchange, South America is thought to have contributed most of the plants found in the vicinity of Barro Colorado Island. All but the commensal insects are also thought to have come from the south. By contrast, the commensals are thought to have come with mammalian hosts from North America. Some of the larger mammals such as tapir, deer, peccaries and the carnivores are thought to have come from

---

[31] Egbert Giles Leigh, Jr., *Tropical Forest Ecology*, p. 47.

[32] There has been much debate about this figure. See for example Aaron O'Dea, Harilaos A. Lessios, Anthony G. Coates, Ron I. Eytan, Sergio A. Restrepo-Moreno, Alberto L. Cione, Laurel S. Collins, Alan de Queiroz, David W. Farris, Richard D. Norris, Robert F. Stallard, Michael O. Woodburne11, Orangel Aguilera, Marie-Pierre Aubry, William A. Berggren, Ann F. Budd, Mario A. Cozzuol, Simon E. Coppard, Herman Duque-Caro, Seth Finnegan, Germán M. Gasparini, Ethan L. Grossman, Kenneth G. Johnson, Lloyd D. Keigwin, Nancy Knowlton, Egbert G. Leigh, Jill S. Leonard-Pingel, Peter B. Marko, Nicholas D. Pyenson, Paola G. Rachello-Dolmen, Esteban Soibelzon, Leopoldo Soibelzon, Jonathan A. Todd, Geerat J. Vermeij and Jeremy B. C. Jackson, 'Formation of the Isthmus of Panama', *Science Advances*, 2:8 (17 August 2016) <http://advances.sciencemag.org/content/2/8/e1600883/tab-pdf>[Accessed 1 June 2017].

North America, while others such as the sloths and monkeys are thought to have come from the south.[33]

As these examples suggest, the species, and so the species diversity of a forest may (also) be a function of more or less ancient migrations resulting in part from the *process* of the land. In Volume 2, I will have a more to say about the geomorphological and biogeographic specificities of Barro Colorado. Here I would like to make a deceptively obvious point: wildlife tourism may be said to be contingent on biogeography in so far as the tourism entails a quest to experience encounters with particular wildlife in particular biomes and ecosystems. Unless one goes to a zoo, or unless one views wildlife via the media of mass communication (or indeed one lives in the area), it is necessary to travel to the region where certain species exist in order to observe them.

The verity of the last point notwithstanding, the mere existence of one or another species in a certain geographic range will of course not be a sufficient motivation for wildlife tourism, or indeed ecotourism; beyond human social factors such as the availability of a tourist infrastructure, it is clear that factors such as dispersion (variations in local density), and ease of access to wildlife 'hot spots' will also typically play a role. For example, even if jaguars (*Panthera onca*) or pumas (*Puma concolor*) were to be relatively easy to spot (which they certainly are not), a visitor to Barro Colorado would be highly unlikely to encounter either species despite being within their geographic range; while both species have occasionally been photographed on the island via camera traps, neither is believed to frequent the island. By contrast, and perhaps for this very reason, the also hard-to-spot ocelot (*Leopardus pardalis*) is thought to have a particularly high density on the island,[34] and so should, theoretically, be more likely to be observed by visitors with suitable knowledge and equipment.

Alas, none of these felids *are* easy to spot even in the most suitable of contexts,[35] and this brings me to the last part of this chapter, in which I would like to offer an account of the *wildlife's* capacity to evade detection. Thus far I have focussed on what might be termed the structural observational conditions in tropical moist forests, but of course, in a tropical forest it is not just the human observer that moves, and is able to detect: so do the wildlife themselves, which not only 'exist' in a location, but do so in ways that improve the animals' chances to sur-

---

[33] Egbert Giles Leigh, Jr., *Tropical Forest Ecology*, p. 8.

[34] See Torrey W. Rodgers, Jacalyn Giacalone, Edward J. Heske, Jan E. Janečka, Christopher A. Phillips and Robert L. Schooley, 'Comparison of Noninvasive Genetics and Camera Trapping for Estimating Population Density of Ocelots (*Leopardus pardalis*) on Barro Colorado Island, Panama', *Tropical Conservation Science*, 7:4(2014), 690-705.

[35] In fact, even scientists devoted to studying ocelots and larger cats have noted that, whilst looking for the creatures, they have later realised that they themselves were under observation by specimens that were in close proximity, but still managed to avoid detection until they triggered camera traps minutes after a scientist left the scene.

vive in a context which, as I noted earlier, is intensely competitive. A key aspect of this adaptation involves taking advantage of the conditions that the forests afford so as to make it less likely for any organism to either detect prey, or to escape detection by prospective predators – including those armed with no more (and no less) than photographic cameras.

To fully understand these capacities, it would be necessary to study each pertinent species's ethology – in the terms that I introduced in Part 3, each species' affections and affects. Here I will merely adopt a general approach based on mainstream ecological discourse – in particular, the typology of defences proposed by the ecologist John A. Endler,[36] which gives a good sense of the multiple ways in which prey attempt to interrupt what Endler describes as the different stages of predation.[37] The stages in question are those of encounter, detection, identification, approach, subjugation and consumption. In so far as wildlife tourists and ecotourists do not generally subjugate or literally consume the creatures they observe, I suggest that Endler's account of the first four stages are the most relevant.[38] The following account lists the pertinent stages and variations, and gives brief descriptions of those defences/counter-defences which are not self-evident and so require some elaboration.

The first phase is what Endler describes as the **encounter**, and involves getting within a distance from which predators can detect prey. Here anti-predator defences include the following:

1) *Rarity*, which reduces the random encounter rate between predator and prey, and reduces the risk still further if predators exhibit apostatic behaviour, which is to say that they prey differentially among most common phenotypes or species. Rarity also makes predator specialisation unlikely.

2) *Apparent rarity*, which involves similar effects to rarity, but with no costs for mate-finding (amongst others). Apparent rarity takes place via a) differences between predator and prey inactivity times or seasons; b) hiding or inconspicuous resting places; c) polymorphism, or the presence of two or more distinct forms (morphs) in the same population of a single species; and d) seasonal changes in colour patterns or other signals.

---

[36] John A. Endler, 'Interactions Between Predators and Prey', in John R. Krebs and Nicholas B. Davies (Eds.) *Behavioural Ecology: An Evolutionary Approach*, 3rd edition, (Oxford: Blackwell Scientific, 1991), 169-196. I am grateful to Hannah ter Hofstede for the reference to Endler's work.

[37] John A. Endler, 'Interactions Between Predators and Prey', p. 169.

[38] John A. Endler, 'Interactions Between Predators and Prey', pp. 170-171.

3) *One-upmanship*, which involves greater detection distance of predators by prey than vice versa.

The second phase is what Endler describes as **detection**, and involves, as the name suggests, the detection of prey as objects which are distinct from the background. In this stage defences include

1) *Immobility*, which makes the animal less visible or even invisible for another whose sensory mode requires motion. This defence may depend upon seasonal changes in colour patterns.

2) *Crypsis*, or the use of 'camouflage' and other similar systems of concealment which reduce the 'signal-to-noise' relation of prey in the predator's sensory field.

3) *Confusion*, which makes detection of a single individual more difficult, or makes it difficult to 'fix' on a single individual for long enough to identify it as edible. This category includes a) random or unpredictable movement, or movement which may shift predator's attention to other prey species; b) movement between contrasting sensory backgrounds; c) random or unpredictable sensory effects, especially when enhanced by, and genetically correlated with, colour patterns; d) extreme abundance: predator saturation, schooling or other concerted behaviour; and e) polymorphism, different forms within a species.

4) *Sensory limits and perception*, which involves the following: a) minimum distance for non-detection of any spot or pattern element; b) minimum distance for detection of colour; c) flicker fusion, which has to do with the threshold at which an intermittent visual stimulus is perceived; d) 'private wavelengths', or wavelengths which only conspecifics can detect; and e) sealed shells to prevent leakage of chemical cues.

The third phase is what Endler describes as **identification**, and involves the identification of prey as profitable and edible, as coupled to the decision to attack. Here Endler distinguishes between the following defences:

1) *Masquerade*, a special resemblance to inedible objects

2) *Confusion*

3) *Aposematism*, conspicuousness associated with distastefulness

4) *Mullerian mimicry*, with less distasteful species strongly resembling the more distasteful species

5) *Batesian mimicry*, with palatable species strongly resembling the distasteful species

6) *Honest signaling of unprofitability.*

The fourth and last phase is what Endler describes as **approach**, which in the case of actual predation involves the attack. Again, wildlife tourists and especially eco-tourists do not generally attack wildlife, but of course the animal under observation would not know that, and might well mistake the guide and tourists' attempts to get as close as possible a prelude to an attack. This being the case, the animal in question might enact the following defences, which might have the effect of interrupting or making more difficult any efforts to observe a specimen or specimens, even if only by causing the visitor to retreat, or even take flight him/herself:

1) *Mode of fleeing*, which includes elements such as a) speed; b) sprint for cover; and c) a different style or medium of movement from that of the   predator (e.g. flying, running, swimming)

2) *Unpredictable behaviour*

3) *Rush for cover* or other predator-inaccessible microhabitat

4) *Startle, bluffing and threat behaviour*

5) *Redirection* of movement (an abrupt change in direction)

6) *Encouraging premature attack*, for example the stotting or pronking movements of gazelles

7) *Aggregation and predator saturation*, i.e. 'strength in numbers'. [39]

The significance of these defences for wildlife observation will hopefully be evident to the reader. Unlike the scenario that unfolds when, say, a visitor enters a zoo, or watches wildlife on TV, when walking along a forest trail a visitor's chances of encountering wildlife will hinge on the forest's floristic structure and dynamics such as the ones described by Endler. In general, any encounter will depend on chance, more or less immediate dynamics of action and reaction, and then again from at least a traditional ethological perspective, the 'laws', if one can call them that, of each species' morphology and behaviour as determined by a combination of long term adaptations and more immediate responses to immedi-

---

[39] John A. Endler, 'Interactions Between Predators and Prey', pp. 170-171.

ate conditions. The latter aspects will depend not only on the species-being of each organism, but on each individual creature's own morphological/ethological capacities, and the immediate surroundings; it might be noted for example that, just as some specimens are particularly successful in predation, some specimens may be more adept at avoiding detection than others. And just as some environments may be particularly well suited to certain defences, on occasion the prey will find itself bereft of its preferred cover. We return by this route to the kind of philosophical issue that I raised via Deleuze in Part 3, and which I will return to in Chapter 14.

Now Endler's account suggests the need to complement my earlier distinction between the understory and the overstory with what might be described as a 'Squirrel *Coda*' (Latin pun not intended): an observer may of course walk around a tree in an attempt to see any object that the tree's trunk or its branches occlude. But when it comes to wildlife observation, this principle may be meaningless if the creature responds, squirrel-like, to the movements of the human observer in order to maintain any protection afforded by occlusion. The observer will also be thwarted if, as happens with spider monkeys and other canopy dwellers, specimens either move up to, or indeed simply *stay* in what I described earlier as the overstory. In such a case, it is unlikely that unaided locomotion, at least amongst relatively inexperienced tourists, is likely to make much of a difference from the all-important visual perspective; the creature will only become visible if there is a 'lucky break', literally or figuratively speaking.

12
Cultural Places and Spaces

If the tourists whose group I joined on Barro Colorado could find spider monkeys foraging high in the canopy of the island's tropical forest, it was because the island had such a forest, and it afforded the conditions in which a substantial group of *A. Geoffroyi* could survive. Then again, if all we could catch was a *glimpse* of the monkeys, it was because that self-same forest, as coupled to the ethology of the species, made it difficult to observe the monkeys. Both aspects justified Chapter 11's focus on the most manifestly physical geography of wildlife observation, and latterly on the defences of the creatures themselves.

In this chapter I will adopt a very different kind approach. Having begun by putting the emphasis on what might be termed physical space, I would now like to turn to social space, or what is perhaps better described as *socio-cultural* space. If physical space is associated with what is conventionally known as nature, socio-cultural space refers to the various ways in which humans structure and inhabit physical space.

The first justification for this shift is that, however important the forest and the affections of its wild denizens, wildlife observation entails techniques of observation that are primarily the result of cultural practice. As I have noted from the start of this study, wildlife observation is not to be confused with wildlife, or indeed with the physical space in which encounters with wildlife occur, however interrelated the different elements may be.

A second justification involves the reductionist conceptions of space and place that pervade positivist approaches – not least, the assumption that place is determined by essentially constitutive objects, or that space is something like an empty vessel that 'contains' those objects. As I began to explain in Chapter 4 via the theory of James Gibson, there is rather more to space and place than positivist conceptions of each category imply.

For these and other reasons that will become clear in due course, this chapter will turn to human geography, and more specifically to a subfield of human geography known as cultural geography. In Chapter 8, it was suggested that 'cultural sociology' was actually a shorthand with which to refer to a set of sociological, anthropological, and philosophical perspectives that analyse the relation between codes, social institutions, and cultural formations as materialised in everyday practice. In this chapter, I propose to employ cultural geography in a similarly broad way. It is possible to suggest that cultural geography started in the late 1980s, when human geographers embraced cultural studies to engage in what came to be known as the 'new' cultural geography.[1] However, in this study I will use the term cultural geography to refer more generally to a form of inquiry that explores the interrelationships between place, space and cultural practice – and this in a manner that at once echoes and contributes to the epistemological and ontological turns associated with cultural sociology.

Why cultural geography, and not, say, any of the other subfields of human geography? The members of the visitor group that I referred to at the start of Part 3 – like all of the rest of the people who partook in my ethnographic fieldwork – were transient visitors in Barro Colorado's forest, and were there in order to engage in tourism. Tourism is a modern cultural leisure practice: one that occurs when an individual or a group temporarily leaves the place known as 'home' and travels elsewhere to experience the pleasures of different places, their events and objects. Such travel is temporary in the sense that the visitors do not migrate to the tourist destinations, expecting to stay there; on the contrary, tourists expect to return home, and to do so within a time scale that is typically framed by the duration of annual leave from work, or shorter breaks made possible by weekends, bank holidays etc. Indeed, according to the sociologists John Urry and Jonas Larsen, tourism is a leisure practice that presupposes, and can only really be understood with reference to its opposite: work that is organised and regulated by an itself modern cultural classification of labour and leisure.[2] A first justification for developing a cultural geographic account of wildlife observation amongst tourists is thus that tourism itself emerges in, and takes place in such a modern *cultural* space, however much it may also involve a temporary immersion in a more or less intact ecosystem.

A second, and closely related justification has to do with the way in which the places, spaces and objects of wildlife observation are tacitly conceived and experienced. Wildlife observation is generally premised on a strong classification that opposes wildlife to livestock and to domesticated animals more generally, and on an equally strong classification of natural and socio-cultural space. But strictly speaking, anyone wishing to observe wildlife need do no more than spend some time observing birds in the local park or garden. Then again, a microscope would probably reveal a fascinating fauna grazing in the fibrous savannahs of carpets; the same instrument applied to the self (if that were possible) would reveal wildlife lurking in the follicular undergrowths of the own eyelashes. Yet even if some garden birds and other not-so-wild wildlife *do* have a devoted following, neither sparrows, nor pyroglyphids, nor staphylococci are likely to be particularly

---

[1] The 'new' in 'new cultural geography' is a recognition that a much earlier form of cultural geography developed in the US in the 1920s. See for example William M. Denevan and Kent Mathewson (Eds.), *Carl Sauer on Culture and Landscape: Readings and Commentaries* (Baton Rouge: Louisiana State University Press, 2009). For an introductory account of cultural geography, see also Mike Crang, *Cultural Geography* (London: Routledge, 1998).

[2] John Urry and Jonas Larsen, *The Tourist Gaze 3.0* (London: Sage, 2013), p. 4.

high on most wildlife tourists' or ecotourists' must-see lists. The very notion of wildlife, like that of nature, is typically premised on a tacit coding orientation (cf. Chapter 8) that encourages wildlife tourists to attend to a rather restricted list of creatures deemed to be 'properly' wild, and to do so in an equally, if not more restricted set of locations – the observed animals' original habitats, and ideally, places that are themselves thought to be, if not *pristine*, then certainly more or less *intact* wildernesses. In so far as it entails this kind of coding orientation, wildlife observation presupposes what are themselves a characteristically modern cultural set of classifications, and with them a cultural disposition with respect to place and space. Just as cultural sociology offers a problematisation of concepts such as discourse, genre and technique, cultural geography critically engages with everyday understandings of place and space, as well as nature, landscape, wilderness and several other associated concepts. The pages that follow will offer an overview of the ways in which such concepts and their associated practices are reconceived, starting with *place*.

To begin with, it may be helpful to go back to the physical geographic perspective on tropical forests (cf. Chapter 11). According to that perspective, tropical forests may be located in certain areas of the planet (a band comprising the equatorial region of the earth, and extending north to the Tropic of Cancer and South to the Tropic of Capricorn). Moreover, what characterises such forests is a certain characteristic climate, floristic structure, and with it the presence of certain types of flora and fauna. This kind of approach is typical for any physical scientific account of a place, or a kind of place. From such a perspective, place may be conceived as a particular point or set of points in space, a location or an area that is circumscribed by more or less clear boundaries, and which contains those objects and processes that provide a place with its identity.

This kind of conceptualisation may seem to be a matter of common sense, and so it is until one starts to question, as cultural geographers do, several of its fundamental assumptions. The first of these is a characterisation of boundary as something akin to a physical *line*. If this is patently problematic in the case of geographic formations with 'fuzzy' perimeters, it is also problematic in those cases where there *is* a relatively clearcut natural boundary (e.g. a river or a large body of water), but where connections nonetheless continue to exist between the delimited place and a beyond that are vital to at least some of the elements within the original place, thereby contradicting any attempt to seal off its boundaries. This issue may be one of the reasons why, historically, tropical *islands* have been such attractive research sites for many scientists.[3] In so far as islands would appear to be, and in some sense really *are* neatly circumscribed, they eliminate the need for a potentially arbitrary geographic demarcation, and with it the ontological ambi-

[3] On this matter see for example Richard H. Grove, *Green Imperialism: Colonial Expansion, Tropical Island Edens and the Origins of Environmentalism, 1600-1860* (Cambridge: Cambridge University Press, 1995).

guities of representation. In Volume 2, I will explain for example that several of Barro Colorado's pioneering scientists and naturalists viewed its character as an island as a positive, indeed desirable feature for precisely these reasons.

As I noted above, even if many scientists would accept the problematic nature of boundaries, few might question the identification of place with the objects that a delimited place is said to contain. Here what is at issue is not so much that x, y or z objects/processes are associated with a certain place, but the suggestion that this association, duly observed and established, defines the essence of the place, and in so doing eliminates any 'subjectivity' in the classification of the place. In fact, any attempt to define place purely with reference to its ostensibly objective contents dissimulates what is, from a cultural geographic perspective, the necessarily arbitrary character of any representation of those contents.

Here it may be helpful to offer some examples. A first example that is pertinent to this part of the study is the debate about whether or not the black-headed spider monkey (*Ateles fusciceps*) should be considered to be a subspecies of Geoffroy's spider monkey (*A. Geoffroyi*), or indeed if all of the currently classified species of spider monkey should be regarded as being subspecies of one single species.[4] Depending on how one interprets this debate, then an at least partly different account of the species (in the singular, or the plural) and its/their *range* will be given.

An even more relevant example involves the ways in which ecologists may attempt to identify the attributes of different kinds of tropical forests. For example, I myself have used Leslie Holdridge's classification of Life Zone Systems (cf. Chapter 11), but any one forest might be classified in a variety of ways contingent on the choice of classificatory schema. This may be illustrated by a comment made by Thomas Croat, the author of one of the most comprehensive studies of Barro Colorado's flora, who noted that the vegetation of the island 'is semi-evergreen moist tropical forest (Knight, 1975a)' but '[i]n terms of the classification system devised by Beard (1944; 1955), the oldest forest on the island would appear to be intermediate between Evergreen Seasonal Forest and Semi-Evergreen Seasonal Forest (Bennet, 1963). And in the Holdridge Life-Zone System… the forest is tropical moist forest'.[5] It might be objected that Croat was writing in the 1970s, but consider this passage in an account of the vegetation in Costa Rica's *La Selva* biological reserve written in the 1990s: 'In different vegetation classification systems, the natural forest of La Selva is known as tropical wet forest (Holdridge 1947), tropical rain forest (Richards 1952), continuously wet tropical rain forest (Walter

---

4 See for example, Andrew C. Collins, 'The Taxonomic Status of Spider Monkeys in the Twenty-First Century', in Christina J. Campbell (Ed.) *Spider Monkeys: Behavior, Ecology and Evolution of the Genus Ateles* (Cambridge: Cambridge University Press, 2008) 50-80.

5 Thomas Croat, *The Flora of Barro Colorado* (Stanford, CA: Stanford University Press, 1978) p. 6. I've left the Harvard-style references to give readers a better sense of what approaches Croat is referring to.

1973), tropical ombrophilous lowland forest (Mueller-Dombois and Ellenberg 1974; UNESCO 1978b), tropical humid forest (Udvardy 1975), tropical lowland evergreen rain forest (Whitmore 1984), and tropical moist forest (Myers 1980). The latter – or an inverted form (moist tropical forest) – is used in a far more general sense than Holdridge's (1947) narrowly defined tropical moist forest'.[6]

Those responsible for the different classification systems would doubtless agree on the location of the *La Selva* forest, and on many if not most of its key characteristics; but of course a similar point could be made about politically contested territories. There is, for example, no discussion about the location of Northern Ireland from the point of view of its physical geography; what is at stake is its political *identity*, viz., whether that part of the island belongs to the United Kingdom (as per the Unionist view), or to Ireland (as per the Republican perspective). Much the same kind of issue is at stake when one tries to argue that a species is better classified as a subspecies, or that a forest is better classified as being this or that type, and not another. The key is in how one characterises the boundaries, and as part of that aspect, the relative weight that one attaches to one or another element, according to a certain pattern of interpretation – in the mentioned examples, a discourse (cf. Chapter 8) – that leads the researchers to construct a physical geography, or one of its elements, in a certain way and not another. The key insight afforded by cultural geography is that the definition and indeed the very experience of place is contingent on such interpretations, as mediated by symbolic forms (or what I prefer to describe as signs), and more generally by the cultural formations that the symbolic forms both shape and are shaped by (cf. Parts 2 and 3).

From the kind of cultural geographic perspective which I am introducing, what is at issue is thereby not just the 'name', but the underlying coding orientation, viz., the *indexing, classification and framing* of the phenomenon or aspects of the phenomenon, as mediated by the subjects' discourse (cf. Chapter 8). Put less technically, the demarcation and identification of place (and as we shall see, of space, landscape and wilderness) entails an unavoidable element of social and cultural 'subjectivity' in the Foucaultian sense of this last term. From a cultural geographic perspective, places are not simply more or less natural locations with a univocal identity whose essence can, and should be determined by way of positivist research into the place's 'contents'; whatever else they are, places are socially and culturally constructed entities whose ascribed identities reflect the social and cultural orientations of all those subjects – individual, but also institutional – who attempt to define the places' natures. Far from being univocal, cultural geographers, like their cultural sociological colleagues, would argue that the social construction of places always reflects the *heteroglossic* character of the parties that

---

[6] Gary S. Hartshorn and Barry E. Hammel, 'Vegetation Thypes and Floristic Patterns', in Lucinda A. McDade, Kamaljit S. Bawa, Henry A. Hespenheide, and Gary S. Hartshorn (Eds.), *La Selva*, 73-89, p. 73.

attempt to define them. Heteroglossia is the literary critic Mikhail Bakhtin's term for the existence of multiple 'voices' within one language, within one novel, or we might add, in the characterisation of the nature of one 'same' place.[7]

This raises a question: if the researcher is not to proceed according to positivist principles, then how else might one conceive and analyse a place? In keeping with what I explained in Part 2, I will argue at a later stage that the nominalist critiques offered by some of the earlier forms of cultural geography do not really address this question adequately. There is, however, at least one proposal which is certainly relevant to the present study. Doreen Massey (1944 - 2016), a leading human geographer, argues in her earlier research that places 'are not so much bounded areas as open and porous networks of social relations… their "identities" are constructed through the specificity of their interaction with other places rather than their counterposition to them… moreover… those identities will be multiple (since the various social groups in a place will be differently located in relation to the overall complexity of social relations and since their reading of those relations and what they make of them will also be distinct)'.[8] Massey is clearly referring to 'social' places, but in so far as biological reserves are themselves (also) such places, then her conception might well be employed to analyse their construction by individuals and by institutions.

One implication of this shift is that it enables the analyst to recognise the possibility that one 'same' place may be differentially constructed by different social groups (we return by this route to heteroglossia). A more subtle implication is that to analyse public understandings of one place, one must refer to other places. This last point seems particularly pertinent to tourism. As I have begun to suggest, one way of characterising tourism is precisely in terms of a quest for a geography of *alterity*, i.e. a place or places that are visited because they are, or *seem* very different from the 'own' everyday life place or places. According to John Urry,[9] and to Urry and Jonas Larsen, such a geography, or such a quest underpins what they describe as the *tourist gaze*:

> The tourist gaze is directed to features of landscape and townscape which separate them off from everyday experience. Such aspects are viewed because they are taken to be in some sense out of the ordinary. The viewing of such tourist sights often involves different forms of social patterning, with a much greater sensitivity to visual elements of landscape or

---

[7] See Mikhail Bakhtin, *The Dialogic Imagination: Four Essays*, 2nd edition edited by Michael Holquist and translated by Caryl Emerson and Michael Holquist (Austin, TX: University of Texas Press, 1982[1975]).

[8] Doreen Massey, *Space, Place and Gender* (Cambridge: Polity Press, 1994), p. 121.

[9] See John Urry, *The Tourist Gaze* (first and second editions, London: Sage 1990 and 2002, respectively). Much the same conceptualisation is offered in John Urry and Jonas Larsen *The Tourist Gaze 3.0.*

townscape than normally found in everyday life. People linger over such a gaze, which is then often visually objectified or captured through photographs, postcards, films, models and so on. These enable the gaze to be reproduced, recaptured and redistributed over time and across space. […] The gaze is constructed through [such] signs, and tourism involves the collection of signs.[10]

While this quote does not make it explicit, it will be clear that what Urry and Larsen characterise as a 'separating off from everyday experience' entails a logic of classification (as per my theory of coding orientations), and with it, a tacit comparison between different places and the objects found there. This being the case, the meaning of one place is, from this identitary perspective, very much reliant on the meaning of another. Put very simply, one of the reasons why tourists may travel to visit tropical forests is precisely that they may regard those forests' geography and wildlife as being quite different from the 'own' geography, the 'own' fauna. Travel from place A to place B occurs precisely because B is thought to be different from A. Ontologically speaking, it is, in this sense, no coincidence that Urry and Larsen's underlying semiotics of tourism relies on the proposals of the literary theorist Jonathan Culler, who himself basis his theory on an application of the semiologist Roland Barthes' *Elements of Semiology*.[11] According to Culler, 'Roland Barthes … writes in his *Elements of Semiology* that … once society exists, every usage is converted into a sign of this usage … By wearing blue jeans, for instance, one signifies that one is wearing blue jeans. This process is crucial, Barthes continues, and *exemplifies the extent to which reality is nothing other than that which is intelligible*. Since it is as signs that our practices have reality, they swiftly become signs, even if signs of themselves'.[12] And this, Culler suggests, is what happens in the case of tourism: 'In their most specifically touristic behavior … tourists are the agents of semiotics: all over the world they are engaged in reading cities, landscapes and cultures as sign systems'.[13] As I explained in Chapter 5, from a Saussurean perspective, the meaning of a sign is always contingent not on the represented object, but on another similarly arbitrary sign, or better yet, a similarly arbitrary *system* of signs. Urry and Larsen establish, in this sense, a certain homology between the valuation of home and away, and the meaning of one (arbitrary) sign and another.

---

[10] John Urry and Jonas Larsen *The Tourist Gaze 3.0*, p. 4.

[11] See Jonathan Culler, 'The Semiotics of Tourism', in *Framing the Sign: Criticism and its Institutions* (Oxford: Blackwell 1988 [1981]), 153-167; and Roland Barthes, *Elements of Semiology*, translated by Annette Lavers and Colin Smith (New York: Hill and Wang, 1973[1964]).

[12] Jonathan Culler, 'The Semiotics of Tourism', p. 154, italics added to the original.

[13] Jonathan Culler, 'The Semiotics of Tourism', p. 155.

Chapter 12 Cultural Places and Spaces

I've said enough about the category of place to give the reader some sense of how cultural geographers problematise it. I would now like to turn to a cultural geographic critique of realist conceptions of *space*. To do so, I will turn to the work of a scholar that seems particularly germane to this part of the study, and who has been influential across both cultural geography and cultural sociology: I refer to the social theorist and philosopher Henri Lefebvre (1901-1991).[14]

Implicit to the anti-essentialist critique of place is a problematisation of the notion that space is somehow 'empty'. The reader may recall that in Part 1 I referred to James Gibson's critique of what he describes precisely as 'empty space'.[15] Lefebvre raises similar issues, but gives his critique a cultural materialist basis. As part of an account of what he describes as social space, Lefebvre reveals the fallacies of empty space, and follows his problematisation with a conception of social space that brings to the foreground the importance of institutions, representations, and lived culture.

Lefebvre's critique starts with what he describes as the 'double illusion'.[16] The first aspect of the double illusion, or what Lefebvre describes as the *illusion of transparency*, makes space appear *luminous*, a substance which, like the air it is apparently modelled on, gives action 'free reign'. Devoid of opacity, space is thereby completely open to the observing mind. The obverse of the illusion of transparency is what Lefebvre describes as the *realist illusion*, the illusion of the 'natural simplicity' of 'things' – things that have a natural reality in a space which then becomes an ordering of those things. According to Lefebvre, each of these illusions 'refers back to the other, reinforces the other, and hides behind the other'.[17] 'The shifting back and forth between the two, and the flickering or oscillatory effect that it produces, are thus just as important as either of the two illusions considered in isolation'.[18]

Of course, scientists have long known that apparently empty spaces around us are in fact filled with gaseous molecules that human perceptual systems are mostly unable to detect; to be sure, from a strictly scientific perspective, space is never truly 'empty' if only because forces are always at work 'on' and 'in' that space (for example, the gravitational). I would nevertheless argue that even now the scientific *modelling* and representation of 'space itself' continues to posit an empty space. What is true for science in general is also true for *ecosystems*.

---

14 Henri Lefebvre, *The Production of Space*, translated by Donald Nicholson-Smith (Oxford: Blackwell, 1991 [1974]).

15 James J. Gibson, *The Perception of the Visual World*, pp. 59-60.

16 Henri Lefebvre, *The Production of Space*, pp. 27-29. The following account builds on the one offered in Nils Lindahl Elliot, *Mediating Nature* (London: Routledge, 2006), pp. 46-47.

17 Henri Lefebvre, *The Production of Space*, p. 27.

18 Henri Lefebvre, *The Production of Space*, p. 30.

Consider an example found in a TV series titled *Deep Jungle* (Granada Wild and New York's Thirteen/WNET, 2004). While much of the series adheres to time-honoured media conventions for representing tropical forests – not least, the very title of the series – its episodes include at least one remarkable innovation: the representation of research about the floristic structure of tropical forests. In one scene, the series portrays aspects of the research being conducted at the time of the series' production by the biologist and mathematician Roman Dial. Dial's research includes efforts to quantify tropical forests' canopy's structure, in particular the space-air gap between trees. To this end, the series shows Dial using climbing equipment to reach the tops of emergent trees, and a laser rangefinder to measure the distance between his own location and different trees in the surrounding canopy. As Judith Cushing *et al.* explain in an article in *Computing in Science and Engineering*,[19] the researchers substitute cylindrical coordinates (r,θ,z) for Cartesian (x, y, z), and employ digital technologies in order to facilitate the acquisition and inscription of the resulting data. The TV series then represents that data by simulating a 'fly-through' of digitised trees, arranged as multiform towers of blocks, even as a voice-over narration explains that 'A laser beam and GPS linked to a computer measures the volume of the jungle's trees and plots the spaces between them. The technique turns complex vegetation into simple blocks. The result of all this: the first ever 3D jungle map'.[20]

Cushing *et al.* might or might not have issues with the TV programme's representation of their research, but even so, this example illustrates perfectly the double illusion, i.e. an 'empty' space 'filled' with objects, in this case trees made out of digital 'blocks'. Arranged in a grid (Cartesian, cylindrical, etc.), the data-based objects can be mapped, and the map can be used to discover hitherto uncharted features of forests (in this case, of forest canopies). Of course, similar principles, methods and technologies are now used to map all kinds of spaces – not least those involving the 'jungles' of advanced capitalism, including 3D maps of countries, cities, buildings, and even of human bodies.

It would be absurd to deny that there is a certain correspondence between such models and their objects – in the Peircian terms that I introduced in Part 2, such models constitute a good example of iconic legisigns. The well-known fact that this kind of approach might miss some objects is not, in and of itself, a reason for dismissing it – if anything it is a reason for improving the delicacy of the analysis. What is at issue is something rather more complex and subtle. The substitution of geometrical space (a space of planes, lines, points, and angles) for a space that entails a mutuality between organisms and environment is arguably a contempo-

---

[19] Judith Bayard Cushing, Nalini Natkarni, Barbara J Bond and Roman Dial, 'How Trees and Forests Inform Biodiversity and Ecosystem Informatics', in *Computing in Science and Engineering*, 5:3 (May 2003), 32-43.

[20] Granada Wild/Thirteen/WNET, *Deep Jungle*, 2004.

rary version of what feminist ecologists and radical environmental historians have long characterised as the discursive 'death of nature'.[21]

Back to Lefebvre: while Lefebvre does not really engage with natural or physical space, he does offer an alternative to the double illusion. Where positivist approaches treat space as empty space, Lefebvre articulates what he describes as *spatial practice*, the *representation of space*, and *representational spaces*.[22] I will explain what each of these concepts means in a moment; first I should clarify that I will exchange this somewhat confusing nomenclature for three terms which arguably clarify what is at stake, without losing too much of the original connotation. I will substitute *instituted* space for spatial practice; *represented* space for the representation of space; and *lived* space for representational spaces.[23] I will also somewhat modify aspects of Lefebvre's characterisation in order to make it more pertinent to the study of wildlife observation and tourism.

How, then, might these categories be employed to characterise wildlife observation amongst tourists? *Instituted space,* or what might also be termed 'the institutional production of space' (spatial practice) includes the particular locations and spatial sets characteristic of a social formation, but as shaped by relations of production and reproduction.[24] Lefebvre understands relations of production from a neo-Marxist perspective – simplifying somewhat, as entailing the material relations that underpin a social order, and which are developed as part of the social production, exchange, and distribution of wealth. We return by this route to what I characterised in Chapter 8 as the cultural-material(ist) turn.

For example, in the case of a tropical forest that is protected as a biological reserve (national park or other), instituted space includes the forest's physical geography from the perspective of its organisation as a reserve, including the demarcation of a territory. It also includes any infrastructure that may exist on the reserve, not least any trails or other facilities designed to cater for, say, rangers, scientists or tourists. However, these aspects are approached not only as so many 'physical features', but as the expression of particular relations of production, and with these relations, the social order that leads to the establishment, and then the ongoing organisation and reorganisation of a particular place.

For instance, the setting aside of Barro Colorado as a 'natural park' in 1923, and the development of a research station on the island, reflected the state of bio-

---

[21] See for example, the analysis offered by Carolyn Merchant, *The Death of Nature: Women, Ecology and Scientific Revolution* (San Francisco, CA: HarperSan Francisco, 1980).

[22] Henri Lefebvre, *The Production of Space*, p. 33.

[23] Lefebvre himself establishes a homology with a different triad which I have also employed in another study: perceived, conceived, and lived space. See Nils Lindahl Elliot, *Mediating Nature*. Ideally, one would speak of institut*ing*, represent*ing*, and liv*ing* space, with the present participle emphasising the ongoing making and remaking of each dimension. I have, however, chosen the -ed suffix for the sake of clarity.

[24] Henri Lefebvre, *The Production of Space*, p. 33.

logical inquiry at the time; but as I will explain in Volume 2, such inquiry was itself intimately related to the US's rising imperial power, and in Barro Colorado's specific case, to the existence of a colonial enclave known as the Panama Canal Zone – itself an instance not only of US imperialism, but of significant changes in the global capitalist order in the first decades of the twentieth century. All these elements resulted not just in the drawing of the boundaries of the new natural park as so many lines, but in particular social relations – amongst them, the exclusion of 'non-scientists' from Barro Colorado, and with this change, the re-classification of local peoples as the equivalent of domestic staff, or indeed as potential trespassers, to not say 'poachers' in the new park. As this example makes clear, instituted space is space interpreted in relation to the predominant social and cultural institutions and fields of interaction (economic, political, scientific, etc.).

For its part, *represented space* (or the representation of space) includes any and all symbolic signs (cf. Part 2), and indeed the *discourses* produced in relation to the instituted space. Simplifying greatly, social space is also shaped by its actors'/subjects' representations of space, but this in a manner that is itself at least partly determined by one or another social order, and the interests of those subjects in relation to particular institutions and fields.

Returning to the example of Barro Colorado, historically the park came not only to be *organised* as a biological reserve in the context of a colonial enclave, but also *represented* via a growing number of genres (scientific, media, etc.) which constructed partly different identities for the island and its ecology, depending at least in part on the producers of the messages, their location in particular fields, and their real and/or imagined audiences. As I will explain in some detail in Volume 2, from an early stage the biological reserve began to be portrayed by scientists with recourse to scientific, but also romantic discourses of nature: as a 'floating laboratory' and as a 'unit of jungle', but also as an 'island ark' and as a place to get away from it all: as the ornithologist and natural history curator Frank M. Chapman put it in 1926, '... on Barro Colorado every prospect pleases and we are the only men. There are no automobiles ... no radios, no jazz, no movies, no "hold-ups" or similar evidences of decadent human nature; in fact, none of the afflictions of modern life, but in their place a soothing, primeval calm such as remains in but few parts of the civilised world.'[25] It might be noted that when Chapman suggests that he/they are the only men, this is not merely a gendered reference to humans; as noted by the historian Pamela M. Henson, while something like a who's-who of the U.S. biological sciences' elite obtained access to the research laboratory on Barro Colorado, in the first decades female researchers were actively excluded from the station.[26]

---

25 Frank M. Chapman, 'An Island Ark: An Unusual Wild Life Refuge near Panama', *World's Work*, 53(November 1926):61-74, p. 61.

26 Pamela M. Henson, 'Invading Arcadia: Women Scientists in the Field in Latin America, 1900-1950', *The Americas*, 58:4 (2002), 577-600.

Chapter 12 Cultural Places and Spaces

The third element of the triad, *lived space* (representational spaces) refers to the manner in which an instituted and represented space is experienced in everyday life. In a more substantial change with respect to Lefebvre's theory, I suggest that this aspect refers to what the cultural historian and ethnographer Michel de Certeau describes as the more or less *tactical* ways in which subjects employ and reemploy, and in so doing 'make do' with the places and spaces they appropriate.[27]

This conceptualisation requires elucidation. De Certeau describes strategy as 'the calculus of force-relationships which becomes possible when a subject of will and power (a proprietor, an enterprise, a city, a scientific institution) can be isolated from an "environment."'[28] As de Certeau explains, a strategy presupposes that a place can be circumscribed as the own, thereby generating a boundary with an exterior (with competitors, targets, etc.). By contrast, a tactic is 'a calculus which cannot count on a "proper" (a spatial or institutional localisation), nor thus on a borderline distinguishing the other as a visible totality. The place of a tactic belongs to the other', and so a tactic 'insinuates itself into the other's place, fragmentarily, without taking it over in its entirety, without being able to keep it at a distance'.[29] When, for example, a tourist visits a biological reserve, the visitor enters a space that has in some sense already been defined, be it by way of the constitution of the park itself, or by way of infrastructural aspects such as trails or observation points, and then again, the rules for the use of these aspects. Some visitors may attempt to transgress the rules, the official trajectories (as set out by trails), or indeed the outer boundaries, but, depending on their number, their capital, and the length of their stay, the transgressions are unlikely to have much effect on the organisation of the park; on some level, merely to visit it, to use its trails, and to view wildlife from its observation points is to experience the park in a tactical manner. Of course, over time, the balance of power may well change if, for example, sufficient visitors transgress the existing rules in such a manner that the organisation of the park has to change. Then again, the strategy-tactic nexus is not itself an essential one, and today's tacticians may become tomorrow's strategists (or vice-versa).

Now Lefebvre's triad must not lead to the conclusion that there is always a neat unity, let alone a relation of linear or mechanical causation between instituted, represented, and lived space. For example, while Lefebvre emphasises the importance of linking social space to relations of production, he is by no means an economic determinist. On the contrary, it is possible to conceive of a context where one 'same' instituted space may be quite differently represented and lived. For instance, by 2007 the Barro Colorado that was administered, represented and

---

[27] Michel de Certeau, *The Practice of Everyday Life,* translated by Steven Rendall (Berkeley, CA: California University Press, 1984).

[28] Michel de Certeau, *The Practice of Everyday Life*, p. xix.

[29] Michel de Certeau, *The Practice of Everyday Life*, p. xix.

lived by scientists was quite different from the one experienced by tourists – which is not to suggest a simple opposition between science and tourism, let alone a homogeneity *within* the two mentioned groups. Indeed, my ethnographic research with groups of visitors on Barro Colorado revealed what were at times important differences across tourist groups, and between some visitors and the organisation of the Smithsonian's Visitors Programme. One way of characterising this difference is with reference to the geographer David B. Weaver's distinction between 'hard' and 'soft' ecotourism (or what I prefer to describe as *unconditional* and *conditional* ecotourism).[30] Simplifying somewhat, unconditional ecotourism is one that, in theory at least, is truly 'environmentally aware' in so far as it involves travel of a kind that presupposes a very active commitment to sustaining beyond human-environments, to the point that the travellers may actually spend significant amounts of time working to try to preserve a biological reserve.[31] By contrast, conditional ecotourism is probably indistinguishable from what some in the industry describe as 'nature tourism', i.e. the kind of tourism whose members are attracted to sites and/or events of outstanding ecological significance, but who require as a condition for travel the kinds of luxuries associated with 'mass' tourism – for example, easy access, a certain standard of accommodation, visitor programmes with translators, etc. For most visitors who engage in conditional ecotourism, there is little or no commitment to 'true' sustainability.

The point I am leading up to is this: while the Smithsonian's Visitors Programme was structured around catering to conditional ecotourism (cf. Volume 2), on occasion visitors arrived on the island that had either been misinformed by tour operators, and/or had no intention of adhering to the proposed (and quite strictly enforced) tour itinerary. In such situations, guides were forced to find ways of mediating the practices of visitors who had no interest in Barro Colorado's ecology (let alone its scientific station), or who seemed determined to make their own way around the island.

Other tourists were closer to the ideals of 'unconditional' ecotourism, and found that the Smithsonian's Visitors Programme catered too much for what was, in effect, a more traditional form of nature tourism. They would have preferred to have stayed for longer on the biological reserve, and in some cases to have participated in scientific research, or conservationist activities.

Examples like these would appear to show the wisdom of the proposals of Doreen Massey, viz., that the concept of *place* must be understood 'not so much [as] bounded areas as open and porous networks of social relations', locales

---

[30] David B. Weaver, 'Ecotourism as Mass Tourism: Contradiction or Reality', *Cornell Hotel and Restaurant Administration Quarterly*, 42 (2001), 104-112.

[31] As I will acknowledge in the postscript, and especially in Volume 2, in the context of catastrophic climate change, one may question any such efforts so long as they still involve avoidable forms of consumption, and with them, avoidable green-house gas emissions.

whose meanings are fixed not by 'nature itself', or even by real or alleged 'locals' (local inhabitants), but by interpretations arising from multiple, and indeed what are often *contested* identities.[32] An analogous point can be made about space, understood as the triadic (instituted, represented, lived) articulation of the social relations by means of which particular places are appropriated. There is no guarantee that social relations will determine how particular places are appropriated.

Until now, my account of cultural geography has focussed on the reconceptualisation of place and space. But what can cultural geographers contribute to an analysis of those places and spaces associated with nature? For many cultural geographers, the academic referent for an analysis of the concept of nature is Raymond Williams, and more specifically, his problematisation of ideas of nature.[33] Williams was one of the founding figures of Cultural Studies, and a cultural sociologist (or as he put it, a 'sociologist of culture') in his own right (cf. Part 3). As I have already presented Williams' critique of ideas of nature (cf. Chapter 8), here I will focus on the way in which insights very much like those proposed by Williams are used in cultural geography to develop critiques of everyday conceptions of landscape, wilderness, and the tropics.

Let's begin with landscape. In the everyday realist perspective, landscape is typically taken to refer to all the visible features of a countryside – in effect, a kind of 'scene' that unfolds like a horizon for the beholder. As such, landscape is typically thought to include not only topographical features, but vegetation, and any structures built by humans. I say any structures, but Williams would note that, at least from the time of the industrialisation of Britain, the notion of landscape came to be used primarily, if not exclusively to refer to rural areas thought to be unchanged by modernity. A key representational source for such a conception may found in the landscape paintings, and at a later stage in the landscape gardens that presented idealised images of agrarian or even Arcadian scenes. This is ironic, for as a number of scholars have noted, at precisely the moment when the English countryside began to be represented in highly idealised ways, it was undergoing the extraordinary upheavals caused by the rise of industrial capitalism.[34]

Partly to pierce the bubble of such idealisation, but also to enable a cultural materialisation of modern sensibilities regarding landscapes, one of the earliest, and most important proposals of the new cultural geographers was to treat landscape as being entirely a matter of social construction. This is particularly evident

---

[32] Doreen Massey, *Space, Place and Gender* (Cambridge: Polity Press, 1994), p. 121.

[33] See Raymond Williams, 'Nature', in *Keywords: A Vocabulary of Culture and Society*, revised edition (London: Fontana, 1983), 219-224. See also Raymond Williams, 'Ideas of Nature', in *Problems in Materialism and Culture* (London: Verso, 1980), 67-85.

[34] See for example the analysis of Anne Bermingham, *Landscape and Ideology: the English Rustic Tradition, 1740-1860* (London: University of California Press, 1986).

in the work of Denis Cosgrove and Stephen Daniels, two scholars who played a key role in the development of cultural geography. According to Cosgrove and Daniels, landscape is

> a cultural image, a pictorial way of representing, structuring or symbolising surroundings. This is not to say that landscapes are immaterial. They may be represented in a variety of materials and on many surfaces – in paint on canvas, in writing on paper, in earth, stone, water and vegetation on the ground. A landscape park is more palpable but no more real, nor less imaginary, than a landscape painting or poem ... And of course, every study of a landscape further transforms its meaning, depositing yet another layer of cultural representation.[35]

The influence of Cosgrove and Daniels, and of like-minded scholars can be noted in the fact that, as late as the early 2010s, sociologists like John Urry and Jonas Larsen still define landscape (which they place in inverted quotation marks) as 'a *human* way of visually forming, through cultivated eyes, skillful techniques and technologies of representing, a physical environment.' '"Landscape" is about how humans take control and possession of, and derive pleasures from, "nature". It is a specific way of relating to "nature" that fuses "reality" with images and representations'.[36] While Urry and Larsen at least acknowledge explicitly that there is a 'physical environment', in their work landscape 'itself' is actually a matter of cultural construction.

This kind of conception of landscape may be contrasted with the kind more likely to be proposed by, say, landscape ecologists. Brian Walker, one such ecologist, defines landscape as a self-organising system of physical/chemical factors which are topographically determined, and which interact with the biological components that occupy them.[37] While Walker acknowledges that 'the resulting patterns of biological communities are strongly influenced by human use and management activities',[38] there is no question but that the landscape, *qua* landscape, has an existence that is wholly independent of its observation by individuals. By contrast, the kind of constructivist critique developed by the first generation of cultural geographers (and cultural sociologists like Urry) insists firstly that humans can only observe what passes for nature via the thick spectacles of sym-

---

35 Denis Cosgrove and Stephen Daniels, 'Introduction: Iconography and Landscape', in Denis Cosgrove and Stephen Daniels (Eds.) *The Iconography of Landscape: Essays on the Symbolic Representation, Design, and Use of Past Landscapes* (Cambridge: Cambridge University Press, 1988), 1-10, p. 1.

36 John Urry and Jonas Larsen *The Tourist Gaze 3.0*, p. 110. Italics in the original.

37 Brian Walker, 'Landscapes and the Biosphere', in Simon A. Levin (Ed.) *The Princeton Guide to Ecology* (Princeton: Princeton University Press, 2009) 423-424, p. 423.

38 Brian Walker, 'Landscapes and the Biosphere', p. 423.

bolic forms; and secondly, that landscape *is itself* a symbolic form, in Peircian terms a symbolic sign. From this perspective, landscapes exist not in the land, but in a group's cultural imaginations as expressed via what Lefebvre describes as represented space. This stance can also be contrasted with that of an ecological psychologist like James Gibson, who speaks of surfaces and objects whose affordances vary from species to species, but which remain surfaces and objects, nonetheless (cf. Chapter 4). In effect, cultural geographers like Cosgrove and Daniels tend to sever anything like an indexical relation between land and landscape, and treat landscape as a more or less entirely representational entity.

Cultural geographers adopt a similar approach with respect to the concept and experience of *wilderness*. Cultural geographers and environmental historians alike have long noted how wilderness entails a discursive interpretation that is analogous to the one employed to represent nature. And indeed, we might say that the concept of wilderness explicitly *territorialises* the tendency of the modern discourses of nature to oppose nature and modern culture. In an analysis that echoes that of Cosgrove and Daniels vis-a-vis landscape, and Raymond Williams vis-a-vis nature, the environmental historian William Cronon argues that, far from being 'the one place on earth that stands apart from humanity', wilderness and its referents are 'quite profoundly a human creation – indeed, the creation of very particular human cultures at very particular moments in human history'. Cronon suggests that this process is nevertheless dissimulated by the fact that '[w]ilderness hides its unnaturalness behind a mask that is all the more beguiling because it seems so natural.'[39] Cronon does not mean to question the existence of what he describes as 'the nonhuman world we encounter in wilderness'.[40] Instead, he questions what it is that motivates people to seek out wilderness. The answer, he argues, lies in cultural invention. 'Go back 250 years in American and European history, and you do not find nearly so many people wandering around remote corners of the planet looking for what today we would call "the wilderness experience"'. On the contrary, as late as the eighteenth century, 'the most common usage of the word "wilderness" in the English language referred to landscapes that generally carried adjectives far different from the ones they attract today. To be a wilderness was to be "deserted", "savage", "desolate," "barren" – in short, a "waste," the word's nearest synonym.'[41]

In the interlude between these early, and often negative accounts, and the kind of idealisation that is expressed by the late nineteenth century, there emerged first in Europe and then in the United States a romantic discourse about nature – one which sublimated wild areas, and in so doing projected onto landscapes idealis-

---

[39] William Cronon, 'The Trouble with Wilderness; or, Getting Back to the Wrong Nature', in William Cronon (Ed.) *Uncommon Ground: Rethinking the Human Place in Nature* (New York: W.W. Norton, 1996) 69-90, p. 69.

[40] William Cronon, 'The Trouble with Wilderness', p. 70.

[41] William Cronon, 'The Trouble with Wilderness', p. 70.

ing images of the kind promoted by poets like William Wordsworth in Britain and essayists like Henry Thoreau and John Muir in the United States. If for the first advocates of this kind of discourse wilderness is both a space and a place where one can experience the sublimity of God and his creation, by the end of the nineteenth century, wilderness becomes a rather more secular entity, e.g. a space and place where one can experience the joys of a nature thought to be more or less devoid of any human intervention – a nature where, for example, many believe they can go back in time to the basics of the kind of 'rugged individualism'[42] long associated with famous explorers and hunters.

It is often forgotten that many of those rugged individuals were the same men (and they were almost always *men*) who made it their business to pursue aboriginal groups – groups who not only regarded 'wilderness' as home long before the Europeans arrived, but who spent millennia modifying those same environments. It is thus doubly ironical that, as Cronon notes, in the United States '[t]he movement to set aside national parks and wilderness areas followed hard on the heels of the final Indian wars'; '[t]he myth of wilderness as "virgin," uninhabited land had always been especially cruel when seen from the perspective of the Indians who had once called that land home'.[43] Indeed, and as I myself have noted elsewhere, what would become the very first US nature park, the Yosemite (the Yellowstone was the first *national* park) was 'discovered' by mercenaries who were paid by gold diggers to evict aboriginal groups from what would become known as the Yosemite Valley during the California Gold Rush[44] – the same rush that, as I will explain in Volume 2, provided a first motivation for the construction of the precursor to the Panama Canal, the trans-isthmian railway known as the Panama Railroad (now Panama Canal Railway). The success of this railway whetted entrepreneurs' appetites for a canal, a work which would itself eventually lead to the eviction of tens of thousands of inhabitants from numerous villages and farms along the Chagres River.[45] We return, by this route, to what I described earlier with Lefebvre as instituted space.

Let's turn, finally, to the geographic category of the tropics, which of course underlies the notion of a 'tropical forest', and plays a key role in positioning subjects in relation to tourism identified by way of travel to 'the tropics'. As I noted in Chapter 11, in physical geographic discourse the tropics have a clear, indeed apparently wholly unambiguous location as a region found between the Tropics of Cancer and Capricorn. By contrast, from a cultural geographic perspective, the very category of the tropics is itself an abstract singular – Raymond Williams'

---

[42] William Cronon, 'The Trouble with Wilderness', p. 77.

[43] William Cronon, 'The Trouble with Wilderness', p. 79.

[44] Nils Lindahl Elliot, *Mediating Nature*, p. 136.

[45] See Nils Lindahl Elliot, 'A Memory of Nature: Ecotourism on Panama's Barro Colorado Island', *Journal of Latin American Cultural Studies*, 19:3(2010), 237-259.

term for a classification that employs a singular name for a real multiplicity of things or processes[46] (cf. Chapter 8) – and one that is charged with a centuries-old history of imperialist and colonialist meanings, values and politics.

This is, broadly, the point made by the historian David Arnold in his critique of what he describes as the *discourse of tropicality*.[47] Arnold argues that, however much the tropics can be defined with reference to objective physical boundaries, such boundaries have, historically, been inextricably entangled with cultural discourses that have projected onto the tropics moralising images and values. As Arnold puts it,

> Calling a part of the globe 'the tropics' (or by some equivalent term, such as the 'equatorial region' or 'torrid zone') became, over the centuries, a Western way of defining something culturally alien, as well as environmentally distinctive, from Europe (especially northern Europe) and other parts of the temperate zone. The tropics existed only in mental juxtaposition to something else – the perceived normality of the temperate lands. Tropicality was the experience of northern whites moving into an alien world – alien in climate, vegetation, people and disease. And this sense of the physical and cultural consequences of moving from one zone to another was more acutely felt in the Atlantic world, where the transition from temperate to 'torrid' was relatively rapid and where it was closely bound up with the Atlantic slave trade, than it was in the Indian Ocean or the Pacific.[48]

Arnold notes that over the centuries during which this discourse emerged, there was an oscillation between two narratives about the tropics. On the one hand, a positive narrative portrayed the tropics as being paradisiacal. The tropics 'came to be seen as lands of great natural abundance, alive with luxuriant vegetation and exotic birds and animals, and blessed with perennially warm climates. Spared the cold and hunger of northern winters, humans in the tropics could enjoy easy, year-round, subsistence in return for minimal labour. The tropics, especially the islands of the Caribbean, the Indian Ocean and latterly the Pacific, conjured up visions of an earthly paradise, a veritable Garden of Eden'.[49]

By contrast, the negative version of the narrative dwelt on what European colonizers regarded as the inherent dangers of the tropics: according to Arnold,

---

[46] See Raymond Williams, 'Nature', in *Keywords: A Vocabulary of Culture and Society*, 219-224. See also Raymond Williams, 'Ideas of Nature', *Problems in Materialism and Culture* (London: Verso, 1980), 67-85, p. 69.

[47] See David Arnold, '"Illusory Riches": Representations of the Tropical World, 1840-1950', *Singapore Journal of Tropical Geography*, 21:1(2000), 6-18. See also David Arnold, *The Problem of Nature: Environment, Culture and European Expansion* (Oxford: John Wiley, 1996).

[48] David Arnold, *The Problem with Nature*, pp. 142-143.

[49] David Arnold, '"Untold Riches"', p. 7.

'Powerfully negative representations of the tropics centred on images of primitiveness, violence and destruction – the speed and fatality of tropical diseases, the destructiveness of tropical storms, the ferocity of tigers and other carnivorous beasts prowling in malarious jungles ... – as well as the detrimental effect of tropical abundance on the moral and physical well-being of human inhabitants, whose easy subsistence bred indolence and provided no stimulus to technological innovation and the arts of civilisation'.[50]

A normative oscillation between the two different narratives – which nonetheless are premised on absolute difference – is especially evident in the context of modern interpretations of tropical *forests*, not least via the figure of the 'jungle', long a trope for the discourse of tropicality more generally. If in narratives like Joseph Conrad's *Heart of Darkness* we find images of the jungle as a 'living hell' – images that are probably based on King Leopold II's colonisation of the Congo region[51] – then as far back as Columbus's first accounts of what are now described as the Americas we find the correlate of the hellish tropical forest in the form in the images of Edenic spaces. As Columbus' journal puts it whilst describing 'Fernandina' (most probably the Bahamian island today known as Long Island) 'It is a very green island, level and very fertile... I saw many trees very unlike those of our country. Many of them have their branches growing in different ways and all from one trunk, and one twig is one form, and another in a different shape, and so unlike that it is the greatest wonder in the world to see the great diversity...'.[52]

However, in that same journal we find a lament that, according to the historian Antonello Gerbi, presages what eventually becomes a discourse about *neo*tropical wildlife amongst colonial explorers: exuberant flora, but meager fauna – in the journal's words, 'no beasts... save parrots'.[53] That Columbus too, should be disappointed in the real or apparent meagerness of the fauna might be taken a kind of historical evidence of biophilia. But of course, Columbus too, was engaging in what might be described as media-based dynamics of transmediation when he embarked on his travels; as a number of scholars have noted, one of Columbus's favoured guides was a text ostensibly authored by the medieval Italian traveller, Marco Polo. Then again, it seems likely that when Columbus *did* encounter more substantial wildlife in the Caribbean, he interpreted it from more generalised popular literatures and narratives. A case in point: one day early in 1493, when

---

[50] David Arnold, '"Untold Riches"', p. 8.

[51] See Adam Hochschild, *King Leopold's Ghost* (New York: Houghton Mifflin Company, 1998).

[52] Christopher Columbus, *The Journal of Christopher Columbus*, translated by Clements R. Markham (London: Elibron Classics, 2005 [1893]), p. 47.

[53] Antonello Gerbi, *Nature in the New World: From Christopher Columbus to Gonzalo Fernández de Oviedo.* Translated by J. Moyle (Pittsburgh: Pittsburgh University Press, 1985), p. 15.

Columbus's journal noted that the Admiral had seen three mermaids, and while they 'rose well out of the sea', 'they [were] not so beautiful as they are painted'[54] – a reference, presumably, to the West Indian Manatee, *Trichechus manatus*.

The cultural critic Candace Slater has noted the extent to which, in more recent times, rain forests, and more specifically those forests found in the Amazonia, have been the subject of Edenic, or 'quasi-Edenic' forms of representation. Describing her experience of watching an IMAX film about the Amazon, Slater offers a critique that shows up the extent to which such representations project onto the region's forests an image that leaves out 'the less-glamorous swamps and brush lands, the big and the little cities'; a similar problem arises when it comes to the representation of the people who live in the region: yes, there are native healers, whirling Indians, and folksy-looking people (in Slater's account of the people shown in the IMAX film), but where are the descendants of black slaves, the Sephardic Jews, the Japanese agricultural workers, the Arab merchants and the mixed-blood rubber trappers 'who have helped create the rich, distinctive cultures of an immense and varied region'?[55]

Slater's reference to the multiculturalism of the Amazon region allows me to turn to the last concept, and problematisation that I will consider in this chapter. Earlier, when discussing Doreen Massey's conception of place, I noted that from a cultural geographic perspective, no place is sealed onto itself; as suggested by Massey, the identities of places are constructed through the specificity of their interaction with other places, and not just by the superposition of places. To this we might add that, if one place is always constructed in relation to other places, some places entail 'within them' a multiplicity, a multi-spatiality which requires a special kind of analysis.

Multi-spatiality can be approached in at least four different ways. In this chapter I will consider three, and in Chapter 14 I will consider a fourth. The kind of multi-spatiality that Slater is referring to may be regarded as a synchronic one, and occurs when, within a territory, there is a more or less explicit multiplicity, for example, an obvious heterogeneity of landscapes, or of cultural groups. If, despite this heterogeneity, a name or a map leads people to imagine that the territory is homogeneous, then we have a recipe for the kind of problem that Raymond Williams points out with respect to abstract singulars. 'Amazonia', or maps showing the supposed distribution of forested areas along the Amazon river and its immense watershed, may be good examples in so far as they omit cities, areas devoted to agriculture, or indeed any cultural heterogeneity.

A second way of approaching multi-spatiality is diachronic, and is particularly pertinent in the case of tourism. As I noted earlier, tourism involves travelling

---

[54] Christopher Columbus, *The Journal of Christopher Columbus*, translated by Clements R. Markham (London: Elibron Classics, 2005 [1893]), p. 154.

[55] Candace Slater, *Entangled Edens: Visions of the Amazon* (Berkeley, CA: California University Press, 2002), p. 3.

from 'home' to an 'away', and so is multi-spatial even in those cases where someone only ever travels once from one kind of space to another.   Then again, those who are able to engage in tourism typically do so throughout much of their lives, and so by adulthood have typically travelled on holiday to numerous places, and with them, potentially very different spaces. Upon reaching any one destination, they thus bring with them a knowledge and experience of what are potentially numerous other sites, but also numerous other spaces. If to this kind of multi-spatiality we add the one obtained via the media of mass communication – so-called 'armchair travelling' – then it becomes clear that the experience of one site (and indeed one *sight*) may be mediated by many others.

The third kind of multi-spatiality is the one that requires a somewhat different kind of analysis. It is a multi-spatiality that occurs when people, or more generally bodies from diverse points of origin come together in one place, and in so doing generate a kind of '*in situ*' multi-spatiality. It is not just that within one *area*, one can designate different places – that, after all, can happen even in one home. Nor is it that someone has 'crossed' different spaces or places to reach one point. It is that the very place constitutes a kind of confluence, a point of convergence for actors/subjects that start from potentially quite different places and spaces. In this chapter I will approach this phenomenon via the Foucaultian concept of heterotopia.

The concept of heterotopia is one that Michel Foucault proposed and most thoroughly described in a lecture that he presented in 1967, but which was only published shortly before Foucault's death in 1984 (as a short text titled 'Of Other Spaces', or in the original French, *Des Espaces Autres*[56]). However, Foucault's first published reference to the concept occurs in the preface of *The Order of Things*,[57] where the concept is treated somewhat differently. In this study as in others,[58] I will stick to the lecture notes, which I believe provide the more useful insights.

Anyone who reads 'Of Other Spaces' will find that the lecture is full of insufficiently developed ideas, and the answers that it offers for some of the questions that Foucault raises are less than satisfactory (but then, it was only a lecture; how many would wish to be able to give such a *lecture*?). From the

---

[56] Michel Foucault, 'Of Other Spaces', translated by Jay Miskowiec, *Diacritics*, 16 (Spring 1986), 22-27.

[57] Michel Foucault, *The Order of Things*: *An Archeology of the Human Sciences*, translated from the French (London: Routledge, 1970 [1966]), p. xviii.

[58] I have employed the concept of heterotopia in several other investigations. See for example Carmen Alfonso and Nils Lindahl Elliot, 'Of Hallowed Spacings', in Jane Arthurs & Iain Grant (Eds.) *Crash Cultures* (London: Intellect, 2002), pp. 153-174. See also my analysis of the eruption of Krakatoa in Nils Lindahl Elliot, *Mediating Nature*, pp. 164-171; and Nils Lindahl Elliot 'Museums and the Challenge of Transmediation: The Case of Bristol's Wildwalk', in Michelle Henning (Ed.) *The International Handbooks of Museum Studies: Museum Media* (Oxford: John Wiley, 2015), 43–68.

perspective of our own times, parts of the text also appear to have sexist, or ageist passages. There are, however, some nuggets that a number of researchers have used to try to theorise a phenomenon that is arguably as ubiquitous as it is modern: I refer to sites, but as we shall see also to *practices* that bring together, for whatever reason, people from different 'walks of life', and this in ways that one might say do 'strange' – or indeed marvellous – things in and to the sites, and by implication to everyone and everything that goes there. Heterotopias, Foucault suggests, are real places that 'are something like counter-sites, a kind of effectively enacted utopia in which the real sites, all the other real sites that can be found within the culture, are simultaneously represented, contested, and inverted. Places of this kind are outside of all places, even though it may be possible to indicate their location in reality'.[59] Examples include holiday villages or resorts, cinemas, brothels, museums, zoos, or cruise ships.

Staying provisionally within a Foucaultian frame, this understanding of heterotopia constitutes a useful way of conceiving mass tourist destinations – places where people travelling from multiple places and spaces converge precisely because the places appear to be, and in some respects really *are turned into* 'utopias' – places that, in Foucault's words, are as 'perfect' as the own everyday place is 'messy, ill constructed, and jumbled', and so work as a space of 'compensation'.[60] This would appear to be a particularly valid insight for destinations devoted to mass *nature* tourism – places where people travel in the hope, however implicit, of experiencing 'nature itself', or wildlife in 'their' habitats – the world before the great fall of industrialisation and consumer societies. We return by this route to Frank Chapman's suggestion, that '... on Barro Colorado every prospect pleases and we are the only men. There are no automobiles ... no radios, no jazz, no movies, no "hold-ups" or similar evidences of decadent human nature...'[61]

Important as this aspect is, Foucault notes that heterotopias have 'the curious property of being in relation with all the other sites', but this 'in such a way as to suspect, neutralize, or invert the set of relations that they happen to designate, mirror, or reflect'.[62] In the case of wildlife tourism or even ecotourism, when individuals travel from afar to visit the 'undisturbed' destinations, the very act of going there, even if for no other reason than to gaze upon the site or some of its objects, introduces a change, a 'disturbance'. In some cases, if the effects of a tourist presence are significant enough to drastically transform the visited place, then the utopian quality of a destination may quickly turn into a dystopian one of the kind associated with overdeveloped resort areas. There a combination of

[59] Michel Foucault, 'Of Other Spaces', p. 24.

[60] Michel Foucault, 'Of Other Spaces', p. 27.

[61] Frank M. Chapman, 'An Island Ark', p. 61.

[62] Michel Foucault, 'Of Other Spaces', p. 24.

social inequality, crowding, and pollution may dramatically reduce the quality of life for residents, and indeed the quality of the leisure space and time for the tourists themselves.

Even when no such overt destruction takes place, part of the 'strangeness' in tourist heterotopias is that objects/bodies that are visited in order to be observed *in situ* are transformed into objects of semeiotic consumption. As I began to explain in Chapter 8, semeiotic consumption may seem like a wholly benevolent, and unproblematic activity. In fact, it is arguably a key part of what drives advanced capitalism, and so in one way or another is bound to have detrimental effects for the environment. Whatever one's view on that matter, semeiotic consumption means that whereas wildlife tourism is ostensibly premised on the firstness of 'being there', of seeing things with one's own eyes, an identitary logic creeps in (cf. Part 3): even as one 'sees with one's own eyes', one searches for what somebody else has seen and represented with *their* eyes, with their observational techniques and technologies. We return by this route to Urry and Larsen's suggestion, that the tourist gaze is 'largely preformed by and within existing mediascapes',[63] and that 'much tourism becomes, in effect, a search for the photogenic'.[64]

Having offered a very brief characterisation of the way in which heterotopia might be used to analyse tourist destinations, I would now like to propose three changes to Foucault's theory that make it more germane to the present study. The first is still quite close to Foucaultian theory: heterotopia is not only a matter of an actual place, an 'empirically verifiable' site of the kind suggested by Foucault's examples. I say this understanding that Foucault was keen to highlight the notion that heterotopias are at once particular locations, and 'all other sites', and so in some sense anti-locations. After acknowledging this point, the first change is that it is also possible and useful to speak of heterotopic *practices*, i.e. ways of doing that produce not only actual (as in geographically locatable), but also 'virtual' heterotopias. Good examples of the latter are TV news shows, or blockbuster wildlife documentary series such as the BBC's *Planet Earth*.

The second change is designed to counter a certain immobility in Foucault's conceptualisation of heterotopia. More than a kind of a-temporal juxtaposition of sites, heterotopias entail the coming together, or the *bringing* together, of multiple *trajectories*, multiple durations, whose relatively sudden convergence accounts in part for the strangeness mentioned earlier. When I say 'relatively sudden' convergence, I mean that there is what the following chapter will describe as a *topological* space.

The third change is designed to counter the anthropocentric inclination that is evident in Foucault despite (or perhaps I should say thanks to) his posthumanist inclinations. Foucault's concept takes for granted a human social frame; even if

---

[63] John Urry and Jonas Larsen, *The Tourist Gaze 3.0* (London: Sage, 2011), p. 179.

[64] John Urry and Jonas Larsen, *The Tourist Gaze 3.0*, p. 178.

Foucault uses the concept of heterotopia to refer to zoos as well, it is clear that it is the *human* sociality that produces heterotopia. Elsewhere I have suggested that it is possible to speak of heterotopias *of nature* in a reference precisely to zoos, natural history museums, and other entities which purportedly bring together in one site – actual or virtual – all the natures of the globe, or indeed the cosmos.[65] But this shift needs to be taken further to acknowledge that there are perhaps also events and places in the natural world that work in ways that are, at the very least, analogous to a human heterotopia. Perhaps the most obvious one is a black hole, though admittedly few would regard black holes as being utopic at least in the conventional sense of the expression (they are u-topic in the most literal sense...). Another example is that of biotic corridors where the flora and fauna from different continents – or merely from the surrounding environs – converge. As suggested in Chapter 11, the Panamanian isthmus has been an important example of such a convergence for millennia, and in Volume 2 I will consider the irony that arises when an ancient 'natural' heterotopia is overcoded by the rather more recent heterotopia constituted by the Panama Canal. Additional examples include so-called salt licks and clay licks (sites where geophagia serves as a dietary supplement, or for pharmacological purposes), a famous example being one on a bend of the Manu River in Perú which attracts large numbers of parrots; and then again, bait balls, which bring together not only large numbers of bait fish, but what is at times an astonishing variety of bait fish predators.

In Chapter 11 I noted that the possible presence of one or another species within a certain geographic range was no guarantee of suitable conditions for wildlife tourism vis-a-vis that species. The presence-in-principle of a species would be meaningless in the absence of an adequate tourist infrastructure, or if the creature was so rarely encountered as to effectively negate any efforts to observe it. One way of conceiving places where wildlife tourism and indeed ecotourism thrive is precisely in places where 'natural' heterotopias, or something very much like them, occur. In such places, one finds something like a meta-heterotopia: the heterotopia constituted by tourists converges on, and observes, the beyond-human heterotopia, or its more or less natural equivalent.

---

[65] See Nils Lindahl Elliot, *Mediating Nature*, pp. 47-49.

13
Actors, Networks and Things

Let's go back to the encounter with Barro Colorado's spider monkeys. A physical geography and ecology such as I outlined in Chapter 11 tells us about the physics of the beyond-human aspects of the island's tropical forest, and about *A. Geoffroyi* regarded as a species. However, it tells us little or nothing about Barro Colorado regarded as a socio-cultural space. This explains why in Chapter 12 I turned to cultural geography. In marked contrast to physical geography, cultural geography tells us about the manifold ways in which particular social and cultural groups may inhabit, construct, and indeed on occasions *de*struct particular places. Note, however, that cultural geography tells us little or nothing about the physics of places such as Barro Colorado. Moreover, at least in its earlier versions, it completely excludes beyond-human animals from the analysis. The landscapes of Cosgrove and Daniels are populated exclusively by symbolic forms, and their producers are exclusively human.

We have, then, a set of mutual exclusions, a set of epistemological and ontological mirrors: for physical geography, there is only really physical space; for cultural geography, at least in its earlier incarnations, there is only really what I described as socio-cultural space. Alas, there is an additional problem. If the division of academic labour that I have just highlighted works to reproduce the nature-culture dualism, it also works to exclude the consideration of all that which cannot be neatly pigeonholed as being nature or culture. The analyses can only proceed by affirming one of two positions: a 'biologist' stance, or in the terms proposed by the philosopher Kate Soper,[1] a 'nature-endorsing' discourse, sociobiological versions of which would insist that everything is ultimately no more (and no less) than an expression of a transcendental nature; or a 'culturalist' stance, or what Soper describes as a 'nature-sceptical' discourse, nominalist versions of which assert the radical discontinuity between symbolic forms and the natural world. In the end, all may agree, staying with Soper, that a nature does exist in the form of material structures and processes that are 'independent of human activity (in the sense that they are not a humanly created product), and whose forces and causal powers are the necessary condition of every human practice'.[2] But at least until the late 1990s, both academic tribes remain committed, however inadvertently, to a dualistic dance. As Soper notes, the cultural reductionists cannot quite rid themselves of the nature-culture dualism: even if nature has become a bad word – or as Soper puts it, little more than a 'convenient, but fairly gestural

---

[1] Kate Soper, *What is Nature? Culture, Politics and the Non-Human* (Oxford: Blackwell, 1995), p. 4.

[2] Kate Soper, *What is Nature?*, pp. 132-133.

concept'[3] – the term is still used to make points about culture. The obverse tendency occurs amongst the biological reductionists, whose writing are often peppered with references to the 'human world'.

This dance may explain why, between the 1980s and 1990s, much of the scholarly effort was put into trying to purify nature of culture, and culture of nature. When it is culture *or* nature, or even culture *and* nature (the latter variation still presupposing a strong boundary between the two entities), there is unlikely to be much room for phenomena that are not one or the other.

This was, and even today remains an unhelpful state of affairs for research about wildlife observation. Anyone who tries to adopt a biologist or a culturalist stance vis-a-vis this practice will soon find that neither 'side' can provide an adequate account of the activities involved. A tourist's encounter with a wild animal in a more or less intact ecosystem is at once a matter of particular genres and techniques of observation – not least, those associated with mass communication and with tourism itself; and an encounter with creatures which may continue to live their lives in much the same ways that their ancestors have for millennia. Wildlife observation must thus be regarded as a nature-culture hybrid *par excellence* – a phenomenon that so blurs at least the conventional boundaries between culture and nature (or the referents of such terms) that it denies the kind of pigeonholing that I mentioned earlier.

I've just referred to a nature-culture hybrid, and beginning the last decade of the twentieth century (in some cases earlier), a number of critical social scientists and philosophers began to use the concept of hybridity to propose new ways of engaging with, and indeed moving beyond the nature-culture dualism. In Part 3 I considered contributions by Merleau-Ponty, Gilles Deleuze, and Félix Guattari that in some ways prepared the ground for this kind of approach; in this chapter I will consider the proposals of two other scholars who have also contributed to what might be described as the 'hybridity turn': first, Bruno Latour,[4] one of the founders of what is known as Actor-Network-Theory (ANT); and second, Sarah

---

[3] Kate Soper, 'Future Culture: Realism, Humanism and the Politics of Nature', *Radical Philosophy*, 102 (July/August 2000) 17-26, p. 17.

[4] See for example Bruno Latour, *We Have Never Been Modern*, translated by Catherine Porter (Cambridge, MA: Harvard University Press, 1993[1991]); Bruno Latour, *The Politics of Nature: How to Bring the Sciences into Democracy*, translated by Catherine Porter (Cambridge, MA: Harvard University Press, 2004); and Bruno Latour, *Reassembling the Social: An Introduction to Actor-Network-Theory* (Oxford: Oxford University Press, 2005).

Whatmore,[5] herself a founder of a field in human geography known as hybrid geographies. Both perspectives are quite closely interrelated in so far as Whatmore's proposals build on the work of Latour (amongst several other sources). This being the case, I will begin by introducing aspects of Latour's research.

Latour is a leading late twentieth/early twenty-first century sociologist, anthropologist, and philosopher. In his earlier research, the semiotician Algirdas Greimas (1917-1992) was an important influence, and by Latour's own account, in more recent times the sociologist Gabriel Tarde (1843-1904) has also provided theoretical and methodological guidance. Latour develops research that is not only constructivist and anti-essentialist, but avowedly relativist, and as the cultural theorist Barbara Herrnstein Smith puts it, thoroughly nominalist.[6] It is nevertheless also neo-empiricist. All this is particularly evident in Latour's proposals regarding Actor-Network-Theory. In general, ANT researchers argue that the meanings, functions, and roles of things are entirely contingent on their place in networks. However, the networks are themselves regarded as a matter of trajectories and inter-relations of things that are always in a state of flux, and so they can never be fully represented, if indeed they can be represented at all. All the researcher can do is to give some sense of the interconnections, and by this means an impression of a whole that vanishes before the analyst has finished describing it (if indeed it ever existed as a 'whole'). ANT may thereby be regarded as a form of high postmodernism, albeit one with a new materialist, neo-empirical emphasis on *things* – a point that I will return to later.

By way of a brief contextualisation, it is possible to trace the start of ANT, or something very much like it in the ethnographies of the scientific workplace that Science and Technology Studies (STS) scholars engaged in the 1970s and 80s. Latour played a central role in these; for example, in 1979 Latour published with Steve Woolgar a study titled *Laboratory Life*,[7] an ethnography of the scientific workplace that studied the practices of researchers working at the Salk Institute for Biological Studies. The research outraged many scientists in so far as Latour and Woolgar made the case that the scientists and the lab technicians at the Salk I

---

[5] See for example Sarah Whatmore and Lorraine Thorne, 'Wild(er)ness: Reconfiguring the Geographies of Wildlife', *Transactions of the Institute of British Geographers*, 23:4 (1998), 435-454; Sarah Whatmore, 'Hybrid Geographies: Rethinking the "Human" in Human Geography', in Doreen Massey, John Allen, Philip Sarre (Eds.) *Human Geography Today* (Cambridge: Polity Press, 1999), 22-39; Sarah Whatmore, *Hybrid Geographies: Natures Cultures Places* (London: Sage, 2002); and Sarah Whatmore, 'Materialist Returns: Practicing Cultural Geography in and for a More-Than-Human World', *Cultural Geographies*, 13 (2006), 600-609.

[6] See for example Barbara Herrnstein Smith, 'Review of Bruno Latour "An Inquiry into Modes of Existence', *Common Knowledge*, 20:3(2014), 491-493, p. 492.

[7] Bruno Latour and Steven Woolgar, *Laboratory Life: The Construction of Scientific Facts* Second Edition (Princeton: Princeton University Press, 1986).

Institute worked to manufacture papers that represented objects on the basis of what was, in effect, an utterly relativist logic. According to *Laboratory Life*, facts are constructed (as opposed to simply 'discovered') by way of technical, material and human resources which work to determine what counts as 'reality'. From this perspective, reality is not so much that which is demonstrated by logic, deduction, and empirical research, but 'the set of statements considered too costly to modify'[8] thanks to the enormous expense of the research, but also the symbolic capital invested in one or another paradigm. Latour and Woolgar's aim was not so much to question the *solidity* of facts arrived by empirical research, as to point out the fundamental importance that a system of production, and with it a social network have in *generating* facts – a production process that, if successful, is eventually erased *as production* by the very notion of a more or less widely accepted facticity.

As Latour and Woolgar put it, 'We ... need to stress the importance of not "reifying" the process by which a substance [in the aforementioned study's case, the Thyrotropin Releasing Factor or TRF immunoassay] is constructed. An object can be said to exist solely in terms of the difference between two inscriptions. In other words, an object is simply a signal distinct from the background of the field and the noise of the instruments'; 'the extraction of the signal and the recognition of its distinctiveness themselves [depend] on the costly and cumbersome procedure for obtaining a steady baseline. This, in turn, [is] made possible by laboratory routine and by the iron hand of the scientist...'.[9]

While the word 'fact' implies that something is unquestionably true, Latour and Woolgar argue that facts are only treated as such within certain contexts, and the same goes for many of the objects referred to via those facts. 'Scientific activity is not "about nature," it is a fierce fight to construct reality. The laboratory is the workplace and the set of productive forces, which makes construction possible'; '[e]very time a statement stabilises, it is reintroduced into the laboratory (in the guise of a machine, inscription device, skill, routine, prejudice, deduction, programme, and so on), and it is used to increase the difference between statements. The cost of challenging the reified statement is impossibly high. Reality is secreted'.[10]

The emergence and formalisation of ANT as a field of inquiry concerned with nature-culture hybridity (amongst other topics) is more recent, and finds an early expression in Latour's ironically titled *We Have Never Been Modern*, which was originally published in the French language in 1991. In this work, Latour offers an

---

[8] Bruno Latour and Steven Woolgar, *Laboratory Life*, p. 243.

[9] Bruno Latour and Steven Woolgar, *Laboratory Life*, pp. 126-127.

[10] Bruno Latour and Steven Woolgar, *Laboratory Life*, p. 243.

extended critique (though today he might not wish to call it that[11]) of what he describes as the 'modern constitution', and with it the characteristically modern opposition of nature and culture. Latour uses the metaphor of a constitution to describe the originating 'laws' of a modern imaginary of nature, and through it, the conceptualisation, but also the *enactment* of certain relations between the objects thought to pertain to nature or culture.

As explained by Latour, this opposition acquired its characteristic form in the work of two seventeenth-century philosophers: Robert Boyle (1627-1691), and Thomas Hobbes (1588-1679). Boyle and Hobbes set about purifying what today we would describe as the categories of Nature (Boyle) and Society (Hobbes). Simplifying somewhat, where Boyle perfects a way of representing nature that appears to be independent of the scientist, Hobbes proposes a way of describing social and political order in terms of an idealised human nature (i.e. a nature that is abstracted from the most material circumstances). Paraphrasing Latour, those in the Boylean tradition attempt to bracket the social in order to produce a faithful representation of the natural world, whereas those in the Hobbesian tradition propose to bracket beyond-human objects in order to generate a faithful representation of the essence of the human condition.[12] The title of Latour's book is meant to remind the reader that, strictly speaking, we have never been Modern, if that means accepting as true-in-itself the opposition between material objects and human essences, and with it the nature-culture dualism that the writings of these scholars, and the many that followed in their footsteps, institute. The last point notwithstanding, Moderns of course *have* been *ever so modern* in so far as they/we accept a Boylean-Hobbesian-style nature-society opposition as a given – an opposition that continues to apply, to this day, in the cases of wildlife tourism and ecotourism.

So what is the conceptual alternative? Already in *We Have Never Been Modern*, Latour anticipates what will be at once a major theme, and a key methodological displacement for Actor-Network-Theory: where the modern constitution suggests that the world can be divided into human and nonhuman constituents (one might even say, *constituencies*), Latour suggests a 'Parliament of Things' that is, by contrast, all about nature-culture *hybrids*. From this perspective, moderns are, like all other cultural groups, immersed in a world of 'mixtures of nature and culture',[13] and it is these 'mixtures' that need to be explained. 'Let us again take up', Latour proposes, 'the two representations and the double doubt about the faithfulness of the representatives, and we shall have defined the Parliament of Things. In its confines, the continuity of the collective is reconfigured. There are no more naked truths [as suggested by physical scientists in the Boylean tradition], but there are

---

[11] See Bruno Latour, 'Why Has Critique Run out of Steam? From Matters of Fact to Matters of Concern', *Critical Inquiry* 30 (Winter 2004), 225-248.

[12] Bruno Latour, *We Have Never Been Modern*, pp. 143-144.

[13] Bruno Latour, *We Have Never Been Modern*, p. 30.

no more naked citizens [as suggested by philosophers in the Hobbesian tradition], either. The mediators have the whole space to themselves...'.[14]

In the expression 'Parliament of Things', the choice of *things* must be understood in relation to what will become an at once increasingly nominalist, and neo-empiricist stance. Latour may be, and indeed *is* highly critical of positivism and the kind of nominalism that *it* partakes in (a rejection of abstract objects, cf. Chapter 1). However, he nevertheless embraces a second kind of nominalism, the one opposed by Charles Sanders Peirce, if not by realist philosophers more generally. Going back to the description that I offered with Peirce of nominalism (cf. Chapter 5), Latour would like to persuade us that, in effect, only names, including the concepts that scientists use, have generality; the world itself is no more (and no less) than agglomerations of things whose relations, and whose very natures are constantly being remade thanks to their shifting positions in networks. Those who engage in ANT thereby adopt a stance vis-a-vis representation which in some respects mirrors that of positivism: far from being the end-all suggested by the new cultural geography, representation is a kind of necessary evil. That said, and in direct opposition to positivism, ANT entails a radical rejection of the rule of phenomenalism (cf. Chapter 1). As I understand it, ANT, and Latour's philosophy rejects out of hand the possibility of anything like indexicality (Peirce, *passim*) when it comes to concepts or symbolic forms more generally.

As I noted above, Latour initially applies this kind of critique to the work of physical scientists. More recently, Latour has turned what increasingly sounds like a nominalist *wrath* on social scientists, berating them/us for essentialising 'the social', and overlooking the importance of the aforementioned 'things'. 'It is always things – and I ... mean this last word literally – which, in practice, lend their "steely" quality to the hapless "society"'.[15] Contrary to what has been suggested by many sociologists (going back, Latour suggests, at least as far as Emile Durkheim's *The Rules of Sociological Method*[16]), social processes are not to be studied in the 'internal constitution' of a group, in its motivations, social ties or social forces, but precisely in all those *things* that Latour repeatedly invokes. As Latour puts it, '[W]hen social scientists add the adjective 'social' to some phenomenon, they designate a stabilized state of affairs, a bundle of ties that, later, may be mobilized to account for some other phenomenon. There is nothing wrong with this use of the word so long as it designates what is *already* assembled together, without making any superfluous assumptions about the *nature* of what is assembled'; if however, '"social" begins to mean a type of material, as if the adjective [is] roughly comparable to other terms like "wooden", "steely", "biological", "economical", "mental", "organizational", or "linguistic"', '[a]t that point, the mean-

---

[14] Bruno Latour, *We Have Never Been Modern*, pp. 143-145.

[15] Bruno Latour, *Reassembling the Social: An Introduction to Actor-Network-Theory*, p. 46.

[16] Emile Durkheim, *The Rules of Sociological Method*, translated Sarah A. Solovay and John H. Mueller (New York: Free Press, 1966[1895]).

ing of the word breaks down since it now designates two entirely different things: first, a movement during a process of assembling; and second, a specific type of ingredient that is supposed to differ from other materials'.[17] The latter meaning Latour dismisses as an exercise in essentialism, and as the second leg of the nature-culture dualism: where physical scientists essentialise the natural, social scientists essentialise the social.

The same problem accrues, he suggests, to any effort to develop a 'social explanation' of some state of affairs. By this account, any researcher's attempt to develop a metalanguage (a vocabulary of theoretical concepts, more or less technical explanations such as the ones presented in this volume) with which to explain that state of affairs is likely to be entirely misleading in so far as it changes what the research subjects believe, or the practices they engage in. This is particularly true for any attempt at critique, i.e. to explain a practice in a way that arrives at conclusions about the nature of an imagined sociality that are not shared by the subjects of the investigation. Such explanations entail, according to Latour, a form of *vampirism*: 'sociology stops being empirical and becomes "vampirical"'.[18]

It follows, or so Latour suggests, that the task of ANT is not to move from description to explanation, but to stay on the level (if we can call it that) of description, and to do so via an 'infra-language' that is as close as possible to the discourse of the actors. As Latour puts it, in ANT 'analysts are allowed to possess only some *infra*-language whose role is simply to help them become attentive to the actors' own fully developed metalanguage, a reflexive account of what they are saying. In most cases, social explanations are simply a superfluous addition that ... dissimulates what has been said...'.[19]

Latour nonetheless clarifies that 'ANT is not the empty claim that objects do things "instead" of human actors: it simply says that no science of the social can even begin if the question of who and what participates in the action is not first of all thoroughly explored, even though it might mean letting elements in which, for lack of a better term, we would call *non-humans*.' 'The project of ANT is simply to extend the list and modify the shapes and figures of those assembled as participants and to design a way to make them act as a durable whole.'[20]

The last curious phrase returns us to Latour's nominalism: there is no durable whole, at least not the kind of whole that may be fixed by language. And the same is true for the actors and the networks referenced by the very name ANT. As La-

---

[17] Bruno Latour, *Reassembling the Social*, p. 1.

[18] Bruno Latour, *Reassembling the Social*, p. 50. For accounts of Latour's aversion to critique, see Bruno Latour, 'Why Has Critique Run out of Steam? From Matters of Fact to Matters of Concern', *Critical Inquiry* 30 (Winter 2004), 225-248. See also the much earlier Bruno Latour, 'The Enlightenment without the Critique: A Word on Michel Serres' Philosophy', *Royal Institute of Philosophy Lectures*, 21(1987), 83-97, p. 85.

[19] Bruno Latour, *Reassembling the Social*, p. 49. Italics in the original.

[20] Bruno Latour, *Reassembling the Social*, p. 72. Italics in the original.

tour puts it, '[a]n "actor" in the hyphenated expression actor-network is not the source of an action but the moving target of a vast array of entities swarming toward it'.[21] Action itself is 'dislocated', as well as 'borrowed, distributed, suggested, influenced, dominated, betrayed, translated. If an actor is said to be an *actor-network*, it is first of all to underline that it represents the major source of uncertainty about the origin of action'.[22] As suggested above, the actor is not necessarily a *human* actor; it could equally be a beyond-human animal, a computer or a kettle, a hurricane or indeed the stench of a factory. In much the way that Norbert Wiener suggested that for cybernetics it matters not whether a message is produced by a human, a non-human animal or a machine (cf. Chapter 3), and that Deleuze suggests that anything is a body/a body is anything (cf. Chapter 10), for ANT the 'sacred' divide between the human and the non-human is replaced with an equally sacred obligation to be thing-oriented, or one might also say object-oriented.

Echoing Wiener in a rather different way, ANT introduces a social version (perhaps one should say 'social' version) of the Gibbsian statistical mechanical principle that the notion of a precise measurement is misleading (what can be measured is the most likely *distribution* of things, cf. Chapter 3). As I have already explained, Latour rejects any suggestion that the meaning of objects is fixed; on the contrary, the meaning(s) or role(s) of an actor (human, machine, etc.) depend(s) on its changing position as part of a network, and indeed Latour borrows from the structuralist semiologist Algirdas Greimas the concept of the *actant* to underscore this point. In the Greimasian theory of narrative, an actant is a *role* played by any person or object which entails a play of continuity and change; the role, as narrative role, remains much the same across different stories (e.g. the subject, the object, the sender, receiver or helper in fairy tales) but the figure changes. As Greimas puts it, 'In the broad sense of the term, an actant can be either the linguistic representation of a human person or the character of a story or again an animal or a machine'.[23] We have only to consider the countless narrative examples where beyond-human animals, or indeed robots and other machines are given human roles. Latour suggests that much the same play occurs in non-literary contexts: one 'same' object can play very different roles in different contexts, and presumably, different objects can play the 'same' role across similar contexts – the point being that the role is utterly relational, and so entirely context-dependent.

I have said something about the actor in Actor-Network-Theory, but what about the *network*? As I began to suggest earlier, Latour does not mean an unchanging, or mechanically transposable matrix of things that have fixed relations

---

[21] Bruno Latour, *Reassembling the Social*, p. 46.

[22] Bruno Latour, *Reassembling the Social*, p. 46. Italics in the original.

[23] Algirdas Julien Greimas, *Narrative Semiotics and Cognitive Discourses*, translated by Paul J. Perron and Frank H. Collins (London: Pinter Publishers, 1990), p. 115.

(as per, say, the image of a computer network). Instead, according to Latour a network is 'a string of actions where each participant is treated as a full-blown *mediator*'[24] – the latter being Latour's term for something that changes something else. In contrast to *intermediaries* – things that '[transport] meaning or force without transformation' – mediators 'transform, translate, distort, and modify the meaning or the elements they are supposed to carry'.[25] '[T]he network does not designate a thing out there that would have roughly the shape of interconnected points ... It is nothing more than *an indicator of the quality of a text* about the topics at hand. It qualifies its objectivity, that is, the ability of each actor to *make* other actors *do* unexpected things. A good text elicits networks of actors when it allows the writer to trace a set of relations defined as so many translations.'[26] Translation, for its part, refers to 'a relation that does not transport causality but induces two mediators into coexisting'.[27]

I will raise some issues with Latour's proposals at the end of this chapter. For now, I would like to turn to the work of the geographer Sarah Whatmore. Whatmore develops a set of proposals with respect to nature-culture hybridity, but now in the context of human geography.[28] This is a paradigm shift that emerged in the late 1990s and early 2000s; while we shall see that Whatmore builds on multiple sources and currents, her analyses suggest close proximity to Latour's variety of ANT.

In her book *Hybrid Geographies*, Whatmore offers not so much a definition, as a series of academic coordinates with which to locate what she means by hybrid geographies. In so far as those coordinates provide a good sense of the manifold theories that inform the work, it is worth quoting them in some length (here as with other authors, I leave Whatmore's Harvard-style references to indicate the sources):

---

[24] Bruno Latour, *Reassembling the Social*, p. 128. Italics added to the original.

[25] Bruno Latour, *Reassembling the Social*, p. 39.

[26] Bruno Latour, *Reassembling the Social*, p. 129. Italics in the original.

[27] Bruno Latour, *Reassembling the Social*, p. 108.

[28] Beyond the writings of Whatmore, see also Paul Cloke and Owain Jones, 'Turning in the Graveyard: Trees and the Hybrid Geographies of Dwelling, Monitoring and Resistance in a Bristol Cemetery', *Cultural Geographies*, 11 (2004) 313-341; D. Demeritt, 'Hybrid Geographies, Relational Ontologies and Situated Knowledges', *Antipode*, 37:4 (2005), 818-23; Steven Hinchliffe, M. B. Kearnes, Monica Degen, and Sarah Whatmore, 'Urban Wild Things: A Cosmopolitical Experiment', *Environment and Planning D: Society and Space* 23 (2005), 643–58; Chris Philo, 'Spacing Lives and Lively Spaces: Partial Remarks on Sarah Whatmore's Hybrid Geographies', *Antipode* 37:4 (2005), 824–33; Leslie Head and Jennifer Atchison, 'Cultural Ecology: Emerging Human-Plant Geographies', *Progress Reports in Human Geography*, 33:2 (2009), 236-245.

'Hybrid geographies' allies the business of thinking space (Crang and Thrift, 2000) to that of thinking through the body (Kirkby [sic] 1997), in other words to apprehend and practice geography as a craft. This enterprise gestures towards Michel Serres' insistence that 'there is a sense in space before the sense signifies'(1991:13) in two ways: by attending simultaneously to the inter-corporeal conduct of human knowing and doing *and* to the affects of a multitude of 'message-bearers' that make their presence felt in the fabric of social life. To map the lively commotion of these worldly associations is to travel in them, negotiating 'modes of access and ways of orienting ourselves to the concrete world we inhabit'(Bingham and Thrift, 2000: 292).[29]

What happens, Whatmore asks, as a consequence of such mappings of knowledge? 'A preliminary response to the question at the outset would be – an upheaval in the binary terms in which the question of nature has been posed and a re-cognition of the intimate, sensible and hectic bonds through which people and plants; devices and creatures; documents and elements take and hold their shape in relation to each other in the fabric-ations of everyday life.'[30]

A detailed account of the different contributions of these influences – or the ways in which Whatmore innovates on their proposals – is beyond the scope of this chapter. It suffices to note that the scholars and fields include those that engage in what one researcher has described as 'non-representational theory'[31] – an unfortunately named theory in human/cultural geography that embraces affect and performance as alternatives to the older emphasis on discourse and representation; Science and Technology Studies (STS), with a special place for the research of Donna Haraway;[32] what Whatmore describes as biophilosophy, and which presumably includes the kinds of theories of organisms, affects and assemblages that I described in relation to Gilles Deleuze and Félix Guattari in Part 3, but also the work of the philosopher Michel Serres (more on his concept of topology, below);[33] new materialist feminist theory with its problematisation of cultural sociological

---

[29] Sarah Whatmore, *Hybrid Geographies*, p. 3.

[30] Sarah Whatmore, *Hybrid Geographies*, p. 3.

[31] See Nigel Thrift, *Non-Representational Theory: Space/Politics/Affect* (London: Routledge, 2008). See also the interview with Thrift in Ben Anderson and Paul Harrison (Eds.) *Taking-Place: Non-Representational Theories and Geography* (Farnham, Surrey: Ashgate, 2010).

[32] See for example, Michel Serres, *Rome: The Book of Foundations*, translated by F. McCarren (Stanford, CA: Stanford University Press, 1991); Michel Serres, *The Five Senses*, translated by M. Sankey and P. Cowley (London: Continuum, 2008); and Michel Serres and Bruno Latour, *Conversations on Science, Culture, and Time*, translated by R. Lapidus (Ann Arbor: University of Michigan Press, 1995).

[33] See for example, Donna Haraway, *Simians, Cyborgs and Women. The Reinvention of Nature* (London: Free Association Books, 1991).

and geographic constructivism, and which includes the work of Anne Kirby;[34] and then again, the emergence, contemporaneous to hybrid geographies, of *animal* geographies,[35] a subfield of cultural geography that shifts scholarly attention to beyond-human animals in the context of the critique of speciesism, and the more general turn towards the problematisation of embodiment[36] (cf. Part 3).

I will return to the question of animality in Chapter 14. Here it may be reiterated that, after recognising the significance of these and other influences, in my view Whatmore's account of hybrid geographies is closest to STS/ANT. This is evident across several aspects of Whatmore's proposals, which in one way or another emphasise the importance of actors (including the beyond-human) and networks conceived in ways that move away from the nature-culture binary by embracing that 'parliament of things' recommended by Latour, and in a manner that emphasises flow and 'positionality' over structure and location. As Whatmore herself describes them, hybrid geographies refuse 'the purifying impulse to fragment living fabrics of association and designate the proper places of "nature" and "society"', and seek instead to countenance the world 'as an always already inhabited achievement of heterogeneous social encounters where, as [STS philosopher] Donna Haraway puts it, "all of the actors are not human and all of the humans are not "us" however defined"'.[37] Or as Whatmore also explains, the essays that make up *Hybrid Geographies* 'follow the interferences of "things", from elephants and soybeans to deeds and patents, in the geographies of social life'.[38]

In this chapter I am particularly keen to consider what Whatmore has to say about beyond-human animals and wilderness. Whatmore suggests that hybrid geographies refuse the *a priori* of a nature that is 'outside', or as Whatmore puts it, 'the choice between word and world'.[39] In so far as the wild is defined as that which lies outside of human civilisation's historical and geographical reach, it 'renders the creatures that live "there" inanimate figures in unpeopled landscapes, removing humans to the "here" of a society from which all trace of animality has been expunged'.[40] Instead of starting from an exterior world of anything like an 'original nature', Whatmore would like to start from the premise that '"wild" animals have been, and continue to be, routinely imaged and organized

---

[34] See for example Vicky Kirby, *Telling Flesh* (London: Routledge, 1997).

[35] See for example, Jennifer Wolch and Jody Emel (Eds.) *Animal Geographies: Place, Politics and Identity in the Nature-culture Borderlands* (London: Verso, 1998).

[36] For an introduction to this shift, see for example Donn Welton (Ed.) *The Body: Classic and Contemporary Readings* (Oxford: Blackwell, 1999).

[37] Sarah Whatmore, *Hybrid Geographies*, p. 3.

[38] Sarah Whatmore, *Hybrid Geographies*, p. 3.

[39] Sarah Whatmore, *Hybrid Geographies*, p. 3.

[40] Sarah Whatmore and Lorraine Thorne, 'Wild(er)ness', pp. 435-436.

within multiple social orderings in different times and places'.[41] These orderings, Whatmore suggests, 'confound the moral geographies of wilderness which pre-suppose an easy co-incidence between the species and spaces of a pristine nature, confining their place to the margins and interstices of the social world'.[42] As Whatmore and Lorraine Thorne put it, the problem is not to start with the wild or with wildlife *as* 'the outside', but to approach it *from* 'the inside', an inside where 'the everyday worlds of people, plants and animals are already in the process of being mixed up'.[43]

As part of this shift, Whatmore proposes to develop a hybrid account of wildlife regarded as 'a relational achievement spun between people and animals, plants and soils, documents and devices in heterogeneous social networks which are performed in and through multiple places and fluid ecologies'[44] – a fluidity that Whatmore seeks to study via the figure of a topology. As Whatmore and Lorraine Thorne put it, they would like to 'reimagine wildlife topologically – as fluid, relational achievements that configure "human" and "animal" categories and lives in intimate, if not necessarily proximate, ways'.[45]

Here a brief excursus on topological space is required to clarify what it is that interests the authors. The philosopher Michel Serres, another scholar that is prominent in Whatmore's writing, offers a marvellously simple way of illustrating the shift from metric to topological space. In a 'conversation' with Bruno Latour, Serres uses a handkerchief to explain the difference: if one takes a handkerchief and spreads it out as if to iron it, one can see in it certain fixed distances and proximities. If, for example, one sketches a circle in one area, one can mark out nearby points and measure far-off distances.

If, however, one takes that same handkerchief and one crumples it by putting it in one's pocket, then suddenly two hitherto distant points will be close, or even superimposed. Or if one tears the handkerchief in certain places, the opposite happens: two points that were close now become distant. Topology is the science of nearness and rifts, while metrical geometry is the science of stable and well-defined distances.[46]

Instead, then, of treating wild animals as entirely *settled* objects of investigation in more or less clearly defined territories – themselves regarded as metric spaces – Whatmore proposes to approach beyond-human animals in ways that acknowl-

---

[41] Sarah Whatmore, *Hybrid Geographies*, p. 14.

[42] Sarah Whatmore, *Hybrid Geographies*, pp. 9-10.

[43] Sarah Whatmore and Lorraine Thorne, 'Wild(er)ness', p. 437.

[44] Sarah Whatmore, *Hybrid Geographies*, p. 14.

[45] Sarah Whatmore and Lorraine Thorne, 'Wild(er)ness: Reconfiguring the Geographies of Wildlife', p. 450.

[46] Michel Serres with Bruno Latour, *Conversations on Science, Culture, and Time*, translated by Roxanne Lapidus (Ann Arbor: University of Michigan Press, 1995 [1990]), p. 60.

edge a certain 'crumpling' and tearing of space – the kind that occurs when humans remove specimens from their original habitats and put them in zoos, or indeed, when humans introduce themselves into habitats that were once relatively undisturbed.

In all of this, there is a risk that the researcher continues to treat animals as being no more than the objects of human interventions, or indeed as the victims of human ecological crimes and misdemeanours. In fact, Whatmore underscores the importance of treating the animals as 'active subjects'. The animals' 'constitutive vitality' must be acknowledged, not in the terms of unitary biological essences, but 'as a confluence of libidinal and contextual forces. Here, the multi-sensual business of becoming antelope or wolf and the inscription of these bodily habits in the categorical and practical orderings of human societies are interwoven in the seamless *performance* of wild-life.'[47] The references to 'becoming' antelope or wolf may be taken to be an echo of the kind of analysis offered by Deleuze and Guattari, which I began to consider in Part 3 and I will return to in Chapter 14. For their part, the references to performance may be interpreted as a sign that Whatmore, like many other scholars in the critical social sciences and humanities by the early 2000s, embraces not only the theatrical metaphors originally proposed by the sociologist Erving Goffman,[48] but more generally the 'performance turn' which plays a particularly important role in 'non-representational' theory.[49]

The theoretical implications of Whatmore's proposals may be illustrated via one of the topologies of wildlife that she develops: a study of the Paignton Zoo in the southwest of England, and in particular the enclosure and display in that zoo of an African Elephant (*Loxodonta africana*) known by the name of 'Duchess'. In *Hybrid Geographies*, Whatmore examines the way in which the elephant is at once managed, positioned and presented by way of what I described in Part 3 as techniques of observation.

One such technique (and indeed technology) is what zoos refer to as ISIS, the International Species Information System – a database of animals held in captivity in zoos across the world, which has the aim of systematising, and making more widely as well as instantly available, data on aspects such as genetic profile, age, sex, parentage, location of birth and circumstances of death. Amongst other things, this technology allows the zoos to try to achieve an 'optimal pairing' of breeding specimens (ISIS works in this sense as an über stud-book for zoos across the world).

Another technique and technology of observation involves the zoo's displays. At the time that Whatmore was conducting the research, the Paignton Zoo was

---

[47] Sarah Whatmore, *Hybrid Geographies*, pp. 14-15. Italics added to the original text.

[48] Erwin Goffman, *The Presentation of Self in Everyday Life*, new edition (London: Penguin Books, 1990 [1959]). Nigel Thrift, one of the geographers often cited by Whatmore, offers a discussion of this metaphor in Nigel Thrift, *Non-Representational Theory*, pp. 124-138.

[49] See Nigel Thrift, *Non-Representational Theory*, p. 32, and pp. 128-132.

undergoing a redevelopment that was funded in part by the European Union, and which entailed transforming the until then traditional zoo into a 'wildlife park' with six 'habitats', including an 'African Savannah'. Even as Duchess' data-based details ensured the hyper-mobility of what might be termed Duchess's *digitised* body, the zoo moved Duchess herself and an Asiatic elephant (*Elephas maximus*) into the zoo's new display area as part of an attempt to blur the boundaries, at least for visitors, between real African savannahs and the 'English Riviera' (the popular name for part of the coast along the south of Devon, which is where the Paington Zoo is located).

Whatmore points out that, to begin with, Duchess and her companion elephant refused to move out of the new building and into the exterior areas of the new display: 'both animals remained within the elephant house, refusing to go outside and "explore" their new enclosure. Ordinarily this act would restrict the elephants from the public gaze, but the new elephant house is architect-designed for internal viewing, with a raised public gallery so that they can see more of what goes on behind the scenes and view the elephants even when are sheltering from the weather'.[50] So even as the elephants initially refused the kind of mobilisation entertained by the zoo, the design of the new display space worked to maintain visibility, and so a certain kind of surveillance.

Whatmore concludes the analysis with a telling point: she notes that, while Duchess certainly belongs 'taxonomically' to *Loxodonta africana*, 'the elephant she has become through her life at Paignton Zoo bears only distant relation to those of her kind at home in the African bush, even as such living spaces are themselves being reconfigured in the same patterning of foresight in which she [Duchess] is caught up'.[51] We come back by this route to the point I raised in Part 3 about the limits of essentialist conceptions of species, and the alternative conception offered by a Deleuzian, and Deleuzian and Guattarian theory of affects and assemblages. Duchess may be formally classified as an elephant, but from the Deleuzian/Guattarian perspective represented by the concept of 'becoming', just as a workhorse may be closer to an ox than to other kinds of horses, there is no essential guarantee that an elephant in a zoo still 'really is' an elephant – at least not the kind of elephant referenced, apparently unproblematically, by the zoos themselves. This is much the same stance, albeit for somewhat different theoretical reasons, adopted by Whatmore, and by ANT more generally: Duchess' identity is the one conferred by the network or networks enacted by the Paignton and other zoos.

Before considering the implications of Latour's and Whatmore's perspectives for my own study, one further quality of Whatmore's account may be noted that also suggests a proximity to ANT: its extraordinary oscillation between meta-language and infra-language (Latour, *passim*). As the quotes I offered at the start of this account show, Whatmore's theoretical discourse is remarkably dense from

---

[50] Sarah Whatmore, *Hybrid Geographies*, pp. 46-47.

[51] Sarah Whatmore, *Hybrid Geographies*, p. 47.

the point of view of its metalanguage, which weaves together a fabric of concepts and meta-concepts, as well as palimpsests of concepts used in other fields, 'topology' being a case in point. But in the actual topological case studies, Whatmore switches to descriptions that might almost have been written by the actors working in the institutions that Whatmore analyses. Like Latour, Whatmore would appear to reject the 'philosophy of critique', and to prefer instead a discourse which is, in Latour's terms, ostensibly merely 'attentive' to the researched subjects' own meta-language (in the case of the Paignton Zoo, its use of the ISIS database, its plans for the redevelopment of the zoo, etc.). As the reader may recall, Latour suggests that to do otherwise risks occluding the original meta-language with the researcher's own obfuscations.

Having provided an indication of the principles, and indeed of an analysis that is close to the principles championed by ANT, what implications might these kinds of approaches have for the present study?

Let's return to Barro Colorado, and to the reintroduced spider monkeys that I referred to in the opening example of Part 4. If we go by the kind of ontology favoured by modern science, then the autoecology of Barro Colorado's spider monkeys should be largely unaffected not only by the reintroduction itself, but by the overall changes to Barro Colorado. An exception to this kind of approach would be any changes generated by the fragmentation of the island, which might reduce the extension of available forest, as well as connections between species and specimens on the mainland. Such issues to one side, the assumption would be that, in the case of the spider monkeys, innate behaviours would provide a kind of template that would, in effect, allow nature to 'resume' in the lives of the monkeys. Informing any such study would be a fundamentally 'metric' (or one might also say Euclidean) assumption regarding the space of the monkeys, if not of the island itself: in so far as the reintroduced spider monkey species once inhabited a certain area defined in linear terms (between, say, points along this or that longitude, this or that latitude); and in so far as Barro Colorado is not only located within the designated territory, but is 'well preserved', then little or nothing should really have changed, geographically and ecologically speaking, for the monkeys. From this perspective, the proof must be in both the monkeys', and the island's puddings: once the infant/juvenile monkeys on Barro Colorado stopped being fed by park wardens, they grew up, and not only survived but successfully bred to establish a thriving group on Barro Colorado – one that would seem to be little or no different from any other groups that have not undergone similar traumas. On the level of the island as a whole, once the hullabaloo of canal-building activities came to a stop, and any imbalances generated by the great flood of the Chagres River Valley came to an end, Barro Colorado itself resumed its existence, in more or less the same place, and we might say at more or less the same ecological *time* from when the ecological clock was momentarily stopped, if indeed it ever was stopped at all.

Chapter 13 Actors, Networks and Things

From a hybrid geographic perspective, it is possible to question both sets of assumptions. Where the monkeys are concerned, whatever the short-term or long-term effects of their reintroduction to Barro Colorado, the individuals *did* undergo a brutal process of incorporation into human networks – we return to Whatmore's suggestion that organisms that are treated as being wild are routinely caught up within multiple networks of human social life. In keeping with the custom of the wild animal traders, the mothers/parents of the infant monkeys would have been killed or otherwise separated from their offspring, and the offspring would have been caged and taken to markets, where they were found by Smithsonian staff. To return to Michel Serres' handkerchief, they were torn from their parents, and crumpled into cages where they would most likely have come into close proximity with other monkeys and creatures that were wholly unrelated. To this trauma – but also the sheer experiential difference vis-a-vis a monkey with no such upheaval – we would have to add being transported to Barro Colorado, and once there, kept under close supervision, as well as being fed, by humans. At the time, it was some scientists' and staff members' habit to treat some wild animals as pets or quasi-pets;[52] even if this did not occur with Barro Colorado's spider monkeys, it is to be expected that the monkeys would have grown up having very different relations to humans from those more accustomed to, say, being hunted for food or for the animal trade. This kind of process would render suspect, to say the least, any suggestion that the monkeys once again became 'truly wild' monkeys, if by that one means a conception of animality that draws a very strong, indeed absolute boundary between the domesticated and the wild animal, culture and nature.

The obvious rejoinder – that the monkeys not only bred successfully, but eventually acquired the same feeding habits of their un-introduced 'cousins' – raises a question as to whether ethological research actually has the tools with which to detect, and correctly interpret subtle differences in observable behaviour. Were we, for example, to apply ethological methods to observing humans, it is unlikely that we could discern signs of neuroticism or of any other mental illnesses that have no ostensive expression. Any such approach is, of course, very much in keeping with the Cartesian notion that beyond-human animals lack mind (cf. Chapter 14). For if New World monkeys *do* have anything like mind, then it is entirely conceivable that at least the first generation of specimens would have had thoughts quite different from those of relatives who were able to avoid hunting, or trans-location. But of course, in so far as the mind-body dualism is 'resolved' entirely in favour of body – that is, a mechanistic conception of body – then mind is not, and *does* not, matter.

Where Barro Colorado itself is concerned, in Volume 2 I will consider in some detail the numerous, if often very subtle transformations introduced to the island's geography and ecology by the flooding of the lower Chagres River Valley,

---

[52] See for example, Frank M. Chapman, *My Tropical Air Castle* (New York: Appleton, 1929).

and also by what by 2007 was the better part of a century of intensive research. I will, of course, also consider the growing ravages of catastrophic climate change. Here it will suffice to point to the transformative effects of the cutting of trails; the more or less interventionist, or even invasive research conducted by generations of scientists; and then again, the 'rewilding' of those parts of the original forest (if one can speak in those terms) that had been cut down by farmers prior to the establishing of the biological reserve. Like any trauma suffered by the first generation of spider monkeys, the transformations of Barro Colorado, both 'macro' and 'micro', may be ignored precisely in so far as a transcendental conception – what might be termed 'essence of tropical forest' – is maintained, over and against the principle of immanence, and with it a realisation that Barro Colorado, no matter how well preserved, must have an ecology that is qualitatively different from what it was until 1910, 1913, or 1924 (to refer to just three particularly significant years in the island's environmental history). That does not mean that it is no longer a wilderness; it does mean that one needs to reconsider what exactly is meant by wilderness, and indeed by 'wild' animals.

There is one other contribution which ANT and Whatmore's hybrid geography can make which I would now like to consider. It concerns the role of 'things' in the research. An investigation like the present one might be tempted to focus on the visitors and the beyond-human animals, or more generally on fields and discourses, geographies and ecologies. Latour and Whatmore make a strong case for the need to incorporate all sorts of other things that go beyond the nature-culture dualism, but also beyond the kinds of purifications encouraged by the unwitting acolytes of Boyle and Hobbes. Accordingly, a study like the present one should consider anything and everything from the organismic bodies, to the kinds of bodies described by Deleuze, but also such apparently banal technical and technological details as the ones that I described with respect to high-definition television in the Prologue. In addition to such 'high' technology we would have to add analyses of 'low' technologies such as canopy walks and trails (themselves technologies of observation); the guidebooks used by some tourists; the signs, leaflets, maps, and any other promotional literatures employed by tourist operators or park authorities; the vehicles used to reach the destinations; the food eaten in park cafeterias; the weather encountered; the amount of effort required to walk along trails... This list is by no means exhaustive, it simply is meant to highlight the need to go beyond – far beyond – the usual objects/subjects of investigation.

I would now like to consider some problems with ANT and hybrid geographies, beginning with Latour's proposals. If we go by Latour's own critique (and yes, it is that, a *critique*), then Latour should receive the title of Vampire-in-Chief. How else to explain that, having accused social scientists of being vampires for arriving at conclusions not shared by the subjects of their research, he has no problem with critiquing the vampires? It seems either contradictory, or condescending, to this 'social scientist' at least, that having used a metalanguage also known as Actor-Network-Theory to critique one or two centuries of social scientific practice (longer in the case of the physical sciences), Latour then feels able to deny the rest of the researchers the right to do exactly the same thing with respect to the actors *they* research. Are we really to accept that when the targets of critique are physical or social scientists, then any meta-language goes, but that the same principle does not apply to so-called 'lay' practitioners?

A second problem may seem like a somewhat surprising one, given Latour's emphasis on 'things' – an emphasis which might easily be mistaken for a kind of realism. In fact, Latour makes it clear that one can adopt an empiricist stance, and still be, to repeat Barbara Herrnstein Smith's expression, a 'thorough nominalist'. As I explained in Chapter 1, positivist philosophy has no trouble reconciling empiricism and nominalism of the kind that rejects abstract objects; Latour reconciles empiricism with nominalism of the kind that rejects universals in so far as his theory appears to draw a very sharp line between concepts and the world itself. This is perhaps in keeping with his debt to Greimas, and it may also help to explain Latour's more recent enthusiasm for research based on the sociologist Gabriel Tarde's (1843-1904),[53] and through him Gottfried Leibniz's (1646-1716) 'monadology'.[54]

As I see it, there are two issues with this stance. The first is that it offers at best a very weak, and at worst no explanation at all of how it is that certain representations may also include at the very least an element of indexicality (in Peircian terms). I return by this route to the kind of point raised in Part 2; there is a difference between, say, writing the sentence 'There is a northerly wind', and a weathervane that is made to point towards 'N' by the wind. In a similar fashion, there is a difference between someone randomly saying 'I believe these apples weigh one kilogramme', and a scale showing that a certain number of apples weigh that much. One may argue that no wind-in-itself ever blows, say, *n-o-r-t-h*, but it would be rather foolhardy to deny that the wind *does* blow in some direction, and that there *is* for that reason an indexical link between the wind and whatever direction the weathervane points to. A similar point can be made with respect to the physical magnetic force that makes compasses point to what is called north (however much the magnetic pole appears to be headed for Siberia...), or indeed the force of the waves and the tides detected by, say, the gravimeter at the Strengbach observatory (where Latour is depicted conversing with scientists as part of an

---

[53] See for example, Gabriel Tarde, *Monadology and Sociology*, edited and translated by Theo Lorenc (Melbourne: Re.press, 2012[1895]).

[54] See for example, Bruno Latour, Pablo Jensen, Tommaso Venturini, Sébastian Grauwin and Dominique Boullier, '"The Whole is Always Smaller than its Parts" – A Digital Test of Gabriel Tardes' Monads', *British Journal of Sociology* 63:4 (2012), 590-615, pp. 597-598. Italics added to the original.

'academic celebrity' interview in the *New York Times*[55]): in each case, there really are forces that modify the representations produced by the machines.

To be sure, in so far as winds blowing in a certain direction (or any other forces) *do* occur repeatedly in multiple places and in ways that can be ascertained objectively (not least from the point of view of their directionality), then I would argue that indexical relations may be more than purely localised relations of action and reaction. Over time, representations may be produced that themselves reflect this generality. This being the case, there really *can* be a *general* continuity between such representations and the represented.

Having made the case for a scholastic realism (cf. Chapter 5), I would nevertheless say that the second issue with nominalism – an issue that it shares with positivist realism – is that it forces one to choose, in effect, between a realist and an idealist stance. From a Peircian perspective, that is a bit like being forced to choose between a glass that is half full or half empty. Or, to use another fluid metaphor, *one never bathes in the same river, until one goes back to bathe in the same river*. Put more technically, there can indeed be a *semeiotic* continuity between dynamical and immediate objects, as per the kind of synechistic, ideal-realist, and scholastic realism of Peircian philosophy, without that forcing the analyst to give up on symbolic signs, or indeed a sceptical stance more generally.

Here as elsewhere in this volume I use the extra 'e' in semeiotic advisedly. Where some nominalists would have us oppose representation to the ineffable flows of the world – hence, perhaps (and ironically), some human geographers' enthusiasm for a 'non-representational' theory – a Peircian semeiotic phenomenology enables one to argue that everything is indeed a matter of firstness (one way of interpreting the later Latour's enthusiasm for Leibniz's monads); but that everything is *also* secondness, *and* thirdness (cf. Part 2). Accordingly, one can acknowledge that while it's true that *every* bath in a river, indeed every *river* is certainly unique (Firstness), the brute forces of *any* flooding river *will* drag a person inadvertently caught in the flow, and will do so again and again (Secondness); though of course, whether the person is saved from drowning may well depend on a social network (Thirdness), a point repeatedly and rightly invoked by those who dispute naive claims that natural disasters are entirely *natural*.

It may further be pointed out that, even as a flooding river drags an individual, the unlucky individual will probably have thoughts on their own plight, and those thoughts will take the form of signs that are at least partly generated by the uniqueness of the own situation *and* by the forces of the river. It is precisely those signs that on occasion may save someone from drowning, or indeed from being devoured by a crocodile (cf. Conclusions to Part 2). If there were not a pretty di-

---

55 Ava Kofman, 'Bruno Latour, the Post-Truth Philosopher, Mounts a Defense of Science', *New York Times Magazine*, 25 October 2018 <https://www.nytimes.com/2018/10/25/magazine/bruno-latour-post-truth-philosopher-science.html>[Accessed 3 December 2019].

rect relation between the different aspects, then it is unlikely that the thoughts would arise, or indeed that a person would be saved from a flood. Peirce, like other realists, would probably argue that it is precisely when one stops believing in such links that the worst kind of relativism sets in – one that is corrosive of whole social orders: if there is no such relation, why bother to call out a warning, or for help? This is increasingly a problem that Latour himself is having to address; ironically, the very science that he worked so hard to relativise, is now under such attack by interested corporations (and the fascist politicians that many such corporations are increasingly allied with) that Latour himself is having to go through some hoops in an effort to explain why his research should actually strengthen faith in science.[56]

A third problem, this time with the hybridity turn, involves precisely the use of the concept of hybridity (which is especially central to Whatmore's writing on the subject, but is also invoked by Latour). This is undoubtedly a useful shorthand with which to refer to the many issues that arise when one questions the nature-culture dualism. Use of the term nonetheless runs the risk of reintroducing the very opposition that the researchers attempt to leave behind. Earlier I quoted Latour when he suggested in *We Have Never Been Modern* the need to consider 'mixtures of nature and culture',[57] and I can think of no better example of the risk I have just referred to. Whatmore is clear that the concept should not be taken to entail an arithmetic logic (nature + culture), let alone a purity of terms. But so long as one continues to use the concept, one cannot quite shake off precisely those connotations. I thereby suspect that the more powerful approach is the one adopted by Deleuze, and by Deleuze and Guattari, who simply refer to bodies, affects, and assemblages – for which, and by means of which 'hybridity' may be explained without having to evoke the dualism again and again.

Let's now turn to some issues arising specifically in Whatmore's hybrid geographies of wildlife. In principle, Whatmore takes a good distance from a Foucaultian-style posthumanism that subsumes wild animals to social orderings (this as part of her commentary on the use of the Foucaultian category of heterotopia, which she seems to approach along the lines presented in *The Order of Things*). Whatmore rightly points out the issues with an externalising conception of wilderness, re-situating the analysis 'inside' – an inside where, as I quoted earlier, 'the everyday worlds of people, plants and animals are already in the process of being mixed up'. I am nonetheless left with the impression that in some places Whatmore's analysis seems closer to a Foucaultian critique of incarceration than it does to a fully hybrid geography, by her own criteria. This may have something to do with the choice of case studies, which in several cases entail animals that are caged or otherwise physically constrained by actual organisations.

---

56 See Ava Kofman, 'Bruno Latour, the Post-Truth Philosopher, Mounts a Defense of Science'.

57 Bruno Latour, *We Have Never Been Modern*, p. 30.

Possibly for this selfsame reason, there appears to be little in the way of a recognition of what may be described, for want of a more sophisticated term, as the varying *degrees* of wildlife incorporation within anthropomorphic networks. It is, for example, one thing to refer to spider monkeys incorporated as part of a simulacrum of a rain forest in a zoo, and another to refer to those animals in a more or less intact rain forest (however fragmented). There is, in this sense, a real risk that the otherwise profound insights afforded by Whatmore (and in a different way, by Cronon) with respect to the externalisation of wilderness may lead the analyst to overlook real differences in the degree of incorporation, and by implication, the degree of hybridity (to use Whatmore and Latour's term).

A more critical alternative might start from a point that Whatmore makes at the very beginning of *Hybrid Geographies,* when she quotes William Cronon in an epigraph as saying that '*wild*ness (as opposed to wilderness) can be found anywhere; in the seemingly tame fields and woodlots of Massachusetts, in the cracks of a Manhattan sidewalk, even in the cells of our own bodies'.[58] As I interpret this statement, the wild is all around, and *within* us if for no other reason than that all organismic bodies have a capacity for action and passion, for affects (cf. the discussion of Spinoza, in Chapter 10, and my reconsideration of this question in Chapter 14).

In so far as we are, from a Peircian perspective, but specks of semeiotic dust in a universe infinitely perfused with signs (or as Deleuze would perhaps describe them, 'modes'), then it would be presumptuous, to say the least, to assume that we are essentially different from everything else in the cosmos. We return by this route to a principle of synechism (cf. Part 2). Even so, this acknowledgement should not prevent researchers from arriving at some kind of *geographic* account that allows them to judge the extent to which some humans have rather illiberally dusted certain places and spaces with signs mostly of their/our own making, they/our own *industry* – and this in ways that curtail, more or less drastically, the opportunity for beyond-human organisms to *be*, or as I will put it with Deleuze and Guattari in Chapter 14, *to become*. Put simply, there are places in the world in which it is far easier for certain kinds of beyond-human organisms to flourish, and to do so relatively free of human interference, and vice-versa: there are also places where any attempt to become will be almost certainly entail anthropogenic destruction.

This being so, it may be helpful to develop something like a notion of *degrees of wild(er)ness* ('wild(er)ness' being an expression by Whatmore and Thorne, which I both borrow and repurpose). We are always immersed in some 'degree' of wild(er)ness; but at least in the Anthropocene, bodies in the planet's surface typically entail at least some degree of anthropomorphic modification. So after acknowledging Soper's 'deep' nature – to repeat her definition once again, a nature whose material structures and processes really are independent of human activity

---

[58] William Cronon, 'The Trouble with Wilderness', p. 89. Italics in the original.

but whose forces and causal powers are the necessary condition of every human practice[59] – the researcher may attempt to develop a characterisation of those 'degrees'. The following is a rough sketch of how such a categorisation might proceed (the numbers are conceived as indicators of qualitative change, as opposed to absolute levels).

Degrees of Wild(er)ness

Degree 5: When a beyond-human formation, and any of the objects that make it up, exist in such a way that there is no contact with humans (or indeed any of the members of the genus *Homo*). By contact, I mean no literal physical contact (e.g. touching), but also no immediate co-presence, no observation (immediate, dynamical, or mediate), no 'human initiated' semeiosis whatsoever. The actual formation and its objects remain so entirely removed from humans or hominids as to be inexistent to them, and perhaps for this reason, they might even be unimaginable. Examples include biomes prior to the evolution and arrival of hominids in them, and of course all those aspects and regions of the cosmos that exist even now, but qualify as per the stated criterion.

Degree 4: Humans come into contact with a beyond-human formation, and thereby at least some of its objects, but the contact is for all practical purposes *rhematic* (cf. Chapter 7) in the sense that the beyond-human objects are left to be what they may be. There is no durable modification of any aspect of the beyond-human bodies in the formation; there is only a fleeting relation of physical co-presence with humans, and/or observation, mutual or not, from a certain distance. Examples include tropical forests in the Americas at the very moment of the first arrival of humans some 13- to 14,000 years ago; or even now, humans observing lava rising from a volcano, or humans observing the birth or death of a star millions of light years away.

Degree 3: Humans come into contact with a beyond-human formation, and at least some of its beyond-human bodies, and do so in ways that have an indexical consequence for one or more of the beyond-human bodies' morphology or ethology, whilst leaving those objects' broader formation largely unchanged. An example is be found in early forays into hitherto unknown and uninhabited ecosystems, in which the presence of humans may lead some creatures to flee or to become otherwise aroused, or even result in the deliberate or inadvertent death of others, but hardly changes the ecosystem, or any of its species, as durable networks. Historically the arrival of human groups to areas that were previously

---

[59] Kate Soper, *What is Nature?* pp. 132-33.

uninhabited would almost certainly have led to a swift transition to from Degree 4 to Degree 3.

Degree 2: Humans have repeated contact with a beyond-human formation and its beyond-human bodies, and these have durable consequences for the beyond-human formation and its networks. Aspects of the formation continue to exist, and do so according to a pre-human logic that itself remains partly intact, and certainly 'wild'. However, as part of this process, many of the beyond-human bodies, and indeed some or many aspects of the network, may be extensively modified or eliminated, in some cases even before they are known to humans. As an example I would point to all those tropical forests that, by the start of the twenty-first century, were still regarded as being more or less intact, but had undergone significant modification, including the loss of species. As this representation makes clear, there may be a very broad range of 'sub-degrees' of modification as part of this category.

Degree 1: Humans attempt to replace a beyond-human formation, or a major part of that formation, with an assemblage that is structured for the purposes of more or less exclusively *human* use. The most obvious example is a city, but other examples are farms, human-made reservoirs, and some might now say the planet itself. Note however that even in such cases, beyond-human objects will continue to make unplanned, and in some cases disruptive or even destructive use of the new formation, as happens when predators attack domestic animals, pests damage crops, or indeed beyond-human organisms simply take up residence in the cultural formation, and/or take advantage of the foraging possibilities opened up by the new habitat, the 'new' body. So even here, a certain wilderness in the conventional sense of the term remains.

I will render this scheme more complex in a moment. First I should explain that with this scheme it is possible to recognise that some geographies may be rather less *hybridised* than others. At least in those cases, it would be wrong to exclude, *a priori*, a degree of separation from human activity – which is not to suggest that we humans are ever 'out' of nature, or that nature only exists 'over there'. Again, what is at issue is the extent to which an approach can recognise the existence of some places that are much less extensively and intensively modified by human activity than others.

The following are some ways of rendering this approach more complex. First, while I speak of 'degrees of wild(er)ness', more than 'states', there are ongoing dynamics. Those dynamics may mean that a tropical forest becomes less intact over time, or on the contrary, it may also undergo a process of rewilding. Of course, 'true' rewilding, as opposed to a purely instrumental intervention at the behest of environmentalists (romantic or not), occurs in all those places that are abandoned by humans, and/or where human occupation is destroyed by a cataclysm (natural or not), eventually enabling beyond-human bodies to reclaim them. Examples include the ancient Maya cities and temples that were claimed back by forests; at least some parts of the Nouabalé-Ndoki National Park in the Republic of Congo, which were once thought to be utterly unchanged by human intervention, modern or ancient;[60] Pompeii when it was buried by Mount Vesuvius in AD 79; or ironically, what has happened in Chernobyl, where species such as the grey wolf (*Canis lupus*), the racoon dog (*Nyctereutes procyonoides*), Eurasian boar (*Sus scrofa*) and the red fox (*Vulpes vulpes*) appear to have made a remarkable comeback.[61] (I would nonetheless note that there is a real danger of sublimation in many of the accounts of this last instance of 'rewilding'.)

Second, as I began to note above, there is bound to be significant variation *within* the 'degrees', with some formations that might otherwise be classified as entailing a similar degree of wild(er)ness having significantly more human intervention – and damage – than others. Conversely, a similar point might be made about 'rewilded' areas, some of which may be much further along a 'line' of succession – a process that might itself be transformed, more or less permanently, by human intervention.

It is also the case that, within one region, indeed *within* one ecosystem, there may be varying degrees of the kind just described (a case in point, the territory comprised by the Barro Colorado Nature Monument, cf. Volume 2), down to the level of the presence of, say, a relatively small patch of forest cleared for a radio-telemetry tower. Then again, and as shown by the remarkable long-term research conducted by William F. Laurance and other researchers about habitat fragmentation,[62] the transformation of tropical forests (and presumably other kinds of more intact ecosystems) entails dynamics that are not best described in the transcendental(ist) terms favoured by, say, Island Biogeography Theory.[63]

An analogous point might be made about the human body itself, which will exhibit varying degrees of humanising intervention, ranging from tattoos to the

---

[60] See Nils Lindahl Elliot, *Mediating Nature*, pp. 15-17.

[61] Sarah C. Webster, Michael E. Byrne, Stacey L. Lance, Cara N. Love, Thomas G. Hinton, Dmitry Shamovich, 'Where the Wild Things Are: Influence of Radiation on the Distribution of Four Mammalian Species Within the Chernobyl Exclusion Zone', Frontiers in Ecology and Environment <https://esajournals.onlinelibrary.wiley.com/doi/full/10.1002/fee.1227>[Accessed 1 July 2018].

[62] See for example, William F. Laurance, José C. Camargo, Philip M. Fearnside, Thomas E. Lovejoy, G. Bruce Williamson, Rita C. G. Mesquita, Christoph F. J. Meyer, Paul E. D. Bobrowiec and Susan G. W. Laurance, 'An Amazonian Rainforest and Its Fragments as a Laboratory of Global Change', *Biological Reviews*, 93:1 (2018), 223-247.

[63] See William F. Laurance, 'Beyond Island Biogeography Theory: Understanding Habitat Fragmentation in the Real World', in Jonathan B. Losos and Robert E. Ricklefs (Eds.) *The Theory of Island Biogeography Revisited* (Princeton: Princeton University Press, 2010), 214-236.

use of birth control or the core destruction generated by, say, a cup of polonium-210 served at an expensive hotel in London. Conversely, even in the most sanitised of bodies, beyond-human organisms not only can, but do and must take hold, and in so doing can prevent death but also take away lives; we return by this route to Cronon, but perhaps also to Latour's suggestion that we have never been modern.

It may be noted finally that modern cultures make the question of 'contact' a very complex one – not least, in so far as such cultures have for some time been able to affect at least all organic beyond-human formations on and above the surface of the planet (if not also those further down, or indeed those millions of miles away), from afar. An early example may be found in the spread of industrial pollution. A more recent phenomenon involves the spread of nuclear radiation. And of course, the most recent and dramatic change is that of anthropogenic climate change, which, amongst many other consequences, may now be affecting the growth rates of different kinds of plants in tropical forests[64] – and this despite the fact that the forests in question may be far removed from the centres of industrial activity that play the biggest roles in driving climate change. So contact is no longer exclusively a matter of *organismic* bodily co-presence. Merely by using a plastic bag and disposing of it improperly, I may well help to kill a pilot whale despite never even having seen one in flesh and blood;[65] countless other examples of the direct but unseen consequences of consumption might be invoked. So the threats posed by modern humans to organisms and assemblages beyond themselves (let alone *to* themselves) are now, and have for some time been subject to temporalities and spatialities that do not require one organismic body to be within physical perceptual distance to another to cause untold damage – the issue raised amongst others by the sociologist Ulrich Beck as part of his theory of qualitatively different kinds of environmental risks.[66]

Issues of these kinds notwithstanding, the point that I would like to reiterate here is that 'wild(er)ness' is something that may and *will* occur as part of *any* of the 'degrees' outlined above (hence stopping at 1, and not at 0 degrees, from the point of view of wilderness). According to the schema I have just outlined, wild(er)ness also occurs in a patch of garden overrun by 'weeds'; a coral reef formed on the wreck of a naval vessel sunk in World War II; or indeed, in the body of someone ravaged by a virus – and of course also someone who is 'perfectly healthy'. As these examples hopefully suggest, wild(er)ness is not best con-

[64] See for example the issues raised in relation to the growth of lianas by Geertje van der Heijden, Jennifer S. Powers, and Stefan A. Schnitzer, 'Lianas reduce carbon accumulation in tropical forests', *Proceedings of the National Academy of Sciences*, 112 (2015): 13267-13271.

[65] Agence France Press, 'Whale Dies From Eating More Than 80 Plastic Bags', *Guardian* online, 3 June 2018 <https://www.theguardian.com/environment/2018/jun/03/whale-dies-from-eating-more-than-80-plastic-bags>[Accessed 4 June 2018].

[66] Ulrich Beck, *Risk Society: Towards a New Modernity* (London: Sage, 1992).

ceived in the mostly all-or-nothing terms that continue to haunt both popular, or until recently even some everyday scientific discourses. Put more technically, it is not best conceived as a kind of unchanging (or unchanged) essence; the weeds in the garden grow there at least in part thanks to the changed circumstances, and of course the same is true for the coral reefs on the sunken ship, or the person that falls ill thanks to the stress generated by a social class position, or indeed thanks to partaking in tourism in a tropical forest. If there *is* an essence, it is the capacity to *become* – the subject of the following chapter.

14

Anthropomorphism and Becoming-Animal

In this, the last full chapter of Volume 1, I'd like to consider the question of animality in greater detail than I have thus far. At least *in situ* wildlife observation is all about observing wild animals; but what exactly do we mean when we refer to wild animals, and before that, to *animals, tout court*?

As the growing number of researchers devoted to Animal Studies[1] have pointed out, scholars in the critical social sciences and humanities have long adopted one of two equally problematic stances with respect to questions like these. The oldest stance, whose roots go at least as far back as Descartes, suggests that beyond-human animals may be defined by opposition to humans: animals are essentially different from humans because they lack mind and the capacity to produce language and culture. The hermeneutic philosopher Martin Heidegger goes so far as to suggest that even the animals' bodies are essentially different. As Heidegger puts it in his famous 'Letter on "Humanism"', published in 1946, 'The human *body* is something *essentially other* than an animal organism'.[2] So it's not just that humans can speak and are aware of the meaning of death; they have limbs, in particular hands, that can engage in actions that are intrinsically different from those performed even by other creatures with a broadly similar morphology (e.g. the great apes).

The second, more recent stance suppresses explicit references to the hierarchy between humans and beyond-humans, but effectively maintains that hierarchy by assuming firstly that symbolic forms are human, if not exclusively human, constructs; and secondly, by ignoring beyond-human animals even when the subject of analysis would suggest the need to acknowledge their vital presence. So, for example, the new cultural geographers were very keen on the analysis of land-

---

[1] See for example Mary Midgley, *Animals and Why They Matter: A Journey Around the Species Barrier* (Harmondworth: Penguin, 1983); Peter Atterton and Matthew Calarco (Eds.) *Animal Philosophy: Ethics and Identity* (London: Continuum, 2004); and Linda Kalof and Amy Fitzgerald (Eds.) *The Animals Reader: The Essential Classic and Contemporary Writings* (Oxford: Berg, 2007). For the 'animal turn' that occurred in human geography in the late 1990s, see the introductions by Jennifer Wolch and Jody Emel (Eds.), *Animal Geographies: Place, Politics and Identity in the Nature-Culture Borderlands* (London: Verso, 1998); and Chris Philo & Chris Wilbert (Eds.) *Animal Spaces, Beastly Places: New Geographies of Human-Animal Relations* (London: Routledge, 2000). For the much broader interest in, and rise of an academic industry ('Animal Studies'), see the useful overview by Cary Wolfe, 'Human, All Too Human: "Animal Studies" and the Humanities', *PMLA*, 124:2(March 2009), 564–575.

[2] Martin Heidegger, 'Letter on "Humanism"', in William McNeill (Ed.) *Pathmarks* (Cambridge: Cambridge University Press, 1995[1946]), p. 267. Italics added to the original.

scapes, but landscapes were treated as being entirely symbolic in nature, and if they had any beyond-human animals in them, they too, were treated as arbitrary, human constructs.

The most recent forms of scholarship, such as ANT and hybrid geographies, go a long ways in addressing the inequities and silences of the aforementioned research. However, as I began to explain in Chapter 13, a combination of nominalism, a lingering focus on entirely human social institutions, and insufficiently nuanced conceptions of hybridity often mean that beyond-human animals continue to be incorporated in human networks by the analysts themselves. When the emphasis is entirely on how natures and their creatures have become hybridised, it becomes difficult to talk about that which is not best described as being 'hybrid'.

In this chapter I will respond to these tendencies as follows. First, I will begin by offering a brief historical contextualisation of the predominant scientific discourse with respect to the nature of animality. This discourse anthropomorphicises animals as machines, and in so doing transforms them into something which, on some level, is quite dead. As the last point begins to suggest, anthropomorphism plays a key role in modern discourses about animals, and so it will be necessary to consider the concept of anthropomorphism in its own right, and this will the second aspect of the chapter. Can humans ever conceive, and indeed relate to beyond-human animals in anything other than anthropomorphic ways?

In the final part of the chapter, I will consider three approaches that not only contest the discourse I have just referred to, but offer provocative alternatives. The approaches are those of the biologist and proto-biosemiotician Jakob von Uexküll; the philosophers Gilles Deleuze and Félix Guattari; and finally the philosopher and STS scholar Donna Haraway.

Let's begin, then, with a brief historical contextualisation. In Chapter 10 I discussed René Descartes' *cogito*. Nestled in Descartes' metaphysics one finds a way of referring to beyond-human animals that establishes an important precedent for modern discourses about beyond-human animals. As far as Descartes is concerned, such animals are, in effect, mindless automatons – though as I will explain later, Descartes qualifies this characterisation in important, and indeed remarkably critical, ways.

What is perhaps Descartes' most famous use of the machine metaphor occurs in the fifth part of the *Discourse on Method*.[3] There Descartes refers to a work that he meant to publish (parts of which survive as an unpublished manuscript, *The Universe*), in which he planned to approach animals as machines. According to Descartes, he aimed to show what structure the nerves and muscles the human body must have to enable the 'animal spirits' to move the body's members. As evidence for such spirits, or what he also describes as a 'corporeal soul', Descartes notes that 'in the case of severed heads ... we can see moving and biting the earth

---

[3] René Descartes, *A Discourse on the Method*, translated by Ian Maclean (Oxford: Oxford University Press, 2006).

shortly after having been cut off, although they are no longer animate.'[4] These and various other aspects of morphology and ethology which Descartes planned to consider (and which he *did* end up considering in the *Discourse on Method*), reveal animals' capacity to engage in actions that are not necessary directed by the will.

> This will not appear at all strange to those who know how wide a range of different automata or moving machines the skill of man can make using only very few parts, in comparison to the great number of bones, muscles, nerves, arteries, veins, and all the other parts which are in the body of every animal. For they will consider this body as a machine which, having been made by the hand of God, is incomparably better ordered and has in itself more amazing movements than any that can be created by men.[5]

So animals may be conceived as machines, albeit with two important provisos: the first is that the nonhuman animal machines 'would never be able to use words or other signs by composing them as we do to declare our thoughts to others'; this even though it is possible, Descartes notes, to 'conceive of a machine made in such a way that it emits words, and even utters them about bodily actions which bring about some corresponding change in its organs (if, for example, we touch it on a given spot, it will ask what we want of it; or if we touch it somewhere else, it will cry out that we are hurting it, and so on)'.[6] However, such a machine would not be able to 'put these words in different orders to correspond to the meaning of things said in its presence, as even the most dull-witted of men can do'.[7]

The second proviso is that, while such machines might do many things as well or even better than humans, they would not be able to act consciously. The reason for this is that the nonhuman animal machines' organs are 'disposed in a certain way' – they are set in a kind of *spatial* aspic which renders them unable to react freely in the way in which human reason can, and which is the result of the fact that reason is mind, i.e. *res cogitans*, as opposed to *res extensa*.[8]

Before leaving Descartes, it might be noted that this position is a rather more reflexive one than some have suggested (Descartes is often treated as the philosophical bogey by animal rights activists). Try as he might, Descartes cannot find any evidence in the bodies of beyond-human animals for anything like true mind; but then he acknowledges that he cannot find that in human body parts either. This continuity, he suggests, can be explained by the fact that humans

------

[4] René Descartes, *A Discourse on the Method*, p. 45.

[5] René Descartes, *A Discourse on the Method*, p. 46.

[6] René Descartes, *A Discourse on the Method*, p. 46.

[7] René Descartes, *A Discourse on the Method*, p. 46.

[8] René Descartes, *A Discourse on the Method*, p. 47.

receive their capacity to think directly from God, and as I just noted, in any case mind is not a matter of extension. But if mind really *isn't* a matter of extension, how do we know that God hasn't given animals mind?

In a passage of a letter sent in 1646 to the Marquess of Newcastle, Descartes reveals the methodical rigour for which he is famous when he acknowledges that the issue is perhaps not entirely settled: 'The most that one can say is that though the animals do not perform any action which shows us that they think, still, since the organs of their body are not very different from ours, it may be conjectured that there is attached to those organs some thoughts such as we experience in ourselves, but of a very much less perfect kind.'[9] To be sure, as I noted above, Descartes believes that many of the activities that we humans engage in are themselves driven by that 'corporeal soul'; so, barring that God-given difference to do with mind (*res extensa* vs. *res cogitans*), Descartes actually does acknowledge significant continuities between humans and at least some kinds of beyond-human animals. Some but not all: as Descartes puts it, 'if they [animals] thought as we do, they would have an immortal soul like us. This is unlikely, because there is no reason to believe it of some animals without believing it of all, and many of them such as oysters and sponges are too imperfect for this to be credible...'[10]

Today it is still possible to find variations of this discourse in scientific discourse, and this is particularly evident in the continuing dominance of the metaphor of mechanism in biology, if not in the physical sciences more generally. This is not a simple metaphor, or one that is advertised up front by biologists. Instead, it is to be found in one of three kinds of understandings, which, as described by Daniel Nicholson, entail one or indeed a combination of the following: 1) mechanism as 'mechanicism', the philosophical thesis that conceptualises living organisms as 'machines that can be completely explained in terms of the structure and interactions of their component parts'; 2) mechanism as 'machine-mechanism', which refers to the internal workings of a machine-like structure; or 3) mechanism as 'causal mechanism', which is a 'step-by-step explanation of the mode of operation of a causal process that gives rise to a phenomenon of interest'.[11]

In the extended quote that I offered in Chapter 3 of Norbert Wiener's account of the synapse, we may find a particularly good example of machine-mechanism: *'The synapse is nothing but a mechanism for determining whether a certain combination of outputs from other selected elements will or will not act as an adequate stimulus for the discharge of the next element, and must have its precise analogue in the computing ma-*

---

[9] René Descartes, Letter to the Marquess of Newcastle, 23 November 1646, in *Descartes: Philosophical Letters*, edited and translated by Anthony Kenny (Oxford: Clarendon, 1970).

[10] René Descartes, Letter to the Marquess of Newcastle.

[11] Daniel J. Nicholson, 'The Concept of Mechanism in Biology', *Studies in History and Philosophy of Biological and Biomedical Sciences,* 43(2012), 152-163, p. 153.

chine. The problem of interpreting the nature and varieties of memory in the animal has its parallel in the problem of constructing artificial memories for the machine'.[12] In that same chapter I considered David Marr's account of visual perception, and this offers a good example of causal mechanism in so far as he characterises the step-by-step process by which the retina and central nervous system encode visual information (cf. Chapter 3). Finally, in Chapter 2 I considered Edward O. Wilson's account of biophilia, and with it Wilson's sociobiological principles. In these we find further examples not just of these forms of mechanism, but of mechani*cism* more generally. In *Consilience*, Wilson's defence of evolutionary psychology, he describes the human brain as follows:

> The brain's true meaning is hidden in its microscopic detail. Its fluffy mass is an intricately wired system of about a hundred billion nerve cells, each a few millionths of a meter wide and connected to other nerve cells by hundreds or thousands of endings. If we could shrink ourselves to the size of a bacterium and explore the brain's interior on foot ... we might eventually succeed in mapping all the nerve cells and tracking all of the electrical circuits. But we could never thereby understand the whole. Far more information is needed. We need to know what the electric patterns mean, and well as how the circuits were put together and, most puzzling of all, for what purpose.[13]

The last three sentences may seem to contradict mechanism, but actually, what Wilson is suggesting is that one day it *will* be possible to explain all of the structure and interactions of the component parts of this machine of machines. To be sure, it may be recalled that, as far as Wilson is concerned, the semiotic aspects of cerebral activities are themselves machine-like 'memes' which are governed by 'epigenetic' rules on a biological 'leash' (cf. Chapter 2).

While their accounts refer specifically to the *human* nervous system and brain, Wilson, Wiener, and Marr (and indeed Descartes, had he had the investigative means) would agree that much the same structure occurs in the brains of other animals. From their perspective, there is little or nothing anthropomorphic about this kind of claim, or the representations I have just considered. However, I would now like to explain why such representations *are* anthropomorphic through and through, and what consequences this has for our understanding of animality (and indeed, any other beyond-human phenomenon one might care to think of). In order to do so, it will be necessary to re-conceptualise anthropomorphism itself.

Now in everyday discourse, the concept of anthropomorphism appears to have a straightforward meaning: it refers to the attribution of human characteristics or behaviour to a nonhuman entity, be it a beyond-human animal, a god, a

---

[12] Norbert Wiener, *Cybernetics: Or Control and Communication in the Animal and the Machine*, 2nd edition (Cambridge, MA: M.I.T. Press, 1961 [1948]) p. 14. Italics added to the original.

[13] Edward O. Wilson, *Consilience: The Unity of Knowledge* (London: Abacus, 1998), p. 106.

machine, or any other body. In the physical sciences, anthropomorphism, thus understood, is one of the cardinal sins, a 'bias' to be avoided by way of critical ways of researching beyond-human animals, and the beyond-human world more generally.

Ironically, even as scientists offer compelling evidence that the difference between human and nonhuman animals (as they would call them) is by no means as great as 'humanists' would have us believe, this conceptualisation of anthropomorphism rests on a very firm human/nonhuman boundary. Presumably this contradiction may be navigated with recourse to the idea that, while human and beyond-human animals may share bodily features, humans bring to the equation thoughts, and with them representations that must be critically filtered in order to prevent representations compromised by particular social or cultural 'values'. It is difficult to see how this is not a return to a version of the mind-body dualism.

Whether or not that is the case, it would appear that the conviction that it *is* possible to avoid anthropomorphism rests on what I described in Chapter 1 as the positivist rules of phenomenalism and objectivism. If scientists are to engage in objective research vis-a-vis nonhuman organisms (and objects more generally), then they must not foist onto them any values. One way of achieving this is by engaging in rigorously inductive observational procedures. Another is to engage in equally rigorously *deductive* procedures based on adequately tested knowledge foundations (foundationalism). If, for example, someone were to say that red-crowned cranes (*Grus japonensis*) doing a mating dance are expressing *love* for one another, then this would have to be regarded as a category error; it is only humans, and indeed perhaps only modern humans, who at least profess to love one another. The same would be true of someone suggesting that a male red-capped manikin (*Ceratopipra mentalis*) engaging in its courtship movements is doing a 'dance' to 'show off'. What is at stake, a sociobiologist might argue, is not love, art, or vanity, but a purely innate behaviour which has a genetically-determined function: for the male and female to reaffirm the sexual bond, to gauge each other's potential to pass on genes, etc.

From a hermeneutic perspective, the naiveté of such a positivist conception of anthropomorphism lies in its advocates' conviction that it really is possible to conceptualise the nonhuman world in wholly, or even 'mainly' nonhuman ways (or at least, ways not vitiated by a human cultural subjectivity). Explicitly or implicitly, this is one of the reasons why modern science has such faith in quantification: by assigning a beyond-human animal – or its actions – a number, and by coding the actions in ways that themselves may be tabulated and statistically manipulated, value-laden words may be avoided in favour of letting the 'facts speak for themselves'. This is the kind of logic that underlies the notion that a species can be described by specifying, for example, a range of dimensions or weights. But in a somewhat different way, it is also the logic behind the binomial system of classification, which has been used since the time of Carl von Linné to identify species in a manner that is itself thought to be entirely objective.

In fact, the description of species by way of the binomial system, or indeed by way of measurements of one or another kind, is by no means beyond the vicissitudes of representation or, to put the matter more generally, *semeiosis*. In Part 2, I explained that any and all signs always entail at least a degree of partiality, a degree of aspectivalism; in so far as this is the case, then a degree of valorisation must inevitably creep into even the most rigorous of observational procedures if only because any one observation, or indeed any set of observations, always entails a choice to represent things in one way, 'from one perspective', and not another.

From the point of view of cultural theory (including cultural sociology, or cultural geography), the concept of *legein* offers another way of demystifying objectivism. As I began to explain in the conclusions to Part 3, for the philosopher and sociologist Cornelius Castoriadis, *legein* entails an identitary and ensemblist logic of sets.[14] In so far as the features of an animal can be quantified, and in so far as the repeated quantification of specimens thought to 'belong' to that species yields a range, then species can be characterised with reference to a mathematical logic: let species A be made up of the set of specimens with features x, y, z... What Castoriadis makes clear is that what may seem to be an utterly objective logic based on 'pure' observation, i.e. measurement, is actually a *cultural* operation. Even pure mathematics, removed from any biological, chemical or physical application – *mathesis* 'itself' – cannot escape *legein*, as Castoriadis illustrates via his critique of the logic of sets. Sets, he notes, presuppose their own positing; they can be constituted only by presupposing that they have already been constituted. The key principle is that a set is a set because *it is defined* as a set: x is part of set A because A is {x, y, z}. Far from being based on a purely indexical or iconic order of reference, *mathesis* is thus itself a matter of an *identitary* logic (let x be...), and so is not just a matter of what I described in Part 2 as thirdness, but of precisely the kind of 'conventional', which is to say *symbolic* thirdness whose influence positivist philosophy attempts either to deny outright, or at least to 'keep in its place'. From this perspective, one way of interpreting the use of numbers to explain the morphology or ethology of beyond-human animals is in terms of a technique that works, however inadvertently, to conceal the ineluctably anthropomorphic quality of *any* identification, any *explanation* whatsoever regarding the beyond-human.

If this analysis is valid, the issue is not whether some account is anthropomorphic or not; *anthropomorphism is unavoidable*. Anthropomorphism results automatically from the use of any symbols in so far as those symbols can be said to be humanly produced. This approach echoes the thought of Charles Sanders Peirce when he suggests that '"Anthropomorphic" is what pretty much all conceptions

---

14 Cornelius Castoriadis, *The Imaginary Institution of Society*, translated by Kathleen Blamey (Cambridge: Polity Press, 1987[1975]). I also refer to this part of Castoriadis' work in Nils Lindahl Elliot *Mediating Nature*, p. 69.

are at bottom; otherwise other roots for the words ... would have to be found'.[15] This perspective might be taken further by suggesting that anthropomorphism may occur even in those cases where no symbolic representation is employed; later in this chapter I will present the biologist Jakob von Uexküll's concept of *umwelt*, which one might use to suggest that every species perceives its surrounding world in ways that are filtered by particular affections. If this account is valid, then it might be argued that anthropomorphism occurs already on the level of perception and the perceptual systems.

Above I suggested that Peirce regards all human semeiosis as being anthropomorphic. But Peirce would nonetheless argue that this is emphatically *not* to say that all anthropomorphisms are equally valid when it comes to theorising the nature of one or another creature, one or another ecological relation or process. Key to this anti-relativist stance is phenomenological secondness, and the logic of indexes. To return to the example of the weathervane (cf. Chapters 7, 13), there is arguably nothing in the wind that resembles the letters N, S, E, or W; the symbols of wind direction on a weathervane are clearly that – wholly, indeed we might say *expressly* anthropomorphic symbols. But the wind *will* cause the weathervane to swing in a certain direction, and this action is, in itself, hardly cultural or conventional in any culturalist sense of the expression. Moreover, assuming that a weathervane has been set up with the aid of an accurate compass, such that 'N' really does point towards the magnetic North, then one cannot say that the 'N' in that weathervane is arbitrary in the way that, say, a child pointing north in whatever direction it pleases is.

To make the same point more generally, indexical signs (or relations very much like them) may and indeed do inform scientific practice, just as they inform the workings of the cosmos itself. One way of understanding the nature of modern science is precisely in terms of a venture that tries to explain the natural world by 'preserving' the indexicality of indexes, i.e. trying to keep them free of symbolic 'contamination'. Such a venture is ultimately an impossible one if only because, as noted in Part 2, indexes do not occur in isolation from other kinds of signs. But the last point should not lead the analyst to lurch back to a constructivist stance that denies indexicality, *tout court*; clearly some ways of recognising, recording, and explaining indexes are better than others – and this not least because the bodies and the events they are associated with do have a certain agency of their own when it comes to producing signs. This being true, they may also have a certain capacity to determine at least aspects of the signs that *humans* themselves produce. Returning to the example with which I opened Part 3 (a man observing a crocodile along the shores of Barro Colorado Island, and interpreting it with reference to TV images of a different species of crocodile in the Serengeti), the fact that someone observes beyond-human animals in a manner that evokes the

---

[15] Charles Sanders Peirce, *CP* 5.47. (For Peircian referencing conventions, see footnote 16, Chapter 5.)

meanings, valorisations, and indeed bodily situations associated with one or another cultural practice (or indeed the human perceptual system) does not simply negate (1) the existence of forms which are not human and (2) the capacity of the human physiological-social-cultural apparatus to perceive at least aspects of those forms in ways that are largely, if not entirely continuous with the original bodies themselves. Put differently, recognising that *legein* does occur – there really *is* the operation of 'distinguish-choose-posit-assemble-count-speak' (or a version of this sequence) – of course does not contradict that dynamical objects, or what I also described in Part 3 as dynamical *bodies* may really be 'there', and may have a say, or at any rate contribute to a certain form to distinguishing, choosing, positing, assembling, counting and speaking. As I noted earlier, by 2018 the primatologist Katharine Milton counted 23 spider monkeys on Barro Colorado; even if one agrees that '23' is little or nothing like 23 spider monkeys, or that the words 'spider monkeys' are little or nothing like the real spider monkeys, unless Milton made a mistake, there really were 'that many' 'creatures' of 'that sort' on 'Barro Colorado Island'; by virtue of their very presence, their very existence, they directly affected the number that Milton arrived at. The use of the Arabic numeral system clearly has consequences for how we may represent the 'twenty-threeness' of the monkeys (as I noted with David Marr in Chapter 3, the choice of the numeral system has consequences for the operations one can engage). But at the risk of stating the obvious, provided that the observer has counted accurately, there is a certain continuity between the number 23, and the number of bodies.

I will return to this issue later in this chapter, when I reconsider the principle of immanence. For now I would like to expand the theory of anthropomorphism by noting that anthropomorphism often goes hand in hand with *cosmomorphism*[16] (also known as 'zoomorphism'), and indeed that the two may work together to produce what I describe as *circuits of anthropomorphism and cosmomorphism*.[17] This notion provides a way of explaining how actors may use anthropomorphisms (however inadvertently or unselfconsciously) to incorporate beyond-human animals and other beyond-human objects in cultural ensembles, or what I described in Part 3 as cultural formations.

As it applies to beyond-human animals, the circuit in question can be said to entail a twofold process. Let's begin with the process of anthropomorphism: a beyond-human animal is represented in a manner that, consciously or more likely unselfconsciously, projects onto it humanising meanings, which is really to say the meanings associated with one or another coding orientation (I will exemplify this process in a moment). In practice and in time, the meanings, and any relations tied to them may come to seem so natural as to be perfectly identified with

---

[16] A term that I borrow from Edgar Morin. See for example Edgar Morin, *The Cinema or the Imaginary Man*, translated by Lorraine Mortimer (Minneapolis: University of Minnesota Press, 2005 [1956]).

[17] Nils Lindahl Elliot, *Mediating Nature*, pp. 42-43.

the animal itself. If so, the agents of anthropomorphism may not regard their own understandings of the animal, or their relations to it, as being anthropomorphic at all: there will be, or seem to be, a kind of entelechy, in the manner defined by Peirce (cf. Chapter 7): object, ground and interpretant seem to be so perfectly identified that there appears to be no sign at all, simply an object presenting itself, or presenting itself via utterly rhematic signs. This may, and indeed is likely to have consequences for cultural practice in so far as an animal, ecosystem or any other beyond-human body comes to be incorporated into the kind of networks theorised by Latour and Whatmore. So, for example, given that great white sharks are, after all, 'man-eaters', it makes good sense from the point of view of human self-preservation if they are pursued to the point of extinction. Or, to offer a more subtle example, given that all wild animals have lives riven by 'natural economies', then it makes perfect sense to suggest that amongst, say, mongooses (*Herpestidae*) 'delayed development of the young occurs in solitary species because they cannot *cover* the energetic *costs* of rapid development in their young'.[18] In this example, as in countless other instances of biological research, it is possible to note how a modern economic discourse pervades the biological explanation.[19] In each example, it is not that there is not some element of the represented, or imagined animal that cannot be used to justify the anthropomorphism; on the contrary, anthropomorphisms typically start from some form, some sign produced by the creature itself. The problem occurs in so far as the scientist is unaware that a certain discourse (in this case an economic one, that structures modern understandings of countless fields), is silently structuring the entire *ecological* account. Ironically we return by this example to the original motivations for opposing anthropomorphism.

Now if anthropomorphism is the outward leg of the circuit, cosmomorphism, the second aspect of the process, is the return leg of the dynamic. In so far as an anthroporphism, or a pattern of anthropomorphism, *is* normalised or naturalised, and in so far as an anthropomorphicised animal (or some aspect of it) seems to express, wholly naturally, some aspect of the human self (the individual human self, the social self, the cultural self, etc.), then that animal may be employed as an ostensibly, and indeed *ostensively* natural symbol. Cosmomorphism occurs when a beyond-human animal (but more generally, a beyond-human body or process, which is why I prefer cosmomorphism to zoomorphism), that has been anthro-

---

[18] T. H. Clutton-Brock, A. Maccoll, P. Chadwick, D. Gaynor, R. Kansky, and J.D. Skinner, 'Reproduction and Survival of Suricates (*Suricata suricatta*) in the Southern Kalahari', *Afr. J. Ecol.*, 37 (1999), 69–80, p. 78. Italics added to the original.

[19] During a presentation about meerkats *(Suricata suricatta)* that I attended at the department of biological sciences of a leading British university, I raised this issue with the presenter, who suggested quite matter-of-factly that there was no metaphor at all: meerkats not only incurred certain costs by engaging in certain actions, but in particular circumstances had to choose between x costs and y benefits...

pomorphicised, is employed to *naturalise* what are themselves at least partly constructed aspects of the self, or of the own practices.

A somewhat obvious but still useful example of cosmomorphism involves the use of animals as heraldic symbols. A more tacit example occurs when a tropical forest, anthropomorphicised as a place characterised by its ruthlessness and lawlessness, is used to describe an economic or social system, as in references to 'jungle capitalism', or to 'the law of the jungle'. Strictly speaking, there is nothing capitalist about tropical forests, and of course jungles are not necessarily 'lawless' places. However, over time such terms have come to be associated according to a kind of popular common sense.

A similar kind of analysis might be offered with respect to a variety of animals that loom large in the popular imaginaries of nature: not least, my earlier example of the great white shark (or sharks more generally), a taxon whose ostensive 'ruthlessness' has long been used to characterise sociopaths (e.g. the 'loan shark'). I've used negative examples, but of course a similar circuit can and does apply to all those cases where certain species are singled out for conservation, and/or become particularly popular – viz., the 'charismatic' fauna. And indeed, on the basis of my account of anthropomorphism, it is possible to specify what precisely I mean by the concept of charisma when it is applied to wild animals.

It should be acknowledged that merely to speak of the 'charisma' of wild animals entails anthropomorphism in so far as charisma has to do with a particular kind of relation of authority amongst *humans*, viz., the type of authority that Max Weber describes as 'resting on devotion to the exceptional sanctity, heroism or exemplary character of an individual person, and of the normative patterns or order revealed or ordained by him'.[20] The last point notwithstanding, if it is useful to refer to the charisma of beyond-human animals, it is precisely because the concept enables the analyst to explain how certain creatures are anthropomorphicised in ways that entail acquiring a following 'of their own', indeed a certain devotion amongst at least some groups. Here again, an ideal-realist approach like Peirce's is desirable to avoid the pitfalls of both biologism (e.g. biophilia) and culturalism (e.g. cultural constructivism). Starting from certain qualities found in the represented animal (what with Peirce we might describe as *qualisigns*, cf. Part 2), the groups in question begin to focus on the creature, and to do so in ways that categorise those qualities according to particular patterns of classification, and relate to them according to particular patterns of framing. Put more technically, the animal comes to be encoded, or the subject of a particular coding orientation (cf. Chapter 8). Of course, a similar thing might be said about *all* coded human interactions with beyond-human animals; what distinguishes

---

[20] Max Weber, *Economy and Society: An Outline of Interpretive Sociology*, edited by Guenther Roth and Claus Wittich, translated by Ephraim Fischoff, Hans Gerth, A. M. Henderson, Ferdinand Kolegar, C. Wright Mills, Talcott Parsons, Max Rheinstein, Guenther Roth, Edward Shils, and Claus Wittich (Berkeley: University of California Press, 1968), p. 215.

the charismatic relation is that those who attribute or project charisma onto an animal do so in ways that involve a more or less explicit element of *attraction*, and with it a certain *ethic*. The two elements, attraction and ethic taken together, produce an *aesthetic*, viz., a suggestion that this or that species is attractive for reasons that implicitly or explicitly establish a certain valorisation, and with it a tacit or overt sense of the good.

For any social period and group, it is always possible to point to certain species that are accorded a charismatic salience. For example, from the 1960s onwards, Common Bottlenose Dolphins (*Tursiops truncatus*) were represented as being exceptionally intelligent, helpful and above all *friendly* creatures – a reading that was no doubt facilitated by the dolphins' 'smiles', but also by the famous TV series (*Flipper*) produced between 1964 and 1967, narrative elements of which were then reproduced via the staged shows produced with actual dolphins captured in the wild and then used in captivity in numerous marine parks across the world.

Or to refer to another example, from the late 1980s, but especially during the 1990s and early 2000s, a type of mongoose (hephestid) known in the English and South African Dutch languages as the Meerkat (*Suricata suricatta*) acquired a charisma of its own, again thanks in no small part to TV programmes. In 1987, a BBC programme titled *Meerkats United* observed meerkats as if they were, in effect, cuddly agrarian socialists capable quite literally of 'standing up for themselves'. Again, at least a part of this interpretation would have started from actual elements of the species' morphology and ethology, in this case, for example, a capacity to stand upright, and no doubt also the so-called altruistic behaviours exhibited by this remarkably social species.

It may, however, be noted that another less popular programme produced by the BBC about a decade later (*Meerkats Divided*, 1996) portrayed the same species as members of urban gangs battling it out in a 'West Side'-style Kalahari – a representation that was not entirely misleading in so far as the members of the species are known to kill the offspring of other individuals to advance their own social positions (that is, at any rate, one interpretation of the practice). Years later, the meerkats were turned into caricatures of themselves when, beginning 2009, the advertising agency VCCP used CGI meerkats (styled after a Russian oligarch, his family, and friends) to promote on US and British commercial television a price comparison website owned by the BGL group. The choice of meerkats deliberately blurred the proximity of 'meerkat' and 'market' (*comparethemeerkat/ comparethemarket*) with what had, by then, become an instantly recognisable, and as I have already explained, charismatic species. Despite painting rather less alluring pictures of the species, both *Meerkats Divided* and the VCCP meerkats endowed the animals with a certain *magnetism*, in much the way that otherwise morally reprehensible characters in overtly human dramas can themselves be made attractive.

And indeed, the notion of charisma as I use it may be employed to describe creatures whose attraction resides as much in so-called 'negative' attributes as in 'positive' attributes. In much the same manner that audiences may still be drawn to the villains in films or in theatre, the charisma of a beyond-human animal will be a function not just of a conventionally moralistic aesthetic, but of a narrative process that positions actors, narrators and narratees in such a way as to encourage audiences to engage in a certain circuit of anthropomorphism and cosmomorphism. Put more simply, a story is told that positions not only the wildlife, but also the audiences, to *follow the plot*, and by extension, its characters, according to certain coding orientations. This is also the kind of following that I had in mind when I described the charismatic fauna as 'having certain following'.

The pragmatic in question has the capacity to turn even potentially dangerous species into objects of desire, e.g. objects that are worth observing not just via the media, but *in situ*. This suggestion offers a way of connecting media, anthropomorphism and wildlife observation considered as a tourist practice. As I noted at the beginning of the volume, wildlife observation as a form of tourism (if not more generally) tends to be closely related to the mediazation of wildlife, and such mediazation has a capacity to generate or mobilise certain desires (as per Deleuze and Guattari's account of desiring-production, cf. Chapter 10). Both aspects point not only to the importance of anthropomorphism regarded as a humanising process, but to the key role that may be played by cultural formations – assemblages of institutions, discourses, genres, techniques and technologies – in that process. If certain species acquire the status of 'charismatic fauna', it is typically because cultural *formations* encourage audiences, or other mass agglomerations to perceive, conceive and interpret them in certain ways.

It may seem that I have strayed a long way from the questions with which this chapter began regarding the nature of animality, and of wild animals in particular. In fact, the approach I've outlined identifies one first very substantial issue when it comes to trying to answer what is a wild animal, and before that, what is an animal: even before one can begin to explain what an animal or a wild animal is, one must acknowledge that anthropomorphism and cosmomorphism are likely to play a major, indeed constitutive role in the explanation. The notion of a *wild* animal, or even of an animal, *tout court*, is from this kind of perspective already an anthropomorphism in the making, and this for reasons that have to do not just with a tacit, and wholly modern opposition – the wild and the domestic, a variant of nature and culture – but with the anthropomorphic character of *any* symbolic form in its own right. No animal is an a-n-i-m-a-l.

From this perspective, we can say that wildlife observation, in tropical forests or in any other biome, entails what is in some respects a human kind of fiction. This might seem like a pleonasm – is not all fiction human? Whatever the answer to that question, I am actually referring to the word's Latin etymology, *fictio*, i.e. entailing a certain 'constructedness', a certain form or contrivance. A wild animal, or any beyond-human animal, is at least partly a fiction generated via the kinds of narratives, and more generally the *discourses* circulated by the BBC or any other modern institution, and so one must begin by questioning the forms, constructions, and contrivances of such discourses.

The last point notwithstanding, as I have already noted the risk with such a stance is that if one takes it too far, then one arrives at the nominalist conclusion that representations of beyond-human animals are, as a rule, *no more* than modern fictions, which is at least as problematic as saying that an animal is no more (and no less) than the kind of animal that might be measurably revealed by way of a mechanistic scientific analysis. So we need to pursue this matter from a different perspective – one that moves away from representation and discourse to make more room for the beyond-human animals themselves. This being the case, I would now like to turn to three approaches that offer promising alternatives to the mechanism of mainstream biological science, and the culturalism of both cultural sociology and geography at least in their earlier variants.

The first that I will consider is the work of a scholar whom I've mentioned in different parts of this volume, and whom I have described as a biologist, and as a 'proto-biosemiotician': Jakob von Uexküll (1864-1944). It is pertinent to begin by underscoring the fact that Uexküll was a practicing scientist well-versed in the empirical methods of scientific research. During the first half of the twentieth century, he nevertheless developed a biological theory that rejected the mechanism of mainstream biology, and with it the derivative status that many continue to accord biology to this day. By derivative status, I mean the idea that physics and chemistry provide the first principles, or a kind of intellectual scaffolding for biology. As Uexküll puts it in the preface to *Theoretical Biology*, 'the old scaffolding, borrowed from chemistry and physics, will suffice no longer. For chemistry and physics do not recognise conformity with plan in Nature. Biology, however, consists in the setting up of a scaffolding of doctrine that takes account of this conformity as the basis of life'.[21]

It should be clarified that the plan that Uexküll refers to is not that of a teleological conception of evolution, but instead a functionalist and vitalist perspective on the nature of each organism. Each organism behaves according to the designs, if one can use the term, of vital semiotic cycles (amongst other works, Uexküll published a work titled *A Theory of Meaning*[22] that concerned itself with what today we would describe as biosemiotics). Such cycles act as a plan that shapes the responses of each organism for a given activity, for a given context. The most famous of Uexküll's examples for this conceptualisation is that of the functional cycles of a tick's (*Ioxidae*) feeding process, which Uexküll describes in detail at the

---

[21] Jakob von Uexküll, *Theoretical Biology*, translated by D. L. MacKinnon (New York: Harcourt Brace, 1926), p. xi

[22] Jakob von Uexküll, 'A Theory of Meaning', in *A Foray Into the Worlds of Animals and Humans* with *A Theory of Meaning*, translated by Josep D. O'Neil (Minneapolis: University of Minnesota Press, 2010 [1934 & 1940, respectively]).

start of what is perhaps his most popular work, *A Foray Into the Worlds of Animals and Humans*.[23] (It is this account that Deleuze refers to as part of his conceptualisation of affect, cf. Chapter 10).

According to Uexküll, the female tick's feeding habits entail just three functional cycles: first, for a tick hanging from the branch of a shrub, the passing of a mammal secreting the odour generated by butyric acid acts as a stimulus for the tick to let go. Once it falls on the mammal, and second, the collision with the hair acts as a tactile stimulus for the tick to start running about, looking for a place to drill into the epidermis of the host. The third cycle kicks in when the greater warmth of a bald patch of skin allows the tick to get past the mat of hair, and to attach itself to begin feeding.

As Uexküll explains, 'This is no doubt a case of three reflexes, each of which is replaced by the next and which are activated by objectively identifiable physical or chemical elements'; however, anyone who stays on that level of analysis 'proves that he [sic] has not seen the real problem at all. It is not a question of the chemical stimulus of the butyric acid any more than it is of the mechanical stimulus (the hair) or of the thermal stimulus (the greater warmth of bald skin). It is only a question of the fact that, among the hundreds of effects that emanate from the mammal's body, only three become feature carriers for the tick. Why these three and no others?'[24]

The kind of answer that Uexküll offers involves at least two theoretical displacements vis-a-vis the mechanism and also the behaviour*ism* that dominates biological approaches in Uexküll's time, if not in our own. First, the reference above to 'feature *carriers*' is a hint that Uexküll treats the stimuli as signs – what he calls perceptive signs, and which he contrasts to effect signs. Given my preference for Peirce, I will not consider Uexküll's semiotic theory in this study. I would, however, like to at least point out its neo-Kantian qualities, and one example which stands in a stark opposition to some of what is argued by James Gibson: in *A Theory of Meaning*, Uexküll explains that when someone throws a stone at an angry dog barking, 'nobody who observed what happened and picked up the stone afterward would doubt that this was the same object, "stone," which initially lay in the street and was then thrown at the dog'; [...] '[n]either the shape, nor the weight, nor the other physical and chemical properties of the stone have changed ... and yet it has undergone a fundamental transformation: it has changed its *meaning*.'[25] So long as the stone is no more (and no less) than a part of the country road, '[i]ts meaning was in its participation in the function of the path. It had, we could say, a "path tone"; upon being thrown at the dog, that changed fundamentally – a 'new meaning was impressed upon in it. It

---

[23] Jakob von Uexküll, *A Foray Into the Worlds of Animals and Humans*.

[24] Jakob von Uexküll, *A Foray Into the Worlds of Animals and Humans*, p. 50.

[25] Jakob von Uexküll, *A Theory of Meaning*, p. 140. Italics in the original.

received a "throwing tone".[26] This approach contrasts markedly with Gibson's theory of affordance, which denies that meaning has anything to do with direct perception (cf. Chapter 4).

I will return to the significance of tonality, or what I describe as the melodic metaphor, in a moment. First it is necessary to explain that the second displacement involves what is a kind of Copernican revolution for theoretical biology, and one that will catch the eye of the early Maurice Merleau-Ponty (cf. Chapter 9): as Uexküll explains, when physiologists treat organisms as machines, they neglect the question of the subject. In *A Foray*, Uexküll explains that '[f]or the physiologist, every living thing is an object that is located in his human world. He investigates the organs of living things and the way they work together just as a technician would examine an unfamiliar machine.' By contrast, the biologist should '[take] into account that each and every living thing is a *subject* that lives in its own world, of which it is the center. It cannot, therefore, be compared to a machine, only to the machine operator who guides the machine'.[27]

This 'machine operator' image – perhaps a concession by Uexküll to the power of mechanism in biology – is really a reference to the subjectivity of organisms, an account of which becomes a central aim of Uexküll's research, and which is encapsulated in the concept of *umwelt*. *Umwelt* – translated variously as 'surrounding-world', 'self-world' or more simply as 'environment' (perhaps unhelpfully, given the last term's current everyday meaning) – refers to the subjective, meaningful, or as the biosemiotician Kalevi Kull also puts it, the 'subjectivised'[28] world of the organism. This is how Uexküll himself introduces the term in *Theoretical Biology*:

> If an observer has before him an animal whose world he wishes to investigate, he must first and foremost realise that the indications that make up the world of this other creature are his own, and do not originate from the mark-signs of the animal's subject, which he cannot know in the least. Consequently, these indications are, one and all, beneath the sway of the laws of our attention; and, as soon as our attention is directed to them, we cannot free them from these laws. ... Since we are not in a position to investigate the appearance-world of another subject, but only that part of our appearance-world surrounding it, we had better speak of the *surrounding-world* [*umwelt*] of the animal. It is only for the observer himself that the surrounding-world and the appearance-world are identical.[29]

---

26 Jakob von Uexküll, *A Theory of Meaning*, p. 140.

27 Jakob von Uexküll, *A Foray*, p. 45. Italics added to the original.

28 Kalevi Kull, 'Jakob von Uexküll: An Introduction', *Semiotica*, 134:1/4 (2001), 1-59, p. 7.

29 Jakob von Uexküll, *Theoretical Biology*, p. 78-79.

Uexküll uses the metaphor of a soap bubble to refer to how each kind of organism is caught up in its own *umwelt*, its own perceptual world. The use of this metaphor is meant above all else to highlight the monadic quality of each *umwelt*: each species, if not each *organism*, lives in a bubble of its own thanks to the fact that it alone can experience the/its world in that particular way.

In *A Foray*, Uexküll employs a propaedeutic device to illustrate how one same room – which includes a bookcase, a small table set with plates and glasses, as well as chairs, an easy chair and settee – would be perceived via the *umwelten* of a human being, a dog, and a fly. In a series of colour plates that illustrate the 'same' room, Uexküll uses colours to highlight which parts of the room would be of significance to each creature. As Uexküll notes in *Theoretical Biology*, we can only try to conceive another organism's *umwelten* via our own, and indeed it is to be supposed that the room in question would 'look' very different to each of the mentioned subjects. Even so, the propaedeutic device of changing colours does highlight what aspects are likely to be significant to each creature. For example, a dog that has learned to sit on a chair when it hears the command 'chair' 'will look [in such a room] for another seat, to wit, another *dog* seat, which needs not at all be suitable for human use. Seats as carriers of meaning for sitting all have the same sitting tone, for they can be exchanged with one another arbitrarily, and the dog will still use them'; most of the rest of the furniture will only have an 'obstacle tone' for that dog.[30]

Here again, is a reference to tonality. For Uexküll, what we may describe as a melodic metaphor provides an alternative to the mechanical, and behaviourist conceptions. As Uexküll defines the terms in *Theoretical Biology*, '[b]y "melody" we understand the orderly sounding of musical notes one after the other; by symphony, their sounding together. When combined, melody and symphony give us harmony'.[31] Nature is precisely a matter of the coming together of 'plans' regarded as so many melodies combining to produce a symphony with harmonies. Or as Uexküll himself puts it,

[e]very Umwelt has its own spatial and temporal dimensions. The Umwelten intersect in many ways without disturbing each other. They do not interact mechanically but are still connected according to a plan as the notes of an oratorio are harmonically connected. It is thus musical and not mechanical laws that we need to study if we want to find out about the laws of Life. [...] As the harmony of the sounds is only a part of the design of the performance of an orchestra, which embraces also the forms and the materials of the instruments, so the sensory perceptions and the intentional impulses of the animals constitute only a part of the design, that is revealed to us most clearly in the bodily forms and movements'; [...] the performances of animals are not

---

[30] Jakob von Uexküll, *A Foray*, p. 142. Italics added to the original.

[31] Jakob von Uexküll, *Theoretical Biology*, p. 29.

products of a harmonic build of the body, it is the harmony of the performance that determines that of the body.[32]

Later in the same text, Uexküll suggests that '[e]very Umwelt of a normal animal is a faultless composition of nature – you only have to understand how to look for its theme and its notes'.[33] An analogous point is made with respect to the relationships between species, which operate in a contrapuntal relationship to each other: the 'melody' of the tick operates as a counterpoint to that of a mammal, the melody of a spider and its web the same again with respect to the fly it tries to trap, and so forth.

As noted by Brett Buchanan,[34] it is this melodic metaphor (as I call it) that the early Merleau-Ponty draws on. Uexküll's metaphor first finds its way into Merleau-Ponty's *The Structure of Behaviour*, Merleau-Ponty's earliest book, in the form of a brief reference: '"Every organism," said Uexkull, "is a melody which sings itself." This is not to say that it knows this melody and attempts to realize it; it is only to say that it is a whole which is significant for a consciousness which knows it, not a thing which rests in-itself [*en soi*].'[35]

As I began to explain earlier, Uexküll's theory is at least partly grounded in a Kantian, or at any rate a neo-Kantian perspective. This aspect, combined with its vitalism, is welcome in so far as it pushes back at mechanistic perspectives on animality. Uexküll's concept of *umwelt* has the added advantage that it permits us to conceive the possibility that in humans there is a form of corporeal-situational anthropomorphism that is in some sense fundamental to the kind of anthropomorphism that is mainly discursive. I refer to an anthropomorphism that results from the kinds of affections that most humans have when it comes to their perceptual systems. On some level, it might be said, we observe the world in the ways that our organismic capacities allow us to, and this constitutes a 'first filter' of the kind described by Uexküll with respect to ticks and other animals.

The risk with such a perspective is, of course, a return to an *identitary* approach grounded in each species' 'plan': the animal is not itself, but its species' plan. Uexküll himself would doubtless agree that the *umwelts* of animals are not themselves fixed, but here I suggest that it is pertinent to (re)turn to the philosophy of Gilles Deleuze, and Deleuze and Félix Guattari. I provided an introduction to aspects of that philosophy in Chapter 10, and I would now like to introduce additional elements, and to consider their implications for the question of animals.

---

[32] Jakob von Uexküll, 'The New Concept of Umwelt: A Link Between Science and the Humanities', *Semiotica*, 134:1/4 (2001), 111-123, p.117.

[33] Jakob von Uexküll, 'The New Concept of Umwelt', p. 120.

[34] Brett Buchanan, *Onto-Ethologies*: *The Animal Environments of Uexküll, Heidegger, Merleau-Ponty and Deleuze* (Albany, NY: Suny Press, 2008).

[35] Maurice Merleau-Ponty, *The Structure of Behaviour*, p. 159.

Perhaps a good way to begin the discussion is to consider several examples that take us back to that Spinozan/Deleuzian/Guattarian mantra, 'no one knows what a body can do'. The first example involves a sloth on Barro Colorado Island. At the time that I was conducting ethnographic research on Barro Colorado, a group of scientists had set up an experimental telemetric tracking system. As part of what was known as the ARTS project (Automatic Radio Telemetry System),[36] scientists and engineers established a network of towers with automated receivers to relay data about the location and movements of animals with VHF transmitters attached by the scientists. At one point, a sloth was having a radio collar fitted, and it reportedly managed to bite one of the researchers. So either the sloth was rather quicker than is suggested by the stereotypical anthropomorphism of the 'slothful' sloth, or the researcher was even slower than the sloth...

A second example, which received some media attention in 2018, was that a group of scientists investigating GPS-collared wolves in northern Minnesota were able to document numerous instances of members of the pack catching and consuming fish in a freshwater spring.[37] Previously some wolves, especially in coastal marine habitats, had been suspected of scavenging on dead fish, but here was incontestable evidence that these 'red meat eaters' were quite capable of learning to fish for themselves.

Two further examples – the ones that Deleuze and Guattari would perhaps most appreciate – are in some sense diametrically opposed to each other. The first involves a 'domestic' animal: in mid-June 2018, a three-year-old cow named 'Betsy' managed to escape from a pen during a rodeo in Achorage, Alaska, and by the end of the year had still not been recaptured by the owner, or indeed by police and other authorities who even used an infrared camera-equipped drone to try to find Betsy after it made its way into Alaska's Far North Bicentennial Park – an area described as a 4,000-acre (approx. 1,600 hectare) expanse of rugged forest, presumably with apex predators such as wolves, grizzly bears and mountain lions. Alas, it seems that every time that a new sighting occurred over the following six or so months, the cow managed to escape 'back into the wild'.[38]

---

36 See for example Roland Kays, Sameer Tilak, Margaret Crofoot, Tony Fountain, Daniel Obando, Alejandro Ortega, Franz Kuemmeth, Jamie Mandel, George Swenson, Thomas Lambert, Ben Hirsch, and Martin Wikelski, 'Tracking Animal Location and Activity with an Automated Radio Telemetry System in a Tropical Rainforest', *The Computer Journal*, 2011, 1-18.

37 See Thomas D. Gable, Steve K. Windels, and Austin T. Homkes, 'Do Wolves Hunt Freshwater Fish in Spring as a Food Source', *Mammalian Biology*, 91 (July 2018), 30-33.

38 See Antonia Noori Farzan, 'A Cow Escaped the Rodeo and Disappeared Into an Anchorage Park. For Six Months, No One Has Been Able to Catch Her', *Washington Post* online, 16 January 2019 < https://www.washingtonpost.com/nation/2019/01/16/cow-escaped-rodeo-disappeared-into-an-anchorage-park-months-no-one-has-been-able-catch-her/>[Accessed 16 January 2019].

The other example involves a 'wild animal' *par excellence*: a grizzly bear, whose subspecies would appear to say it all (*Ursus arctos horribilis*). Yet anyone who looks up the subject of 'bear trainers' on the internet will find numerous videos showing *U. arctos horribilis* interacting with such trainers, including one video produced by Barcroft TV, which shows Doug Seus, described as a *professional* bear trainer, 'wrestling' with 'Bart', a male grizzly bear weighing close to 700 kilograms (1,300 pounds) and the better part of 2.5 meters in height when standing.[39] Of course, Bart would have been brought up in a humanising environment from the time it was a cub, and most likely would have been subjected to what might well be a ruthless regime of operant conditioning. However, the very fact that this might result in a bear able to play with humans in the manner depicted must nonetheless give pause to anyone who would seek to essentialise wildness as a kind of uncontrollable instinct to remain completely separate from any kind of anthropocentric or anthropomorphic formation.

Now there is hardly anything new about stories such as the ones I've just presented. On the contrary, there is a very established and *wildly* popular genre in the new media that specialises in revealing videos of animals doing surprising things. There is also a long history of narratives across various 'old' media devoted to chronicling the wholly unexpected affections (and their often disastrous consequences) amongst both wild and domestic animals. To be sure, the narratives in question are not merely concerned with the affections of *beyond-human* animals; in 2019, the media reported on the extraordinary events surrounding Travis Kauffman, a 'Colorado man' who was running along a trail in the foothills outside of Fort Collins when he was attacked by what was described as a 'young' mountain lion (and so presumably not a fully grown specimen). The mountain lion effectively botched the ambush in so far as it made a noise before it could reach Kauffman, who turned, saw it, yelled, failed to stop the attack, but did manage to protect himself by throwing up his arms. The lion then made a second mistake by latching on to Kauffman's wrist; in the wrestling match that ensued over the next ten or so minutes, Kauffman managed to kill the creature by suffocating it.[40]

The first issue that Deleuzian philosophy brings into focus is the extent to which these and countless other examples cannot really be explained by mechanistic, or more generally ontologies of identity for which fixing the nature of bodies (beyond-human or human) via essentialising concepts is the philosophical staple. To return to the points raised in Chapter 10, it is only really

---

[39] See Barcroft TV, 'Wrestling A Grizzly Bear in My Garden', YouTube, <https://www.youtube.com/watch?v=9nSWc43TLaI >[Accessed 1 May 2019].

[40] For a transcript of an interview with Lulu García-Navarro, see National Public Radio, 'Colorado Man Tells His Mountain Lion Attack Story', 17 February 2019, < https://www.npr.org/templates/transcript/transcript.php?storyId=695536837&t=1551367772209>[Accessed 21 February 2019].

possible to recognise what *a* body can do by fully acknowledging the importance of modes, and by adopting a flat ontology. This point is a deceptively simple, and so easily misunderstood one. When, for example, I used the example of the sloth bite in a conference presentation, a biologist in the audience quite legitimately pointed out that exceptional events such as this occur all the time in the course of ecological research, and that they mean little or nothing to the researchers. The gist of the response was, every now and again a sloth might bite a scientist, but so what? If we were to approach the matter statistically, it could almost certainly be shown that the balance of probability would tally with a scientific conception of the capabilities of the animal regarded as a member of a species. From this kind of perspective, 'statistical flukes' are either just that, or obey a situational logic which, examined more carefully, would probably prove not to be so unexpected or surprising after all.

What this kind of perspective leaves out is an ethic, in the form of an ontological *choice* which Deleuze and Guattari seek to bring to the foreground. This ethic is grounded partly in the kind of philosophy that I introduced in Chapter 10. However, it is also grounded in advanced mathematics and 'hard' (sic) science. In Part 1 I mentioned the importance of Willard Gibbs' statistical mechanics to Norbert Wiener's cybernetic theory, and indeed it will be recalled that Wiener contended that in so far as no physical measurements can ever be absolutely precise, then the calculation of displacements involving machines and other dynamic phenomena must be approached as a matter of their most likely *distribution*. The problem is not simply to arrive at the explanation of phenomena on the basis of ostensibly immutable laws, as per the Newtonian (or indeed Comtian) perspective, but to use maths to try to arrive at some control over uncertainty and contingency, or what Wiener liked to characterise as a problem of *entropy*. Given the particular problems that Wiener set out to solve as part of his cybernetic perspective, he was keen to use this kind of perspective as part of an effort to control human animals and their interface with machines.

Deleuze and Guattari might well agree with at least a part of Wiener's cosmology – in effect, that of quantum mechanics[41] – but they would be horrified at his effort to use it as a tool for control. Indeed, simplifying greatly, for Deleuze and Guattari the problem is to *recognise* the flux, the fluidity, not to fix it (as in stop it, symbolically or otherwise). Put differently, the philosophers might argue that a recognition of quantum mechanics is pointless if one regresses, however implicitly, to a Newtonian physics, let alone a Newtonian *Politics* (I use the capital 'P' quite deliberately; it is perhaps not well known that, late in his life, Sir Isaac Newton became the Master of the newly established Royal Mint in London, and

---

[41] For an analysis of this aspect of Deleuze's work, see Wim A. Christiaens, 'The Deleuzian Concept of Structure and Quantum Mechanics', in Diederik Aerts, Sven Aerts, and Christian de Ronde, *Probing the Meaning of Quantum Mechanics: Physical, Philosophical and Logical Perspectives* (Brussels: World Scientific Papers, 2014), 189-208.

beyond reorganising the hitherto chaotic production of coins, he took it upon himself to pursue counterfeiters [or 'coiners', most of whom were impoverished women] with a particular zeal[42]).

In keeping with their perspective, Deleuze and Guattari would like to consider what sorts of politics become possible when it is recognised, in the most fundamental way possible, that we are not only what or whom we are *told* we are, or indeed what we tell *ourselves* that we are; and of course, the same applies for beyond-human animals: the animals are not simply what they are represented as being. Indeed, Deleuze and Guattari would argue that adhering, however implicitly or unselfconsciously, to an identitary and transcendental conception of *species* leads precisely to the view that unexpected variations in morphology or ethology are a matter of exception, and not the very stuff of the species. From a Deleuzian perspective, Charles Darwin and Alfred Russel Wallace's genial discoveries regarding evolution are systematically undermined by anyone – perhaps even by the great Charles Darwin himself – who fails to put sufficient emphasis on the 'fluidity' of species-being. Species are not, by evolutionary theory's own standards, fixed entities; on the contrary, *that was the great insight* of Darwin and Wallace.

If this is the case, how could those who favour adaptationist perspectives regress to a certain immobilism? There is no easy answer to this question, but part of the problem probably involves the temporality of evolutionary theory, which typically presupposes changes across hundreds of thousands of years, and so may ironically leads scientists to forget change, and also non-evolutionary *causes* of change in much shorter time scales. Another motivation may be that, in so far as any conception of change is theorised quite literally along the *lines* of a mechanistic conception of causation, then mechanism takes over.

With these examples and this discussion in mind, I would now like to introduce the Deleuzian concept of *becoming* to the present analysis. To understand this concept it is necessary to engage in a relatively detailed explanatory excursus regarding the thought of another of the great influences in Deleuze's, and by extension Deleuze and Guattari's, philosophy: the philosopher Henri Bergson (1859-1941). In his introduction to Deleuzian philosophy, Todd May explains the importance of Bergson – the second in what Todd describes as Deleuze's 'holy trinity' (Spinoza, Bergson, and Nietzsche) – as follows: if the concept of immanence is not to regress into a mindless sameness, it must have recourse to the concept of expression. 'But expression', May notes, 'is a temporal concept. Expression happens temporally; for Deleuze, expression is so inseparable from temporality that we might think of expression and temporality as the same thing, seen from two different angles. What Spinoza creates with his

---

[42] For a brief account of Newton's role as the warden, and then the master of the Royal Mint, see Nils Lindahl Elliot, *Mediating Nature*, pp. 77-79.

concept of expression and what Bergson creates with his concept of temporality are a seamless weave in Deleuze's philosophical perspective.'[43]

To understand that weave, it is possible to draw a parallel between the critique of 'empty space' (cf. James Gibson, cf. Chapter 4; see also the account of the double illusion by Henri Lefebvre in Chapter 12), and Bergson's critique of a quantitative conception of time. Conceived as 'empty space', the boundaries of a certain space can be measured, and so can the positions of the objects found within it. This is a quintessentially Cartesian space, with each of the elements occupying a clear and distinct position. A similar conceptual transformation may be achieved with respect to time if it is spatialised, and in so doing, regarded as a container in its own right (with things happening 'in' time), or indeed spatialised as a line that is composed of a succession of points that extend both backwards and forwards in time. Any point on a line like this can be given a number (say, 2019), and indeed the line's points can be divided into segments (say, 2010-2019). From this kind of perspective it may be inferred, as the pre-Socratic philosopher Zeno of Elea (c. 495 - 430 BC) did in his famous paradoxes, that Achilles will never catch up with the tortoise, or indeed that it is impossible to reach a destination: Achilles will only ever be arriving to a point the tortoise has just left, and in so far as a distance is infinitely divisible in halves, then each time someone sets out, they will always fail to reach their destination because at each point they will still have another half of the distance to cover. If motion involves distance and time thus conceived, it must be an illusion.

Of course, the real illusion is that an action as it unfolds in time is divisible, and it is this problem that Bergson addresses by way of his concept of duration (*durée*). A first way to approach this concept is by way of a psychological conception of time (this is not ultimately how Bergson himself understands duration, but it serves a preliminary propaedeutic purpose). In so far as linear time objectifies and indeed spatialises time as a kind of line, it overlooks the importance of what might be termed experienced or lived time. For any person, time is anything *but* objective in as much as it involves a lived experience that, if anything, is utterly *subjective*. So it is that in some circumstances time may seem to fly, whilst in others it can slow to a crawl. How one lives this or that moment is not a matter of a divisible succession of instants, but of what kind of action one is engaged in, and how one perceives, conceives and interprets that action. Clearly, this kind of experience unfolds in a continuous manner; it is, in this sense, a kind of continuous *becoming*.

Bergson was aware of this kind of time, but was also aware of the danger of 'phenomenologising' time, and in so doing of simply shifting from the kind of objectivist conception that underpins linear time to a subjectivist understanding of the kind that underpins psychological time (with its emphasis on

---

[43] Todd May, *Gilles Deleuze: An Introduction* (Cambridge: Cambridge University Press, 2005), p. 41.

'consciousness'). What he proposed instead is a conceptualisation of duration that characterises it as *qualitative* multiplicity, as distinct from *quantitative* multiplicity.

To explain this difference it may be helpful to turn to an example with animals that Bergson himself employed. When someone counts the sheep in a flock and finds that there are 50 of them, the counting exercise neglects the sheep's 'individual differences' and only takes into account what they have in common, despite the fact that each sheep is different – indeed, each sheep may well be recognised by the farmer. As Bergson puts it, 'the idea of number implies the simple intuition of a multiplicity of parts or units, which are absolutely alike.'[44] And yet, as Bergson notes, the sheep must be somehow distinct from each other; otherwise they would merge into a single unit. One way to distinguish between them is to note the position they occupy in space. Space may provide criteria for differentiation even if we disregard the physical location of the sheep, as happens if, say, we count fifty times in succession the image of a single sheep. It may seem that such a series lies in duration and not in space. But this is not in fact the case; as Bergson puts it, 'if we picture to ourselves each of the sheep in the flock in succession and separately, we shall never have to do with more than a single sheep. In order that the number should go on increasing in proportion as we advance, we must retain the successive images and set them alongside each of the new units which we picture to ourselves: now, it is in space that such a juxtaposition takes place and not in pure duration.'[45]

Whichever way one tries to turn the matter, counting, and by implication dividing things implies *quantitative* multiplicity, a multiplicity which involves space – or put more precisely, it necessarily involves a representational operation of spatialisation. By contrast, what Bergson describes as *qualitative* multiplicity is all about indivisibility. Take for example the case of a single one of those 50 sheep being born, growing up, and dying. According to the logic of *quantitative* multiplicity, this is just one sheep among 50, it is 1, 2, 3 or more years old, weighs so many kilograms, etc. The validity of these measurements *as measurements* notwithstanding, no one measurement, or even a set of measurement, is the sheep, and this not least because throughout the duration of its existence, this or any other sheep will go through numerous changes, and will do so indivisibly as it grows.

This insight reveals the false premises of Zeno's paradoxes. As Bergson puts it, 'the truth is that each of Achilles' steps is a simple indivisible act, and that, after a given number of these acts, Achilles will have passed the tortoise. The mistake of the Eleatics arises from their identification of this series of acts, each of which is of a definite kind and indivisible, with the homogeneous space which underlies

---

44 Henri Bergson, *Time and Free Will: An Essay on the Immediate Data of Consciousness*, translated by F. L. Pogson (London: George Allen & Unwin, 1910 [1889]), p. 76.

45 Henri Bergson, *Time and Free Will*, p. 77.

them'.[46] Even as each of Achilles' and of the tortoise's steps are indivisible, they are *also* of a different magnitude in space. Together, these aspects mean that Achilles will of course overtake the tortoise. And the indivisibility of movement regarded from the point of view of duration means that of course we *will* reach a destination, and not simply travel endlessly by halves.

It is now possible to go back to Deleuze's concept of *becoming*, and Deleuze and Guattari's *becoming-animal*. Let's begin with the former term. Put simply, the Bergsonian concept of duration, and with it qualitative multiplicity, are what allow Deleuze to give expression a conception of temporality that is in keeping with the principle of immanence that Deleuze borrows from Spinoza, and translates into his own ontology. It is, to be sure, not just Spinoza and the principle of immanence that are at stake: as I began to note earlier, like Bergson himself, Deleuze is keenly aware of the significance of the change from classical mechanics to statistical mechanics (cf. Chapter 3), from Newtonian to quantum mechanics. In this context, the problem is no longer to try to say what something is, as per the classical metaphysic, but what it is becoming.

This shift begins to explain why in *A Thousand Plateaus*, Deleuze and Guattari suggest that the multiplicities that most interest them are 'micromultiplicities', as opposed to the 'macromultiplicities'. The latter are 'extensive, divisible, and molar; unifiable, totalisable, organisable; conscious or preconscious', whereas the former are 'libidinal, unconscious, molecular, intensive multiplicities composed of particles that do not divide without changing in nature, and distances that do not vary without entering another multiplicity'.[47]

Far from being set in the aspic of molar formations (where molar refers to the continuities, representational or other, that result from the consolidation or indeed *coagulation* of particular assemblages), the individuals of micromultiplicity are like particles that engage in endless random motion and constant collisions with other like particles. Their movements are, in this   sense, Brownian[48] (as in Brownian motion). From this perspective, 'becoming' is *becoming particular*, in both the conventional and the Brownian/quantum/Bergsonian sense of particular: an individual undergoing a constant displacement and change, however minute or infinitesimal.

Now along with the conceptual pairs macro/micro, molar/molecular, Deleuze and Guattari refer to a process of territorialisation, deterritorialisation, and reterritorialisation – what I will describe, for brevity's sake as 'trans-territorialisation'. While a biology and ontology of the kind advocated by, say, an Edward O. Wilson would put all of the emphasis on a more or less fixed territory – it is no coincidence, from this point of view, that Wilson would like to set aside

---

46 Henri Bergson, *Time and Free Will*, p. 113.

47 Gilles Deleuze and Félix Guattari, *A Thousand Plateaus*, p. 33.

48 Gilles Deleuze and Félix Guattari, *A Thousand Plateaus*, p. 33.

half of the world, or the equivalent thereof, to prevent another mass extinction[49] – Deleuze and Guattari put the emphasis on a *process* that leads to the constant reformation of territory – we might speak in this sense of becoming territory. If bodies are undergoing constant change, so are 'their' territories. Their territories are constantly being gained, lost, and eventually reconstituted not just because there is constant motion, but because all the bodies that make up the space are undergoing constant change. Here again, becoming stands in contrast (or perhaps I should say *moves* in contrast) to the older ontological emphasis on being: to reiterate the earlier point, for Deleuze, and for Deleuze and Guattari, ontology is not so much about 'what things are' (as in the Cartesian quest for the clear and distinct, and for the fundamental substances) but on becoming.

Of course, it may be objected that this is a false opposition. In fact, Deleuze and Guattari do not oppose becoming to being, any more than they think that macromultiplicity and micromultiplicity stand in a dualistic relation of the kind that might be advocated by, say, structuralists of any ilk. This is made evident by their concept of *strata*, which Deleuze and Guattari describe as 'phenomena of thickening on the Body of the earth, *simultaneously* molecular and molar: accumulations, coagulations, foldings'.[50] Admitting that it is a rather conventional distinction, Deleuze and Guattari propose that there are three major strata, the physiochemical, organic, and anthropomorphic or 'alloplastic'.[51] Each of these has very diverse forms and substances (we come back to Hjelmslev, cf. Appendix 2), as well as a variety of codes and milieus. However, as Deleuze and Guattari put it, each stratum, and its many possible subdivisions, is 'extremely mobile. One stratum is always capable of serving as a *substratum* of another, or of colliding with another, independently of any evolutionary order'; '[s]tratification is like the creation of the world from chaos, a continual, renewed creation'.[52] Returning to Michel Serres, one might say that strata are better conceived along the lines of topological space.

Perhaps a good way of illustrating what is at stake in 'becoming' is by referring to something that seems completely immobile. Here I have in mind a wonderful example offered by the geographer Doreen Massey as part of her redevelopment of her theory of space and place. The reader will recall that I introduced Massey as part of the first wave of cultural geography, in and for which an identitary and cultural materialist approach to space prevailed. By the mid-2000s, Massey's work shifted to embrace a notion of space that is not only about multiplicity (as was in some respects already the case in Massey's earlier

---

[49] See Edward O. Wilson, *Half Earth: Our Planet's Fight For Life* (London: Liveright, 2016).

[50] Gilles Deleuze and Félix Guattari, *A Thousand Plateaus*, p. 502. Italics added to the original.

[51] Gilles Deleuze and Félix Guattari, *A Thousand Plateaus*, p. 502.

[52] Gilles Deleuze and Félix Guattari, *A Thousand Plateaus*, p. 502.

work on *place*), but about becoming. In *For Space*, Massey puts forward three deceptively simple propositions: First, that space should be recognised as the product of interrelations. Second, that space should be understood as 'the sphere of the possibility of the existence of multiplicity in the sense of contemporaneous plurality; as the sphere in which distinct trajectories coexist; [and] as the sphere therefore of coexisting heterogeneity'.[53] As Massey explains, '[w]ithout space, no multiplicity; without multiplicity, no space. If space is indeed the product of interrelations, then it must be predicated upon the existence of plurality', and multiplicity and space are co-constitutive.[54] Finally, Massey further suggests that space is always under construction, i.e. space is always in the process of being made, it is never finished or closed.[55]

Perhaps the most obvious implication of this kind of approach is an expressly political one: if space is conceived in this manner, it is not possible to justify, other than on utterly ideological grounds, the notion that people 'belong' to certain places and not others, one of the fundamental presuppositions of nativism and racism. On the basis of Massey's conceptualisation, it is possible to normalise the geographies of the rapidly changing societies now found across much of the world, where a combination of migration, rapid technological change, and economic transformations are producing constant changes in the social relations.

It might be objected that this is all fine and well with what Deleuze and Guattari describe as the alloplastic stratum, but what about the physiochemical and the organic? Can Massey's approach – or indeed Deleuze and Guattari's – offer anything to someone studying not a multicultural neighbourhood, or the commute between central London and Milton Keynes (as Massey does in some of her research), but, say, tropical forests and the organisms found there? Massey herself raises this question with reference to the comments of a friend, who notes that 'That's all right [Massey's approach is] when you talk about human activity and human relations. I can understand and relate to it then: the interconnectivity, the essential transience ... but I live in Snowdonia [a national park in northern Wales] and my sense of place is bound up with the mountains'.[56]

Massey responds by way of an example that I mentioned in the introduction to this volume, and which is drawn from another set of mountains, those found around the town of Keswick, in the English Lake District – a place where Massey and her sister once stayed, presumably on holiday. Massey notes that it would be easy to analyse the town itself in the terms of her theory. But she notes that the same is possible with respect to Skiddaw, 'a massive block of a mountain, over 3000 feet high, grey and stony', 'not pretty, but impressive' and indeed seemingly

[53] Doreen Massey, *For Space* (London: Sage, 2005), p. 9.

[54] Doreen Massey, *For Space*, p. 9.

[55] Doreen Massey, *For Space*, p. 9.

[56] quoted in Doreen Massey, *For Space*, p. 131.

immovable and timeless[57] – precisely the kind of conception that many tourists might have of countless other landscapes visited for their heritage value, or for their geographic or ecological significance.

In fact, when Massey considers in detail the geography and especially the geology of the mountain, it becomes clear that her three propositions also have something to offer anyone interested in analysing physical geographic spaces. Much of the landscape around Keswick was etched and moulded by the glaciers of the ice ages that retreated 10,000 years ago. Then again, the rocks of Skiddaw were laid down in a sea which existed some 500 million years ago, the result of the erosion of still older lands. To the south of Skiddaw, there is also evidence, for those who can read the landscape, of more recent but still very ancient volcanic activity, with its own transformational effects. Finally, the 'unmovable' and 'timeless' Skiddaw, a part of today's northern England in the 'British' imaginary, was actually once south of the equator. As Massey puts it, '[t]he bit that we know today as the slates of Skiddaw crossed the equator about 300 million years ago'.[58] This being the case, Massey argues that Skiddaw's 'timeless shape is no such thing. Nor has it been "here" for ever. Nor again is this a matter of past history alone, for the movement of continents of course continues'.[59] Indeed, the rocks of Skiddaw 'are *immigrant* rocks, just passing through here, like my sister and me only rather more slowly, and changing all the while. Places as heterogeneous associations. If we can't go "back" home, in the sense that it will have moved on from where we left it, then no more, and in the same sense, can we, on a weekend in the country, go back to nature. It too is moving on.'[60] (I might note in passing that a better example of a 'progressive' circuit of anthropomorphism and cosmomorphism would be hard to find.)

What is true for that 'block of a mountain' would be even more true for tropical forests – and not least, those undergoing accelerated, and indeed accelerat*ing* changes produced by fragmentation, let alone catastrophic climate change. The verity of this kind of insight may be visualised by time-lapse photography, which reveals how plants thought not to be, and in some sense clearly *not* capable of engaging in locomotion nevertheless produce motions that are invisible to the naked eye. Elsewhere I have argued that wildlife documentary representations of these kinds of movement tend to 'animalise' plants and so constitute a kind of visual anthropomorphism;[61] but perhaps the more fundamental point is an ontological one in the sense that it acknowledges that plants and forests too, are in a constant

---

[57] Doreen Massey, *For Space*, p. 131.

[58] Doreen Massey, *For Space*, p. 133-134.

[59] Doreen Massey, *For Space*, p. 135-137.

[60] Doreen Massey, *For Space*, p. 137. Italics added to the original.

[61] See Nils Lindahl Elliot, 'Signs of Anthropomorphism: the Case of Natural History Television Documentaries', in *Social Semiotics*, 11(2001), 289-305.

process of flux, and this no doubt is what the producers of a series like *The Private Life of Plants* (BBC, 1995) in some way wished to convey.

I think that I've said enough about 'becoming' to provide a sense of its place within the Deleuzian, and Deleuzian/Guattarian ontology. What, then, do Deleuze and Guattari mean by 'becoming-animal', and what does the concept offer anyone asking the question (or at any rate, some questions) about animality? If Deleuze and Guattari's concepts are generally difficult to pin down – no doubt precisely because the authors wish to be consistent with their own philosophy – becoming-animal is perhaps one of the concepts that poses the greatest challenge, and so requires the greatest hermeneutic effort. A first way to engage with the concept is by considering the tripartite distinction that Deleuze and Guattari propose for animals in *A Thousand Plateaus*.

First, there are 'individuated animals, family pets, sentimental, Oedipal animals each with its own petty history, "my" cat, "my" dog'; according to Deleuze and Guattari's withering assessment, these are animals that 'invite us to regress, draw us into a narcissistic contemplation, and they are the only kind of animal psychoanalysis understands'.[62]

I will return to the critique of psychoanalytic representations of animality below. First it is pertinent to explain that the second kind of animal refers to animals 'with characteristics or attributes; genus, classification, or State animals; animals as they are treated in the great divine myths, in such a way as to extract from them series or structures, archetypes or models', with the Jungian analysis of animals offering more profound insights than those of Freud.[63] This type of animal is the kind that I myself was referring to earlier when I spoke of a circuit of anthropomorphism and cosmomorphism, e.g. of meerkats that are treated as benign 'agrarian socialists' or as heroic/malevolent 'street gangs', of animals that are counted, species that are measured. We return to an identitary logic: meerkat is x, this is sheep no. 48.

The third kind of animal involves what Deleuze and Guattari describe as 'demonic animals, pack or affect animals that form of a multiplicity, a becoming, a population, a tale...'.[64] In this clause we find one of countless examples of Deleuze and Guattari being naughty, seemingly setting out to deliberately confuse the reader. Are not animals in tales those of the second kind? And is the very notion of population – redolent as it is of sociobiological connotations – not itself a sign of a 'molar' animal, i.e. one conceived by more or less fixed type, and not one constantly in flux, i.e. a 'molecular' animal?

Deleuze and Guattari clarify that any one animal may be treated in all three ways: 'There is always the possibility that a given animal, a louse, a cheetah or an

---

[62] Gilles Deleuze and Félix Guattari, *A Thousand Plateaus*, p. 240.

[63] Gilles Deleuze and Félix Guattari, *A Thousand Plateaus*, pp. 240-241.

[64] Gilles Deleuze and Félix Guattari, *A Thousand Plateaus*, p. 241.

elephant, will be treated as a pet, my little beast. And at the other extreme, it is also possible for any animal to be treated in the mode of the pack or swarm', and conversely there may be a 'favourite', or indeed a leader, in the pack.[65] This notion of the pack is itself not an easy one, and so also requires elucidation.

In *A Thousand Plateaus*, a first indication of what the authors mean is provided by their trenchant critique of Freud's interpretation of the now-famous dream of a patient he described as Wolf-Man (*der Wolfsmann*), and who was subsequently identified as Sergei Pankejeff. The details of this case are not unimportant, but go beyond the scope of this chapter; it suffices to explain that Freud gave Pankejeff the 'Wolf-Man' pseudonym (designed to protect Pankejeff's identity in Freud's writings) thanks to Pankejeff's dream as a young child that there was a tree full of white wolves outside his bedroom window – a dream from which he woke up in a state of extreme distress, and which Freud interpreted in terms of his patient witnessing his parents having sex *a tergo* (or the displacement of this kind of copulation seen amongst animals to Pankejeff's parents).

Freud eventually claimed to have cured Pankejeff, a claim that Pankejeff disputed. One way or the other, the case achieved a certain notoriety in psychoanalytic circles, with post-Freudian scholars questioning key aspects of Freud's interpretation of Pankejeff's dream. As I noted in Chapter 10, in *Anti-Oedipus* Deleuze and Guattari acknowledge Freud's theory of the unconscious, but reject its Oedipal turn; in *A Thousand Plateaus*, the philosophers return to their earlier critique: '[n]o sooner does Freud discover the greatest art of the unconscious, this art of molecular multiplicities, than we find him tirelessly at work bringing back molar unities, reverting to his familiar themes of *the* father, *the* penis, *the* vagina, Castration with a capital C ... (On the verge of discovering a rhizome, Freud always returns to mere roots)'.[66] Or as they put in the following page, 'For Freud, when the thing splinters and loses its identity, the word is still there to restore that identity or invent a new one. Freud counted on the word to reestablish a unity no longer found in things. Are we not witnessing the first stirrings of a subsequent adventure, that of *the* Signifier...?'[67] Clearly, this is Deleuze and Guattari having a go at Lacanian psychoanalysis. In what may be interpreted both as a symbolic gesture to oppose this kind of unification, but also as part of the turn to duration and to qualitative multiplicity (or *micro*multiplicity), Deleuze and Guattari seize on the fact that Freud reduces the pack of wolves outside of Pankejeff's window to a single wolf, *the* Father. In the case of the 'Wolf-Man' as in other cases, Deleuze and Guattari argue that Freud fails to comprehend the importance of the mass, of the crowd: 'Freud tried to approach crowd phenomena from the point of view of the unconscious, but he did not see clearly, he did not see that the unconscious itself

[65] Gilles Deleuze and Félix Guattari, *A Thousand Plateaus*, p. 241.

[66] Gilles Deleuze and Félix Guattari, *A Thousand Plateaus*, p. 27.

[67] Gilles Deleuze and Félix Guattari, *A Thousand Plateaus*, p. 28.

was fundamentally a crowd.'[68] The crowd that the philosophers are referring to is a (social) collectivity, presumably along the Marxist lines I referred to in Chapter 10, but also along the lines of qualitative multiplicity. We can say that each individual is one, but from the perspective of duration, one is in fact many. While Freud does recognise the multiplicity of libidinal currents in Wolf-Man, he reduces the unconscious to 'the One', in this case, the impact/signification of the one wolf, the one and only Father. According to Deleuze and Guattari, instead of trying to unify or fix the flux, what analysts (psychoanalysts, biologists, sociologists, *et al.*) might do instead is to approach matters from the point of view of *intensity* (i.e. duration, qualitative multiplicity, *indivisibility*): 'the Wolf is the pack, in other words, the multiplicity instantaneously apprehended as such insofar as it approaches or moves away from zero, each distance being nondecomposable. Zero is the body without organs of the Wolf-Man.'[69] When Deleuze and Guattari speak of a 'body without organs' ('BwO'), they do not refer to a body that is literally without organs, but rather to what we may now describe as the 'body-becoming', the body of duration, the body of micromultiplicity and so *not* the body of molarity – not a body dissected into its various parts, but equally, not a body identified by species. One might say a body is zero – no value, or at any rate *nothing else* – in so far as it is becoming itself.

In this context, 'becoming-animal' may be interpreted as a way of linking up duration, micromultiplicity or molecularity, *and territory* – or rather, *territorialisation*. As Deleuze and Guattari put it, 'The wolf, as the instantaneous apprehension of a multiplicity in a given region, is not a representative, a substitute, but an *I feel*. I feel myself becoming wolf, one wolf amongst others, on the edge of the pack. A cry of anguish, the one Freud hears: Help me not become wolf (or the opposite, Help me not fail in this becoming)'.[70] This process is not a matter of representation or of signifiers, *per se*; '[t]he wolf, the wolves, are intensities, speeds, temperatures, nondecomposable variable distances. A swarming, a wolfing.'[71] 'Lines of flight or of deterritorialization, becoming-wolf, becoming-inhuman, deterritorialized intensities: that is what multiplicity is. To become wolf [or to become anything else] is to deterritorialize oneself following distinct but entangled lines'.[72]

In this account, it may seem that becoming- and becoming-animal are entirely anthropocentric concepts. However, in keeping with Deleuze and Guattari's overall philosophical approach, this is not the case. Perhaps the best example that doesn't involve a human (or indeed a critique of psychoanalysis) is the one in

---

[68] Gilles Deleuze and Félix Guattari, *A Thousand Plateaus*, p. 29.

[69] Gilles Deleuze and Félix Guattari, *A Thousand Plateaus*, p. 31.

[70] Gilles Deleuze and Félix Guattari, *A Thousand Plateaus*, p. 32.

[71] Gilles Deleuze and Félix Guattari, *A Thousand Plateaus*, p. 32.

[72] Gilles Deleuze and Félix Guattari, *A Thousand Plateaus*, p. 32.

which the philosophers describe the relation between certain orchids and wasps. Although no species is identified (perhaps in keeping with the authors' philosophical aversion to that second type of animal, the animal 'with characteristics'), one may surmise that the philosophers refer to any of the numerous genera and species of orchids with flowers whose shape encourages pseudocopulation by one or more usually *several* species of hymenopterans.   In *A Thousand Plateaus*, Deleuze and Guattari use such orchids and wasps to exemplify trans-territorialisation: 'The orchid deterritorializes by forming an image, a tracing of a wasp; but the wasp reterritorializes on that image. The wasp is nevertheless deterritorialized, becoming a piece in the orchid's reproductive apparatus. But it reterritorializes the orchid by transporting its pollen. Wasp and orchid, as heterogeneous elements, form a rhizome.'[73]

Two points require clarification here: on the one hand, it will be clear that Deleuze and Guattari are conceptualising territory in a manner that is itself at once existential and mobile – perhaps a version of body-as-situation (cf. Chapter 9). On the other hand, the reference to the rhizome invokes one of the key theoretical and political themes – or perhaps I should say *banners* – of *One Thousand Plateaus*. Deleuze and Guattari begin this work by berating modern cultures for adopting the tree and the conventional image of a root system as one of the dominant metaphors for explanation, and for life itself. As examples we might refer to the 'family tree', or to a genealogy that traces a social 'branch' 'back to its roots', or indeed to the 'tree of life' so often invoked by environmentalists and ecologists alike. What bothers Deleuze and Guattari is that this kind of explanation goes back not only to a 'transcendentalist' ontology, but one that proceeds in such a way as to divide the world according to a wholly hierarchical, and molar logic. We have only to think of the way in which some very wealthy and established families might use a representation of a 'family tree' to explain, literally and figuratively, 'who goes where', or who *comes* from where. The model is a vertical one in more ways than one.

To this, Deleuze and Guattari oppose the rhizome, a continuously growing underground stem that puts out, as the Oxford Dictionary puts it, lateral stems and 'adventitious' roots at intervals. Where the tree and roots are all about verticality, rhizomes are (mainly) about horizontality. As anyone who has grown mint will know, they are also in some sense irrepressible, and no doubt this too, makes them a better metaphor for two scholars concerned with stopping the fix, with freeing the flow from the tyranny of mechanism and representationalism. The following quote at once illustrates this aspect, and sheds additional light (perhaps one should say, fosters rhizomatic growth) with respect to the notion of the 'pack': 'a rhizome as subterranean stem is absolutely different from roots and radicals. Bulbs and tubers are rhizomes. Plants with roots or radicals may be rhizomorphic on other respects altogether: the question is whether plant life in its specificity is

---

[73] Gilles Deleuze and Félix Guattari, *A Thousand Plateaus*, p. 10.

not entirely rhizomatic. Even some animals are, in their pack form. Rats are rhizomes. Burrows are too, in all their functions of shelter, supply, movement, evasion, and breakout ... The rhizome includes the best and the worst: potato and couchgrass, or the weed. Animal and plant, couchgrass is crabgrass'.[74]

If 'rhizomaticity' offers an apt metaphor for the kinds of horizontal entanglements and ceaseless becomings that Deleuze and Guattari are so keen to theorise, it also says something about their take on semiotics. As I've already noted, Deleuze and Guattari declare war on semiology, and on any understanding of semiotics that privileges representation. They are, however, quite happy to think of semiotic relations along rhizomatic lines: 'A rhizome ceaselessly establishes connections between semiotic chains, organizations of power, and circumstances relative to the arts, sciences, and social struggles. A semiotic chain is like a tuber agglomerating very diverse acts, not only linguistic, but also perceptive, mimetic, gestural, and cognitive...'.[75]

Back to the orchid and the wasp: Deleuze and Guattari suggest that, however much the orchid imitates the wasp (and indeed ecologists would refer to this as an example of sexual *mimicry*), the most important thing going on, as far as they are concerned, is not to do with imitation or resemblance. It is 'a capture of code ... an increase in valence, a veritable becoming, a becoming-wasp of the orchid and a becoming-orchid of the wasp', and with it, 'an exploding of two heterogeneous series on the line of flight composed by a common rhizome that can no longer be attributed to or subjugated by anything signifying.'[76] One may question, as I did in the Appendix 2, if this is Deleuze and Guattari going too far down the road of anti-representation, but they are certainly onto something when they emphasise the easily taken-for-granted independence of the two creatures – indeed, the philosophers quote the biologist and entomologist Rémy Chauvin approvingly when he speaks of '"the *aparallel evolution* of two beings that have absolutely nothing to do with each other"'[77] – but at the same time the way in which orchid and wasp nevertheless *become each other* during 'pseudocopulation'. I put pseudocopulation in inverted commas because of course, from the perspective of either species (or rather, either individual organism), it is not *pseudo*copulation at all.

As may be clear, in this interpretation, becoming-animal is emphatically *not* about exchanging one molar state or category for another. On the contrary, it is a matter of something akin to a 'direct' *molecular* change, and exchange: the wasp takes away pollen thanks to the fact that the pollen will bond, however tem-

---

74 Gilles Deleuze and Félix Guattari, *A Thousand Plateaus*, p. 7.

75 Gilles Deleuze and Félix Guattari, *A Thousand Plateaus*, p. 7.

76 Gilles Deleuze and Félix Guattari, *A Thousand Plateaus*, p. 10.

77 Rémy Chauvin, in Gilles Deleuze and Félix Guattari, *A Thousand Plateaus*, p. 10. Italics in *A Thousand Plateaus*.

porarily, with a part of its body, and the wasp will leave sperm on the orchid's labellum. In the case of humans, one 'becomes-animal' in so far one gives up the molar certainties (or the 'certainties of molarity') for molecular flux with any other body. But this is not mostly a matter of a voluntary, conscious change or choice; it is a matter of affect, in the sense of a relationality that is not in itself instrumental (which is not to say that certain bodies might not try to render the relation instrumental, as occurs day in and day out with advertising, or with populist politics).

Before considering the significance of this conceptualisation for wildlife observation, and linking it to my analysis of anthropomorphism, I would like to refer to one strongly dissenting perspective vis-a-vis Deleuze and Guattari's characterisation of animals: that of the scientist and feminist philosopher Donna Haraway. Doing so has the double benefit that it not only tests Deleuze and Guattari's arguments, but also introduces new dimensions to the research.

It is pertinent to begin by noting that Haraway's research is associated with STS and with ANT. In the 1980s and 90s, she was best known for her remarkable history of primatology (and biological science more generally).[78] Haraway was, and remains centrally concerned with the kinds of issues raised by the so-called posthumanities, and this thanks not least for her feminist conceptualisation of cyborgs.[79] In the mid-2000s, her research took a surprising turn in so far as it started to focus on issues arising from what Haraway describes as the 'companion species'. In the first lines of *When Species Meet*,[80] Haraway describes her interest as follows: '(1) Whom and what do I touch when I touch my dog? and (2) How is "becoming with" a practice of becoming worldly? I tie these questions together in expressions I learned in Barcelona from a Spanish lover of French bulldogs, *alter-globalisation* and *autre-mondialisation*.'[81]

Three points leap up in this characterisation: first, the fact that Haraway refers to dogs, indeed to 'my dog', i.e. the pets for which Deleuze and Guattari reserve some of their most scathing discourse; second, the fact that Haraway speaks not of 'knowing' animals, or even 'gazing at them', but of *touching* them – thereby putting the focus on haptic 'contact zones' which were once almost literally overlooked by any animal studies; and finally, that in her account, it is not just becoming (Deleuze and Guattari, *passim*) but 'becoming *with*' – a preposition that serves at once to mark a distance from the Deleuzian perspective, and to announce an emphasis on companionship.

---

[78] See for example Donna Haraway, *Primate Visions: Gender, Race, and Nature in the World of Modern Science* (London: Routledge, 1989).

[79] Donna Haraway *Simians, Cyborgs, and Women* (London: Free Association Books, 1991).

[80] See Donna Haraway, *When Species Meet* (Minneapolis: Minnesota University Press, 2008).

[81] Donna J. Haraway, *When Species Meet*, p. 3.

In this context, Haraway suggests that 'companion species' is more a 'pointer' than a 'category' – a pointer for a form of relationality that she describes by referring to the etymologies and meanings of 'companion' and 'species': '*Companion* comes from the Latin *cum panis*, "with bread." Messmates at table are companions. Comrades are political companions. A companion in literary contexts is a vade mecum or handbook' (amongst other meanings); as a verb, the term also refers '"to consort, to keep company," with sexual and generative connotations always ready to erupt'.[82]

Where species is concerned, 'The Latin *specere* is at the root of things here, with its tones of "to look" and "to behold." In logic, *species* refers to a mental impression or idea, strengthening the notion that thinking and seeing are clones'; however, species is also 'about the dance linking kin and kind', with the ability to interbreed reproductively being the 'rough and ready requirement for members of the same species'.[83] Then again, other meanings of species include the ones figured in environmental/conservationist circles (as in 'endangered species'), but also in racist discourse: '[t]he discursive tie between the colonized, the enslaved, the noncitizen, and the animal – all reduced to type, all Others to rational man ... is at the heart of racism and flourishes, lethally, in the entrails of humanism'.[84] Woven into these meanings is the reductionist version of the typological female, conceived from the perspective of a reproductive function. As Haraway puts it, '*Species* reeks of race and sex; and where, and when species meet, that heritage must be untied and better knots of companion species attempted within and across differences. Loosening the grip of analogies that issue in the collapse of all of man's others into one another, companion species must instead learn to live intersectionally.'[85] The last being, in some sense, the aim of *When Species Meet* – if, as Haraway quotes Anna Tsing saying, '"Human nature is an interspecies relationship"', then the realisation of that relationship constitutes a path towards the *autre-mondialisation* mentioned at the start of the book.

Much of what I have just described here would appear to fit well with Deleuze and Guattari's approach. But Haraway expresses, if anything, a fury with the two philosophers. 'The making each other available to events that is the dance of "becoming with" has no truck with the fantasy wolf-pack version of "becoming-animal" figured in Gilles Deleuze and Félix Guattari's famous section of *A Thousand Plateaus*'; '[m]undane, prosaic, living wolves have no truck with that kind of wolf pack...'; '... I want to explain why writing in which I had hoped to find an ally for the tasks of companion species instead made me come as close as I

---

82 Donna J. Haraway, *When Species Meet*, p. 17.

83 Donna J. Haraway, *When Species Meet*, p. 17.

84 Donna J. Haraway, *When Species Meet*, p. 18.

85 Donna J. Haraway, *When Species Meet*, p. 18.

get to announcing, "Ladies and Gentlemen, behold the enemy!"'.[86] In the text that follows, Haraway effectively accuses Deleuze and Guattari of contradicting their own, or at any rate Deleuze's, 'flat ontology'. From Haraway's perspective, Deleuze and Guattari's conceptualisation of animals is 'a symptomatic morass for now *not* to take earthly animals – wild or domestic – seriously'.[87]

It is difficult not to interpret Haraway's response at least partly in terms of the affections [in the older psychological sense] of someone who clearly loves her dogs – but then the obverse point might also be made: the writing of Deleuze and Guattari clearly reflects *their* affections in the form of an unmistakable disgust for pet ownership. Whatever each party's inclinations, Haraway raises an important problem by suggesting that Deleuze and Guattari – like Jacques Derrida and other philosophers – fail to 'meet the gaze' of real animals, and to engage with a 'simple obligation' of companion species: to become curious about what a cat (or indeed, beyond-human animals more generally) 'might actually be doing, feeling, thinking, or perhaps making available'[88] to them (Haraway's example of the cat is a reference to the celebrated analysis by Jacques Derrida[89]). We return, by this route, to the issues I raised vis-a-vis Barro Colorado's trans-territorialised spider monkeys.

As part of this approach, Haraway references the philosopher, lawyer, and social reformer Jeremy Bentham (1748-1832), who in the 1820s established a kind of early road map for animal welfarism when he suggested that utilitarian principles ought also to apply to beyond-human animals. Simplifying somewhat, early utilitarian philosophers like Bentham propose that the guiding moral principle for undertaking any kind of action ought to be to produce the most good for an individual, or a population. Put differently, the fundamental ethic ought to be to produce *happiness* and pleasure in the greatest number of individuals, and so to spare those individuals from any avoidable pain. Across Bentham's works, those individuals are people, which is to say, humans. But in a now famous footnote to *An Introduction to the Principles of Morals and Legislation* (first published in 1789), Bentham notes how arbitrary it is to exclude animals from these principles. The question about nonhuman animals, he argues, is not, 'Can they *reason*? nor, Can they *talk*?' – a reference if not directly to Descartes, then certainly to a Cartesian variety of humanism – 'but, Can they *suffer*?'[90]

---

[86] Donna J. Haraway, *When Species Meet*, p. 27.

[87] Donna J. Haraway, *When Species Meet*, p. 29.

[88] Donna J. Haraway, *When Species Meet*, p. 20.

[89] Jacques Derrida, 'The Animal That Therefore I am (More to Follow)', translated by David Wills, *Critical Inquiry*, 28(2)(2002), 369-418.

[90] Jeremy Bentham, *An Introduction to the Principles of Morals and Legislation* (New York: Hafner Press, 1948), p. 311.

I might note in passing that, in the England of the first half of the nineteenth century (and far beyond), the last question would have had a particular resonance. At the very point in England's industrialisation when some kinds of animals had become the victims of a social viciousness – the preeminent example being that of the whipping of horses, which were quite literally failing to keep up with the increasing speed of industrial capitalism – Bentham asked a question that dramatically undermined the widely accepted boundary between the human and the beyond-human.

A pretty direct line of intellectual descent can be traced between Bentham and, fully 150 years later, the proposals of Peter Singer[91] and the late twentieth century animal rights movement. As Singer notes, Bentham went so far as to entertain the idea that beyond-human animals might one day themselves have rights. However, Singer also notes that, while Bentham and other utilitarian thinkers were influential in the establishment of laws that prohibited gross cruelty to animals, there was no question but that the priorities of humans should come first.[92] And indeed, in a perhaps less often quoted passage of *Principles of Morals and Legislation*, Bentham argues, extraordinarily disingenuously, that there 'is a very good reason why we should be allowed to eat such non-human animals as we like to eat: we are the better for it, and they are never the worse.'[93]

Returning to Haraway, the philosopher is keen to address many additional questions which neither Bentham nor most zoologists raise, but which in her view may be equally important: 'Can animals play? Or work? And even, can I learn to play with *this* cat? Can I, the philosopher, respond to an invitation or recognize one when it is offered? ... And what if the question of how animals engage *one another's* gaze *responsively* takes center stage for people?'[94]

For Haraway as for many other writers in the burgeoning field of animal studies, these are not questions that might only be asked about pets, or indeed by philosophers with pets; similar issues may be raised about the research of ethologists with wild animals, which as Haraway shows, 'engage animals as objects of their vision, not as beings who look back and whose look their own intersects, with consequences for all that follows'.[95] In an extended account of the research of the zoologist (now bioanthropologist) Barbara Smuts, Haraway reveals how Smuts tried to write her PhD by following standard-for-the-time (the

---

[91] Peter Singer, *Animal Liberation* (New York: Harper-Collins, 1975).

[92] Peter Singer, 'Preface', in Peter Atterton and Matthew Calarco (Eds.) *Animal Philosophy: Ethics and Identity* (London: Continuum, 2004), xi-xiii, p. xi.

[93] Jeremy Bentham, *An Introduction to the Principles of Morals and Legislation* (New York: Hafner Press, 1948), p. 311.

[94] Donna J. Haraway, *When Species Meet* (Minneapolis: University of Minnesota Press, 2008), p. 22. Italics in the original.

[95] Donna J. Haraway, *When Species Meet*, p. 21.

1970s) scientific advice that, when studying baboons in the Great Rift Valley, she should be as 'neutral as possible, to be like a rock, to be unavailable, so that eventually the baboons would go on about their business in nature as if data-collecting humankind were not present'.[96] In fact, it was not until Smuts learned the semiotics of the baboons' social cues and responded appropriately that the baboons began to accept her, and act 'normally' around her.

From Haraway's perspective, Deleuze and Guattari fail to acknowledge these kinds of issues and possibilities; they also fail to say anything about actual wolves (or most other animals). 'I know that D&G [Deleuze and Guattari] set out to write not a biological treatise but rather a philosophical, psychoanalytic, and literary one requiring different reading habits for the always nonmimetic play of life and narrative. But no reading strategies can mute the scorn for the homely and the ordinary in this book'.[97]

Worse, according to Haraway, their privileging of individual animals makes their philosophy guilty of a 'sublime ecstasy'. Haraway quotes Deleuze and Guattari as saying '"Wherever there is multiplicity, you will find also an exceptional individual, and it is with that individual that an alliance must be made in order to become-animal"' – and that, says Haraway, 'is a philosophy of the sublime, not the earthly, not the mud; becoming-animal is not an autre-mondialisation'.[98] Haraway also quotes Deleuze and Guattari as suggesting that *anyone who likes cats or dogs is a fool* (italics in Deleuze and Guattari's own text), and she even accuses them of misogyny because they suggest that '"Ahab's Moby Dick is not like the little cat or dog owned by an elderly woman who honors and cherishes it"'.[99] As Haraway puts it, 'The old, female small, dog- and cat-loving: these are who and what must be vomited out by those who will become-animal. Despite the keen competition, I am not sure I can find in philosophy a clearer display of misogyny'.[100]

If one accepts Haraway's assessment, then it is easy to understand why Haraway (or any other critical scholar: pet lover, pet hater, or none of the above) *would* be incensed not only by Deleuze and Guattari's take on animality, but by their philosophy more generally. Yet, after acknowledging the valuable contribution that Haraway makes with respect to questions involving the gaze, ethology, and what might be termed 'everyday' animals, it seems to me that her reading fails to see the wood for the trees, or perhaps I should say the animals for the pets. Anyone who accuses especially Deleuze, but also Deleuze and Guattari of a sublime ecstasy has either not understood Deleuze's ontology of immanence,

---

[96] Donna J. Haraway, *When Species Meet*, p. 23-24.

[97] Donna J. Haraway, *When Species Meet*, p. 29.

[98] Donna J. Haraway, *When Species Meet*, p. 28.

[99] Donna J. Haraway, *When Species Meet*, p. 30.

[100] Donna J. Haraway, *When Species Meet*, p. 30.

or is making the rather bold claim that Deleuze and Guattari flatly contradict that ontology by privileging individuals or certain kinds of animals. The equivalent would be for this volume to suggest that Deleuze and Guattari must hate trees and be 'anti-environmentalist' because they are against what they describe as arborescent culture, and in favour of rhizomes.

Evidence of a misreading, or misinterpretation is particularly evident in Haraway's claim that Deleuze and Guattari believe that all dog- or cat-lovers are fools. Deleuze and Guattari do appear to be contemptuous with respect to the very idea of pet ownership. But if one returns to the text that Haraway cites, one finds that Deleuze and Guattari are not actually claiming that all pet lovers are fools. The full text is this: 'These animals [individuated animals, family pets, sentimental, Oedipal animals] invite us to regress, draw us into a narcissistic contemplation, and they are the only kind of animal psychoanalysis understands, the better to discover a daddy, a mommy, a little brother behind them (when psychoanalysis talks about animals, animals learn to laugh): *anyone who likes cats or dogs is a fool*. And then there is a second kind...'[101]

My own interpretation – one that I believe is consistent with Deleuze and Guattari's use of italics – is that the authors are suggesting that *psychoanalysis* treats patients who like cats or dogs as fools. Deleuze and Guattari are explicitly critical of the way in which Freud interprets Wolf-Man's dream, and this not least on the grounds that his interpretation takes the patient for a kind of fool.

The text cited above – and the passages I cited earlier in these pages – make clear two further points which would confirm my sense that Haraway's otherwise insightful analysis of 'companion species' somewhat looses the plot when it vents its anger at Deleuze and Guattari. First, Deleuze and Guattari are clear that they are critical not just of keeping pets, but of *individuated* (as distinct from *individual*) animals more generally; this is part of the reason why they raise the question of the 'pack', the 'swarm'. In this context, it seems an embarrassing misunderstanding at best, and a reading in bad faith at worst to accuse Deleuze and Guattari of falling for a kind of romantic-liberal discourse of the sublime. An alternative interpretation is that, having insisted on the importance of the 'pack', Deleuze and Guattari are nonetheless keen to acknowledge that relations *will* and indeed *must* involve individuals – even individual organisms. To deny that, or to appear to deny that, would be to fall for an absurd collectivism, and to contradict the principle of immanence.

Second, and as I noted earlier, Deleuze and Guattari make it clear that any one animal *can be all three kinds of animal* (individuated, 'molar', and 'demonic' or we might also say 'molecular'), and this complicates any suggestion that the philosophers are being dichotomous or dualistic. If a dog or a cat or any other creature can be treated as a 'demonic' animal – or as Deleuze and Guattari also make clear, if an elephant can be treated as a pet and of course, both as molar

---

[101] Gilles Deleuze and Félix Guattari, *A Thousand Plateaus*, p. 240.

animals – then any suggestion that the authors are simply in the business of sublimating animality becomes rather difficult to sustain.

As regards Haraway's claim of misogyny, certainly the opposition between Ahab in Moby Dick and an 'elderly woman' with a pet suggests a typically sexist, and indeed *ageist* opposition. Questions can and also *have* been raised about Deleuze and Guattari's suggestion that feminist politics should move beyond the molarity of a specifically female subjectivity.[102] But to accuse Deleuze and Guattari of the 'clearest' instance of misogyny in philosophy is, as the colloquial English expression puts it, over the top.

Back to pets: Haraway herself arguably runs the risk of sublimating the issues involving *ownership* – the last being a term which raises the question of the fundamental asymmetry that really does render keeping pets immensely problematic, and this even if *of course* pet ownership, like the ownership of all sorts of other 'things' (in Latour's parlance), is not simply reducible to a capitalist logic. As Haraway makes clear, there can be a liveliness about relationships with 'companion species' that is not only meaningful in its own right, but which provides deep insights to human/beyond-human animal relations more generally. That much is not in question, and is a valuable contribution to the discussion. What *is* in question, but what Haraway does not give enough weight to, is the extent to which the 'surplus of affect' that is revealed by pet ownership is likely to rest on a huge differential in power between the pet, and the pet owner. To say the last is not to automatically assume an identitary relation (animals *as* pets), let alone the *inferiority* of the animals as pets; to return to an earlier Deleuzian/Spinozan theme, nobody knows what a body can do, and indeed there are countless examples of animals that have rescued their owners, detected hitherto unnoticed threats, or indeed have attacked and even killed their owners. The last points not withstanding, the inverse order is the one that typically prevails: it is the owners that more often than not attack the animals, and the owners that kill them, even if late in a pet's life this action is performed, or at least justified, as an act of kindness.

Where Haraway is on better critical ground – and where, in a different way, a certain romanticism may be attributed to Deleuze and Guattari – is in her suggestion that, when it comes to animals, Deleuze and Guattari don't seem very interested in the most 'ordinary' creatures (their many references to rats nothwithstanding). And indeed, a reader might be forgiven for being left with the impression that Deleuze and Guattari's emphasis on becoming, on deterritorialisation, lines of flight, and the molecular almost invites one to adopt a totalising ethic against being, territorialisation, the molar, or indeed, 'staying put'. If a certain masculinism *is* to be attributed, perhaps it would entail a celebration

---

[102] See for example the analysis offered by Claire Colebrook in the introduction to Ian Buchanan and Claire Colebrook, *Deleuze and Feminist Theory* (Eds.) (Edinburgh: Edinburgh University Press, 2000), 1-17.

of nomadism – though of course, precisely the opposite argument might also be made: that anyone who simply equates nomadism with masculinism runs the risk of implicitly associating femininities with a sedentary, which is potentially also to say a *domestic*, nature. One way or the other, it is of course one thing to point out and problematise an identitary logic, and another to suggest that one can ever really leave it behind.

I would like to conclude this chapter by returning to the question of the natures of animality, and more specifically to wild animality: what might the theories/philosophies presented in this chapter tell us about wild animality? Is there such a thing as a wild animal, and by implication, 'wildlife observation'? To answer this question, it may be useful to return to the example of 'Bart', the bear trained by Doug Seus. When it comes to wild animals, few animals, at least for an Anglo-American or European imaginary (if one can so crudely equate imaginaries with regions) offers a better example than the brown bear, and more specifically, the subspecies that goes by the name of the grizzly. And yet, clearly it is possible to break down the ostensibly natural barrier that separates humans from this archetypal wild animal. This point is not actually contradicted by the fact that such bears have been known to kill trainers (which indeed they have[103]). What such deaths illustrate is that the animals have the capability to kill humans; but then the same might be said about pet dogs, provided that there is a sufficiently big differential in power between the specific dog and a would-be victim. An argument can be made that, in the case of some species, there is an affection that is likely to be particularly damaging or even lethal if anything 'goes wrong'. With Uexküll we might add that some *umwelts* may also be particularly resistant to domestication. But that is different from saying that there is such a thing as an essential difference, based on the criterion of 'wildness', between say, a grizzly bear and a domestic cat, a cow and an elk. Let a cat become feral, and after one or two generations those cats may themselves need 'cat trainers'. Let cows 'roam free', and soon it might prove to be quite difficult to milk them. In each case, what is really at stake is a combination of the individual affections, and the assemblage that the creature forms, or *becomes* a part of. From this perspective, what matters is not only the 'species-being', let alone an essentialised conception of species, but what might be termed the *species-becoming*. Put enough effort and time in, and wolves may become dogs. But mistreat even a pet dog sufficiently and the dog may become at least as dangerous as any wolf supposedly is. Or to use a somewhat more positive example, let a cow into an Alaskan wilderness, and if it is wily enough, it may elude even the most determined of *human* hunters.

Does this mean that 'wildlife observation' is a misnomer, and so does not, strictly speaking, exist? Not at all. In so far as some animals in some

---

[103] See for example, Steve Gorman, 'Hollywood Grizzly Bear Kills Trainer', *Reuters* online, 23 april 2008, <https://www.reuters.com/article/us-bear-attack-idUSN2343261920080423> [Accessed 1 May 2019].

circumstances are indeed more independent of human cultural formations than others, and in so far as some territories are themselves less affected by human intervention than others (we return to the point made in Chapter 13), then certainly it is legitimate to represent this difference with reference to categories such as the wild animal, and indeed, a wilderness. However, the key lies not in some itself *essential* distance between human and beyond-human bodies (e.g. the conventional dualism of the 'domestic' vs. 'wild'), but in situations and trajectories. From this perspective, any body has wild(er)ness merely by virtue of having a potential for becoming. It may seem that this principle only applies to organismic bodies (the biotic), but just as mountains move, a book will become mouldy with damp, and indeed most materials left to their own devices will transform over time. Of course, with some materials this process may be a lot slower than with others, and perhaps the ultimate sign of 'humanisation' is a material like plastic that resists becoming in the way that many other everyday materials do not. The current obsession with plastic is a surprisingly belated recognition of this kind of difference – and one that acts as a cautionary note to all those who would use a Deleuzian/Guattarian philosophy to dismiss any kind of 'humanism' whatsoever.

Conclusions to Part 4

Let's return to the encounter with the spider monkeys. The kind of spider monkey that most tourists would perhaps have liked to encounter was the *represented* spider monkey, the monkey 'with characteristics' seen on TV or in other media – the monkey whose remarkably long limbs, and quasi-human face invites relatively explicit forms of anthropomorphism and cosmomorphism. The monkey that, shown in close-ups, dramatically extends the affections of the human observer in such a manner as to enable that observer to 'be there', *right up there in the forest canopy*, dangling from branches with the electronic equivalent of a prehensile tail, sharing the most intimate aspects of its life. Alas, as I explained at the beginning of this part of the volume, what those of us on the ground got instead was at first no sighting whatsoever, and then a glimpse – almost literally a (monkey) *line of flight*.

That this observational differential existed reveals the discontinuity between two sets of affections and affects which, like all affections and affects, are 'context dependent'. Had the tourists never seen TV or close-up photos of non-human animals, doubtless their expectations would have been quite different. Then again, had the spider monkeys been, say, olive baboons (*Papio anubis*) in the eastern African savannah, or even in relatively open woodland, things might also have been very different, albeit for a different set of reasons. If there was a conflict between the two assemblages (that of the spider monkeys foraging in the tropical forest, and that of monkeys or other creatures represented via wildlife TV), it was at least in part because the two entailed very different spatialities. Barring aids to observation of the kind that some biological reserves make available (e.g. 'canopy walks'), a tourist on the ground, unlike a prospective tourist watching TV, is constrained by the kinds of conditions that I described in Chapters 4 and 11. Those conditions can be theorised, initially at least, via a combination of physical geography (including landscape ecology), an ecological account of the kind proposed by John Endler, and indeed the kind of ecology of perception described by James Gibson. On one level Barro Colorado's forest and topography really *do* constitute something like a surface with certain barriers and obstacles which tourists must overcome if they hope to see wildlife. In keeping with this perspective, Endler's characterisation of different predator-prey relations certainly does allow us to better understand why tourists (or anyone else, for that matter) might find that, in already difficult visual circumstances, certain species' morphologies and/or ethologies might make it even less likely that an encounter will occur. And even if it *were* to occur, the encounter might well be as fleeting as the one I described at the beginning of Part 4. Put simply, there is an ecological and geographic materiality which is not only undeniable as a bio-physical materiality, but which can be accurately and usefully characterised via positivist ecological, physical geographic, and ecological-psychological theories.

There are, however, good reasons for going beyond such perspectives when it comes to developing a critical account of wildlife observation amongst tourists visiting tropical forests. The first returns us to a point I first made in Part 2: positivist research tells us little or nothing about the semiotic *intríngulis* of encounters with wildlife. The 'new' cultural geographers were right, in this sense, when they foregrounded the importance of symbolic forms in relation to socio-cultural experiences of landscape, and of place and space more generally. How one interprets the flora and fauna found in a tropical forest does entail a semiotic dimension, even it is not one that is best conceived along the Saussurean, or post-Saussurean lines long favoured by cultural geographers.

Then again, there are also the more-than-semiotic, which is to say socio-cultural materialities that are underscored by the earlier cultural sociologists, and also by analyses of the kind developed by Doreen Massey or Henri Lefebvre. In the vocabulary of Henri Lefebvre, the three-fold articulation of social space along instituted, represented, and lived dimensions points to the importance of recognising that space is not only *not* 'empty', but is constituted by objects, processes and relations which are mediated by institutions, discourses and beyond-linguistic representations, as well as by everyday life itself. What is true for the explicitly human social spaces that Lefebvre mostly refers to is also true for the biological reserves which most tourists visiting tropical forests travel to, and which may themselves be described with reference to a triadic logic (of instituted, represented, and lived space).

And yet, STS and ANT scholars such as Bruno Latour and Sarah Whatmore rightly warn us of the importance of not straying too far in the direction of a representationalist and culturalist perspective – one that overlooks the significance not just of 'things' (Latour, *passim*), but of *hybrid*, and *topological* spaces – spaces and with them places where bodies and trajectories that have hitherto been quite separate suddenly become 'folded' onto one another as in Michel Serres' marvellous example of the handkerchief. These kinds of insights seem particularly important in the case of wildlife tourism, ecotourism, or nature tourism more generally – forms of tourism which typically entail the equivalent of parachuting into a 'wilderness', and staying there relatively briefly. In such contexts, it is not legitimate to concentrate entirely, as the earlier cultural geographers and sociologists did, on socio-cultural space. Nor is it legitimate to simply oppose that space to nature. On the contrary, it is necessary to propose ways of avoiding the purification of culture from nature, and of nature from culture (or all that is conventionally associated with each concept), and that is precisely what ANT and the hybrid geographers try to do. As I explained in Chapter 13, hybrid geographies invite us to reject what Whatmore describes as 'the purifying impulse to fragment living fabrics of association'; instead, the researcher must seek, in Whatmore's terms, to understand the world 'as an always already inhabited achievement of heterogeneous social encounters'.[1]

Latour's and Whatmore's proposals seem particularly germane to the present study when it comes to recognising the ways in which wildernesses and wild animals may get caught up in networks which render them something different from the sublimated image of animals and places wholly undisturbed by humans. Across this volume, most of the examples I have offered have been drawn from my ethnographic research on Barro Colorado, a site that perhaps illustrates particularly strongly the wisdom of the emphasis on hybridity. However, a similar point might be made about wildlife observation in virtually any tropical forest, or indeed any other kind of biome elsewhere in the world. To be sure, merely to observe wildlife must constitute a hybrid practice in its own right if only because it entails precisely the kind of 'co-production' that Whatmore highlights via her work.

The above strengths notwithstanding, as I explained towards the end of Chapter 13 hybrid geographies, like the critical conception of hybridity itself, have certain theoretical tendencies which are themselves problematic for someone interested in studying wildlife observation in tropical forests. First, the very notion of hybridity may inadvertently promote a dualistic form of writing in so far as its terms of engagement continue to be that of the 'mixture', the 'combination' of two elements that are conceived as being more or less separate. Both Latour and Whatmore work against any such tendency, noting that it is not a case, as Whatmore puts it, of a 'one plus one', or 'mixture of two pure forms'.[2] But it is difficult to get past the separation when the very term one uses on some level reiterates that separation.

Second, Latour and Whatmore arguably under-theorise the question of animality, and indeed their analyses continue to put the emphasis on human sociality, notwithstanding especially Whatmore's suggestions to the contrary. Latour seems far more comfortable studying the work of scientists in cutting edge laboratories, and Whatmore's study still seems very close to a Foucaultian analysis of disciplinary techniques – no bad thing in itself, but something of a disappointment given her suggestions that she is moving on from that kind of research. This being so, both authors' proposals may go too far in the direction of denying a place and space for a wilderness that is not already subsumed within human networks.

It is for reasons like these that in Chapter 13 I sketched an alternative conception of wilderness, or to borrow Whatmore and Thorne's expression, wild(er)ness. And in Chapter 14 I considered the proposals of Uexküll, and then Deleuze, Guattari and Haraway. Especially Deleuze, and Deleuze and Guattari's

---

[1] Sarah Whatmore, *Hybrid Geographies*, p. 3.

[2] Sarah Whatmore, *Hybrid Geographies*, p. 3. See also Bruno Latour, *Pandora's Hope: Essays on the Reality of Science Studies* (Cambridge, MA: Harvard, 1999).

proposals regarding body, affect and assemblage, as articulated via the Spinozan/ Deleuzian principle of immanence, allow us to understand how it is that sloths 'in general' may be very slow, but one can bite a scientist; that a cow may not only survive, but even elude high tech human attempts to recapture it, in the Alaskan wilderness; that Bart the grizzly bear can learn to wrestle with Doug the grizzly bear trainer; or indeed, that Barro Colorado's spider monkeys may seem identical to any non-reintroduced spider monkey, and yet be quite different for historical reasons.

To reiterate a point made in Part 3 via Manuel DeLanda, the regularities of species, regarded as molarities, are not what is at issue; to deny their existence would be absurd. It would also be absurd to deny that many of the regularities *do* have a basis in evolutionary processes extending back millennia. What *is* at issue is the ontology that one employs to explain the nature of individuals. The conventional ontology starts from essentialised types, and renders any bodies that do not fit with such types as being exceptional; by contrast, the principle of immanence is inclusive in so far as it starts from the 'body itself', which by Deleuzian definition, cannot be an exception. Exception is, from this perspective, the rule.

In Parts 2 and 3, I developed interdisciplinary versions of this approach via the concepts of immediate, dynamical and mediate modes of observation (semeiotic-phenomenology), and via the concepts of body, affect, and assemblage (bio-cultural). Given the analysis developed in this part of the volume, how might such an approach be expressed ecologically and geographically? I propose that the concepts of trajectory, encounter, and heterotopia (the last term suitably adapted) provide homologous eco-geographic tools with which to analyse wildlife observation amongst tourists visiting tropical forests. This being the case, I would like to say something about each of these concepts now.

Let's begin with trajectory. It may be acknowledged from the start that the concept has a predominantly physical (or 'physicist') connotation, i.e. trajectory as the path that a body with mass, and in motion follows through space as a function of time. Certainly the bodies that partake in wildlife observation (not least, those of the observer and the observed) have trajectories in this sense of the expression, which involves a quantitative multiplicity. From the moment of birth, each tourist will quite literally have travelled along a certain path, starting from the place of birth to the spot where an encounter with wildlife occurs. This aspect of trajectory is of course a significant one, and any ethnographic research into tourist experiences would need to find out about it, if only to obtain a sense of where else a visitor has been before reaching any given destination (a point I will consider in Volume 2).

That said, as developed in this study, the concept of trajectory takes us back to Bergson's notions of duration and qualitative multiplicity, and also to Maurice Merleau-Ponty and Simone de Beauvoir's notion of a body *as* situation (cf. Chapter 9). What is involved is not just a series of 'points' travelled by a body regarded as a machine-like vehicle. What there is instead is a vital entity that emerges, develops, and has a motility that is charged with a certain intentionality. This intentionality need not necessarily be conceived along the lines of a consciousness of the kind privileged by the earlier Merleau-Ponty, or in a different way, by the earlier forms of cultural sociology and geography. What is at stake is more generally conceived as a cross between Uexküll's 'plan', and an individuality understood along the lines of Spinozan/Deleuzian principle of immanence (which is of course *not* to deny that there may be a certain consciousness as well). Put succinctly, a body's trajectory is a *dynamical situation*, the ongoing and changing 'first co-ordinates' that Merleau-Ponty described, 'the anchoring of the active body in an object, the situation of the body in face of its tasks'.[3] But this bodily space, this bodily situation entails a multiplicity understood from the perspective of duration, or to use the Deleuzian concept, a becoming. Duration and becoming are also what make the trajectory, like the body itself, dynamical.

As Bergson would point out, trajectory conceived in this manner is indivisible. One only has to think, in this sense, of an individual person who is born, grows up, decides to travel for reasons to do with tourism and wildlife observation, and encounters one or another animal in a tropical forest. The person is no more 'divisible' than is the other animal in the encounter. This conception may suggest a kind of seamless flow, a trajectory that is entirely unified in the way that a gesture, or that jumping from one tree to another, are. Put differently, it may be assumed that a trajectory entails crossing one same space, e.g. a physical space conceived along the lines of the double illusion, or indeed 'empty space'. I hope that it will be clear by now why this would entail a mistaken understanding of trajectory. From the perspective of Henri Lefebvre, space is never empty. And from the perspective of Michel Serres, however much a line might be traced across a metric space, the sudden folds and rips of modern cultures would virtually guarantee a 'topological line', which is in some sense to say no *line* at all. To insist on the indivisibility of trajectory as trajectory is not necessarily imply a uniformity of space – especially not in the case of tourists' trajectories.

I've said enough to give some sense of how I understand trajectory; let's now turn to the concept of encounter. The reader may recall that, as part of his account of the stages of predation, Endler distinguishes between encounter, detection, identification, approach, subjugation and consumption (cf. Chapter 11). In my reference to Endler, I ignored subjugation and consumption, arguing that tourists don't generally subjugate, let alone literally consume, the wildlife they observe. However, I have observed instances where guides have themselves caught one or another creature so as to be able to show it at close range to visitors. Moreover, I have also noted that tourists do engage in the *semeiotic* consumption of wildlife.

---

[3] Maurice Merleau-Ponty, *Phenomenology of Perception*, translated by Colin Smith (London: Routledge, 2002 [1945]), p. 115.

These kinds of issues to one side, in general, 'encounter' as I conceive it involves a combination of what Endler describes as encounter, detection, identification and approach – but all these understood as an outcome of the crossing of at least two trajectories, at least two *durations*; and as a matter of a logic of action and reaction, affect, and with these a chiasmatic relation between the human and the beyond-animals.

The last points require clarification. While Endler distinguishes between several different stages of predation, when it comes to wildlife observation, the encounter occurs as soon as the human observer becomes aware of, and reacts by paying attention to, one or another creature. This is not to suggest that the observed animal may not itself be an observer; a creature may also detect the presence of the human observer, and indeed, may well do its best to flee, or to adopt any of the defences described by Endler (though again, what exactly happens will be contingent on both the species-related affects, and the haecceity of the individuals). In many cases tourists might themselves be detected by beyond-human animals, without the *visitors* being aware of it. Strictly speaking, the latter scenario also constitutes an encounter, but from an anthropocentric *tourist* perspective, it will not *count* as encounter; the encounter will occur, and will matter only if the *visitor* becomes aware of, and pays attention to, the animal. This 'paying attention to' will eventually always entail what I described in Part 2 as a mediate mode of observation, i.e. one that entails mediation by primarily symbolic forms, and thereby the kind of interpretation that is of interest to hermeneuticians, and indeed to those who specialise in so-called 'heritage interpretation', i.e. formal or informal ways of teaching and learning about nature.

As this account begins to make clear, there is a degree of intentionality and purposiveness about encounters – one which may be more or less strongly guided by discourse. That said, it is necessary to add at least three caveats to this account. First, encounters always entail the element of chance. Second, while one might say that all encounters 'tend' towards mediateness, some encounters may be more a matter of immediacy, or dynamical modes of observation. Then again, even a 'controlled' approach to a 'known' animal, preceded by a guide's indexes and introductory commentary, may suddenly change if, for example, the observed animal unexpectedly turns and attacks the guide or the visitors.

The third caveat links chance to assemblage and thereby to a broader context. Had the visitors in Part 4's opening example been trying to observe spider monkeys on the Panamanian mainland, they might well have found that the monkeys fled their presence in the assumption that any human might well be a prospective hunter. Then again, had the tourists been on a tour to what is locally known as 'Monkey Island' (a part of the Panama Canal that is to the south of Barro Colorado), an encounter would almost have been guaranteed because private tour operators have effectively trained at least capuchin monkeys (*Cebus capuchinus*) to board boats, so that they may be fed by tourists. From the perspective developed throughout this volume, wildlife encounters obey at once a logic of the haecceity of this or that dynamical body, its affective relations with other bodies, but then again, the kinds of assemblages which work at once to enable and constrain the encounters. The assemblages in question include, of course, ecologies of the kind I described in Chapter 11, but also the ones generated by tour operators (and human actors more generally).

Let's now turn to the final concept in the triad, that of heterotopia. It is relevant to preface the comments that follow by returning to Doreen Massey's suggestion that space should be recognised as the product of interrelations, as 'the sphere of the possibility of the existence of multiplicity in the sense of contemporaneous plurality; as the sphere in which distinct trajectories coexist; [and] as the sphere therefore of coexisting heterogeneity'.[4] 'Without space, no multiplicity; without multiplicity, no space. If space is is indeed the product of interrelations, then it must be predicated upon the existence of plurality', and multiplicity and space are co-constitutive.[5] From this perspective, heterogeneity is not the exception, it is the *rule* when it comes to space. However, as I began to suggest in Chapter 12, there are different kinds of multi-spatiality, and here I would add, different kinds of spatial heterogeneity and multiplicity. It seems to me that what characterises the spatiality of wildlife observation in the context of tourism is not just the coming together of multiple trajectories per se, but much the kind of heterotopic logic described by Foucault: there is a kind of coming together in one site, or via one practice, of all the worlds, or at any rate, a 'sample' of such worlds, with important consequences for the nature of the site, and for the experience of the site.

This kind of space at once reflects, is made possible by, and produces the kind of topological space described by Serres. To continue with Serres' analogy, it is not just that one handkerchief is folded over or crumpled; it is that several handkerchiefs are crumpled together 'in one place', enabling the relatively sudden juxtaposition of an extraordinary cross section, and interconnection of strata (Deleuze and Guattari, *passim*). It may well be that this kind of multi-spatiality has long been the rule in modern cultures; however, in destinations that attract tourists seeking wildlife encounters, the heterotopic space acquires a hybrid heterogeneity that is quite different from the one found in modern, everyday urban places. It may well be that in such places there is a kind of secret, or 'nocturnal' hybrid heterotopicality which eludes observation by all but the most specialised or determined of observers. By contrast, in many of the most popular destinations for wildlife tourism, there is an overtly hybrid heterotopicality: visitors and beyond-human animals congregate, and do so quite publicly, in one place.

---

[4] Doreen Massey, *For Space*, p. 9.

[5] Doreen Massey, *For Space*, p. 9.

Conclusions to Volume 1

Wildlife Observation from a
Geosemeiotic Perspective

Our guide suddenly stops and, turning towards us, points to her nose. A pungent smell fills the air – a smell that, to this observer at least, appears to combine something like a human body odour with a stew well advanced in its preparation. A second later we hear a crash in the undergrowth and catch a glimpse of several plumply hirsute bodies quite literally turning tail and disappearing into the forest. When the commotion dies down, our guide explains that we've just surprised a herd of collared peccary (*Pecari tajacu*). Peccary, she says, are wild relatives of pigs. In so far as peccary have a keen sense of smell but relatively poor eyesight, it's likely that we inadvertently surprised the herd by approaching it from downwind. On Barro Colorado, she further explains, peccary can often be found eating fruit discarded by monkeys high up in the island's trees. While the *species* is not considered to be critically endangered, in Panama as in other parts of Central America, a combination of hunting and habitat loss have led to a drastic reduction in numbers, and even to local extinction in areas where peccary were once common.[1]

I later learn that in the US, peccary are sometimes referred to as musk hogs in recognition of their capacity to produce the strong odour that I've just referred to. This capacity involves a prominent skin gland, known by scientists as the dorsal gland, which is located in peccaries' rumps, and opens via a nipple-like papilla. Peccary can use this gland either to rub their secretions onto surfaces, or when alarmed, to forcibly eject the glandular fluids in jet-like streams. The fluid contains a chemical cocktail that includes geranylgeraniol (a diterpene alcohol), squalene (a 30-carbon organic compound), and isomers of springene.[2]

Several different theories have been proposed regarding the function of the ejection of these compounds by peccaries. Some scientists believe they may act to repel predators, while others suggest that the main function involves pheromonal activity[3] (presumably these hypotheses are not mutually exclusive). Whatever the precise ethology, the effect is to fill the environment with what to 'us humans' is a certain *redolence*, and it is this that we, or perhaps I should say that *some* of us, perceive during the encounter. And indeed, if I put scare quotes around 'us humans', it is because somewhat astonishingly, in discussions with visitors after the

---

[1] The species was put on the IUCN's Red List of Endangered Species in 2011, four years after I began the research for this book. See http://www.iucnredlist.org/details/41777/0 [Last accessed 28 October 2016].

[2] John S. Waterhouse, Jia Ke, John A. Pickt and Paul J. Weldon, 'Volatile Components in Dorsal Gland Secretions of the Collared Peccary, Tayassu tajacu (Tayassuidae, Mammalia)', *Journal of Chemical Ecology*, 22:7 (1996), 2459-69. Springene is a diterpene homolog of *ß*-farsene; intriguingly, *ß*-farsene is itself a compound that is released by aphids as an alarm pheromone. Note: Since the publication of Waterhouse et al.'s article, the collared peccary has been reclassified as *Pecari tajacu*.

[3] John S. Waterhouse, Jia Ke, John A. Pickt and Paul J. Weldon, 'Volatile Components in Dorsal Gland Secretions of the Collared Peccary', p. 2460

event, some members of the group explain that they didn't really notice the smell until the guide mentioned it.

* * *

In the introduction to this volume, I suggested that the concluding chapter would draw the different perspectives presented across the work into a single approach – what I described as a *geosemeiotic* approach. The following statements will do just that by articulating the semeiotic-phenomenological, bio-cultural, and eco-geographic perspectives. In these conclusions I will do no more than to provide a very general account that will then be specified in Volume 2 (which offers a more concrete analysis involving visitors touring Barro Colorado Island). While I have explained that the overall research is concerned with tourists visiting tropical forests, I hope that the generality of the presentation which follows will make it easier for researchers interested in wildlife observation in other biomes (and perhaps even in the theory of observation more generally) to consider whether any of what follows may also apply in/to their own research contexts.

1. *Wildlife observation entails encounters with/between dynamical bodies.* The bodies are dynamical initially in the sense that Peirce gives to dynamical objects. This is, ontologically speaking, a principle of realism. It is a way of saying that the bodies in question exist, in and of themselves, whether someone thinks about them or not, whether someone *represents* them or not (in this statement it might be more accurate to say *some one*; of course, if enough tourists observe wildlife, or one particularly destructive individual does so, all may change). But it is also a way of saying that the bodies have at least a degree of agency – an agency which may help to determine how one body perceives, and *observes* another.

In the example this chapter began with, the peccaries of course exist whether the tourists I referred to above observe them or not, whether they take pictures of them or not. That much ought to be uncontroversial. I am, however, also suggesting that the peccary begin to determine how the visitors observe and represent them; that is part of what I mean by agency. This suggestion should itself be regarded as a matter of good common sense; but it is something that is arguably denied, however implicitly, by both structuralist and poststructuralist approaches to semiotics, many if not most of which have traditionally assumed that signs are human constructs, and also entirely *arbitrary* constructs, which implies that the signs are entirely unaffected by the objects they may re/present.

2. *The bodies of wildlife observation are, phenomenologically speaking, three-dimensional ('3D') bodies.* Here 'three-dimensional' does not mean that the bodies are endowed with depth, width, and length – though they may be that too. Instead, '3D' refers to bodies with, and of, firstness, secondness, and thirdness. To any observer, any beyond-human animal and indeed any organic or inorganic body

found in the proximate context of an encounter between tourists and wildlife is at once itself and only itself; itself in a relation of action and reaction with another body; and that relation in turn mediated by additional bodies, and this potentially *ad infinitum.*

To offer a very simple example, in the case of the encounter described above, each of the peccaries are at once themselves; themselves reacting to the presence of tourists; but also, themselves reacting to the tourists in a context with additional elements that play a mediating role in the process: for instance, if the peccary can make themselves scarce, it is because there is a forest with a floristic structure that allows the peccary to hide from the visitors. A similar principle applies, of course, to any of the humans that participate in any such encounter; if the visitors lose sight of the peccaries, it is because the visitors cannot see past the boles and the foliage more generally.

This account immediately raises the question: why three dimensions, and not two, or indeed four or five or more? It is not *two* dimensions because Charles Peirce makes a compelling case that phenomenological logic, and with it semeiotic relations involve mediation, and mediation is not itself reducible to dyadic relations. As to whether we should speak of four- or more-dimensional bodies, Peirce argues that all relations of mediation can be accounted for by way of a triadic logic. This is probably a simplification, if only because Peirce's phenomenology says nothing about more elaborate sociological, anthropological, or ecological mediations which go beyond semeiosis, and which may not be reduced to a semeiotic logic. It may thus be necessary to speak of 3D+.

3. *The bodies involved in wildlife observation can only be perceived, conceived and interpreted thanks to the fact that they become signs.* With Peirce, signs may be regarded as triadic entities that associate an object (or in Deleuzian/Guattarian terms, an individual or a body), a ground, and an interpretant. Signs are what allow objects to express themselves (in the Deleuzian sense of expression) by producing/finding a form. But they also require a co-semeiotic interaction with another body, which is required to 'complete' the first sign by way of the production of one or more interpretants. From this perspective, bodies *become* signs. On the one hand, they are not simply entities fixed in a semeiotic aspic; they are in a constant, if at times infinitesimally tiny process of transformation (tiny, that is, to a human *umwelt*). On the other hand, strictly speaking no object is a sign in itself. Signs must always be co-produced in the 'technical' sense that one sign requires another to become a sign. It might also be suggested that, in so far as bodies become signs, one body requires another to become a sign.

In the encounter with the peccary, the first sign of the ungulates' presence for at least some of the members of our group was the pungent smell. But of course, for that smell to become a *smell*, the visitors had to attend to it, and interpret it with additional signs. The same logic applies to the peccaries; before they could turn tail and run off, they too, had to perceive the presence of the visitors, and this they could only do by detecting the signs of that presence, and then interpreting them as signs of danger (i.e. additional signs, or interpretants). Both sets of actions occurred in the midst of a semeiosphere quite literally replete with other kinds of bodies and 'their' signs.

4. *All the bodies in wildlife observation entail a principle of immanence.* In Part 3, I discussed in some detail what is a fundamental shift proposed by Deleuzian philosophy: what I describe, echoing Deleuze himself, as a 'principle of immanence' that suggests that bodies are themselves, before they are anything else, and that bodies express themselves, as opposed to being an expression of something else. Via Spinoza, Deleuze insists that things are what they are (or what they are becoming), and not what they are not (a principle of identity: *x* is *y*, as in x is a peccary). This notion has at once philosophical, ethical, and practical implications. The principle's most obvious philosophical implication is that it establishes a flat ontology. The ethical implication is that the world (or in this case the observational process) is not split up according to a hierarchy that establishes the superiority of one or another of the parts (e.g. mind over body, or human over nonhuman animal). The practical implication is that it recognises wildlife observation for what it is: an activity that involves individuals whose haecceity plays a key role when it comes to determining who will encounter what creatures when and where, and how each and all of the different bodies will respond. The point is not to deny the importance that certain *strata* may have in generating contexts where, say, some kinds of creatures are more likely to be encountered, and to be encountered in certain ways. Instead, it is to introduce a 'first principle', and at the same time, a principle of firstness – that the bodies involved in wildlife observation are always themselves, even as they partake in a relation with another body. That much is or should be obvious (though again, it may be a verity overlooked by certain philosophical conceptions). Less obvious, perhaps, is the notion that this same principle serves to undermine the everyday scientific assumption that individual organisms are *species*, conceived as essential types with relatively fixed ethologies, before they are themselves.

Two very concrete examples begin to explain why an essentialist conception of species-being may be utterly misleading. In earlier chapters I referred to crested guans' (*Penelope purpurascens*) habit of erupting into flight with loud calls and the equally loud beating of wing feathers, generating surprise amongst visitors and guides alike. But in Costa Rica's *La Selva* biological reserve and field station, I came across a flock of guans resting quietly on a branch less than three meters above one of the station's trails; as we walked under the branch where they were roosting, the birds had no ostensive reaction to our presence. A similar point might be made about a herd of peccary that was itself resting in the gardens of that research station; where the peccary on Barro Colorado fled tourists' presence, leaving behind their odorous compounds (living up to their reputation as 'musk hogs'), those in *La Selva* relaxed on the lawns of said gardens, taking advantage of a break between thunderstorms to have a siesta in the sun even as scientists and station staff walked past them. Here were 'specimens' of the 'same' species with what seemed like completely different responses to humans... Clearly, the haecceity of the different creatures' trajectories had consequences for their ethology – an itself obvious point, but one that is easily overlooked when the emphasis, both explanatory and philosophical, is on determining types.

5. *All the bodies in wildlife encounters always have affective relations with other bodies.* To emphasise the principle of immanence is not to suggest that bodies exist in splendid isolation. On the contrary, bodies are always becoming bodies in relation to other bodies. Phenomenologically speaking, a first way to theorise the interrelation is by way of secondness: any one object, any one body is always acting and reacting with another. If any of the peccary could rest on the grass in the *La Selva* station, it was because the force of gravity was almost literally holding them down. In Peirce's phenomenological terms, every action of any one musk hog is always a reaction to something else.

From the perspective of the Deleuzian/Spinozan ontology, this kind of relationality may be theorised via the concept of affect. Affect goes beyond recognising dynamics of action and reaction to insist that what one body is becoming is always a matter of a relation of enablement or disablement with other bodies. Deleuze uses Uexküll's account of the relation between ticks and their hosts to illustrate this kind of interrelation, this conceptualisation of affect, which is different from the one traditionally invoked by psychologists. What Uexküll describes as the tick's three functional cycles, Deleuze describes as three affects: to quote his example once again, 'the first has to do with light (climb to the top of a branch); the second is olfactive (let yourself fall onto the mammal that passes beneath the branch); and the third is thermal (seek the area without fur, the warmest spot). A world with only three affects, in the midst of all that goes on in the immense forest. An optimal threshold and a pessimal threshold in the capacity for being affected: the gorged tick that will die, and the tick capable of fasting for a very long time. Such studies as this, which define bodies, animals, or humans by the affects they are capable of, founded what is today called *ethology*'.[4] This approach, Deleuze explains, can also be applied to human beings, though 'no one knows ahead of time the affects one is capable of; it is a long affair of experimentation, requiring a lasting prudence, a Spinozan wisdom that implies the construction of a plane of immanence or consistency'.[5]

Does this not contradict the earlier underlining of the importance of the haecceity of individuals? Not at all; to return to the example of the peccary in *La Selva*, those peccary did not flee from humans because their affective relation with peo-

---

4 Gilles Deleuze, *Spinoza: Practical Philosophy*, translated by Robert Hurley (San Francisco: City Light Books, 1988 [1970]), p. 124-125.

5 Gilles Deleuze, *Spinoza: Practical Philosophy*, p. 124-125.

ple was such that flight was not a response learned from short or long-term experience.

Where, one might further ask, does this leave Endler's account of the predator-prey relations? That account is an attempt to come up with a general list of the kinds of actions/reactions that species engage as prey to avoid capture. Endler in effect conceives the defences as something akin to types of secondness. No doubt this tabulation succeeds in representing a great many such relations, precisely as it is meant to. But two points need to be made here: the first is that the predator-prey relations always involve the 'element' of haecceity, and the 'chance' of firstness if only because, strictly speaking, there is not a relation of affect between *species*; there is a relation of affect between *individuals,* however much those individuals may be, and will be, the outcome of a more or less shared morphogenesis. We return to Sartre's existence precedes essence, and to Ghiselin's/DeLanda's notion of species as individuals as opposed to transcendental types.

The second point is that of course, one cannot divorce the defences or counter-defences from thirdness, or assemblage, and so a broader context. Depending on the individuals and their circumstances, certain defences or counter-defences will work for/with some individuals, but not others; were this not the case, all defences would always be, *ceteris paribus*, either successful or unsuccessful.

6. *All the bodies in wildlife encounters (and beyond such encounters) both constitute, and are part of, assemblages. Indeed, wildlife observation must itself be regarded as a matter of assemblage.* To reiterate an earlier point, if the peccary could run for cover, it was because there *was* cover to run towards (the forest). Or returning to the example I used in Part 3, if the tourist interpreted the crocodile as if it were a specimen of *C. niloticus* along the Mara River, it was thanks in no small part to the mediation of the media of mass communication, and more specifically, the kinds of wildlife documentaries produced by the BBC and other nature media organisations. In each case, we may point, in Peircian terms, to the importance of thirdness to all manner of relations, but this in a way that is not exclusively symbolic. Even if the relation is primarily a matter of what I described in Part 2 as immediate or dynamical modes of observation, a body and its form must 'join up' with another body and *its* form to produce a relation; in Peircian terms, there must always be an interpretant.

From the perspective of the Deleuzian/Guattarian ontology, thirdness, like mediation more generally, entails a logic of assemblage. The key insight afforded by the concept of assemblage is that it interprets the mediation as a matter of relations of exteriority, and not interiority (Manuel DeLanda, *passim*). Put differently, the part-whole relation is not an organic one, or even a matter of a gestalt; it is one of bodies that are at least partly independent of each other, which is to say *inter-independent*. The bodies act and react in ways whose logic is not simply down to a mechanically conceived causation, as per a Barlowesque conception of neuronal systems, or even a 'magnetic' field, as per Bourdieu/Wacquant's conception of field.

The last point notwithstanding, it may be admitted that some bodies have a greater degree of autonomy, or what one biosemiotician has described as a 'semiotic freedom'[6] to act and react in ways that are more varied. In some cases that semiotic freedom may be such that, depending on the context, a body may choose to more or less consciously ignore another body, or alternatively to make it the centre of attention. This is the kind of difference that I raised in the conclusions of Part 1, where I sketched a situation in which a researcher might, or might not attend to bullet ants, depending on what the researcher was actually looking for – and this despite having not only the requisite *umwelt* (Uexküll, *passim*), but also the ants squarely in the field of view in each of the three described scenarios. Or to offer another example, visitors to a tropical forest might be – and according to my research, generally *were* – so preoccupied with *seeing animals* that they ignored the flora surrounding them. In some cases they probably also ignored non-visual signs of those animals' presence.

This last point provides part of the justification for my choice of the encounter with the peccary for the opening example of these conclusions. As I noted above, not all of the visitors detected, or at least *acknowledged* detecting, the smell of the peccary. This difference presents a particularly interesting challenge to anyone seeking to develop a theoretical approach to wildlife observation. If one adopts an apparently empiricist perspective of the kind favoured by adaptationist or neuro-scientific approaches, then assuming that all of the visitors had intact olfactory systems, the most plausible way to explain this difference is to suggest that some of the visitors either weren't actually exposed to the odorants, or weren't exposed to a sufficiently high number of the molecules to trigger the action potential in the relevant receptor neurons. We come back, albeit via an olfactory route, to the kind of model proposed by Horace Barlow (cf. Chapter 3). If the human olfactory system – or what the ecological psychologist James Gibson describes as the taste-smell perceptual system (cf. Chapter 4) – works in the same way that Barlow seemed to show that the vision of intact frogs does, then all of our noses should have leapt at, or perhaps I should say leapt *away* from, the ungulates' odour.

The most obvious problem with this interpretation is that even the visitors who claimed not to have detected the smell *did* acknowledge the odour when someone pointed it out to them – 'Oh yes...'. This being the case, unless those visitors were latecomers to the ligands (or vice-versa), an alternative interpretation is that what was actually at stake was *attention*: those who did not smell were not paying attention to their sense of smell, or possibly *pretended* not have done so. In either case, it is tempting to reconsider the kind of interpretation offered by the sociologist Zygmunt Bauman (1925-2017), who famously suggested that moderni-

---

[6] Jesper Hoffmeyer, *Signs of Meaning in the Universe*, translated by Barbara J. Haveland (Bloomington: Indiana University Press, 1995 [1993]), p. 61.

ty declared something like a discursive war on smell. In an oft-quoted essay, Bauman argues that 'Scents had no room in the shiny temple of perfect order modernity set out to erect. And no wonder, as scents are the most obstreperous, irregular, defiantly ungovernable of all impressions. They emerge all on their own, and by doing so they betray what one would rather keep secret: that not everything is under control and not all is ever likely to be'. 'Odours', Bauman adds, 'do not respect borderlines and do not fear border guards; they travel freely between spaces which – if order is to be preserved – have to be kept strictly apart. They cannot remain unnoticed, however hard one tries...'.[7]

It may well be that odours really cannot be 'kept apart' – at least not from the unprotected olfactory organs – but I would argue that someone focussing fiercely on the visual field might well ignore at least some signals sent by the own olfactory receptors. Then again, it is of course possible to *pretend* that one has not noticed a smell, and in some contexts, there may be good social reasons for doing just that. In a wonderful essay on smell, the historian Mark S. R. Jenner notes that, even now, '[t]o remark upon odors often violates decorum; to relish them is seen as characteristic of that stage of childhood in which the young person recognizes codes of politeness but has not yet fully internalized them' – an attitude that he says has deep historic roots, and which associates olfaction with the primitive and the childish.[8] From this perspective, at least some visitors may have decided, however consciously or unselfconsciously, to discipline, if not that which, according to Bauman, cannot be disciplined – smells – then that which *can* be: a public response to the smells.

Whatever the precise motivations, one thing is clear: there *were* so-called 'individual differences', however much there may have been affects and ensembles, both 'ungulated' and modern, working to produce a 'universal' response. As I also suggested in Part 1 via Mario Bunge, even if something like direct (or 'mechanical') causation was involved – which is bound to have been the case with the peccaries' volatile compounds – that does not mean that the humans, or indeed the beyond-humans, must act and react in the manner of Cartesian automatons. To say this is not to suggest that the individuals will have so-called 'free will' (semiotic 'freedom' may, in this sense, be a misleading choice of expression). It *does* mean that they will have a certain volition, a certain purposiveness, a certain *agency*, conscious or not. We return by this route to the principle of immanence. Any 'system', any 'network' will work thanks in part to this logic of exteriority, and not despite it.

The more general point is that wildlife observation must itself be regarded as a matter of an *assemblage* of bodies, including the organismic, but also the less-than,

---

[7] Zygmunt Bauman, 'The Sweet Smell of Decomposition', in Chris Rojek and Bryan Turner (Eds.) *Forget Baudrillard* (London: Routledge, 1993), 22-46, p. 24.

[8] Mark S. R. Jenner, 'Follow Your Nose? Smell, Smelling, and Their Histories', *American Historical Review*, 116:2 (April 2011), 335-351, p. 337.

or more-than organismic bodies, all of which come together during acts of perception in ways that entail not only a multiplicity of actors/actants (to echo Latour), but actors/actants acting according to logics that are not themselves exclusively organismic or indeed organic. To return to Manuel DeLanda's definition, an assemblage involves component parts that 'may be detached from it and plugged into a different assemblage in which its interactions are different. In other words, the exteriority of relations implies a certain autonomy for the terms they relate ... Relations of exteriority also imply that the properties of the component parts can never explain the relations which constitute a whole'.[9]

7. *Wildlife observation entails a duration, and so do any participating bodies. But at the same time, wildlife observation is always situated, and situation entails more than 'being in a certain place'.* Wildlife observation takes time. Or, to put it differently, whether they partake in observational practices or not, all '3D' bodies have a temporality. They have a duration, understood from a Bergsonian perspective as an ongoing process of transformation, of flux, and as part of this process, a play of sameness and difference, continuity and discontinuity. From this perspective, even before one approaches wildlife observation as entailing a multiplicity of bodies, it is necessary to recognise the multiplicity of any one body – albeit a multiplicity that is itself indivisible: it is, to return to Bergson's vocabulary, a *qualitative* multiplicity.

This point might seem so obvious as to not even require restatement. But of course, to recognise duration, and to do so in this way, is to render a potentially static account dynamical, processual. Every body involved in the process of wildlife observation – and it is that, a *process* – is undergoing a constant dynamic of change, however minute. The human body is always changing, but so are its circumstances, and the same is true for the rest of the bodies involved in any encounter with wildlife – not least, those that seem most immobile. Here Doreen Massey's account of how the 'timeless' Skiddaw is no such thing offers perhaps the best possible example: even mountains move!

Of course, if this point is taken too far it is not possible to account for the relative stasis, for those 'coagulations' that Deleuze and Guattari describe as strata (cf. number 11, below). For now it is nonetheless more important to underscore the importance of change, as part of a more general effort to 'un-fix' the categories that tend to reify the participants of wildlife observation into literally or figuratively immobile entities.

A related point can and be made about the geography of wildlife observation. Throughout this volume, I have used the expression *in situ* repeatedly to clarify that I was referring to the practice of wildlife observation as it took place in a tropical forest itself. This was, and remains a useful distinction, not least because I explained the importance of recognising the difference between observing the

---

[9] Manuel DeLanda, *A New Philosophy of Society: Assemblage Theory and Social Complexity* (London: Bloomsbury, 2006), pp. 10-11.

wildlife of a tropical forest as represented in, say, a TV screen, and observing wildlife more directly, in an actual tropical forest.

It is, however, also important to recognise that, strictly speaking, any observational practice is *always* 'in situ' in two fundamental ways. The first is the one referred to by Simone de Beauvoir and by Maurice Merleau-Ponty when they refer to the body *as* situation – indeed one might speak of the body as situ-ation (both expressions, *in situ* and situation, derive from the Latin *situs*, or 'site').

The second one takes us back to Henri Bergson, and the notion of duration, or qualitative multiplicity, and with it to the Deleuzian and Guattarian notion of becoming. The body as situation is always the body becoming a different situation, however microscopically. Or to use different Deleuzian/Guattarian terms, there is an ongoing process of de- and re-territorialisation (what I refer to as trans-territorialisation). If it is true that Skiddaw never really stands still, the same is true, only that much more quickly, for any animal; our own, like all other bodies, are always on the move, and so producing a trajectory.

8. *The duration of the bodies of wildlife observation is expressed spatially by way of a trajectory.* In Part 4 I mentioned the ARTS surveillance system which scientists used to try to track the activities of hard-to-observe species on Barro Colorado Island. If we were to apply a similar technology to the tourists and to the peccaries I mentioned at the start of this chapter, then it would be possible to draw a line made up of all the points visited (or at any rate, travelled past) by the human or beyond-human animals. This kind of conception, of trajectories as a kind of linear displacement, is one that is based on what Bergson would describe as *quantitative* multiplicity. As such, it can be segmented or otherwise divided in whatever ways the social or physical scientist deems suitable for the purposes of research.

Certainly this kind of perspective has its uses – not least, the kind of cybernetic uses described with reference to Norbert Wiener and Claude Shannon in Chapter 3. One can, for example, imagine that park authorities would find it easier to manage tourism and the impact of tourism on trails if all visitors could be tagged, and if a log could be kept of how many bodies have walked along this or that trail. In zoological research, scientists can often only really find out what certain species are up to by such means.

I am not actually proposing that this should be done with people or with any other kinds of animals (although of course that is precisely what Google and Facebook have been doing for some time, using mobile phones and other devices as the equivalent of radio collars). On the contrary, in keeping with what I explained about duration, my own understanding of trajectory is one that is not about quantitative, but *qualitative* multiplicity, and this entails a profound shift in perspective; it is, as I described in Part 3, *dynamic situation*. Moreover, the reader may recall that the early Merleau-Ponty (cf. Chapter 9) suggested that what he calls *motility* should be understood 'as basic intentionality';[10] any movement to-wards an object contains 'a reference to the object, not as an object represented, but as that highly specific thing towards which we project ourselves, near which we are, in anticipation, and which we haunt. Consciousness is being-towards-the-thing through the intermediary of the body'.[11] Even if the later Merleau-Ponty abandons the emphasis on perception and consciousness, the more general notion still stands: at any point in the trajectory of a person, the displacement is caused, motivated, or otherwise associated with a certain purposiveness and so a certain agency, however conscious or unselfconscious, which is not in itself a matter of measurable units.

If human trajectories can be said to be purposive, the same is true for beyond-human animals, albeit according to a logic based on at least partially different *umwelten* (Uexküll, *passim*). The peccary turned tail and disappeared into the for-est in order to make themselves scarce (though that may not be how they con-ceived the action). The green and black poison-dart frog I mentioned at the start of Part 2 hopped away for similar or perhaps very different reasons, and the same is true for the specimen of *C. acutus* (Part 3). In the case of the Geoffroy's spider monkeys that we glimpsed as they crossed from one tree to another (Part 4), it is more likely that the purpose was to move to another tree in which to forage for fruit; but of course, this too, was purposive movement. As some ethologists have protested with respect to telemetric systems, simply knowing where one creature is, or has been, cannot replace a proper ethological study (or an anthropologist would say an *ethnographic* study) that considers what the animals are actually 'up to', what they are doing and why they are doing it.

9. *Wildlife observation occurs during encounters, which themselves take place when the trajectories of human observers and beyond-human animals cross – but this such that the human observers attend to the beyond-human animals, and do so in a mediate manner.* I've just suggested that encounters occur when the trajectories of the different an-imals cross. However, from a tourist perspective, an encounter only counts as an encounter if the person or persons who are in the tropical forest not only have co-presence with the beyond-human animal, but are aware of that co-presence, and react to it by attending to the animal for what is itself a certain duration. In any one moment, a tourist in a tropical forest is likely to be surrounded by organisms, many if not most of which will be ignored (they will not become the objects of even cursory attention). Some of those organisms may react to the visitors' pres-ence without being detected by the visitor. However, wildlife observation, *qua* wildlife observation amongst tourists, only occurs when there is an encounter in which the tourist reacts to the animal, becomes aware of its presence, attends to it,

---

[10] Maurice Merleau-Ponty, *Phenomenology of Perception*, translated by Colin Smith (London: Routledge, 2002 [1945]), p. 158-159.

[11] Maurice Merleau-Ponty, *Phenomenology of Perception*, p. 159-160.

and indeed observes it in a more or less sustained fashion, according to one or another coding orientation, one or another technique of observation, viz., what I have described as a *mediate* mode of wildlife observation. The last point should not lead us to forget the importance of what I have also described as the immediate, and the dynamical modes; rather, it is to underscore the importance of the inter-relation between observation, attention, and a certain purposiveness.

10. *Wildlife observation among tourists entails heterotopia.* Even if an encounter involves the crossing of at least two trajectories, it of course also involves a much wider assemblage of bodies, themselves with trajectories, and with varying consequences for, or relations to the observer and the observed. It is, in this sense, necessary to recognise the verity of the conceptualisation of space proposed by Massey, viz., that space should be recognised as the product of interrelations, as 'the sphere of the possibility of the existence of multiplicity in the sense of contemporaneous plurality', as the sphere in which distinct trajectories coexist, and therefore as the sphere of coexisting heterogeneity.[12]

If spatial heterogeneity, thus conceived, acts as a kind of conceptual and practical baseline, in the case of wildlife observation of the kind engaged by tourists, scientists and others who travel to visit tropical forests, we have also the kind of heterogeneity associated with heterotopias, or heterotopic practices. If space is already inherently heterogeneous in its own right, to this heterogeneity we must add the kind that results when individuals from 'all' other places converge in one site, and generate the kinds of events that I described via Foucault in Part 4. To echo Massey once again, on one level there is nothing 'strange' about this in so far as the conceptualisation of space, and indeed of trajectory necessarily implies a multiplicity of 'origins' and 'destinations', a multiplicity of 'beings' (in both the sense of individuals, but also, becomings). To be sure, as I began to note in Part 4, in modern as in other complex civilisations, 'heterotopicality' is probably the rule. But that should not lead us to overlook the peculiarity of the situations that arise when trajectories, considered along the lines of metric space, converge according to what becomes, quite suddenly, a *topological* logic of the kind that I also described in Part 4. When this happens, there may be a collision, or at any rate a rather sudden juxtaposition of potentially very different strata. Topology, Serres suggests, is the science of nearness and rifts; by contrast, metrical geometry is the science of stable and well-defined distances.[13] The heterotopic quality of wildlife observation of the kind engaged by tourists involves precisely such rifts, and this generates conditions that *do* involve a certain strangeness, in the sense that an event is unusual, surprising, or unsettling of a certain order: multiple actors, who might normally have absolutely nothing to do with each other, may suddenly find

---

12 Doreen Massey, *For Space* (London: Sage, 2005), p. 9.

13 Michel Serres with Bruno Latour, *Conversations on Science, Culture, and Time*, translated by Roxanne Lapidus (Ann Arbor: University of Michigan Press, 1995 [1990]), p. 60.

themselves in close proximity, and may have to negotiate the absence of anything like common coding orientations, or indeed, 'life genres'. The point is not to suggest that this is inherently negative (as per a reactionary, and nativist discourse), or inherently positive (as per a liberal, and multiculturalist perspective). It is, rather, to suggest that the resulting assemblage may well be conducive to different forms of trans-territorialisation, and in so doing may transform strata.

11. *Every instance of wildlife observation is unique, and each involves signs that are in some sense unique. But wildlife observation always involves strata, or better yet, is constantly undergoing a process of stratification/re-stratification, territorialisation/trans-territorialisation.* Thus far, the analysis has emphasised mobility and flux. This has been necessary to correct a tendency, which remains in much mainstream theory, to 'fix the flux'. Such a tendency is not only understandable, but in some sense inevitable, even amongst those who try to do otherwise; just as one does not approach Skiddaw expecting it to run away, it is perfectly legitimate, and indeed necessary to point to certain regularities in species, ecosystems, biomes, and the cosmos more generally. Those regularities not only exist, but have real consequences for all manner of phenomena, including wildlife observation. Indeed, simply to speak of 'wildlife observation' presupposes that there is a practice which can be identified as such, and that suggests some kind of regularity. Moreover, to speak of wildlife observation in *tropical forests* presupposes in turn that there are such forests. And while I have rendered complex the notion of the wild, it should be clear, by virtue of the criteria presented in Part 4, that some animals, some places, can be said to entail more wild(er)ness than others. From this perspective, the fluidity invoked by the notion of becoming signs, or indeed becoming, *tout court,* should not lead the analyst to overlook the fact that certain forces really do attempt to, and clearly *do succeed* in generating those 'phenomena of thickening on the Body of the earth', those accumulations, coagulations, and foldings that Deleuze and Guattari refer to as strata. Or, to put the matter more sharply, for any developing set of relations, for any occurring set of actions and reactions and their associated assemblages, some strata *will* work to 'slow things down', to block certain lines of flight, and so to conserve, or to attempt to conserve, a certain order – sociological, biological, or combinations and recombinations of the two and of others which escape any such classification. Clearly, any attempt to communicate, to engage in semeiosis not only presupposes such coagulations, but cannot do without them.

Deleuze and Guattari distinguish between the physiochemical, the organic, and the alloplastic strata. However, according to the philosophers, any stratum is 'extremely mobile. One stratum is always capable of serving as a *substratum* of another, or of colliding with another, independently of any evolutionary order'; '[s]tratification is like the creation of the world from chaos, a continual, renewed creation'.[14] While some bodies may be closer to the kinds of coagulations/accumulations associated with one or another strata, there is, in the practice of wildlife observation as in all other human activities, always an element of assemblage that crosses the strata, and in so doing refuses efforts to 'peg' it as being entirely a matter of one or another stratum. As I noted above, wildlife observation occurs when the trajectories of humans and beyond-human animals cross, and in such encounters, the certainties of everyday *umwelten*, everyday strata may become undone – there may be, indeed there is *bound* to be a dynamic of transmediation now understood as trans-territorialisation. For example, a cocktail of geranylgera-niol, squalene, and isomers of springene, emanating from the rumps of several peccaries, may find its way into the olfactory cavities of visitors, who may then translate the chemicals into a series of symbolic signs, at least some of which may take the form of delomes (arguments, discourse). Conversely, even as guides and visitors animately produce those delomes, their voices will produce sound waves, and their bodies more generally will expel their own 'volatile compounds', both of which may be detected by peccaries and other creatures with a suitably affective olfactory capacity, and which will themselves be translated via creature- and species-specific forms of semeiosis. At each step, there must be *both* change and continuity; understood critically, the very notion of translation presupposes both stratification and destratification, territorialisation and deterritorialisation.

12. *From a semeiotic-phenomenological perspective, one way of recognising the circulation across strata during wildlife observation amongst tourists is to at once distinguish between, and interrelate immediate, dynamical, and mediate modalities of wildlife observation.* If one adopts, as so many poststructuralist scholars once did, the principle that there is nothing in the human world that is not already discursive in one or another way (what might be called discursive constructivism, or simply, *discursivism*); and if one also adheres to the principle that everything in the human world entails culture (culturalism), then it is not possible to admit that there may be forms of wildlife observation, or encounters with beyond-human animals that are not about discourse and/or culture. Or to state this problem from a different perspective, so long as one adopts an identitary logic and treats it as the only ontological show in town, then one cannot really explain the 'becomingness' of certain encounters with beyond-human animals, let alone dynamics of trans-territorialisation that in some sense remake the major stratifications.

In the conclusions to Part 2, I illustrated this point via the example of the crocodile attack, but of course it applies far more generally: even if it is admitted that we can only perceive, conceive, and interpret the world via signs, there are very different *kinds* of signs, some of which are, in a manner of speaking, closer to firstness than to thirdness (despite being signs), and vice-versa: some signs are

---

[14] Gilles Deleuze and Félix Guattari, *A Thousand Plateaus: Capitalism and Schizophrenia*, translated by Brian Massumi (London: Continuum, 1988[1980]), p. 502.

closer to thirdness, despite entailing organismic bodies. Some signs may also have a very different relation to 'their' objects than others, such that there may be, for instance, an indexical as opposed to a symbolic relation of representation, or indeed no *representation* whatsoever.

To acknowledge these kinds of differences, I have proposed that it is possible to distinguish between immediate, dynamical, and mediate modalities of wildlife observation, where each of the categories is determined by the extent to which the observational relation hinges more strongly on firstness, secondness, or thirdness, respectively. Peirce's phenomenology makes it clear that *all* aspects are always present, at least to some degree. This being so, there is no such thing as non- or pre-semeiotic observation; there is, however, always the possibility of pre- or indeed more-than *symbolic* observation, viz. observation that is not primarily driven by, or structured around symbols, legisigns or delomes.

The encounter with the peccaries can illustrate these points quite clearly. For someone entirely unfamiliar with the smell of the ungulates, the first impression, the first 'intake' of smell would most probably be mostly a matter of immediate, and dynamical modes: simplifying greatly, a peculiar 'feeling' (immediate mode), followed by a reaction of surprise (dynamical mode), followed in turn by a questioning mode with informal hypotheses (mediate mode). Conversely, somebody familiar with the peccary and their characteristic odour would doubtless identify the provenance of the smell straight away (mediate mode), but would still have to rely on the most overtly physical (Peirce would say brute) reaction of their perceptual systems in order to be able to detect the smell (dynamical mode). Even in such cases, the sheer haecceity of the encounter and each of its elements would still guarantee an element of firstness, and with it potential to become something else (immediate mode). Just as nobody knows what a body can do, nobody knows just what might happen during an encounter.

This three-way distinction is designed to make room, as part of the theoretical approach, for forms of embodiment, and of embodied encounter which are not themselves already entirely molar, and so may at least partly escape the coagulations of stratification. In wildlife observation such forms may play a particularly important role when the wildlife is unknown, when wildlife appears unexpectedly, or when the encounter involves circumstances which, in one way or another, undermine the apparent certainties associated with a certain molarity – be it a visitor's sense of the animal's expected behaviour, the anticipated character of the encounter as encounter, etc.

13. *Not all the bodies involved in wildlife observation are organismic bodies, or indeed bodies with an overtly physical extension. Even the non-organismic bodies may play a key role in wildlife observation.* Thus far, this account's emphasis has been largely on 'organismic bodies'. But lest we forget, wildlife observation includes not just those bodies that are 'obviously' embodied – e.g. the human and beyond-human animal bodies, the flora, and geomorphological features – but also bodies of

knowledge, technological bodies, and bodies of representation (amongst many others). The latter also have a bearing on wildlife observation, even if they lack an ostensive physical materiality, and with it an ostensive physical extension. Perhaps the best evidence in favour of not relegating such bodies to a secondary status is the extent to which human observers, like their beyond-human co-observers, may completely disregard some manifestly physical bodies that are in close proximity to them whilst becoming utterly absorbed by others that are physically absent.

The example of the person observing a crocodile along the shoreline of Barro Colorado as if it were a specimen of *C. niloticus* in the Serengeti (cf. the start of Part 3) is a good one in that it makes very clear how important even representational bodies (or bodies of representation) may be to 'actual' encounters. But a similar point might be made about the encounter with the peccary: while it was the peccaries' volatile emanations that first alerted the guide to the presence of the ungulates, at least some visitors do not seem to have attended to those emanations. One possibility is that those visitors were so focussed on visual cues that they ignored their own olfactory sense – the one in charge of sensory cues that Bauman would have us believe are inescapable. In Volume 2 I will consider arguments that suggest that one reason for this kind of attentive focus may have to do with the kind of multimodality that is characteristic of the modern cultural emphasis on so-called 'visual culture'.

What I have just explained allows us to come full circle to a question posed at the end of the Prologue: perhaps few readers, I suggested, would disagree with the proposition that we humans do not look at the fauna in rain forests via two-dimensional, 16:9 screens with expertly intensified scenes of the kind made available by TV shows. Or do 'we'?

On the basis of a geosemeiotic approach such as has been outlined in this volume, it is possible to argue that if we are to respect the haecceity of each observer, and if we are to adhere to what has been described via Deleuze as a principle of immanence, then it would be a mistake to establish any simple identity between a person's ways of looking (or otherwise sensing) and the techniques of observation deployed by one or another medium of mass communication. To this philosophical point we might add the kind of practical considerations offered by James Gibson as part of his critique of the realism of what he calls 'drawings', and 'progressive pictures' (cf. Chapter 4). Clearly no one *literally* walks around gazing upon the world through a rectangle with a 16x9 aspect ratio. In keeping with this perspective, throughout this volume I have noted the discontinuities between the forms of observation that unaided visitors on the ground can engage in, and those made possible by the cameras, lenses, editing suites and other technologies of the nature media.

However, this kind of analysis should itself not be taken to such a literal extreme as to adopt a naively naturalistic conception of space, let alone a naively biologistic conception of the workings of what Gibson describes as perceptual systems. Eleanor Gibson devoted her career to studying what she described as perceptual learning,[15] and provided that one acknowledges that such learning can also occur via media such as TV or photography, then it is apparent that any individual's observational capabilities may well be modified by the media in question.

Combining Peircian and cultural sociological concepts, it may be suggested that the media offer interpretants – in the fullest sense of that expression – that may affect how a media user expects to index, classify, and frame a tropical forest and its wild denizens. The media generate signs that may affect what visitors want to observe, how they categorise what they observe, and how they engage, or expect to engage, in actual observational practices.

I have just suggested the media may affect visitors, and indeed in Deleuzian/Guattarian terms, and more generally from what I have described as a bio-cultural perspective, it is possible to conceive of the interrelation between the media and observers as a matter of affect, i.e. a certain capacity to increase or decrease another body's power to act; in this case, to engage in certain forms of observation, where observation is regarded as an assemblage involving organs, 'perceptual systems', and the entire interpretive apparatus made available by diverse cultural formations.

It might be added that, from the perspective of Latour and Whatmore (and more generally what I have described as an eco-geographic perspective), the nature media render even more complex the topological space generated by the act of travelling as a tourist to a tropical forest. If tourism itself entails topology and heterotopia, visitors whose gaze is shaped by the nature or other media can be said to produce something like a hyper-topological, and meta-heterotopic space: the folds of sudden travel are themselves enfolded in the instant voyages afforded by the media, and the congregation of tourism brings with it the archiving globalism of 'planetary' TV series such as the ones produced by the BBC.

It should, however, be noted that any resulting affections are highly ambiguous ones from the point of view of the ethics of media representations as these relate to *in situ* wildlife observation amongst tourists. On the one hand, it may be argued that the nature media do indeed alert visitors as to the existence of particular species, and reveal aspects of their morphology and ethology. Understood in this way, the representations may be regarded, in Spinozan terms, as a call to observational *action*. The nature media may channel and/or generate a desire to partake in certain observational experiences, and this may culminate in the organisation of a trip to a tropical forest (or any other biome depicted by the media).

On the other hand, this selfsame process may, again in Spinozan terms, result in *passion* if and when a tourist discovers that it is not possible to view or other-

---

[15] See Eleanor J. Gibson, *Principles of Perceptual Learning and Development* (New York: Century Psychology Series, 1969). See also Eleanor J. Gibson and Anne D. Pick, *An Ecological Approach to Perceptual Learning and Development* (Oxford: Oxford University Press, 2000).

wise experience wildlife in tropical forests in the manner suggested by televisual or other nature media's techniques of observation. This is not just a matter of the sheer mobility and proximity afforded by telephoto lenses or edited sequences; as I began to suggest in the Prologue, especially the mainstream nature media go to extreme lengths to present at once highly intensified, and utterly idealised images of species and ecosystems. By comparison, actual encounters – even relatively close encounters such as the one that I described with respect to the parrot snake and the turnip-tail gecko – may seem prosaic. We return, by this route, to the kinds of issues raised by Donna Haraway with respect to the neglect of everyday forms of animality.

It may be deduced from this account – and from my incipient critique of se-meiotic consumption – that the nature media have negative consequences from the point of view of *in situ* wildlife observation amongst the media users. But this is not necessarily the case; if we are to be faithful to the principle of immanence, and to the importance of becoming, then one cannot simply assume, *a priori*, what the affective relation will be: once again, *nobody knows what a body can do*. Some media users may well find ways of *mediating the media* in surprising ways, and what is true for visitors is even more true for guides, who may play a crucial me-diating role in their own right.

That said, it would be an exercise in idealism to suggest that the interpretive field is entirely open. If the ethnographic research which I referred to in the intro-duction to this volume is anything to go by, for a majority of first time visitors to tropical forests – and especially those who are frequent users of the nature media– the nature media's characteristic techniques of observation are more likely to gen-erate a sense of *lack* – a sense that a forest such as Barro Colorado's has relatively little to offer.

14. *The different perspectives and dimensions of wildlife observation should not be con-ceived as a matter of an organic explanatory system; they are better understood as a mat-ter of an explanatory assemblage, which articulates a series of homologies between differ-entially situated concepts.* The different perspectives outlined in this chapter may be articulated as a series of homologies – a set of theoretical correspondences or 'parallels' that express the multidimensionality of the approach. These homolo-gies can be represented as follows (see Table 4, below):

PERSPECTIVE	Monadic	Dyadic	Triadic +
Semeio-Phenomenological	Object/Firstness	Ground/Secondness	Interpretant/Thirdness
Socio-Philosophical	Body	Affect	Assemblage
Eco-geographic	Trajectory	Encounter	Heterotopia

Table 4: Dimensions of Wildlife Observation

What I am suggesting is that it is possible to articulate relations between the different approaches to wildlife observation according to a 'triadic +' model (as I explained earlier, it may be necessary to acknowledge that there is more than thirdness, Peirce's defence of thirdness notwithstanding). The last point to one side, in each 'level', in each dimension, it is possible to go from the individual to a more or less 'raw' relation between two individuals, and from there to a more or less elaborate relation of mediation. I say to 'go' from one to the other, but all three remain actual (actually, all three remain), and there is, strictly speaking, no 'progression'.

Accordingly, the Peircian semeiotic phenomenology at once differentiates, and interrelates firstness, secondness, and thirdness, and object, ground, and interpretant. In so doing it recognises objects in themselves, dyadic relations of action and reaction, and triadic, which is really to say semeiotic relations of mediation. The last may themselves be more or less a matter of firstness, of secondness, or thirdness, but all involve all three phenomenological elements.

A Deleuzian/Guattarian socio-philosophical (or one might also say bio-cultural) approach to body/mode suggests that things are first and foremost what they are in themselves, they are the *cause* of themselves, and so are not the result of something else (principle of immanence). But of course, to say this is not to deny that things are what they are always *in relation* to something else; on the contrary, Deleuze interprets Spinoza's concept of *affect* precisely along the lines of the capacities of bodies always being/becoming in a relation with, always being/becoming strengthened or weakened by, other bodies. To be sure, it is never just two bodies acting in relation to each other in some kind of 'closed circuit'; it is, Deleuze and Guattari tell us, *assemblages* of bodies, all working at once as themselves, and in relation to other bodies, and those bodies in relation to others, but always on the basis of relations of exteriority (DeLanda, *passim*). What Peirce explains according to a semeiotic-mathematical logic, Deleuze, and later Deleuze and Guattari explain as a matter of relationships between bodies.

Conclusions to Volume 1

From an eco-geographic perspective, a similar point might be made about the concepts of trajectory, encounter, and heterotopia. Each body has a unique trajectory; the trajectory not only 'involves' a changing set of situations, but *is* dynamic situation. For their part, encounters are about the coming together of at least two trajectories in ways that may or may not involve what I described in Part 3 as the dialectic of attention and observation (after all, bodies may pass each other like ships in the night). Finally, heterotopia, and more generally the kind of heterogeneity theorised by Massey, entails a plurality of trajectories with/of bodies that converge in one place. To reiterate the earlier point, here as in the two earlier triads, the 'third element' necessarily entails the second, and the second the first: just as there is no thirdness without objects acting and reacting in relation to each other, and so of course, no action and reaction without objects; and just as there is no assemblage without affects, and no affects without bodies, there is no heterotopia without encounters produced by the crossing of trajectories. Yet here too, the 'third element' entails something more than just 1+1.

The different lines of analysis should be treated as per the logic of assemblage, in their own right. The semeiotics of wildlife encounters are not simply determined by ethology or geography, any more than they are by cultural formations. Yet clearly the different dimensions are inter-independent, and so the task of the researcher is to tease out those inter-independencies as they occur among particular groups of tourists visiting specific locations. That will be the task of Volume 2, which as I have noted, will be devoted to the analysis of tourist practices on Barro Colorado Island.

Postscript

Of Hecatombs and Rhizomes

Over a decade ago, as part of an analysis of *The Day After Tomorrow* (a 2004 film which dramatised catastrophic climate change), I predicted that, given the way things were going, sooner or later the realities of climate change would overtake even 'exaggerated' films like that one.[1] Given this suggestion, and given what has happened in the years since I made that rather easy prediction, a *mea culpa* is required. *Mea culpa* is originally an expression used in the Catholic prayer of sinfulness, the *Confiteor*. My sin was, and remains to have lived the past decade as if I did not believe my own analysis of *The Day After Tomorrow*. For ten or more years, I have continued to write, and more generally to live everyday life as if nothing had fundamentally changed in the world.

In 2018, narratives such as the one found in *The Day After Tomorrow* started to seem rather less far-fetched to many people across the globe. In that year, extreme heatwaves, wildfires, but also extraordinary snowfalls and other weather-related events occurred on a scale never before recorded in history. In 2019, my own country (the UK) experienced hitherto unheard of temperatures for February, with some parts reaching over 20 C – by the old standards, a fine day, temperature-wise, in the middle of a good summer. Yet across the Atlantic, the opposite scenario unfolded, with a polar vortex enveloping much of North America with low temperatures so extreme that they might well be interpreted as a precursor to *The Day After Tomorrow's* apocalyptic scenes of a winter snowfall that rendered much of North America uninhabitable. In the Caribbean, another devastating Category 5 hurricane suggested that many of that region's countries might already be becoming uninhabitable. Then again, far to the north, scientists discovered that the permafrost in the Canadian High Arctic is experiencing rates of degradation which are already exceeding those projected to occur by *2090* according to the 4.5 version of the Representative Concentration Pathway (RCP) greenhouse gas concentration predictions adopted by Intergovernmental Panel on Climate Change (IPCC) in its fifth Assessment Report.[2]

Each new event appears to confirm a trend that anybody who bothers to do even cursory research on the subject of climate change will soon notice: that if the estimates of the IPCC have always been overly conservative (shaped as they are

---

[1] Nils Lindahl Elliot, *Mediating Nature* (London: Routledge 2006), pp. 233-234.

[2] See Louise M. Farquharson, Vladimir E. Romanovsky, William L. Cable, Donald A. Walker, Steven Kokelj and Dimitry Nicolsky, 'Climate Change Drives Widespread and Rapid Thermokarst Development in Very Cold Permafrost in the Canadian High Arctic', *Geophysical Research Letters*, 10 June 2019, <https://agupubs.onlinelibrary.wiley.com/doi/10.1029/2019GL082187> [Accessed 15 June 2019].

by the politics of climate change denial), even within mainstream climate science itself, what have started out as 'fringe' predictions – i.e. worst-case scenarios that have, to begin with, seemed too dire to seem plausible – have moved steadily to the mainstream, only to be overtaken by even grimmer models.

I am, of course, not the only one to have pretended that, in the words of the fourteenth century anchoress Mother Julian of Norwich (to continue with the religious imagery, though I myself am not that kind of believer), *all shall be well, and all shall be well*. On the contrary, most of the world continues to partake in a greenhouse gas-emitting jamboree, and as I write, that jamboree is becoming even more frenetic. The period from 2017 to 2019 not only saw a crescendo of anthropogenic climate change denials by the usual suspects, but also new rises in atmospheric CO2 emissions. Indeed, after a hiatus between 2014 and 2016, 2017 and 2018 saw increases in the mentioned emissions,[3] and in May 2019 it was announced that CO2 levels in the atmosphere had recently reached 415 parts per million for the first time in three million years, with levels accelerating at an unprecedented rate.[4] By September 2019 the precise reasons for this new jump were unclear, but it was difficult not to conclude that, sooner or later, a discourse of climate change denial supported by Donald Trump and several other far-right wing leaders across the world would legitimise a return to care-less forms of production and consumption, and with them a renewed onslaught on hitherto protected environments.

As if to prove this conclusion, late in the boreal summer of 2019 it was confirmed that under Brazil's president Jair Bolsonaro, the numbers of humanly (and mostly illegally) produced forest fires have shot up in the Brazilian Amazon.[5] To be fair, fires set by people seeking to claim tropical forest for agricultural and other purposes are a phenomenon that predates the arrival of Bolsonaro, and are by no means confined to Brazil. It nevertheless seems very clear that the huge jump in the number of fires in Brazil is directly related to Bolsonaro's avowed policy of 'opening up' the Amazon to development. It appears that there is now a concerted push amongst the far right-wing politicians to quite overtly finish the environmental *job* that was begun, however covertly or passively, by their more liberal predecessors.

It is, of course, very easy to point fingers at individuals such as Trump, Bolsonaro, or now our very own Boris Johnson (himself with a dire environmental

---

[3] As reported by the International Energy Agency's CO2 Emissions Statistics for 2018. See <http://www.iea.org/statistics/co2emissions/ [Accessed 10 May 2019].

[4] Scott Waldman, 'Global CO2 Nears Troubling Benchmark', *Climatewire*, 7 May 2019 <https://www.eenews.net/climatewire/stories/1060286896>[Accessed 10 May 2019].

[5] See for example, Lisandra Paraguassu, 'Amazon Burning: Brazil Reports Highest Forest Fires [sic] Since 2010', Reuters, 20 August 2019 https://www.reuters.com/article/us-brazil-environment-wildfires/amazon-burning-brazil-reports-record-forest-fires-idUSKCN1VA1UK [Accessed 24 August 2019].

policy record). But I would argue that the more important agents, the ones ultimately responsible for what is now a veritable global environmental *emergency*, are the vast and immensely powerful institutions that have for centuries made it their business to produce, and/or promote the burning, of fossil fuels. I refer to the results of the remarkable research conducted single-handedly by Richard Heede, which show that by 2010 nearly two-thirds of anthropogenic carbon emissions produced since the Industrial Revolution were the result of the activities of just 90 companies and government-run industries. Amongst these, eight companies – the so-called carbon majors such as Royal Dutch Shell, ExxonMobil, and BP – accounted for 20% of world carbon emissions from fossil fuels and cement production.[6] According to research made public in 2015, the managers of at least one of these companies were aware of the threat of anthropogenic global warming long before it became a public issue; after considering 'going green', they opted instead for the path of disinformation.[7] Heede's statistics, and the discovery of this latest twist in the conspiracy to normalise climate change constitute a cautionary note for anyone overly keen to relativise, or indeed dismiss as 'conspiracy theorising' any attempts to lay the blame for catastrophic climate change at the corporate door of organisations such as the ones cited by Heede. Some individuals, and individual organisations clearly must take the lion's share of the blame for the environmental emergency, and in a more democratic world, they would long since have been brought to heel and made to pay for the damage.

I will return to this idea in a moment. First I should recognise that now the real tragedy arguably involves the inaction, to not say paralysis on the part of all those of us who publicly acknowledge the reality of anthropogenic climate change. Staying within the realm of 'the establishment', a good, which is really to say terrible example of this paralysis was enacted by Britain's House of Commons on February 28, 2019. After not debating climate change for years – itself an indictment of the supposedly *centre*-right governments of David Cameron and Theresa May – 40 out of 650 MPs showed up for a debate on the subject, with just seven out of 313 Conservative MPs visible on the government side of Westminster's not-as-green-as-they-look benches (Labour Party MPs did not do that much better). Caroline Lucas, a leading member of Britain's Greens and herself that party's sole

---

[6] Douglas Starr, 'Just 90 Companies are to Blame for Most Climate Change, This "Carbon Accuntant" Says', *Science*, 25 August 2016 <http://www.sciencemag.org/news/2016/08/just-90-companies-are-to-blame-most-climate-change-carbon-accountant-says>. For the original research, see Richard Heede, 'Tracing Anthropogenic Carbon Dioxide and Methane Emissions to Fossil Fuel and Cement Producers, 1854-2010', *Climatic Change*, 122(1-2) (2014), 229-241.

[7] See Neela Banerjee, Lisa Song and David Hasemyer, 'Exxon's Own Research Confirmed Fossil Fuels' Role in Global Warming Decades Ago', *Inside Climate News*, 15 September 2015 <https://insideclimatenews.org/news/15092015/Exxons-own-research-confirmed-fossil-fuels-role-in-global-warming>[Accessed September 16, 2019].

MP, noted the absenteeism and pointed out that 'since 2010 [when New Labour was defeated and the Conservative Party returned to power], this government has built a bonfire out of the measures designed to cut emissions. Zero carbon homes targets have been scrapped. Onshore wind has been effectively banned. Solar power has been shafted. The Green Investment Bank has been flogged off. Fracking has been forced on communities who have rejected it.'[8]

Unfortunately, much the same *laissez-faire* logic is evident in the lives of all of those of us who may be comparatively powerless to change national or even local policies, but who still make no major changes even in our own everyday practices. Here I would refer once again, albeit critically, to my own research. In this volume's Prologue, as in the study more generally, I've analysed wildlife observation in relation to tourism, mass communication and other cultural formations, all but ignoring what by late 2016 had already become the burning question even amongst the traditionally conservative nature media producers. How could the BBC Natural History Unit, the National Geographic Society's equivalent, and any other nature media organisations continue to tip-toe around the growing environmental catastrophe?

As I noted in Chapter 8, this is an issue that was raised specifically with respect to the example with which I began this volume – *Planet Earth II* – on the very day that its last episode was first broadcast in the UK (1 January 2017). Shortly after that episode appeared, Martin Hughes-Games, himself the producer of wildlife TV series for the BBC, tore up a decades-long consensus amongst the natural history film producers concerning the ostensibly conservationist credentials of wildlife TV. He did so via an article that he penned for the *Guardian*. While the article began by praising *Planet Earth II* for its 'glorious, spectacular and fascinating' representations, and by professing the 'greatest admiration' for the teams that made the series, in the rest of the piece Hughes-Games let rip:

> I fear this series, and others like it, have become a disaster for the world's wildlife. These programmes are pure entertainment, brilliantly executed but ultimately a significant contributor to the planet-wide extinction of wildlife we're presiding over. [...] The justification, say the programme makers, is that if people (the audience) become interested in the natural world they will start to care about the natural world, and will be more likely to want to get involved in trying to conserve it. Unfortunately the scientific evidence shows this is nonsense.

Citing research which showed that the period during which Attenborough's blockbusters were shown coincided with a 58% decline of vertebrate population

---

8 Caroline Lucas, 'Parliament Must Declare a Climate Emergency - Not Ignore It', *Guardian* online, 4 March 2019, <https://www.theguardian.com/commentisfree/2019/mar/04/climate-change-emergency-westminster>[Accessed 4 March 2019].

abundance worldwide, Hughes-Games went on to suggest that the TV pro-grammes were nevertheless still being made

> as if this worldwide mass extinction is simply not happening. The produc-ers continue to go to the rapidly shrinking parks and reserves to make their films – creating a beautiful, beguiling fantasy world, a utopia where tigers still roam free and untroubled, where the natural world exists as if man had never been. [...] By fostering this lie they are lulling the huge worldwide audience into a false sense of security. "If David Attenborough is still mak-ing these sorts of wonderful shows then it can't be that bad, can it?"[9]

It might be noted that even in this critique, global warming was only one in a list of several other causes of the current mass extinction – a stance that is accurate, and can certainly be defended ethically, but which perhaps shows how, as late as early 2017, what was already catastrophic climate change was still not treated with the urgency that it is finally starting to generate now.

The point I am leading up to is that a criticism like Hughes-Games' might well be levelled at this study: if anyone is still writing books about wildlife observation and tourism, it can't be that bad, can it? But of course it *is* bad – so bad that we may have already reached a point of no return. In this context, to write a study about wildlife observation and tourism is itself arguably a form of climate change denialism. At the risk of stating the obvious, international tourism is an entirely avoidable leisure practice that is reliant on 'carbon-outrageous' forms of trans-portation. In the case of wildlife tourism in tropical forests, the practice also relies on forms of accommodation which are frequently air-conditioned, and have re-quired, at one point or another, the clearing of forest to make space for structures which are often made with cement, a substance whose manufacture is a terrible source of greenhouse gas emissions in its own right. And that is to say nothing about issues of environmental justice, or of impacts on ecosystems that arise with the development of the infrastructure which is so often justified on the back of meeting the demand for tourism.

I am of course aware that it has been argued that one possible benefit of so-called ecotourism and wildlife tourism is that both genres of tourism may actual-ly save species and their habitats from destruction. I will have more to say about this kind of claim in Volume 2, which will offer a history of the present forms of tourism and mass communication in the case of Barro Colorado (including a con-sideration of the question: should biological reserves run by scientists still accept visits from international tourists?). Here I would simply reiterate a point made in this volume: that in so far as ecotourism and wildlife tourism are driven by se-

---

[9] Martin Hughes-Games, 'The BBC's Planet Earth II Did Not Help The Natural World', *Guardian* online, <https://www.theguardian.com/commentisfree/2017/jan/01/bbc-planet-earth-not-help-natural-world> 1 January 2017 [Accessed 2 January 2017].

meiotic consumption, and in so far as such consumption is itself reliant on fossil fuel-based and thereby extractive activities, then the mentioned genres cannot be validly described as being benevolent, whatever their real or alleged comparative advantages over more traditional forms of tourism.

Again, the more general point is that if there ever was a valid justification for transnational wildlife tourism and ecotourism (a big if, full of often overlooked social issues), these activities, like all other avoidable forms of greenhouse gas-producing consumption, must stop, and must do so more or less immediately. Alas, the chances of international tourism coming to an end any time soon are exactly 0. Some would argue, and not without reason, that the economic consequences for some social groups of the end of tourism would be worse than climate change, at least in the short-term. But that kind of argument says nothing about the possibility that something like a Marshall Plan, paid for mostly if not entirely by the carbon majors, could be enacted in the affected areas to achieve a transition away from the kind of political economy that is devastating the planet. Again, the chances of such a plan being put into effect at the present time are themselves exactly 0. So what are the alternatives, and before that, why the global paralysis? In the remainder of this postscript I will offer some thoughts that are based in part on the theoretical approach I've presented in this volume.

By way of an introduction, I'd like to refer to a forthcoming book which addresses global climate change, albeit by way of an analysis of an event that occurred in Colombia in the 1980s: the Armero Disaster of 1985, which took place after the snowcapped *Nevado del Ruiz* volcano (some 5,400 meters high) produced what was, at least by volcanological criteria, a relatively minor eruption. The eruption nevertheless set off pyroclastic surges and flows which melted approximately 10% of the volcano's icecap. The resulting volumes of water combined with volcanic ejecta and liquefied volcanic deposits (the result of volcanic tremors) to trigger *lahars* which literally roared down the *Cordillera Central's* canyons, following the paths of rivers that started from the snow line and finished in the lowlands of the Magdalena and Cauca River valleys, some *5,000 metres* below the crater. As the lahars made their way down the steep mountain slopes, they accumulated more and more debris, including gargantuan boulders. The immense momentum of the lahars meant that the debris flows reached towns that were not only far below, but in some cases the better part of 100 killometres (60 miles) away from the volcano's crater. The town that was worst hit was Armero, where successive lahars covered over 75% of its perimeter in several meters of mud and stone, and killed over 20,000 of the town's 30,000 inhabitants.

As I explain in *Chronicle of a Disaster Foretold: The Lessons of Armero*, the tragedy of Armero was not that its folk were caught unawares in their homes late at night (which many were), or that they were hit by what some in the Colombian media still describe as a wholly unforeseen 'avalanche' travelling at '300 kph' (the lahars were actually thought to travel at about 40 to 50 kph in the steeper gradients). It was that the disaster was specifically predicted, that earlier that day many townsfolk had expressed grave concern and sought responsible leadership, and that the lahars took over two hours to reach Armero after the volcano first erupted at about 9pm on November 13, 1985. This being the case, most if not all of those who were killed could have been saved simply by travelling a distance of one or two kilometres to the nearby hillsides. The real killer was thus *not* the volcano, but the inaction of various state agencies and their leaders. Starting with Colombia's president, but continuing with the governor of the Tolima province (who reportedly continued to play pool in a bar after the eruption started), and ending with the mayor of Armero (who was himself killed during the disaster), none took the steps to ensure the safety of the people of Armero. On the contrary, in the afternoon when the eruption was starting and heavy ash falls covered Armero in grey, the local vicar used the tannoy in his church's bell tower to tell the town that all was well, that all *would* be well. The *hecatomb,* as the Colombian media aptly described the mass deaths, still stands as the worst lahar-related disaster in recorded history.

This raises the question as to why the authorities failed to act. I argue that it was thanks to an at once political, cultural, geographic and geological logic of *assemblage* – in this case, mutually *canceling* relations of exteriority – which conspired mostly without conspiring to immobilise the people who should have mobilised their fellow townsfolk, their fellow citizens. Some 35 years after the event, I will suggest that much the same logic, but now on a global scale, is prevailing in the context of efforts to stop even more catastrophic climate change. We face a hecatomb of hecatombs, but once again there is a seemingly unstoppable logic of assemblage – in this case, a conspicuously Capitalist assemblage – albeit now on a global scale.

As I have already noted, in the case of the climate change emergency there certainly *is* a conspiracy to try to occlude, lessen the gravity, and above all stop the enactment of the policies required to prevent further greenhouse gas emissions. I say *conspiracy,* but actually, today we know exactly who it that has been, and *still is* plotting, increasingly overtly, to stop the measures required to stop climate change. As I began to suggest earlier, a case might be made that if we lived in a democratic world, then the current emergency would either never have reached this stage, or the present situation could be quite rapidly addressed by drawing up a list of, say, the 10,000 most directly responsible politicians and corporate leaders (including not only the top echelons of the carbon majors, but also those of most if not all of the major media/social media companies). The individuals on the list would then be put on trial for the equivalent of crimes against humanity. In this imagined world, the evidence of wrongdoing (including crimes of omission) would be so compelling that most likely the individuals in question would be imprisoned for life if one goes by today's increasingly noxious logic of incarceration. According to this liberal fantasy (for that is what it is), the incarceration of the first 10,000 or so heinous climate criminals would inflict such fear in the rest of the climate-change-denying conspirators – or their numerous enablers –

that the great floodgates of climate change policy would swing wide open, and the world would be saved, if it still can be saved.

One problem with this, as with so many other liberal fantasies, is that it delegates to the state, or at any rate to an imagined judicial entity (for example, one of the kind used to judge war criminals in The Hague) the capacity to deliver justice. Yet perhaps today more clearly than ever before, the US, that liberal nation-state *par excellence*, reveals the pitfalls of believing that at least liberal democracies, even those with a seemingly unshakeable division of powers, can withstand the combined onslaught of corporate corruption and a rising fascism. I for one (perhaps I should say I for *many*) am convinced that corruption is not only 'endemic' to advanced capitalism, but an absolutely integral part of its workings. Yet whatever one's view on that question, since when was state-sponsored fascism stopped with courts, or indeed with democracy?

Perhaps the more serious problem nonetheless lies in the assumption that, if one can get rid of a few bad men (and the worst offenders *are* mostly, if not entirely, *men*), then, to return to Mother Julian, all shall be well, and all *shall* be well. A Deleuzian/Guattarian counter-image would be that of a gargantuan assemblage of assemblages, or a machine made of countless machines that are not only expanding in every direction, but doing so more and more rapidly with every year that passes, and this despite not being 'plugged in' to each other; on the contrary, the 'system' grows at least partly because it is *not* a system – at least not an organic one. Each assemblage, each 'machine', each individual entails/embodies/*is* what I have described as a principle of immanence, and so must be, at the very least, partly independent. Indeed, each 'machine' is ready, if not to spawn a new machine or indeed a new *kind* of machine, then to link up with another similarly inclined machine (though those machines may not seem at all disastrous to their 'operators'). As I began to explain in Chapter 10, the machines in question include not just individual humans, but also farting cows.[10]

The last thing anyone should do is to deny the historic role of a tiny but extraordinarily powerful alliance of Rupert Murdoch-esque figures who, each time that the machine-of-machines appears ready to contract or adapt, have injected more fuel – literal, symbolic, affective – to ensure that the expansion recommences with added vigour. Removing these climate change villains would doubtless make a difference (dare I use the concept). However, in and of itself, such removal would do little to address a world full of desiring bodies – bodies that, by the early 2000s, if not long before, more often than not inhabited the aforementioned as-

---

[10] Gilles Deleuze and Félix Guattari, *A Thousand Plateaus: Capitalism and Schizophrenia*, translated by Brian Massumi (London: Continuum, 1988[1980]). This is Deleuze and Guattari being perhaps their most mischievous selves, seemingly echoing, but in fact turning the mechanism, or *mechanicism* of the dominant modern cultural formations against themselves.

semblage-of-assemblages according to a model of at once the fiercest, but also the least recognised, of addictions.

I should acknowledge straight away that no doubt Deleuze and Guattari would suggest unconventional ways of conceiving all that which goes by the name of addiction, in the same way that they reconceptualise schizophrenia. Addiction is not about a mechanical relation of cause and effect; nor is there a neatly organic relation, i.e. a version of what Manuel DeLanda would describe as relations of interiority. (If either kind of explanation were valid, then all users would become hopelessly addicted to one or another 'addictive substance'.)

All that said, addiction *does* entail something very much like an indexical relation between a substance, and one or more bodies. But here substance is not to be interpreted along the lines of either a Cartesian dualism, or an abject empiricism. By the early twenty-first century, the substance that was most abused was typically semeiotic-symbolic in character, and its consequences defied any simple characterisation along the lines of an opposition of realism and idealism. I am referring to a seemingly permanent need to use mobile phones and other digital technologies to produce, but actually mainly to produce the *consumption* of, symbolic forms.

Unlike alcohol or other explicitly drug-based addictions, this is not an addiction that is likely to destroy one or another organ (though people's necks and thumbs might well be affected). It is instead a version of what might be described as Shannon's Syndrome, or perhaps one should say Shannon's Disease, in honour of the great Claude Shannon (cf. Chapter 3), viz., an absolutely compelling, indeed uncontrollable desire, however unselfconscious or even *un*conscious, to become-bit, to *become-algorithm*.

It would probably be a good idea to distinguish between the syndrome, and the disease versions. The syndrome is the not-necessarily-negative desire to be 'in touch' by digital means (here 'digital' refers to both kinds of digits, but 'in touch' is nonetheless something of a misnomer). The disease is caused by the uses given to that desire by corporate leaders, at least some of whom were said to be forbidding their own children from using their companies' products. As ever, *affect* was/is key, but as I noted in Part 3, affect and with it desire are not best understood as a product of a transcendental force of the kind theorised in their different ways by sociobiologists or psychoanalysts; affect and desire work in tandem with the assemblages-*du-jour*, in this case the kinds of assemblages associated with a form of capitalism that is now late in every sense of the word.

What to do in this mind-bending (dare I use the expression) situation? It is difficult not to employ Deleuzian and Guattarian insights as the basis for an infinite pessimism, and to shift from a *mea culpa* (and a catholic self-flagellation) to an increasingly fatalist, to not say suicidal disposition (Deleuze, who suffered from a life-long illness, took his own life when his affliction became unmanageable). It is not just that there seems to be no way out; it's that any way out would require the efforts of billions, even as it would continue to be undermined by a far right that seems determined to hate till death do us part.

In Britain the marvellously colloquial expression 'It's like herding cats' is often used to represent the difficulties of getting a group of people to agree to something contentious, but in the case of climate change it's like herding all of the world's tigers, lions, leopards and jaguars all at once. Anyone who tries to engage with the Trumps, the Putins, or the Bolsonaros – or those whose interests the politicians both represent and mould – is likely to be mauled even before any serious attempt at 'herding' has even begun. This is one of the functions of the so-called *social* media, whose owners work not only to generate further addiction, but to enable, however inadvertently, a kind of linguistic (and to be sure, not just *linguistic*) shooting range, with veritable battalions of both macro- and micro-fascists just waiting for the next environmental do-gooder to raise their head above the digital parapet. The harassment – a poor word for what are actually attempts to completely destroy individuals and groups – makes apparent the foolishness of anyone who still subscribes to that old positivist English-language nursery rhyme, 'sticks and stones may break my bones but words will never hurt me'.

As this account begins to suggest, another 'function' of the social media is to promote difference of an identitary kind. Twitter and other similar media effectively encourage their users to identify with a certain category (1), and to treat others as being not that category (0) – and viceversa: much of the hatred directed at feminists by misogynist commentators start from the premise that feminists are only and entirely that (1), whilst the misogynist has absolutely nothing to do with anything defended with feminism (0). Exactly the same principle applies to those who espouse homophobia, climate change denialism, let alone the various nationalist causes.

It might be suggested that 'it was always thus' with ideological hatreds, but one key difference in the context of the new media is that identitarianism may be espoused in more or less complete anonymity, and to audiences that may number in the billions, even as 'feeds' channel back similarly identitary voices to those that espouse such views. Thanks in part to this anonymous quality, Twitter and other similar platforms may have effects that are not unlike those of driving a car whilst under the effects of road rage; an otherwise *perfectly reasonable* individual may become a demented curser, a killer on wheels, or in this case, a killer on algorithms.

Yet in so far as the new media's identitary juggernaut generates responses in kind on the part of those who might normally refuse such terms of engagement, then one and all may end up being sucked into a veritable culture war: a war that, like all true wars, is fought on the basis of the most violent reduction of difference, viz. to an absolutely binary us versus them. In such a context, passion, in the Spinozan sense of the word, will reign supreme; all that will matter will be to drastically diminish the power of acting of anyone labelled as Enemy. It goes almost without saying that this is a necessary condition of the rise of fascism, and if this analysis is correct, then the director-owners of the social media are playing a central role in its increasingly state-sponsored return. Suggesting this does not contradict the principle of immanence; the social media constitute an assemblage *par excellence*, and clearly each angry user is responsible for their actions; but such actions would not be possible without the existence, and a particular deployment of, the technologies.

Almost a decade before the risks of anthropogenic climate change became widely known, Deleuze and Guattari wrote something which sums up the difficulty of engaging with the aforementioned machine-of-desiring-machines. As part of a discussion in *A Thousand Plateaus* of 'what movement, what impulse, sweeps us outside the strata',[11] the philosophers suggest that '[e]very undertaking of destratification ... must ... observe concrete rules of extreme caution: a too-sudden destratification may be suicidal, or turn cancerous. In other words, it will sometimes end in chaos, the void and destruction, and sometimes lock us back into the strata, which become more rigid still, losing their degrees of diversity, differentiation, and mobility'.[12] It is not entirely clear from this passage whether Deleuze and Guattari are opposing or linking the two outcomes, but the combination of increasingly catastrophic climate change and the current return to state-sponsored fascism shows that, actually, both phenomena can co-occur: as things are going, it is difficult not to believe that the world will end in utter chaos, but ruled by figures who will make the current fascistoid leaders seem almost quaintly authoritarian.

And yet... a Deleuzian/Guattarian perspective might also be employed to express optimism: once again, *nobody knows what a body can do*, and to this we might add, nobody knows what two or more bodies can do together (affect, *sensu non stricto*). If there is an enemy of critical progressive thought and action, it is teleology, or even worse, teleology coupled to fatalism. And indeed, in 2019 there were many signs that the fiendish concatenation of conspiracy, 'everyday' neoliberalism and emergent fascism was being resisted by activists working across a variety of contexts. Early 2019 saw school strikes by children as young as 10 or 11. In London (and then in other cities across the British isles), the 'Extinction Rebellion' showed how easy it would be for people exercising their democratic right to protest to put a very significant spanner in the works, despite decades of the Tories' and New Labour's enthusiastic embrace of panopticism (London was reportedly one of the major cities with the most CCTV cameras in the world). Even in the US, several states and cities challenged, via local and congressional policy-makers, the forces represented by the US presidency, and promulgated policies designed to redevelop local economies along greener lines. Perhaps most surprisingly, by 2019 even the seemingly perpetually prudent Sir David Attenborough,

---

[11] Gilles Deleuze and Félix Guattari, *A Thousand Plateaus*, p. 502.

[12] Gilles Deleuze and Félix Guattari, *A Thousand Plateaus*, p. 503.

once the 'face' of a nocturnal form of environmental disaster denialism, was warning of looming catastrophe.

What, then, can we do? A first step is to recognise that there both is, and there is not, a 'we'. We are all on the same planet and so forced to share the terrible burden of burgeoning global catastrophe. But of course, concealed by the environmentalist we/one are all sorts of differing interests, not least those arising from huge inequalities in wealth, and with them what are also huge inequalities in the per capita emissions of green house gases. But even if one ignores such fundamental divides, it is by no means clear how large, heterogeneous societies might, could or should mobilise to effect vast changes – and changes on so many levels in so short a time. Amongst those who have devoted their lives to researching developmental and grass-roots politics, it has long been axiomatic that grass-roots movements alone cannot effect major changes; the efforts of billions must be organised and led. Then again, top-down efforts to change cultural processes will only go so far.

Despite the risk of seeming to depoliticise the issues, at least in the time of climate change emergency it seems clear that billions of people across the world *do* have it in their power to make a drastic change, 'if only they can cure their addictions'. From that perspective, what is now required is a form of *abstinence*: we must all stop engaging in avoidable acts of consumption, beginning with semeiotic consumption of the kind I have described in this volume. This will, no doubt, be the hardest thing to do (but here I do not use the word 'thing' as Latour does; what is at stake is a process). As with all addictions, what is at stake is not simply a personal psychology or a purely organismic dependency, but what at times seems like a perfectly matched bio-sociality – body, affect, and assemblage looking to quench a terrible thirst in seemingly perfect synchrony (it is actually *not* a perfect synchrony, it never is).

This postscript is not the place for detailed proposals regarding this problem of problems, but I will at least suggest this: as an integral part of the now desperate need for a veritable economic *revolution* – the term 'transition' is a problematic one in so far as it suggests a gradual process, and in so far as it fails to address questions of both climate and social justice – we need movements all over the world that can not only reduce existing greenhouse-gas emitting forms of consumption to an absolute minimum, but can at the same time encourage people to engage in what I describe as a progressive production of polity, and through it, the development of alternative forms of 'ludism'. Both of these elements will, in turn, require a third: a pedagogy of affect(ion).

Briefly: the production of polity entails generating assemblages that promote what are, at least to start with, relatively local relations of solidarity amongst individuals and groups that normally don't have such relations – but this typically on the basis of overtly, one might even say *intensely* practical issues. Compare and contrast the difference between, say, well-meaning environmental activists producing a video that is distributed via the media of mass communication (see for example the marvellous video produced in 2019 by Greta Thunberg and George Monbiot, and distributed via *The Guardian* online[13]); and a group of neighbours who don't know each other, but who come together on their street to address a problem with, say, excessive and overly fast through-traffic (they would like to generate a safer space for their children to play), or indeed, a group of people who would like to plant more trees (these are, admittedly, in some respects rather suburban, and middle class examples). If I speak of the production of polity, it is precisely to suggest that people may/*must* come together to form new 'societies'. Welcome as they may be, videos such as Thunberg's and Monbiot's cannot replace the latter kind of practice, and on the contrary a case can be made that today part of the problem is that so many of us feel more inclined to consume representations than to 'go out and *do* something'. Even now many are bewitched by the spectacle of our own demise.

Ludism is not luddism (though perhaps a degree of the latter may now also be required). Ludism is the doctrine that play and humour may help to break molarities, and in so doing, unsettle strata via local relations of solidarity (we return to polity). There is, of course, a long history of the carnivalesque, and what I described in Chapter 12 as the grotesque realism subverting the Cartesian ontology. Humour, play, and partying are dead serious when it comes to transforming unequal relations (though of course it can also go the other way; we return by this route to addiction; the powers that become, and stop becoming would probably quite like it if we were all now to party to death).

For its part, the pedagogy of affect(ion) entails at least two aspects: first, making explicit the workings of what I described in the conclusions to Part 3 as non-formal pedagogic modes. The idea would be to show how certain habits, or better yet certain *habituses* and *hexis* (Bourdieu, *passim*) are taught without being taught, and learned without being learned. The parents and grandparents who express love by feeding children sugar, or buying them toys; the parents that are constantly using their mobile phones, tablets or laptops whilst in the presence of children... these are but simple examples of the kind of fundamental pedagogies that must be both recognised, and challenged.

This kind of interruption must, however, be more than matched by a positive pedagogy, one which is devoted to finding forms of expressing desire, and engaging in ludic activities (as per my invocation of 'ludism') that are not premised on further consumption – at least not the kind of profligate, greenhouse gas-rich consumption encouraged by contemporary forms of play – and that, I might add, is to say nothing about the 'old', but now even more valid critique of commodity fetishism. The more general point is that it will not suffice to simply try to stamp

---

13 The video may be accessed at Damian Carrington, 'Greta Thunberg: "We Are Ignoring Natural Climate Solutions"', *The Guardian* online, 19 September 2019, https://www.theguardian.com/environment/2019/sep/19/greta-thunberg-we-are-ignoring-natural-climate-solutions [Accessed 19 September 2019].

on desire; new pleasures, new *leisures* will need to be identified – or ancient ones resurrected – and this communally.

Little or nothing of what I have just suggested is new. What *would* be new is if those who have channeled their vital energies into writing books, or conducting research were to devote a similar effort to partnering up with activists who have the gift of everyday charisma in order to promote the mentioned activities, and to help them grow to the point that they become mass movements of the kinds that are now so desperately required. Here the Deleuzian-Guattarian model of the rhizome is no doubt a more useful one than the archetypal environmental tree; in the times of environmental hecatomb, we need rhizomes.

Index of Key Concepts
and Authors

Index of Key Concepts and Authors

Index of Key Concepts and Authors

Notes

Notes

Notes

Notes

Notes

Notes

Printed in Poland
by Amazon Fulfillment
Poland Sp. z o.o., Wrocław

54878370R00271